SCHAUM'S OUTLINE OF

Theory and Problems of
COLLEGE MATHEMATICS

THIRD EDITION

Algebra
Discrete Mathematics
Precalculus
Introduction to Calculus

FRANK AYRES, Jr., Ph.D.

*Formerly Professor and Head
Department of Mathematics, Dickinson College*

PHILIP A. SCHMIDT, Ph.D.

*Program Coordinator, Mathematics and Science Education
The Teachers College, Western Governors University
Salt Lake City, Utah*

Schaum's Outline Series

McGRAW-HILL
New York   Chicago   San Francisco   Lisbon   London
Madrid   Mexico City   Milan   New Delhi   San Juan
Seoul   Singapore   Sydney   Toronto

The McGraw·Hill Companies

FRANK AYRES, Jr., Ph.D., was formerly Professor and Head of the Department of Mathematics at Dickinson College, Carlisle, Pennsylvania. He is the author or coauthor of eight Schaum's Outlines, including *Calculus*, *Trigonometry*, *Differential Equations*, and *Modern Abstract Algebra*.

PHILIP A. SCHMIDT, Ph.D., has a B.S. from Brooklyn College (with a major in mathematics), an M.A. in mathematics, and a Ph.D. in mathematics education from Syracuse University. He is currently Program Coordinator in Mathematics and Science Education at The Teachers College at Western Governors University in Salt Lake City, Utah. He is also the reviser of Barnett Rich's *Geometry*.

1 2 3 4 5 6 7 8 9 10 11 12 13 14 15 16 17 VLP VLP 0 9 8 7 6 5 4 3

ISBN 0-07-140227-6

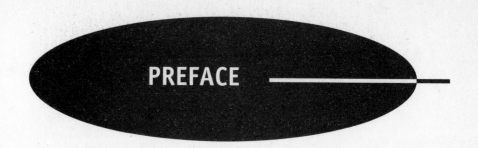

PREFACE

In the Third Edition of College Mathematics, I have maintained the point-of-view of the first two editions. Students who are engaged in learning mathematics in the mathematical range from algebra to calculus will find virtually all major topics from those curricula in this text. However, a substantial number of important changes have been made in this edition. First, there is more of an emphasis now on topics in discrete mathematics. Second, the graphing calculator is introduced as an important problem-solving tool. Third, material related to manual and tabular computations of logarithms has been removed, and replaced with material that is calculator-based. Fourth, all material related to the concepts of locus has been modernized. Fifth, tables and graphs have been changed to reflect current curriculum and teaching methods. Sixth, all material related to the conic sections has been substantially changed and modernized. Additionally, much of the rest of the material in the third edition has been changed to reflect current classroom methods and pedagogy, and mathematical modeling is introduced as a problem-solving tool. Notation has been changed as well when necessary.

My thanks must be expressed to Barbara Gilson and Andrew Littell of McGraw-Hill. They have been supportive of this project from its earliest stages. I also must thank Dr. Marti Garlett, Dean of the Teachers College at Western Governors University, for her professional support as I struggled to meet deadlines while beginning a new position at the University. I thank Maureen Walker for her handling of the manuscript and proofs. And finally, I thank my wife, Dr. Jan Zlotnik Schmidt, for putting up with my frequent need to work at home on this project. Without her support, this edition would not have been easily completed.

PHILIP A. SCHMIDT
New Paltz, NY

CONTENTS

v

PART I

REVIEW OF ALGEBRA

Chapter 1

Elements of Algebra

IN ARITHMETIC the numbers used are always known numbers; a typical problem is to convert 5 hours and 35 minutes to minutes. This is done by multiplying 5 by 60 and adding 35; thus, $5 \cdot 60 + 35 = 335$ minutes.

In algebra some of the numbers used may be known but others are either unknown or not specified; that is, they are represented by letters. For example, convert h hours and m minutes into minutes. This is done in precisely the same manner as in the paragraph above by multiplying h by 60 and adding m; thus, $h \cdot 60 + m = 60h + m$. We call $60h + m$ an *algebraic expression*. (See Problem 1.1.)

Since algebraic expressions are numbers, they may be added, subtracted, and so on, following the same laws that govern these operations on known numbers. For example, the sum of $5 \cdot 60 + 35$ and $2 \cdot 60 + 35$ is $(5 + 2) \cdot 60 + 2 \cdot 35$; similarly, the sum of $h \cdot 60 + m$ and $k \cdot 60 + m$ is $(h + k) \cdot 60 + 2m$. (See Problems 1.2–1.6.)

POSITIVE INTEGRAL EXPONENTS. If a is any number and n is any positive integer, the product of the n factors $a \cdot a \cdot a \cdots a$ is denoted by a^n. To distinguish between the letters, a is called the *base* and n is called the *exponent*.

If a and b are any bases and m and n are any positive integers, we have the following laws of exponents:

(1) $\quad a^m \cdot a^n = a^{m+n}$

(2) $\quad (a^m)^n = a^{mn}$

(3) $\quad \dfrac{a^m}{a^n} = a^{m-n}, \qquad a \neq 0, \quad m > n; \qquad \dfrac{a^m}{a^n} = \dfrac{1}{a^{n-m}}, \quad a \neq 0, \quad m < n$

(4) $\quad (a \cdot b)^n = a^n b^n$

(5) $\quad \left(\dfrac{a}{b}\right)^n = \dfrac{a^n}{b^n}, \qquad b \neq 0$

(See Problem 1.7.)

LET n BE A POSITIVE INTEGER and a and b be two numbers such that $b^n = a$; then b is called an *n*th *root of a*. Every number $a \neq 0$ has exactly n distinct nth roots.

If a is imaginary, all of its nth roots are imaginary; this case will be excluded here and treated later. (See Chapter 35.)

If a is real and n is odd, then exactly one of the nth roots of a is real. For example, 2 is the real cube root of 8, $(2^3 = 8)$, and -3 is the real fifth root of $-243[(-3)^5 = -243]$.

If a is real and n is even, then there are exactly two real nth roots of a when $a > 0$, but no real nth roots of a when $a < 0$. For example, $+3$ and -3 are the square roots of 9; $+2$ and -2 are the real sixth roots of 64.

THE PRINCIPAL nth **ROOT OF** a is the positive real nth root of a when a is positive and the real nth root of a, if any, when a is negative. The principal nth root of a is denoted by $\sqrt[n]{a}$, called a *radical*. The integer n is called the *index* of the radical and a is called the *radicand*. For example,

$$\sqrt{9} = 3 \qquad \sqrt[6]{64} = 2 \qquad \sqrt[5]{-243} = -3$$

(See Problem 1.8.)

ZERO, FRACTIONAL, AND NEGATIVE EXPONENTS. When s is a positive integer, r is any integer, and p is any rational number, the following extend the definition of a^n in such a way that the laws (1)-(5) are satisfied when n is any rational number.

<u>DEFINITIONS</u>	<u>EXAMPLES</u>
(6) $\quad a^0 = 1, a \neq 0$	$2^0 = 1, \left(\frac{1}{100}\right)^0 = 1, (-8)^0 = 1$
(7) $\quad a^{r/s} = \sqrt[s]{a^r} = (\sqrt[s]{a})^r$	$3^{1/2} = \sqrt{3}, (64)^{5/6} = \left(\sqrt[6]{64}\right)^5 = 2^5 = 32, 3^{-2/1} = 3^{-2} = \frac{1}{9}$
(8) $\quad a^{-p} = 1/a^p, a \neq 0$	$2^{-1} = \frac{1}{2}, 3^{-1/2} = 1\sqrt{3}$

[NOTE: Without attempting to define them, we shall assume the existence of numbers such as $a^{\sqrt{2}}, a^{\pi}, \ldots$, in which the exponent is irrational. We shall also assume that these numbers have been defined in such a way that the laws (1)–(5) are satisfied.] (See Problem 1.9–1.10.)

Solved Problems

1.1   For each of the following statements, write the equivalent algebraic expressions: (*a*) the sum of x and 2, (*b*) the sum of a and $-b$, (*c*) the sum of $5a$ and $3b$, (*d*) the product of $2a$ and $3a$, (*e*) the product of $2a$ and $5b$, (*f*) the number which is 4 more than 3 times x, (*g*) the number which is 5 less than twice y, (*h*) the time required to travel 250 miles at x miles per hour, (*i*) the cost (in cents) of x eggs at 65¢ per dozen.

  (*a*) $x + 2$ (*d*) $(2a)(3a) = 6a^2$ (*g*) $2y - 5$
  (*b*) $a + (-b) = a - b$ (*e*) $(2a)(5b) = 10ab$ (*h*) $250/x$
  (*c*) $5a + 3b$ (*f*) $3x + 4$ (*i*) $65(x/12)$

1.2   Let x be the present age of a father. (*a*) Express the present age of his son, who 2 years ago was one-third his father's age. (*b*) Express the age of his daughter, who 5 years from today will be one-fourth her father's age.

  (*a*) Two years ago the father's age was $x - 2$ and the son's age was $(x - 2)/3$. Today the son's age is $2 + (x - 2)/3$.

  (*b*) Five years from today the father's age will be $x + 5$ and his daughter's age will be $\frac{1}{4}(x + 5)$. Today the daughter's age is $\frac{1}{4}(x + 5) - 5$.

1.3　A pair of parentheses may be inserted or removed at will in an algebraic expression if the first parenthesis of the pair is preceded by a $+$ sign. If, however, this sign is $-$, the signs of all terms within the parentheses must be changed.

(a)　$5a + 3a - 6a = (5 + 3 - 6)a = 2a$　　　　(b)　$\frac{1}{2}a + \frac{1}{4}b - \frac{1}{4}a + \frac{3}{4}b = \frac{1}{4}a + b$

(c)　$(13a^2 - b^2) + (-4a^2 + 3b^2) - (6a^2 - 5b^2) = 13a^2 - b^2 - 4a^2 + 3b^2 - 6a^2 + 5b^2 = 3a^2 + 7b^2$

(d)　$(2ab - 3bc) - [5 - (4ab - 2bc)] = 2ab - 3bc - [5 - 4ab + 2bc]$

$$= 2ab - 3bc - 5 + 4ab - 2bc = 6ab - 5bc - 5$$

(e)　$(2x + 5y - 4)3x = (2x)(3x) + (5y)(3x) - 4(3x) = 6x^2 + 15xy - 12x$

(f)　$5a - 2$　　　　　　　　(g)　$2x - 3y$　　　　　　　　(h)　$3a^2 + 2a - 1$

$\quad\ \ \dfrac{3a + 4}{15a^2 - 6a}$　　　　　　　　$\quad\ \ \dfrac{5x + 6y}{10x^2 - 15xy}$　　　　　　　　$\quad\ \ \dfrac{2a - 3}{6a^3 + 4a^2 - 2a}$

$(+)\dfrac{20a - 8}{15a^2 + 14a - 8}$　　　　$(+)\dfrac{12xy - 18y^2}{10x^2 - 3xy - 18y^2}$　　　　$(+)\dfrac{-9a^2 - 6a + 3}{6a^3 - 5a^2 - 8a + 3}$

(i)

$$
\begin{array}{r}
x^2 + 4x - 2 \\
x - 3\,\overline{)\,x^3 + x^2 - 14x + 6} \\
(-)\ \underline{x^3 - 3x^2} \\
4x^2 - 14x \\
(-)\ \underline{4x^2 - 12x} \\
-2x + 6 \\
(-)\ \underline{-2x + 6}
\end{array}
$$

(j)

$$
\begin{array}{r}
x^2 - 2x - 1 \\
x^2 + 3x - 2\,\overline{)\,x^4 + x^3 - 9x^2 + x + 5} \\
(-)\ \underline{x^4 + 3x^3 - 2x^2} \\
-2x^3 - 7x^2 + x \\
(-)\ \underline{-2x^3 - 6x^2 + 4x} \\
-x^2 - 3x + 5 \\
(-)\ \underline{-x^2 - 3x + 2} \\
3
\end{array}
$$

$$\frac{x^3 + x^2 - 14x + 6}{x - 3} = x^2 + 4x - 2 \qquad \frac{x^4 + x^3 - 9x^2 + x + 5}{x^2 + 3x - 2} = x^2 - 2x - 1 + \frac{3}{x^2 + 3x - 2}$$

1.4　The problems below involve the following types of factoring:

$ab + ac - ad = a(b + c - d)$　　　　　　　　　　$a^2 \pm 2ab + b^2 = (a \pm b)^2$

$a^3 + b^3 = (a + b)(a^2 - ab + b^2)$　　　　　　　$a^2 - b^2 = (a - b)(a + b)$

$acx^2 + (ad + bc)x + bd = (ax + b)(cx + d)$　　　$a^3 - b^3 = (a - b)(a^2 + ab + b^2)$

(a)　$5x - 10y = 5(x - 2y)$　　　　　　　　　(e)　$x^2 - 3x - 4 = (x - 4)(x + 1)$

(b)　$\frac{1}{2}gt^2 - \frac{1}{2}g^2t = \frac{1}{2}gt(t - g)$　　　　　　　(f)　$4x^2 - 12x + 9 = (2x - 3)^2$

(c)　$x^2 + 4x + 4 = (x + 2)^2$　　　　　　　(g)　$12x^2 + 7x - 10 = (4x + 5)(3x - 2)$

(d)　$x^2 + 5x + 4 = (x + 1)(x + 4)$　　　　　(h)　$x^3 - 8 = (x - 2)(x^2 + 2x + 4)$

(i)　$2x^4 - 12x^3 + 10x^2 = 2x^2(x^2 - 6x + 5) = 2x^2(x - 1)(x - 5)$

1.5　Simplify.

(a)　$\dfrac{8}{12x + 20} = \dfrac{4 \cdot 2}{4 \cdot 3x + 4 \cdot 5} = \dfrac{2}{3x + 5}$　　　　(d)　$\dfrac{4x - 12}{15 - 5x} = \dfrac{4(x - 3)}{5(3 - x)} = \dfrac{4(x - 3)}{-5(x - 3)} = -\dfrac{4}{5}$

(b)　$\dfrac{9x^2}{12xy - 15xz} = \dfrac{3x \cdot 3x}{3x \cdot 4y - 3x \cdot 5z} = \dfrac{3x}{4y - 5z}$　　　(e)　$\dfrac{x^2 - x - 6}{x^2 + 7x + 10} = \dfrac{(x + 2)(x - 3)}{(x + 2)(x + 5)} = \dfrac{x - 3}{x + 5}$

(c)　$\dfrac{5x - 10}{7x - 14} = \dfrac{5(x - 2)}{7(x - 2)} = \dfrac{5}{7}$　　　　(f)　$\dfrac{6x^2 + 5x - 6}{2x^2 - 3x - 9} = \dfrac{(2x + 3)(3x - 2)}{(2x + 3)(x - 3)} = \dfrac{3x - 2}{x - 3}$

(g)　$\dfrac{3a^2 - 11a + 6}{a^2 - a - 6} \cdot \dfrac{4 - 4a - 3a^2}{36a^2 - 16} = \dfrac{(3a - 2)(a - 3)(2 - 3a)(2 + a)}{(a - 3)(a + 2)4(3a + 2)(3a - 2)} = -\dfrac{3a - 2}{4(3a + 2)}$

1.6　Combine as indicated.

(a)　$\dfrac{2a + b}{10} + \dfrac{a - 6b}{15} = \dfrac{3(2a + b) + 2(a - 6b)}{30} = \dfrac{8a - 9b}{30}$

(b)　$\dfrac{2}{x} - \dfrac{3}{2x} + \dfrac{5}{4} = \dfrac{2 \cdot 4 - 3 \cdot 2 + 5 \cdot x}{4x} = \dfrac{2 + 5x}{4x}$

(c) $\dfrac{2}{3a-1} - \dfrac{3}{2a+1} = \dfrac{2(2a+1)-3(3a-1)}{(3a-1)(2a+1)} = \dfrac{5-5a}{(3a-1)(2a+1)}$

(d) $\dfrac{3}{x+y} - \dfrac{5}{x^2-y^2} = \dfrac{3}{x+y} - \dfrac{5}{(x+y)(x-y)} = \dfrac{3(x-y)-5}{(x+y)(x-y)} = \dfrac{3x-3y-5}{(x+y)(x-y)}$

(e) $\dfrac{a-2}{6a^2-5a-6} + \dfrac{2a+1}{9a^2-4} = \dfrac{a-2}{(2a-3)(3a+2)} + \dfrac{2a+1}{(3a+2)(3a-2)}$

$\qquad\qquad = \dfrac{(a-2)(3a-2)+(2a+1)(2a-3)}{(2a-3)(3a+2)(3a-2)} = \dfrac{7a^2-12a+1}{(2a-3)(3a+2)(3a-2)}$

1.7 Perform the indicated operations.

(a) $3^4 = 3\cdot3\cdot3\cdot3 = 81$ (f) $2^6\cdot2^4 = 2^{6+4} = 2^{10} = 1024$ (k) $a^{10}/a^4 = a^{10-4} = a^6$

(b) $-3^4 = -81$ (g) $\left(\tfrac{1}{2}\right)^3\left(\tfrac{1}{2}\right)^2 = \left(\tfrac{1}{2}\right)^5$ (l) $a^4/a^{10} = 1/a^{10-4} = 1/a^6$

(c) $(-3)^4 = 81$ (h) $a^{n+3}a^{m+2} = a^{m+n+5}$ (m) $(-2)^8/(-2)^5 = (-2)^3 = -8$

(d) $-(-3)^4 = -81$ (i) $(a^2)^5 = a^{2\cdot5} = a^{10}$ (n) $a^{2n}b^{5m}/a^{3n}b^{2m} = b^{3m}/a^n$

(e) $-(-3)^3 = 27$ (j) $(a^{2n})^3 = a^{6n}$ (o) $36^{x+3}/6^{x-1} = 6^{2x+6}/6^{x-1} = 6^{x+7}$

1.8 Evaluate.

(a) $81^{1/2} = \sqrt{81} = 9$ (d) $(-27)^{1/3} = \sqrt[3]{-27} = -3$

(b) $81^{3/4} = \left(\sqrt[4]{81}\right)^3 = 3^3 = 27$ (e) $(-32)^{4/5} = \left(\sqrt[5]{-32}\right)^4 = (-2)^4 = 16$

(c) $\left(\tfrac{16}{49}\right)^{3/2} = \left(\sqrt{\tfrac{16}{49}}\right)^3 = \left(\tfrac{4}{7}\right)^3 = \tfrac{64}{343}$ (f) $-400^{1/2} = -\sqrt{400} = -20$

1.9 Evaluate.

(a) $4^0 = 1$ (c) $(4a)^0 = 1$ (e) $4^{-1} = \tfrac{1}{4}$ (g) $125^{-1/3} = 1/125^{1/3} = \tfrac{1}{5}$

(b) $4a^0 = 4\cdot1 = 4$ (d) $4(3+a)^0 = 4\cdot1 = 4$ (f) $5^{-2} = \left(\tfrac{1}{5}\right)^2 = \tfrac{1}{25}$ (h) $(-125)^{-1/3} = -\tfrac{1}{5}$

(i) $-\left(\tfrac{1}{64}\right)^{5/6} = -\left(\sqrt[6]{\tfrac{1}{64}}\right)^5 = -\left(\tfrac{1}{2}\right)^5 = -\tfrac{1}{32}$

(j) $-\left(-\tfrac{1}{32}\right)^{4/5} = -\left(\sqrt[5]{-\tfrac{1}{32}}\right)^4 = -\left(-\tfrac{1}{2}\right)^4 = -\tfrac{1}{16}$

1.10 Perform each of the following operations and express the result without negative or zero exponents:

(a) $\left(\dfrac{81a^4}{b^8}\right)^{-1/4} = \dfrac{3^{-1}a^{-1}}{b^{-2}} = \dfrac{b^2}{3a}$ (b) $\left(a^{1/2}+a^{-1/2}\right)^2 = a+2a^0+a^{-1} = a+2+\dfrac{1}{a}$

(c) $\left(a-3b^{-2}\right)\left(2a^{-1}-b^2\right) = 2a^0 - ab^2 - 6a^{-1}b^{-2} + 3b^0 = 5 - ab^2 - 6/ab^2$

(d) $\dfrac{a^{-2}+b^{-2}}{a^{-1}-b^{-1}} = \dfrac{(a^{-2}+b^{-2})(a^2b^2)}{(a^{-1}-b^{-1})(a^2b^2)} = \dfrac{b^2+a^2}{ab^2-a^2b}$

(e) $\left(\dfrac{a^2}{b}\right)^7\left(-\dfrac{b^2}{a^3}\right)^6 = \dfrac{a^{14}\cdot b^{12}}{b^7\cdot a^{18}} = \dfrac{b^5}{a^4}$

(f) $\left(\dfrac{a^{1/2}b^{2/3}}{c^{3/4}}\right)^6\left(\dfrac{c^{1/2}}{a^{1/4}b^{1/3}}\right)^9 = \dfrac{a^3b^4}{c^{9/2}}\cdot\dfrac{c^{9/2}}{a^{9/4}b^3} = a^{3/4}b$

Supplementary Problems

1.11 Combine.

(a) $2x+(3x-4y)$ (c) $[(s+2t)-(s+3t)] - [(2s+3t)-(-4s+5t)]$

(b) $5a+4b-(-2a+3b)$ (d) $8x^2y-\{3x^2y+[2xy^2+4x^2y-(3xy^2-4x^2y)]\}$

1.12   Perform the indicated operations.

(a) $4x(x - y + 2)$ (c) $(5x^2 - 4y^2)(-x^2 + 3y^2)$ (e) $(2x^3 + 5x^2 - 33x + 20) \div (2x - 5)$

(b) $(5x + 2)(3x - 4)$ (d) $(x^3 - 3x + 5)(2x - 7)$ (f) $(2x^3 + 5x^2 - 22x + 10) \div (2x - 3)$

1.13   Factor.

(a) $8x + 12y$ (e) $16a^2 - 8ab + b^2$ (i) $(x - y)^2 + 6(x - y) + 5$

(b) $4ax + 6ay - 24az$ (f) $25x^2 + 30xy + 9y^2$ (j) $4x^2 - 8x - 5$

(c) $a^2 - 4b^2$ (g) $x^2 - 4x - 12$ (k) $40a^2 + ab - 6b^2$

(d) $50ab^4 - 98a^3b^2$ (h) $a^2 + 23ab - 50b^2$ (l) $x^4 + 24x^2y^2 - 25y^4$

1.14   Simplify.

(a) $\dfrac{a^2 - b^2}{2ax + 2bx}$ (d) $\dfrac{16a^2 - 25}{2a - 10} \times \dfrac{a^2 - 10a + 25}{4a + 5}$

(b) $\dfrac{x^2 + 4x + 3}{1 - x^2}$ (e) $\dfrac{x^2 + xy - 6y^2}{2x^3 + 6x^2y} \times \dfrac{8x^2y}{x^2 - 5xy + 6y^2}$

(c) $\dfrac{1 - x - 12x^2}{1 + x - 6x^2}$

1.15   Perform the indicated operations.

(a) $\dfrac{5x}{18} + \dfrac{4x}{18}$ (c) $\dfrac{3a}{4b} - \dfrac{4b}{3a}$ (e) $x + 5 - \dfrac{x^2}{x - 5}$ (g) $\dfrac{2x + 3}{18x^2 - 27x} - \dfrac{2x - 3}{18x^2 + 27x}$

(b) $\dfrac{3a}{x} + \dfrac{5a}{2x}$ (d) $\dfrac{2a - 3b}{a^2 - b^2} + \dfrac{1}{a - b}$ (f) $\dfrac{a + 2}{2a - 6} - \dfrac{a - 2}{2a + 6}$

1.16   Simplify.

(a) $\dfrac{a}{2 - \dfrac{3}{a}}$ (b) $\dfrac{4 - \dfrac{x}{3}}{\dfrac{x}{6}}$ (c) $\dfrac{x - \dfrac{4}{x}}{1 - \dfrac{2}{x}}$ (d) $\dfrac{\dfrac{1}{x} + \dfrac{1}{y}}{\dfrac{x + y}{y} + \dfrac{x + y}{x}}$

ANSWERS TO SUPPLEMENTARY PROBLEMS

1.11   (a) $5x - 4y$ (b) $7a + b$ (c) $t - 6s$ (d) $xy(y - 3x)$

1.12   (a) $4x^2 - 4xy + 8x$ (c) $-5x^4 + 19x^2y^2 - 12y^4$ (e) $x^2 + 5x - 4$

(b) $15x^2 - 14x - 8$ (d) $2x^3 - 13x^2 + 31x - 35$ (f) $x^2 + 4x - 5 - 5/(2x - 3)$

1.13   (a) $4(2x + 3y)$ (e) $(4a - b)^2$ (i) $(x - y + 1)(x - y + 5)$

(b) $2a(2x + 3y - 12z)$ (f) $(5x + 3y)^2$ (j) $(2x + 1)(2x - 5)$

(c) $(a - 2b)(a + 2b)$ (g) $(x - 6)(x + 2)$ (k) $(5a + 2b)(8a - 3b)$

(d) $2ab^2(5b - 7a)(5b + 7a)$ (h) $(a + 25b)(a - 2b)$ (l) $(x - y)(x + y)(x^2 + 25y^2)$

1.14   (a) $\dfrac{a - b}{2x}$ (b) $-\dfrac{x + 3}{x - 1}$ (c) $\dfrac{4x - 1}{2x - 1}$ (d) $\frac{1}{2}(4a^2 - 25a + 25)$ (e) $\dfrac{4y}{x - 3y}$

1.15   (a) $\frac{1}{2}x$ (b) $\dfrac{11a}{2x}$ (c) $\dfrac{9a^2 - 16b^2}{12ab}$ (d) $\dfrac{3a - 2b}{a^2 - b^2}$ (e) $-\dfrac{25}{x - 5}$

(f) $\dfrac{5a}{a^2 - 9}$ (g) $\dfrac{8}{12x^2 - 27}$

1.16   (a) $\dfrac{a^2}{2a - 3}$ (b) $\dfrac{24 - 2x}{x}$ (c) $x + 2$ (d) $\dfrac{1}{x + y}$

Functions

A VARIABLE IS A SYMBOL selected to represent any one of a given set of numbers, here assumed to be real numbers. Should the set consist of just one number, the symbol representing it is called a *constant*.

The *range* of a variable consists of the totality of numbers of the set which it represents. For example, if x is a day in September, the range of x is the set of positive integers $1, 2, 3, \ldots, 30$; if x (ft) is the length of rope cut from a piece 50 ft long, the range of x is the set of numbers greater than 0 and less than 50.

Examples of ranges of a real variable, together with special notations and graphical representations, are given in Problem 2.1

FUNCTION. A correspondence (x, y) between two sets of numbers which pairs to an arbitrary number x of the first set exactly one number y of the second set is called a *function*. In this case, it is customary to speak of y as a *function* of x. The variable x is called the *independent variable* and y is called the *dependent variable*.

A function may be defined

(a) By a table of correspondents or table of values, as in Table 2.1.

Table 2.1

x	1	2	3	4	5	6	7	8	9	10
y	3	4	5	6	7	8	9	10	11	12

(b) By an equation or formula, as $y = x + 2$.

For each value assigned to x, the above relation yields a corresponding value for y. Note that the table above is a table of values for this function.

A FUNCTION IS CALLED *single-valued* if, to each value of y in its range, there corresponds just one value of x; otherwise, the function is called *multivalued*. For example, $y = x + 3$ defines y as a single-valued function of x while $y = x^2$ defines y as a multivalued (here, two-valued) function of x.

At times it will be more convenient to label a given function of x as $f(x)$, to be read "the f function of x" or simply "f of x." (Note carefully that this is not to be confused with "f times x.") If there are two functions, one may be labeled $f(x)$ and the other $g(x)$. Also, if $y = f(x) = x^2 - 5x + 4$, the statement "the value of the function is -2 when $x = 3$" can be replaced by "$f(3) = -2$." (See Problem 2.2.)

Let $y = f(x)$. The set of values of the independent variable x is called the *domain* of the function while the set of values of the dependent variable is called the *range of the function*. For example, $y = x^2$ defines a function whose domain consists of all (real) numbers and whose range is all nonnegative numbers, that is, zero and the positive numbers; $f(x) = 3/(x - 2)$ defines a function whose domain consists of all numbers except 2 (why?) and whose range is all numbers except 0. (See Problems 2.3–2.8.)

A VARIABLE w (dependent) is said to be a function of the (independent) variables $x, y, z, \ldots$ if when a value of each of the variables $x, y, z, \ldots$ is known, there corresponds exactly one value of w. For example, the volume V of a rectangular parallelepiped of dimensions x, y, z is given by $V = xyz$. Here V is a function of three independent variables. (See Problems 2.9–2.10.)

ADDITIONAL TERMINOLOGY If the function $y = f(x)$ is such that for every y in the range there is one and only one x in the domain such that $y = f(x)$, we say that f is a one-to-one correspondence. Functions that are one-to-one correspondences are sometimes called *bijections*. Note that all functions of the form $ax + by + c = 0$ are bijections. Note that $y = x^2$ is not a bijection. Is $y = x^3$ a bijection? (Answer: Yes!)

Solved Problems

2.1   Represent graphically each of the following ranges:

(a) $x > -2$

(b) $x < 5$

(c) $x \leqslant -1$

(d) $-3 < x < 4$

(e) $-2 < x < 2$ or $|x| < 2$

(f) $|x| > 3$

(g) $-3 \leqslant x \leqslant 5$

(h) $x \leqslant -3, x \geqslant 4$

2.2   Given $f(x) = x^2 - 5x + 4$, find

(a) $f(0) = 0^2 - 5 \cdot 0 + 4 = 4$

(b) $f(2) = 2^2 - 5 \cdot 2 + 4 = -2$

(c) $f(-3) = (-3)^2 - 5(-3) + 4 = 28$

(d) $f(a) = a^2 - 5a + 4$

(e) $f(-x) = x^2 + 5x + 4$

(f) $f(b + 1) = (b + 1)^2 - 5(b + 1) + 4 = b^2 - 3b$

(g) $f(3x) = (3x)^2 - 5(3x) + 4 = 9x^2 - 15x + 4$

(h) $f(x + a) - f(a) = [(x + a)^2 - 5(x + a) + 4] - (a^2 - 5a + 4) = x^2 + 2ax - 5x$

(i) $\dfrac{f(x + a) - f(x)}{a} = \dfrac{[(x + a)^2 - 5(x + a) + 4] - (x^2 - 5x + 4)}{a} = \dfrac{2ax - 5a + a^2}{a} = 2x - 5 + a$

2.3   In each of the following, state the domain of the function:

(a) $y = 5x$

(b) $y = -5x$

(c) $y = \dfrac{1}{x + 5}$

(d) $y = \dfrac{x - 2}{(x - 3)(x + 4)}$

(e) $y = \dfrac{1}{x}$

(f) $y = \sqrt{25 - x^2}$

(g) $y = \sqrt{x^2 - 9}$

(h) $y = \dfrac{1}{16 - x^2}$

(i) $y = \dfrac{1}{16 + x^2}$

Ans. (a), (b), all real numbers; (c) $x \neq -5$; (d) $x \neq 3, -4$; (e) $x \neq 0$; (f) $-5 \leqslant x \leqslant 5$ or $|x| \leqslant 5$; (g) $x \leqslant -3, x \geqslant 3$ or $|x| \geqslant 3$; (h) $x \neq \pm 4$; (i) all real numbers.

2.4   A piece of wire 30 in. long is bent to form a rectangle. If one of its dimensions is x in., express the area as a function of x.

   Since the semiperimeter of the rectangle is $\frac{1}{2} \cdot 30 = 15$ in. and one dimension is x in., the other is $(15 - x)$ in. Thus, $A = x(15 - x)$.

2.5   An open box is to be formed from a rectangular sheet of tin 20×32 in. by cutting equal squares, x in. on a side, from the four corners and turning up the sides. Express the volume of the box as a function of x.

   From Fig. 2-1. it is seen that the base of the box has dimensions $(20 - 2x)$ by $(32 - 2x)$ in. and the height is x in. Then

$$V = x(20 - 2x)(32 - 2x) = 4x(10 - x)(16 - x)$$

2.6   A closed box is to be formed from the sheet of tin of Problem 2.5 by cutting equal squares, x cm on a side, from two corners of the short side and two equal rectangles of width x cm from the other two corners, and folding along the dotted lines shown in Fig. 2-2. Express the volume of the box as a function of x.

   One dimension of the base of the box is $(20 - 2x)$ cm; let y cm be the other. Then $2x + 2y = 32$ and $y = 16 - x$. Thus,

$$V = x(20 - 2x)(16 - x) = 2x(10 - x)(16 - x)$$

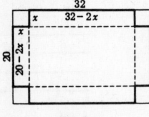

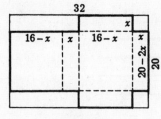

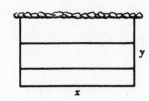

Fig. 2-1   Fig. 2-2   Fig. 2-3

2.7   A farmer has 600 ft of woven wire fencing available to enclose a rectangular field and to divide it into three parts by two fences parallel to one end. If x ft of stone wall is used as one side of the field, express the area enclosed as a function of x when the dividing fences are parallel to the stone wall. Refer to Fig. 2-3.

   The dimensions of the field are x and y ft where $3x + 2y = 600$. Then $y = \frac{1}{2}(600 - 3x)$ and the required area is

$$A = xy = x \cdot \frac{1}{2}(600 - 3x) = \frac{3}{2}x(200 - x)$$

2.8   A right cylinder is said to be inscribed in a sphere if the circumferences of the bases of the cylinder are in the surface of the sphere. If the sphere has radius R, express the volume of the inscribed right circular cylinder as a function of the radius r of its base.

   Let the height of the cylinder be denoted by $2h$. From Fig. 2-4, $h = \sqrt{R^2 - r^2}$ and the required volume is

$$V = \pi r^2 \cdot 2h = 2\pi r^2 \sqrt{R^2 - r^2}$$

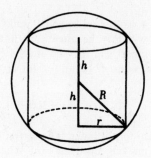

Fig. 2-4

2.9 Given $z = f(x, y) = 2x^2 + 3y^2 - 4$, find

   (a) $f(0,0) = 2(0)^2 + 3(0)^2 - 4 = -4$ (b) $f(2,-3) = 2(2)^2 + 3(-3)^2 - 4 = 31$

   (c) $f(-x,-y) = 2(-x)^2 + 3(-y)^2 - 4 = 2x^2 + 3y^2 - 4 = f(x,y)$

2.10 Given $f(x,y) = \dfrac{x^2 + y^2}{x^2 - y^2}$, find

   (a) $f(x,-y) = \dfrac{x^2 + (-y)^2}{x^2 - (-y)^2} = \dfrac{x^2 + y^2}{x^2 - y^2} = f(x,y)$

   (b) $f\left(\dfrac{1}{x}, \dfrac{1}{y}\right) = \dfrac{(1/x)^2 + (1/y)^2}{(1/x^2) - (1/y)^2} = \dfrac{1/x^2 + 1/y^2}{1/x^2 - 1/y^2} = \dfrac{y^2 + x^2}{y^2 - x^2} = -f(x,y)$

Supplementary Problems

2.11 Represent graphically each of the following domains:

   (a) $x > -3$ (c) $x \geqslant 0$ (e) $|x| < 2$ (g) $-4 \leqslant x \leqslant 4$

   (b) $x < 5$ (d) $-3 < x < -1$ (f) $|x| \geqslant 0$ (h) $x < -3, x \geqslant 5$

2.12 In the three angles, A, B, C of a triangle, angle B exceeds twice angle A by 15°. Express the measure of angle C in terms of angle A.

   Ans. $C = 165° - 3A$

2.13 A grocer has two grades of coffee, selling at \$9.00 and \$10.50 per pound, respectively. In making a mixture of 100 lbs, he uses x lb of the \$10.50 coffee. (a) How many pounds of the \$9.00 coffee does he use? (b) What is the value in dollars of the mixture? (c) At what price per pound should he offer the mixture?

   Ans. (a) $100 - x$ (b) $9(100 - x) + 10.5x$ (c) $9 + 0.015x$

2.14 In a purse are nickels, dimes, and quarters. The number of dimes is twice the number of quarters and the number of nickels is three less than twice the number of dimes. If there are x quarters, find the sum (in cents) in the purse.

   Ans. $65x - 15$

2.15 A and B start from the same place. A walks 4 mi/hr and B walks 5 mi/hr. (a) How far (in miles) will each walk in x hr? (b) How far apart will they be after x hr if they leave at the same time and move in opposite directions? (c) How far apart will they be after A has walked $x > 2$ hours if they move in the same direction but B leaves 2 hr after A? (d) In (c), for how many hours would B have to walk in order to overtake A?

   Ans.   (a) $A, 4x; B, 5x$ (b) $9x$ (c) $|4x - 5(x - 2)|$ (d) 8

2.16   A motor boat, which moves at x mi/hr in still water, is on a river whose current is $y < x$ mi/hr. (a) What is the rate (mi/hr) of the boat when moving upstream? (b) What is the rate of the boat when moving downstream? (c) How far (miles) will the boat travel upstream in 8 hr? (d) How long (hours) will it take the boat moving downstream to cover 20 mi if the motor dies after the first 15 mi?

   Ans.   (a) $x - y$ (b) $x + y$ (c) $8(x - y)$ (d) $\dfrac{15}{x + y} + \dfrac{5}{y}$

2.17   Given $f(x) = \dfrac{x - 3}{x + 2}$, find $f(0)$, $f(1)$, $f(-3)$, $f(a)$, $f(3y)$, $f(x + a)$, $\dfrac{f(x + a) - f(x)}{a}$.

   Ans. $-\frac{3}{2}$, $-\frac{2}{3}$, 6, 0, $\dfrac{a - 3}{a + 2}$, $\dfrac{3y - 3}{3y + 2}$, $\dfrac{x + a - 3}{x + a + 2}$, $\dfrac{5}{(x + 2)(x + a + 2)}$

2.18   A ladder 25 ft long leans against a vertical wall with its foot on level ground 7 ft from the base of the wall. If the foot is pulled away from the wall at the rate 2 ft/s, express the distance (y ft) of the top of the ladder above the ground as a function of the time t seconds in moving.

   Ans. $y = 2\sqrt{144 - 7t - t^2}$

2.19   A boat is tied to a dock by means of a cable 60 m long. If the dock is 20 m above the water and if the cable is being drawn in at the rate 10 m/min, express the distance y m of the boat from the dock after t min.

   Ans. $y = 10\sqrt{t^2 - 12t + 32}$

2.20   A train leaves a station at noon and travels east at the rate 30 mi/hr. At 2 P.M. of the same day a second train leaves the station and travels south at the rate 25 mi/hr. Express the distance d (miles) between the trains as a function of t (hours), the time the second train has been traveling.

   Ans. $d = 5\sqrt{61t^2 + 144t + 144}$

2.21   For each function, tell whether it is a bijection:

   (a) $y = x^4$
   (b) $y = \sqrt{x}$
   (c) $y = 2x^2 + 3$

   Ans.   (a) No   (b) Yes   (c) No

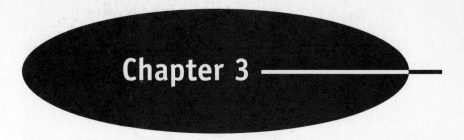

Graphs of Functions

A FUNCTION $y = f(x)$, by definition, yields a collection of pairs $(x, f(x))$ or (x, y) in which x is any element in the domain of the function and $f(x)$ or y is the corresponding value of the function. These pairs are called *ordered pairs*.

EXAMPLE 1.   Obtain 10 ordered pairs for the function $y = 3x - 2$.

The domain of definition of the function is the set of real numbers. We may choose at random any 10 real numbers as values of x. For one such choice, we obtain the chart in Table 3.1.

Table 3.1

x	-2	$-\frac{4}{3}$	$-\frac{1}{2}$	0	$\frac{1}{3}$	1	2	$\frac{5}{2}$	3	4
y	-8	-6	$-\frac{7}{2}$	-2	-1	1	4	$\frac{11}{2}$	7	10

(See Problem 3.1.)

THE RECTANGULAR CARTESIAN COORDINATE SYSTEM in a plane is a device by which there is established a one-to-one correspondence between the points of the plane and ordered pairs of real numbers (a, b).

Consider two real number scales intersecting at right angles in O, the origin of each (see Fig. 3-1), and having the positive direction on the horizontal scale (now called the *x axis*) directed to the right and the positive direction on the vertical scale (now called the *y axis*) directed upward.

Let P be any point distinct from O in the plane of the two axes and join P to O by a straight line. Let the projection of OP on the x axis be $OM = a$ and the projection of OP on the y axis be $ON = b$. Then the pair of numbers (a, b) in that order are called the plane rectangular Cartesian coordinates (briefly, the rectangular coordinates) of P. In particular, the coordinates of O, the *origin* of the coordinate system, are $(0, 0)$.

The first coordinate, giving the directed distance of P from the y axis, is called the *abscissa* of P, while the second coordinate, giving the directed distance of P from the x axis, is called the *ordinate* of P. Note carefully that the points $(3, 4)$ and $(4, 3)$ are distinct points.

The axes divide the plane into four sections, called   *quadrants*. Figure 4-1 shows the customary numbering of the quadrants and the respective signs of the coordinates of a point in each quadrant. (See Problems 3.1–3.4.)

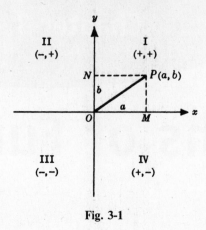

Fig. 3-1

THE GRAPH OF A FUNCTION $y = f(x)$ consists of the totality of points (x, y) whose coordinates satisfy the relation $y = f(x)$.

EXAMPLE 2. Graph the function $3x - 2$.

After plotting the points whose coordinates (x, y) are given in Table 3.1, it appears that they lie on a straight line. See Fig. 3.2. Figure 3-2 is not the complete graph since (1000, 2998) is one of its points and is not shown. Moreover, although we have joined the points by a straight line, we have not proved that every point on the line has as coordinates a number pair given by the function. These matters as well as such questions as: What values of x should be chosen? How many values of x are needed? will become clearer as we proceed with the study of functions. At present,

(1)   Build a table of values.
(2)   Plot the corresponding points.
(3)   Pass a smooth curve through these points, moving from left to right.

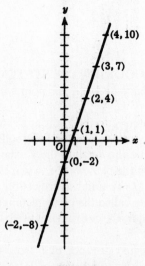

Fig. 3-2

It is helpful to picture the curve in your mind before attempting to trace it on paper. If there is doubt about the curve between two plotted points, determine other points in the interval.

ANY VALUE OF x for which the corresponding value of function $f(x)$ is zero is called a *zero* of the function. Such values of x are also called *roots* of the equation $f(x) = 0$. The real roots of an equation $f(x) = 0$ may be approximated by estimating from the graph of $f(x)$ the abscissas of its points of intersection with the x axis. (See Problems 3.9–3.11.)

Algebraic methods for finding the roots of equations will be treated in later chapters. The graphing calculator can also be used to find roots by graphing the function and observing where the graph intersects the x axis. See Appendix A.

Solved Problems

3.1 (*a*) Show that the points $A(1, 2)$, $B(0, -3)$, and $C(2, 7)$ are on the graph of $y = 5x - 3$.

(*b*) Show that the points $D(0, 0)$ and $E(-1, -2)$ are not on the graph of $y = 5x - 3$.

(*a*) The point $A(1, 2)$ is on the graph since $2 = 5(1) - 3$, $B(0, -3)$ is on the graph since $-3 = 5(0) - 3$, and $C(2, 7)$ is on the graph since $7 = 5(2) - 3$.

(*b*) The point $D(0, 0)$ is not on the graph since $0 \neq 5(0) - 3$, and $E(-1, -2)$ is not on the graph since $-2 \neq 5(-1) - 3$.

3.2 Sketch the graph of the function $2x$. Refer to Table 3.2.

Table 3.2

x	0	1	2
$y = f(x)$	0	2	4

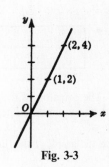

Fig. 3-3

This is a linear function and its graph is a straight line. For this graph only two points are necessary. Three points are used to provide a check. See Fig. 3-3. The equation of the line is $y = 2x$.

3.3 Sketch the graph of the function $6 - 3x$. Refer to Table 3.3.

Table 3.3

x	0	2	3
$y = f(x)$	6	0	-3

See Fig. 3-4. The equation of the line is $y = 6 - 3x$.

3.4   Sketch the graph of the function x^2. Refer to Table 3.4.

Table 3.4

x	3	1	0	-2	-3
$y = f(x)$	9	1	0	4	9

See Fig. 3-5. The equation of this graph, called a *parabola*, is $y = x^2$. Note for $x \neq 0$, $x^2 > 0$. Thus, the curve is never below the x axis. Moreover, as $|x|$ increase, x^2 increases; that is, as we move from the origin along the x axis in either direction, the curve moves farther and farther from the axis. Hence, in sketching parabolas sufficient points must be plotted so that their U shape can be seen.

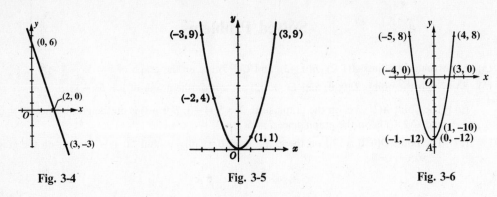

Fig. 3-4   Fig. 3-5   Fig. 3-6

3.5   Sketch the graph of the function $x^2 + x - 12$. Refer to Table 3.5.

Table 3.5

x	4	3	1	0	-1	-4	-5
$y = f(x)$	8	0	-10	-12	-12	0	8

The equation of the parabola is $y = x^2 + x - 12$. Note that the points $(0, -12)$ and $(-1, -12)$ are *not* joined by a straight line segment. Check that the value of the function is $-12\frac{1}{4}$ when $x = -\frac{1}{2}$. See Fig. 3-6.

3.6   Sketch the graph of the function $-2x^2 + 4x + 1$. Refer to Table 3.6.

Table 3.6

x	3	2	1	0	-1
$y = f(x)$	-5	1	3	1	-5

See Fig. 3-7.

3.7   Sketch the graph of the function $(x + 1)(x - 1)(x - 2)$. Refer to Table 3.7.

Table 3.7

x	3	2	$\frac{3}{2}$	1	0	-1	-2
$y = f(x)$	8	0	$-\frac{5}{8}$	0	2	0	-12

This is a *cubic* curve of the equation $y = (x + 1)(x - 1)(x - 2)$. It crosses the x axis where $x = -1, 1$, and 2. See Fig. 3-8.

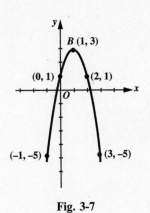

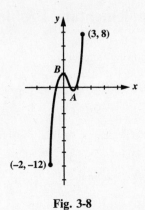

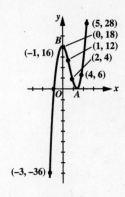

Fig. 3-7   Fig. 3-8   Fig. 3-9

3.8   Sketch the graph of the function $(x + 2)(x - 3)^2$. Refer to Table 3.8.

Table 3.8

x	5	4	$\frac{7}{2}$	3	2	1	0	-1	-2	-3
$y = f(x)$	28	6	$\frac{11}{8}$	0	4	12	18	16	0	-36

This cubic crosses the x axis where $x = -2$ and is tangent to the x axis where $x = 3$. Note that for $x > -2$, the value of the function is positive except for $x = 3$, where it is 0. Thus, to the right of $x = -2$, the curve is *never* below the x axis. See Fig. 3-9.

3.9   Sketch the graph of the function $x^2 + 2x - 5$ and by means of it determine the real roots of $x^2 + 2x - 5 = 0$. Refer to Table 3.9.

Table 3.9

x	2	1	0	-1	-2	-3	-4
$y = f(x)$	3	-2	-5	-6	-5	-2	3

The parabola cuts the x axis at a point whose abscissa is between 1 and 2 (the value of the function changes sign) and at a point whose abscissa is between -3 and -4.

Reading from the graph in Fig. 3-10, the roots are $x = 1.5$ and $x = -3.5$, approximately.

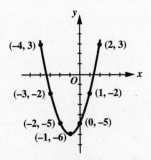

Fig. 3-10

Supplementary Problems

3.10 Sketch the graph of each of the following functions:

(a) $3x - 2$ (c) $x^2 - 1$ (e) $x^2 - 4x + 4$ (g) $(x-2)(x+1)^2$

(b) $2x + 3$ (d) $4 - x^2$ (f) $(x+2)(x-1)(x-3)$

3.11 From the graph of each function $f(x)$ determine the real roots, if any, of $f(x) = 0$.

(a) $x^2 - 4x + 3$ (b) $2x^2 + 4x + 1$ (c) $x^2 - 2x + 4$

Ans. (a) 1,3   (b) $-0.3, -1.7$ (c) none

3.12 If A is a point on the graph of $y = f(x)$, the function being restricted to the type considered in this chapter, and if all points of the graph sufficiently near A are higher than A (that is, lie above the horizontal drawn through A), then A is called a *relative minimum point* of the graph. (a) Verify that the origin is the relative minimum point of the graph of Problem 3.4. (b) Verify that the graph of Problem 3.5 has a relative minimum at a point whose abscissa is between $x = -1$ and $x = 0$ (at $x = -\frac{1}{2}$), the graph of Problem 3.7 has a relative minimum at a point whose abscissa is between $x = 1$ and $x = 2$ (approximately $x = 1.5$), and the graph of Problem 3.8 has $(3, 0)$ as relative minimum point. Also, see Chapter 48 for a more sophisticated discussion of minima.

3.13 If B is a point on the graph of $y = f(x)$ and if all points of the graph sufficiently near B are lower than B (that is, lie below the horizontal drawn through B), then B is called a *relative maximum point* of the graph. (a) Verify that $(1, 3)$ is the relative maximum point of the graph of Problem 3.6. (b) Verify that the graph of Problem 3.7 has a relative maximum at a point whose abscissa is between $x = -1$ and $x = 1$ (approximately $x = -0.2$), and that the graph of Problem 3.8 has a relative maximum between $x = -1$ and $x = 0$ (at $x = -\frac{1}{3}$). See Chapter 48 for additional work on extrema.

3.14 Verify that the graphs of the functions of Problem 3.11 have relative minimums at $x = 2$, $x = -1$, and $x = 1$, respectively.

3.15 From the graph of the function of Problem 2.4 in Chapter 2 read that the area of the rectangle is a relative maximum when $x = \frac{15}{2}$.

3.16 From the graph of the function of Problem 2.7 in Chapter 2 read that the area enclosed is a relative maximum when $x = 100$.

3.17 Use a graphing calculator to locate the zeros of the function $y = x^2 + 3$.

3.18 Use a graphing calculator to graph $y = x^2$, $y = x^4$, and $y = x^6$ on the same axes. What do you notice?

3.19 Repeat Problem 3.18 using $y = x^3$, $y = x^5$ and $y = x^7$.

Chapter 4

Linear Equations

AN EQUATION is a statement, such as (a) $2x - 6 = 4 - 3x$, (b) $y^2 + 3y = 4$, and (c) $2x + 3y = 4xy + 1$, that two expressions are equal. An equation is linear in an unknown if the highest degree of that unknown in the equation is 1. An equation is quadratic in an unknown if the highest degree of that unknown is 2. The first is a *linear* equation in one unknown, the second is a *quadratic* in one unknown, and the third is linear in each of the two unknowns but is of degree 2 in the two unknowns.

Any set of values of the unknowns for which the two members of an equation are equal is called a *solution* of the equation. Thus, $x = 2$ is a solution of (a), since $2(2) - 6 = 4 - 3(2)$; $y = 1$ and $y = -4$ are solutions of (b); and $x = 1, y = 1$ is a solution of (c). A solution of an equation in one unknown is also called a *root* of the equation.

TO SOLVE A LINEAR EQUATION in one unknown, perform the same operations on both members of the equation in order to obtain the unknown alone in the left member.

EXAMPLE 1. Solve: $2x - 6 = 4 - 3x$.

Add 6: $\qquad 2x = 10 - 3x \qquad$ Check: $2(2) - 6 = 4 - 3(2)$

Add $3x$: $\qquad 5x = 10 \qquad\qquad\qquad\quad -2 = -2$

Divide by 5: $\quad x = 2$

EXAMPLE 2. Solve: $\frac{1}{3}x - \frac{1}{2} = \frac{3}{4}x + \frac{5}{6}$.

Multiply by LCD $= 12$: $\quad 4x - 6 = 9x + 10 \qquad$ Check: $\frac{1}{3}\left(-\frac{16}{5}\right) - \frac{1}{2} = \frac{3}{4}\left(-\frac{16}{5}\right) + \frac{5}{6}$

Add $6 - 9x$: $\qquad\qquad\qquad\quad -5x = 16$

Divide by -5: $\qquad\qquad\qquad\quad x = -\frac{16}{5} \qquad\qquad\qquad\qquad -\frac{47}{30} = -\frac{47}{30}$

(See Problems 4.1–4.3.)

An equation which contains fractions having the unknown in one or more denominators may sometimes reduce to a linear equation when cleared of fractions. When the resulting equation is solved, the solution *must* be checked since it may or may not be a root of the original equation. (See Problems 4.4–4.8.)

RATIO AND PROPORTION. The ratio of two quantities is their quotient. The ratio of 1 inch to 1 foot is 1/12 or 1:12, a pure number; the ratio of 30 miles to 45 minutes is $30/45 = 2/3$ mile per minute.

The expressed equality of two ratios, as $\frac{a}{b} = \frac{c}{d}$, is called a *proportion*. (See Problems 4.11–4.12.)

VARIATION. A variable y is said to vary *directly* as another variable x (or y is proportional to x) if y is equal to some constant c times x, that is, if $y = cx$.

A variable y is said to vary *inversely* as another variable x if y varies directly as the reciprocal of x, that is, if $y = c/x$.

Solved Problems

Solve and check the following equations. The check has been omitted in certain problems.

4.1 $x - 2(1 - 3x) = 6 + 3(4 - x).$

$$x - 2 + 6x = 6 + 12 - 3x$$
$$7x - 2 = 18 - 3x$$
$$10x = 20$$
$$x = 2$$

4.2 $ay + b = cy + d.$

$$ay - cy = d - b$$
$$(a - c)y = d - b$$
$$y = \frac{d - b}{a - c}$$

4.3 $\dfrac{3x - 2}{5} = 4 - \dfrac{1}{2}x.$

Multiply by 10: $6x - 4 = 40 - 5x$ Check: $\dfrac{3(4) - 2}{5} = 4 - \dfrac{1}{2}(4)$

$$11x = 44 \qquad\qquad\qquad\qquad 2 = 2$$
$$x = 4$$

4.4 $\dfrac{3x + 1}{3x - 1} = \dfrac{2x + 1}{2x - 3}.$ Here the LCD is $(3x - 1)(2x - 3)$.

Multiply by LCD: $(3x + 1)(2x - 3) = (2x + 1)(3x - 1)$

$$6x^2 - 7x - 3 = 6x^2 + x - 1$$
$$-8x = 2$$
$$x = -\frac{1}{4}$$

Check: $\dfrac{3(-\frac{1}{4}) + 1}{3(-\frac{1}{4}) - 1} = \dfrac{2(-\frac{1}{4}) + 1}{2(-\frac{1}{4}) - 3},$

$$\frac{-3 + 4}{-3 - 4} = \frac{-2 + 4}{-2 - 12}, \qquad -\frac{1}{7} = -\frac{1}{7}$$

4.5 $\dfrac{1}{x-3} - \dfrac{1}{x+1} = \dfrac{3x-2}{(x-3)(x+1)}$.   Here the LCD is $(x-3)(x+1)$.

$$(x+1) - (x-3) = 3x - 2$$
$$-3x = -6$$
$$x = 2$$

Check: $-1 - \dfrac{1}{3} = \dfrac{6-2}{-3}$, $-\dfrac{4}{3} = -\dfrac{4}{3}$

4.6 $\dfrac{1}{x-3} + \dfrac{1}{x-2} = \dfrac{3x-8}{(x-3)(x-2)}$.   The LCD is $(x-3)(x-2)$.

$$(x-2) + (x-3) = 3x - 8$$
$$2x - 5 = 3x - 8$$
$$x = 3$$

Check: When $x = 3, \dfrac{1}{x-3}$ is without meaning. The given equation has no root. The value $x = 3$ is called *extraneous*.

4.7 $\dfrac{x^2 - 2}{x - 1} = x + 1 - \dfrac{1}{x-1}$.

$$x^2 - 2 = (x+1)(x-1) - 1$$
$$= x^2 - 2$$

The given equation is satisfied by all values of x except $x = 1$. It is called an identical equation or *identity*.

4.8 $\dfrac{1}{x-1} + \dfrac{1}{x-3} = \dfrac{2x-5}{(x-1)(x-3)}$.

$$(x-3) + (x-1) = 2x - 5$$
$$2x - 4 = 2x - 5$$

There is no solution. $2x - 4$ and $2x - 5$ are unequal for all values of x.

4.9 One number is 5 more than another and the sum of the two is 71. Find the numbers.

Let x be the smaller number and $x + 5$ be the larger. Then $x + (x + 5) = 71$, $2x = 66$, and $x = 33$. The numbers are 33 and 38.

4.10 A father is now three times as old as his son. Twelve years ago he was six times as old as his son. Find the present age of each.

Let $x = $ the age of the son and $3x = $ the age of the father. Twelve years ago, the age of the son was $x - 12$ and the age of the father was $3x - 12$.

Then $3x - 12 = 6(x - 12), 3x = 60$, and $x = 20$. The present age of the son is 20 and that of the father is 60.

4.11 When two pulleys are connected by a belt, their angular velocities (revolutions per minute) are *inversely* proportional to their diameters; that is, $\omega_1 : \omega_2 = d_2 : d_1$. Find the velocity of a pulley 15 cm in diameter when it is connected to a pulley 12 cm in diameter and rotating at 100 rev/cm.

Let ω_1 be the unknown velocity; then $d_1 = 15$, $\omega_2 = 100$, and $d_2 = 12$. The given formula becomes

$$\frac{\omega_1}{100} = \frac{12}{15} \qquad \text{and} \qquad \omega_1 = \frac{12}{15}(100) = 80 \text{ rev/min}$$

4.12 Bleaching powder is obtained through the reaction of chlorine and slaked lime, 74.10 kg of lime and 70.91 kg of chlorine producing 127.00 kg of bleaching powder and 18.01 kg of water. How many kg of lime will be required to produce 1000 kg of bleaching powder?

Let x = the number of kg of lime required. Then

$$\frac{x(\text{kg of lime})}{1000(\text{kg of powder})} = \frac{74.10(\text{kg of lime})}{127(\text{kg of powder})} \qquad 127x = 74\,100 \qquad \text{and} \qquad x = 583.46 \text{ kg}$$

Supplementary Problems

4.13 Solve for x and check each of the following:

(a) $2x - 7 = 29 - 4x$

(c) $\dfrac{x+3}{x-3} = 3$

(e) $\dfrac{2x+1}{4} - \dfrac{1}{x-1} = \dfrac{x}{2}$

(b) $2(x-1) - 3(x-2) + 4(x-3) = 0$

(d) $\dfrac{4}{x-4} = \dfrac{2}{2x-5}$

(f) $a(x+3) + b(x-2) = c(x-1)$

Ans.   (a) 6   (b) $\dfrac{8}{3}$ (c) 6   (d) 2   (e) 5   (f) $\dfrac{2b-3a-c}{a+b-c}$

4.14 A piece of wire $11\frac{2}{3}$ m long is to be divided into two parts such that one part is $\frac{2}{3}$ that of the other. Find the length of the shorter piece.

Ans. $4\frac{2}{3}$ m

4.15 A train leaves a station and travels at the rate of 40 mi/hr. Two hours later a second train leaves the station and travels at the rate of 60 mi/hr. Where will the second train overtake the first?

Ans.   240 mi from the station

4.16 A tank is drained by two pipes. One pipe can empty the tank in 30 min, and the other can empty it in 25 min. If the tank is $\frac{5}{6}$ filled and both pipes are open, in what time will the tank be emptied?

Ans. $11\frac{4}{11}$ min

4.17 A man invests $\frac{1}{3}$ of his capital at 6% and the remainder at 8%. What is his capital if his total income is $4400?

Ans. $60,000

4.18 *A* can do a piece of work in 10 days. After he has worked 2 days, *B* comes to help him and together they finish it in 3 days. In how many days could *B* alone have done the work?

Ans.   6 days

4.19 When two resistances R_1 and R_2 are placed in parallel, the resultant resistance R is given by $1/R = 1/R_1 + 1/R_2$. Find R when $R_1 = 80$ and $R_2 = 240$.

Ans.   60

4.20 How soon after noon are the hands of a clock together again?

Ans.   1 hr, $5\frac{5}{11}$ min

4.21 How much water will be produced in making the 1000 kg bleaching powder in Problem 4.12?

Ans.   141.81 kg

4.22 The reaction of 65.4 g of zinc and 72.9 g of hydrochloric acid produces 136.3 g of zinc chloride and 2 g of hydrogen. Find the weight of hydrochloric acid necessary for a complete reaction with 300 g of zinc and the weight of hydrogen produced.

Ans.   334.4 g, 9.2 g

4.23   How much water must be used to prepare a 1:5000 solution of bichloride of mercury from a 0.5-g tablet?

   Ans.   2500 g

4.24   Newton's law of gravitation states that the force F of attraction between two bodies varies jointly as their masses m_1 and m_2 and inversely as the square of the distance between them. Two bodies whose centers are 5000 mi apart attract each other with a force of 15 lb. What would be the force of attraction if their masses were tripled and the distance between their centers was doubled?

   Ans. $33\frac{3}{4}$ lb

4.25   If a body weighs 20 lb on the earth's surface, what would it weigh 2000 mi above the surface? (Assume the radius of the earth to be 4000 mi.)

   Ans. $8\frac{8}{9}$ lb

Chapter 5

Simultaneous Linear Equations

TWO LINEAR EQUATIONS IN TWO UNKNOWNS.   Let the system of equations be

$$\begin{cases} a_1x + b_1y + c_1 = 0 \\ a_2x + b_2y + c_2 = 0 \end{cases}$$

Each equation has an unlimited number of solutions (x, y) corresponding to the unlimited number of points on the locus (straight line) which it represents. Our problem is to find all solutions common to the two equations or the coordinates of all points common to the two lines. There are three cases:

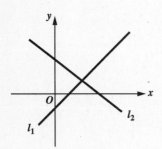

Fig. 5-1

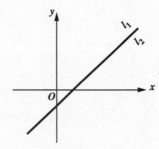

Fig. 5-2

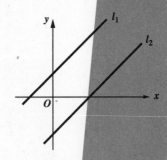

Fig. 5-3

(1)   The system has one and only one solution; that is, the two lines have one and only one point in common. The equations are said to be *consistent* (have common solutions) *and independent*. See Fig. 5-1, indicating two distinct intersecting lines.

(2)   The system has an unlimited number of solutions; that is, the two equations are equivalent or the two lines are coincident. The equations are said to be *consistent and dependent*. See Fig. 5-2, indicating that the two equations represent the same line.

(3)   The system has no solution; that is, the two lines are parallel and distinct. The equations are said to be *inconsistent*. See Fig. 5-3, indicating that the two equations result in two parallel lines.

GRAPHICAL SOLUTION.   We plot the graphs of the two equations on the same axes and scale off the coordinates of the point of intersection. The defect of this method is that, in general, only approximate solutions are obtained. (See Problem 5.1.)

ALGEBRAIC SOLUTION.   A system of two consistent and independent equations in two unknowns may be solved algebraically by eliminating one of the unknowns.

EXAMPLE 1.   Solve the system

$$\begin{cases} 3x - 6y = 10 \\ 9x + 15y = -14 \end{cases}$$

$$(5.1)$$
$$(5.2)$$

ELIMINATION BY SUBSTITUTION

Solve (5.1) for x:

$$x = \tfrac{10}{3} + 2y \qquad (5.3)$$

Substitute in (5.2):

$$9\left(\tfrac{10}{3} + 2y\right) + 15y = -14$$
$$30 + 18y + 15y = -14 \qquad 33y = -44 \qquad y = -\tfrac{4}{3}$$

Substitute for y in (5.3):

$$x = \tfrac{10}{3} + 2\left(-\tfrac{4}{3}\right) = \tfrac{2}{3}$$

Check: Using (5.2),

$$9\left(\tfrac{2}{3}\right) + 15\left(-\tfrac{4}{3}\right) = -14$$

EXAMPLE 2.   Solve the system

$$\begin{cases} 2x - 3y = 10 \\ 3x - 4y = 8 \end{cases}$$

$$(5.4)$$
$$(5.5)$$

ELIMINATION BY ADDITION

Multiply (5.4) by -3 and (5.5) by 2:

$$-6x + 9y = -30$$
$$\underline{6x - 8y = 16}$$
Add: $\qquad\qquad y = -14$

Substitute for x in (5.4):

$$2x + 42 = 10 \quad \text{ or } \quad x = -16$$

Check: Using (5.5),

$$3(-16) - 4(-14) = 8$$

(See Problems 5.2–5.4.)

THREE LINEAR EQUATIONS IN THREE UNKNOWNS. A system of three consistent and independent equations in three unknowns may be solved algebraically by deriving from it a system of two equations in two unknowns. (The reader should consult Chapter 21 for a thorough discussion of this topic.)

EXAMPLE 3. Solve the system

$$\begin{cases} 2x + 3y - 4z = 1 & (5.6) \\ 3x - y - 2z = 4 & (5.7) \\ 4x - 7y - 6z = -7 & (5.8) \end{cases}$$

We shall eliminate y.

$$\begin{array}{ll} \text{Rewrite } (5.6): & 2x + 3y - 4z = 1 \\ 3 \times (5.7): & \underline{9x - 3y - 6z = 12} \\ \text{Add:} & 11x \quad\; - 10z = 13 \end{array} \qquad (5.9)$$

$$\begin{array}{ll} \text{Rewrite } (5.8): & 4x - 7y - \; 6z = -7 \\ -7 \times (5.8): & \underline{-21x + 7y + 14z = -28} \\ \text{Add:} & -17x \quad\quad + 8z = -35 \end{array} \qquad (5.10)$$

Next, solve (5.9) and (5.10).

$$\begin{array}{ll} 4 \times (5.9): & 44x - 40z = \quad 52 \\ 5 \times (5.10): & \underline{-85x + 40z = -175} \\ \text{Add:} & -41x \quad\quad\; = -123 \\ & \qquad\quad x = 3 \end{array}$$

From (5.9): $\quad 11(3) - 10z = 13 \quad z = 2$
From (5.6): $\quad 2(3) + 3y - 4(2) = 1 \quad y = 1$
Check: Using (5.7),

$$3(3) - 1 - 2(2) = 4$$

(See Problems 5.5–5.6.)

SOLUTIONS OF LINEAR SYSTEMS USING DETACHED COEFFICIENTS.

In Example 4 below, a variation of the method of addition and subtraction is used to solve a system of linear equations. On the left the equations themselves are used, while on the right the same moves are made on the rectangular array (called a *matrix*) of the coefficients and constant terms. The numbering (1), (2), (3), ... refers both to the equations and to the rows of the matrices.

EXAMPLE 4. Solve the system

$$\begin{cases} 2x - 3y = \; 2 \\ 4x + 7y = -9 \end{cases}$$

USING EQUATIONS		USING MATRICES
$2x - 3y = \; 2$	(1)	$\begin{pmatrix} 2 & -3 & \vert & 2 \\ 4 & 7 & \vert & -9 \end{pmatrix}$
$4x + 7y = -9$	(2)	

Multiply (1) by $\frac{1}{2}$ and write as (3). Multiply (1) by -2 and add to (2) to obtain (4).

$$\begin{aligned} x - \tfrac{3}{2}y &= 1 \\ 13y &= -13 \end{aligned} \qquad \begin{aligned} &(3) \\ &(4) \end{aligned} \qquad \begin{pmatrix} 1 & -\tfrac{3}{2} & \bigg| & 1 \\ 0 & 13 & \bigg| & -13 \end{pmatrix}$$

Multiply (4) by $\tfrac{3}{2}/13 = \tfrac{3}{26}$ and add to (3) to obtain (5). Multiply (4) by $\tfrac{1}{13}$ to obtain (6).

$$\begin{aligned} x &= -\tfrac{1}{2} \\ y &= -1 \end{aligned} \qquad \begin{aligned} &(5) \\ &(6) \end{aligned} \qquad \begin{pmatrix} 1 & 0 & \bigg| & -\tfrac{1}{2} \\ 0 & 1 & \bigg| & -1 \end{pmatrix}$$

The required solution is $x = -\tfrac{1}{2}$, $y = -1$.

EXAMPLE 5. Solve, using matrices, the system

$$\begin{cases} 2x - 3y + 2z = 14 \\ 4x + 4y - 3z = 6 \\ 3x + 2y - 3z = -2 \end{cases}$$

The matrix of the system

$$\begin{pmatrix} 2 & -3 & 2 & \bigg| & 14 \\ 4 & 4 & -3 & \bigg| & 6 \\ 3 & 2 & -3 & \bigg| & -2 \end{pmatrix}$$

is formed by writing in order the coefficients of x, y, z and the constant terms.

There are, in essence, three moves:

(a) Multiply the elements of a row by a nonzero number. This move is used only to obtain an element 1 in a prescribed position.

(b) Multiply the elements of a row by a nonzero number and add to the corresponding elements of another row. This move is used to obtain an element 0 in a prescribed position.

(c) Exchange two rows when required.

The first attack must be planned to yield a matrix of the form

$$\begin{pmatrix} 1 & * & * & \bigg| & * \\ 0 & * & * & \bigg| & * \\ 0 & * & * & \bigg| & * \end{pmatrix}$$

in which only the elements of the first column are prescribed.

Multiply first row by $\tfrac{1}{2}$:
Multiply first row by -2 and add to second row:
Multiply first row by $-\tfrac{3}{2}$ and add to third row:

$$\begin{pmatrix} 1 & -\tfrac{3}{2} & 1 & \bigg| & 7 \\ 0 & 10 & -7 & \bigg| & -22 \\ 0 & \tfrac{13}{2} & -6 & \bigg| & -23 \end{pmatrix}$$

The second attack must be planned to yield a matrix of the form

$$\begin{pmatrix} 1 & 0 & * & \bigg| & * \\ 0 & 1 & * & \bigg| & * \\ 0 & 0 & * & \bigg| & * \end{pmatrix}$$

in which the elements of the first two columns are prescribed.

Multiply second row by $\tfrac{3}{20}$ and add to first row:
Multiply second row by $\tfrac{1}{10}$:
Multiply second row by $-\tfrac{13}{20}$ and add to third row:

$$\begin{pmatrix} 1 & 0 & -\tfrac{1}{20} & \bigg| & \tfrac{37}{10} \\ 0 & 1 & -\tfrac{7}{10} & \bigg| & -\tfrac{11}{5} \\ 0 & 0 & -\tfrac{29}{20} & \bigg| & -\tfrac{87}{10} \end{pmatrix}$$

The final attack must be planned to yield a matrix of the form

$$\begin{pmatrix} 1 & 0 & 0 & \bigg| & * \\ 0 & 1 & 0 & \bigg| & * \\ 0 & 0 & 1 & \bigg| & * \end{pmatrix}$$

in which the elements of the first three columns are prescribed.

Multiply third row by $-\frac{1}{29}$ and add to first row:

Multiply third row by $-\frac{14}{29}$ and add to second row:

Multiply third row by $-\frac{20}{29}$:

$$\begin{pmatrix} 1 & 0 & 0 & | & 4 \\ 0 & 1 & 0 & | & 2 \\ 0 & 0 & 1 & | & 6 \end{pmatrix}$$

The solution is $x = 4$, $y = 2$, $z = 6$.

SOLUTIONS USING THE GRAPHING CALCULATOR.   Systems of equations such as

$$\begin{cases} 2x - 3y = 2 \\ 4x + 7y = -9 \end{cases}$$

are easily solved using the graphing calculator. See Appendix A for graphing calculator directions. Also, software packages, such as Maple, provide broad capabilities for this topic.

Solved Problems

5.1   Solve graphically the systems

(a) $\begin{cases} x + 2y = 5 \\ 3x - y = 1 \end{cases}$,   (b) $\begin{cases} x + y = 1 \\ 2x + 3y = 0 \end{cases}$,   (c) $\begin{cases} 3x - 6y = 10 \\ 9x + 15y = -14 \end{cases}$.

(a) $x = 1$, $y = 2$ (b) $x = 3$, $y = -2$ (c) $x = 0.7$, $y = -1.3$ (See Fig. 5-4.)

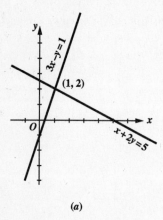

(a)

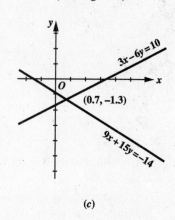

(b)

(c)

Fig. 5-4

5.2   Solve algebraically:

(a) $\begin{cases} x + 2y = 5 & (1) \\ 3x - y = 1 & (2) \end{cases}$ (b) $\begin{cases} 3x + 2y = 2 & (1) \\ 5x + 6y = 4 & (2) \end{cases}$ (c) $\begin{cases} 2x + 3y = 3 & (1) \\ 5x - 9y = -4 & (2) \end{cases}$

(a)   Rewrite (1): $x + 2y = 5$

Multiply (2) by 2: $\underline{6x - 2y = 2}$

Add: $7x \qquad = 7$

$x = 1$

Substitute for x in (1):

$$1 + 2y = 5, \quad y = 2$$

Check: Using (2),

$$3(1) - 2 = 1$$

(b)   Multiply (1) by -5: $-15x - 10y = -10$

Multiply (2) by 3: $\underline{15x + 18y = 12}$

Add: $8y = 2$

$y = \tfrac{1}{4}$

Substitute for y in (1):

$$3x + 2\left(\tfrac{1}{4}\right) = 2, \quad x = \tfrac{1}{2}$$

Check: Using (2),

$$5\left(\tfrac{1}{2}\right) + 6\left(\tfrac{1}{4}\right) = 4$$

(c)   Multiply (1) by 3: $6x + 9y = 9$

Rewrite (2): $5x - 9y = -4$

Add: $11x = 5$

$x = \tfrac{5}{11}$

Substitute in (1):

$$3y = 3 - 2\left(\tfrac{5}{11}\right) = \tfrac{23}{11}, \quad y = \tfrac{23}{33}$$

Check: Using (2),

$$5\left(\tfrac{5}{11}\right) - 9\left(\tfrac{23}{33}\right) = -4$$

5.3   If the numerator of a fraction is increased by 2, the fraction is $\tfrac{1}{4}$; if the denominator is decreased by 6, the fraction is $\tfrac{1}{6}$. Find the fraction.

Let $\tfrac{x}{y}$ be the original fraction. Then

$$\frac{x+2}{y} = \frac{1}{4} \quad \text{or} \quad 4x - y = -8 \quad (1)$$

$$\frac{x}{y-6} = \frac{1}{6} \quad \text{or} \quad 6x - y = -6 \quad (2)$$

Subtract (1) from (2):

$$2x = 2 \quad \text{and} \quad x = 1$$

Substitute $x = 1$ in (1):

$$4 - y = -8 \quad \text{and} \quad y = 12$$

The fraction is $\tfrac{1}{12}$.

5.4   Solve the system

$$\begin{cases} x - 5y + 3z = 9 & (1) \\ 2x - y + 4z = 6 & (2) \\ 3x - 2y + z = 2 & (3) \end{cases}$$

Eliminate z.

Rewrite (1): $x - 5y + 3z = 9$

Multiply (3) by -3: $\underline{-9x + 6y - 3z = -6}$

Add: $-8x + y = 3 \quad (4)$

$$\text{Rewrite } (2): \qquad 2x - y + 4z = 6$$
$$\text{Multiply } (3) \text{ by } -4: \quad -12x + 8y - 4z = -8$$
$$\underline{\qquad\qquad\qquad\qquad\qquad}$$
$$\text{Add:} \quad -10x + 7y \qquad = -2 \quad (5)$$

$$\text{Multiply } (4) \text{ by } -7: \quad 56x - 7y = -21$$
$$\text{Rewrite } (5): \qquad \underline{-10x + 7y = -2}$$
$$\text{Add:} \quad 46x \qquad = -23$$
$$x = -\tfrac{1}{2}$$

Substitute $x = -\tfrac{1}{2}$ in (4):

$$-8(-\tfrac{1}{2}) + y = 3 \quad \text{and} \quad y = -1$$

Substitute $x = -\tfrac{1}{2}$, $y = -1$ in (1):

$$-\tfrac{1}{2} - 5(-1) + 3z = 9 \quad \text{and} \quad z = \tfrac{3}{2}$$

Check: Using (2), $2(-\tfrac{1}{2}) - (-1) + 4(\tfrac{3}{2}) = -1 + 1 + 6 = 6.$

5.5   A parabola $y = ax^2 + bx + c$ passes through the points $(1,0)$, $(2,2)$, and $(3,10)$. Determine its equation.

Since $(1,0)$ is on the parabola:

$$a + b + c = 0 \quad (1)$$

Since $(2,2)$ is on the parabola:

$$4a + 2b + c = 2 \quad (2)$$

Since $(3,10)$ is on the parabola:

$$9a + 3b + c = 10 \quad (3)$$

Subtract (1) from (2):

$$3a + b = 2 \quad (4)$$

Subtract (1) from (3):

$$8a + 2b = 10 \quad (5)$$

Multiply (4) by -2 and add to (5):

$$2a = 6 \quad \text{and} \quad a = 3$$

Substitute $a = 3$ in (4):

$$3(3) + b = 2 \quad \text{and} \quad b = -7$$

Substitute $a = 3$, $b = -7$ in (1):

$$3 - 7 + c = 0 \quad \text{and} \quad c = 4$$

The equation of the parabola is $y = 3x^2 - 7x + 4$.

5.6   Solve, using matrices, the system

$$\begin{cases} x - 5y + 3z = 9 \\ 2x - y + 4z = 6 \\ 3x - 2y + z = 2 \end{cases} \quad \text{(See Problem 5.4.)}$$

Begin with the matrix:

$$\begin{pmatrix} 1 & -5 & 3 & | & 9 \\ 2 & -1 & 4 & | & 6 \\ 3 & -2 & 1 & | & 2 \end{pmatrix}$$

Rewrite first row (since first element is 1):
Multiply first row by -2 and add to second row:
Multiply first row by -3 and add to third row:

$$\begin{pmatrix} 1 & -5 & 3 & | & 9 \\ 0 & 9 & -2 & | & -12 \\ 0 & 13 & -8 & | & -25 \end{pmatrix}$$

Multiply second row by $\frac{5}{9}$ and add to first row:
Multiply second row by $\frac{1}{9}$.
Multiply second row by $-\frac{13}{9}$ and add to third row:

$$\begin{pmatrix} 1 & 0 & \frac{17}{9} & | & \frac{7}{3} \\ 0 & 1 & -\frac{2}{9} & | & -\frac{4}{3} \\ 0 & 0 & -\frac{46}{9} & | & -\frac{23}{3} \end{pmatrix}$$

Multiply third row by $\frac{17}{46}$ and add to first row:
Multiply third row by $-\frac{1}{23}$ and add to second row:
Multiply third row by $-\frac{9}{46}$:

$$\begin{pmatrix} 1 & 0 & 0 & | & -\frac{1}{2} \\ 0 & 1 & 0 & | & -1 \\ 0 & 0 & 1 & | & \frac{3}{2} \end{pmatrix}$$

The solution is $x = -\frac{1}{2}$, $y = -1$, $z = \frac{3}{2}$.

5.7   Solve, using matrices, the system

$$\begin{cases} 2x + 2y + 3z = 2 \\ 3x - y - 6z = 4 \\ 8x + 4y + 3z = 8 \end{cases}$$

Begin with the matrix:

$$\begin{pmatrix} 2 & 2 & 3 & | & 2 \\ 3 & -1 & -6 & | & 4 \\ 8 & 4 & 3 & | & 8 \end{pmatrix}$$

Multiply first row by $\frac{1}{2}$:
Multiply first row by $-\frac{3}{2}$ and add to second row:
Multiply first row by -4 and add to third row:

$$\begin{pmatrix} 1 & 1 & \frac{3}{2} & | & 1 \\ 0 & -4 & -\frac{21}{2} & | & 1 \\ 0 & -4 & -9 & | & 0 \end{pmatrix}$$

Multiply second row by $\frac{1}{4}$ and add to first row:
Multiply second row by $-\frac{1}{4}$.
Subtract second row from third row:

$$\begin{pmatrix} 1 & 0 & -\frac{9}{8} & | & \frac{5}{4} \\ 0 & 1 & -\frac{21}{8} & | & -\frac{1}{4} \\ 0 & 0 & \frac{3}{2} & | & -1 \end{pmatrix}$$

Multiply third row by $\frac{3}{4}$ and add to first row:
Multiply third row by $-\frac{7}{4}$ and add to second row:
Multiply third row by $\frac{2}{3}$:

$$\begin{pmatrix} 1 & 0 & 0 & | & \frac{1}{2} \\ 0 & 1 & 0 & | & \frac{3}{2} \\ 0 & 0 & 1 & | & -\frac{2}{3} \end{pmatrix}$$

The solution is $x = \frac{1}{2}$, $y = \frac{3}{2}$, $z = -\frac{2}{3}$.

Supplementary Problems

5.8   Solve graphically the systems

(a) $\begin{cases} x + y = 5 \\ 2x - y = 1 \end{cases}$ (b) $\begin{cases} x - 3y = 1 \\ x - 2y = 0 \end{cases}$ (c) $\begin{cases} x + y = -1 \\ 3x - y = 3 \end{cases}$

Ans.   (a) (2, 3)   (b) $(-2, -1)$ (c) $(\frac{1}{2}, -\frac{3}{2})$

5.9   Solve algebraically the systems

(a) $\begin{cases} 3x + 2y = 2 \\ x - y = 9 \end{cases}$ (c) $\begin{cases} 3x - y = 1 \\ 2x + 5y = 41 \end{cases}$ (e) $\begin{cases} 1/x + 2/y = 2 \\ 2/x - 2/y = 1 \end{cases}$

(b) $\begin{cases} 3x - 5y = 5 \\ 7x + y = 75 \end{cases}$ (d) $\begin{cases} x + ay = b \\ 2x - by = a \end{cases}$ (f) $\begin{cases} 1/4x + 7/2y = \frac{5}{4} \\ 1/2x - 3/y = -\frac{5}{14} \end{cases}$

Hint: In (e) and (f) solve first for $1/x$ and $1/y$.

    Ans.   (a) $x = 4,\ y = -5$ (c) $x = \frac{46}{17},\ y = \frac{121}{17}$ (e) $x = 1,\ y = 2$

          (b) $x = 10,\ y = 5$ (d) $x = \dfrac{b^2 + a^2}{2a + b},\ y = \dfrac{2b - a}{2a + b}$ (f) $x = 1,\ y = \frac{7}{2}$

5.10 A and B are 30 km apart. If they leave at the same time and travel in the same direction, A overtakes B in 60 hours. If they walk toward each other, they meet in 5 hours. What are their rates?

    Ans.   A, $3\frac{1}{4}$ km/hr; B, $2\frac{3}{4}$ km/hr

5.11   Two trains, each 400 ft long, run on parallel tracks. When running in the same direction, they pass in 20 s; when running in the opposite direction, they pass in 5 s. Find the speed of each train.

    Ans.   100 ft/s, 60 ft/s

5.12   One alloy contain 3 times as much copper as silver; another contains 5 times as much silver as copper. How much of each alloy must be used to make 14 kg in which there is twice as much copper as silver?

    Ans.   12 kg of first, 2 kg of second

5.13   If a field is enlarged by making it 10 m longer and 5 m wider, its area is increased by 1050 square meters. If its length is decreased by 5 m and its width is decreased by 10 m, its area is decreased by 1050 square meters. Find the original dimensions of the field.

    Ans.   80 m × 60 m

5.14   Solve each of the following systems:

(a) $\begin{cases} x + y + z = 3 \\ 2x + y - z = -6 \\ 3x - y + z = 11 \end{cases}$ (c) $\begin{cases} 3x + y + 4z = 6 \\ 2x - 3y - 5z = 2 \\ 3x - 4y + 3z = 8 \end{cases}$ (e) $\begin{cases} 4x - 3y + 3z = 8 \\ 2x + 3y + 24z = 1 \\ 6x - y + 6z = -1 \end{cases}$

(b) $\begin{cases} 4x + 4y - 3z = 3 \\ 2x + 3y + 2z = -4 \\ 3x - y + 4z = 4 \end{cases}$ (d) $\begin{cases} 2x - 3y - 3z = 9 \\ x + 3y - 2z = 3 \\ 3x - 4y - z = 4 \end{cases}$ (f) $\begin{cases} 6x + 2y + 4z = 2 \\ 4x - y + 2z = -3 \\ 7x - 2y - 3z = 5 \end{cases}$

    Ans.   (a) $x = 1,\ y = -3,\ z = 5$ (c) $x = \frac{3}{2},\ y = -\frac{1}{2},\ z = \frac{1}{2}$ (e) $x = -\frac{3}{2},\ y = -4,\ z = \frac{2}{3}$

          (b) $x = 2,\ y = -2,\ z = -1$ (d) $x = 3,\ y = 2,\ z = -3$ (f) $x = \frac{2}{3},\ y = \frac{2}{3},\ z = -\frac{5}{3}$

5.15   Find the equation of the parabola $y = ax^2 + bx + c$ which passes through the points (1,6), (4,0), (3,4). Check your result using a graphing calculator.

    Ans. $y = -x^2 + 3x + 4$

5.16   Solve the systems in Problem 5.14 above using a computer software package such as Maple.

5.17   Repeat Problems 5.8 and 5.9, solving the systems using a graphing calculator.

Chapter 6

Quadratic Functions and Equations

THE GRAPH OF THE QUADRATIC FUNCTION $y = ax^2 + bx + c,\ a \neq 0$, is a parabola. If $a > 0$, the parabola opens upward (Fig. 6-1); if $a < 0$, the parabola opens downward (Fig. 6-2). The lowest point of the parabola of Fig. 6-1 and the highest point of the parabola of Fig. 6-2 are called *vertices*. The abscissa of the vertex is given by $x = -b/2a$. (See Problem 6.1.). See Chapter 36 for a full discussion of parabolas.

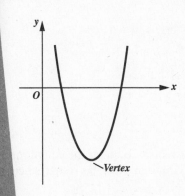

$$y = ax^2 + bx + c,\ a > 0$$

Fig. 6-1

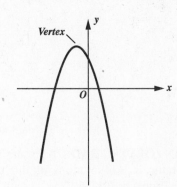

$$y = ax^2 + bx + c,\ a < 0$$

Fig. 6-2

A QUADRATIC EQUATION in one unknown x is of the form

$$ax^2 + bx + c = 0 \qquad a \neq 0 \tag{6.1}$$

Frequently a quadratic equation may be solved by *factoring*. (See Problem 6.2.)

Every quadratic equation (*6.1*) can be solved by the following process, known as *completing the square*:

(*a*)   Substract the constant term c from both members.

(*b*)   Divide both member by a, the coefficient of x^2.

33

(c) Add to each member the square of one-half the coefficient of the term in x.
(d) Set the square root of the left member (a perfect square) equal to $\pm$ the square root of the right member and solve for x.

EXAMPLE 1. Solve $3x^2 - 8x - 4 = 0$ by completing the square.

(a) $3x^2 - 8x - 4$, (b) $x^2 - \dfrac{8}{3}x = \dfrac{4}{3}$,

(c) $x^2 - \dfrac{8}{3}x + \dfrac{16}{9} = \dfrac{4}{3} + \dfrac{16}{9} = \dfrac{28}{9}$, $\left[\dfrac{1}{2}\left(-\dfrac{8}{3}\right)\right]^2 = \left(-\dfrac{4}{3}\right)^2 = \dfrac{16}{9}$,

(d) $x - \dfrac{4}{3} = \pm\dfrac{2\sqrt{7}}{3}$. Then $x = \dfrac{4}{3} \pm \dfrac{2\sqrt{7}}{3} = \dfrac{4 \pm 2\sqrt{7}}{3}$.

(See Problem 6.3.)

Every quadratic equation (6.1) can be solved by means of the quadratic formula

$$x = \frac{-b \pm \sqrt{b^2 - 4ac}}{2a}$$

(See Problems 6.4–6.5.)

It should be noted that it is possible that the roots may be complex numbers.

EQUATIONS IN QUADRATIC FORM. An equation is in *quadratic form* if it is quadratic in some function of the unknown. For example, if the unknown is z, the equation might be quadratic in z^2 or in z^3.

EXAMPLE 2. Solve $x^4 + x^2 - 12 = 0$. This is a quadratic in x^2.
    Factor:

$$x^4 + x^2 - 12 = (x^2 - 3)(x^2 + 4) = 0$$

Then

$$x^2 - 3 = 0 \qquad x^3 + 4 = 0$$
$$x = \pm\sqrt{3} \qquad x = \pm 2i$$

(See Problems 6.11–6.12).

EQUATIONS INVOLVING RADICALS may sometimes reduce to quadratic equations after squaring to remove the radicals. All solutions of this quadratic equation *must* be tested since some may be extraneous. (See Problems 6.13–6.16.)

THE DISCRIMINANT of the quadratic equation (6.1) is, by definition, the quantity $b^2 - 4ac$. When a, b, c are rational numbers, the roots of the equation are

Real and unequal if and only if $b^2 - 4ac > 0$.
Real and equal if and only if $b^2 - 4ac = 0$.
Complex if and only if $b^2 - 4ac < 0$. (See Chapter 39.)

(See Problems 6.17–6.18.)

SUM AND PRODUCT OF THE ROOTS. If x_1 and x_2 are the roots of the quadratic equation (6.1), then $x_1 + x_2 = -b/a$ and $x_1 \cdot x_2 = c/a$.

A quadratic equation whose roots are x_1 and x_2 may be written in the form

$$x^2 - (x_1 + x_2)x + x_1 \cdot x_2 = 0$$

THE GRAPHING CALCULATOR. The graphing calculator can easily be used to find the roots of a quadratic equation. See Appendix A for graphing calculator directions.

Solved Problems

6.1. Sketch the parabolas: (a) $y = x^2 - 2x - 8$, (b) $y = 2x^2 + 9x - 9$. Determine the coordinates of the vertex $V(x, y)$ of each.

Vertex:

(a)
$$x = -\frac{b}{2a} = -\frac{-2}{2 \cdot 1} = 1 \qquad y = 1^2 - 2 \cdot 1 - 8 = -9$$

hence, $V(1, -9)$. See Fig. 6-3(a).

(b)
$$x = -\frac{9}{2(-2)} = \frac{9}{4} \qquad y = -2\left(\frac{9}{4}\right)^2 + 9\left(\frac{9}{4}\right) - 9 = \frac{9}{8}$$

hence, $V\left(\frac{9}{4}, \frac{9}{8}\right)$. See Fig. 6-3(b).

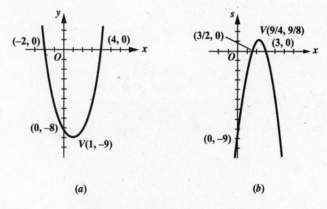

(a)   (b)

Fig. 6-3

6.2. Solve by factoring:

(a) $4x^2 - 5x = x(4x - 5) = 0$ (e) $4x^2 + 20x + 25 = (2x + 5)(2x + 5) = 0$

(b) $4x^2 - 9 = (2x - 3)(2x + 3) = 0$ (f) $6x^2 + 13x + 6 = (3x + 2)(2x + 3) = 0$

(c) $x^2 - 4x + 3 = (x - 1)(x - 3) = 0$ (g) $3x^2 + 8ax - 3a^2 = (3x - a)(x + 3a) = 0$

(d) $x^2 - 6x + 9 = (x - 3)(x - 3) = 0$ (h) $10ax^2 + (15 - 8a^2)x - 12a = (2ax + 3)(5x - 4a) = 0$

Ans.   (a) $0, \frac{5}{4}$ (c) $1, 3$ (e) $-\frac{5}{2}, -\frac{5}{2}$ (g) $a/3, -3a$

           (b) $\frac{3}{2}, -\frac{3}{2}$ (d) $3, 3$ (f) $-\frac{2}{3}, -\frac{3}{2}$ (h) $-\frac{3}{2}a, 4a/5$

6.3. Solve by completing the square: (a) $x^2 - 2x - 1 = 0$, (b) $3x^2 + 8x + 7 = 0$.

(a) $x^2 - 2x = 1$; $x^2 - 2x + 1 = 1 + 1 = 2$; $x - 1 = \pm\sqrt{2}$; $x = 1 \pm \sqrt{2}$.

(b) $3x^2 + 8x = -7$; $x^2 + \frac{8}{3}x = -\frac{7}{3}$; $x^2 + \frac{8}{3}x + \frac{16}{9} = -\frac{7}{3} + \frac{16}{9} = -\frac{5}{9}$;

$x + \frac{4}{3} = \pm\sqrt{\frac{-5}{9}} = \pm\frac{i\sqrt{5}}{3}$; $x = \frac{-4 \pm i\sqrt{5}}{3}$.

6.4. Solve $ax^2 + bx + c = 0$, $a \neq 0$, by completing the square.

Proceeding as in Problem 6.3, we have

$$x^2 + \frac{b}{a}x = -\frac{c}{a}, \quad x^2 + \frac{b}{a}x + \frac{b^2}{4a^2} = \frac{b^2}{4a^2} - \frac{c}{a} = \frac{b^2 - 4ac}{4a^2},$$

$$x + \frac{b}{2a} = \pm\sqrt{\frac{b^2 - 4ac}{4a^2}} = \pm\frac{\sqrt{b^2 - 4ac}}{2a}, \quad \text{and} \quad x = \frac{-b \pm \sqrt{b^2 - 4ac}}{2a}$$

6.5. Solve the equations of Problem 6.3 using the quadratic formula.

(a) $x = \dfrac{-b \pm \sqrt{b^2 - 4ac}}{2a} = \dfrac{-(-2) \pm \sqrt{(-2)^2 - 4(1)(-1)}}{2 \cdot 1} = \dfrac{2 \pm \sqrt{4+4}}{2} = \dfrac{2 \pm 2\sqrt{2}}{2} = 1 \pm \sqrt{2}$

(b) $x = \dfrac{-(8) \pm \sqrt{8^2 - 4 \cdot 3 \cdot 7}}{2 \cdot 3} = \dfrac{-8 \pm \sqrt{64 - 84}}{6} = \dfrac{-8 \pm \sqrt{-20}}{6} = \dfrac{-8 \pm 2i\sqrt{5}}{6} = \dfrac{-4 \pm i\sqrt{5}}{3}$

6.6. An open box containing 24 cm^3 is to be made from a square piece of tin by cutting 2 cm squares from each corner and turning up the sides. Find the dimension of the piece of tin required.

Let $x =$ the required dimension. The resulting box will have dimensions $(x - 4)$ by $(x - 4)$ by 2, and its volume will be $2(x - 4)(x - 4)$. See Fig. 6-4. Then

$$2(x - 4)^2 = 24, \qquad x - 4 = \pm 2\sqrt{3}, \qquad \text{and} \qquad x = 4 \pm 2\sqrt{3} = 7.464,\ 0.536$$

The required square of tin is 7.464 cm on a side.

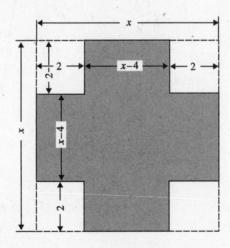

Fig. 6-4

6.7. Two pipes together can fill a reservoir in 6 hr 40 min. Find the time each alone will take to fill the reservoir if one of the pipes can fill it in 3 hr less time than the other.

Let $x =$ time (hours) required by smaller pipe, $x - 3 =$ time required by larger pipe. Then

$$\frac{1}{x} = \text{part filled in 1 hr by smaller pipe} \qquad \frac{1}{x - 3} = \text{part filled in 1 hr by larger pipe}$$

Since the two pipes together fill $\dfrac{1}{\frac{20}{3}} = \dfrac{3}{20}$ of the reservoir in 1 hr,

$$\frac{1}{x} + \frac{1}{x - 3} = \frac{3}{20}, \qquad 20(x - 3) + 20x = 3x(x - 3),$$

$$3x^2 - 49x + 60 = (3x - 4)(x - 15) = 0, \quad \text{and} \quad x = \tfrac{4}{3}, 15$$

The smaller pipe will fill the reservoir in 15 hr and the larger pipe in 12 hr.

6.8. Express each of the following in the form $a(x - h)^2 \pm b(y - k)^2 = c$.

(a) $x^2 + y^2 - 6x - 9y + 2 = 0$.

$$(x^2 - 6x) + (y^2 - 9y) = -2 \qquad (x^2 - 6x + 9) + (y^2 - 9y + \tfrac{81}{4}) = -2 + 9 + \tfrac{81}{4} = \tfrac{109}{4}$$
$$(x - 3)^2 + (y - \tfrac{9}{2})^2 = \tfrac{109}{4}$$

(b) $3x^2 + 4y^2 + 6x - 16y - 21 = 0$.

$$3(x^2 + 2x) + 4(y^2 - 4y) = 21 \qquad 3(x^2 + 2x + 1) + 4(y^2 - 4y + 4) = 21 + 3(1) + 4(4) = 40$$
$$3(x + 1)^2 + 4(y - 2)^2 = 40$$

6.9. Transform each of the following into the form $a\sqrt{(x - h)^2 + k}$ or $a\sqrt{k - (x - h)^2}$.

(a) $\sqrt{4x^2 - 8x + 9} = 2\sqrt{x^2 - 2x + \tfrac{9}{4}} = 2\sqrt{(x^2 - 2x + 1) + \tfrac{5}{4}} = 2\sqrt{(x - 1)^2 + \tfrac{5}{4}}$

(b) $\sqrt{8x - x^2} = \sqrt{16 - (x^2 - 8x + 16)} = \sqrt{16 - (x - 4)^2}$

(c) $\sqrt{3 - 4x - 2x^2} = \sqrt{2} \cdot \sqrt{\tfrac{3}{2} - 2x - x^2} = \sqrt{2} \cdot \sqrt{\tfrac{5}{2} - (x^2 + 2x + 1)} = \sqrt{2} \cdot \sqrt{\tfrac{5}{2} - (x + 1)^2}$

6.10. If an object is thrown directly upward with initial speed v ft/s, its distance s ft above the ground after t s is given by

$$s = vt - \tfrac{1}{2}gt^2$$

Taking $g = 32.2 \, \text{ft/s}^2$ and initial speed 120 ft/s, find (a) when the object is 60 ft above the ground, (b) when it is highest in its path and how high.

The equation of motion is $s = 120t - 16.1t^2$.

(a) When $s = 60$:

$$60 = 120t - 16.1t^2 \quad \text{or} \quad 16.1t^2 - 120t + 60 = 0$$

$$t = \frac{120 \pm \sqrt{(120)^2 - 4(16.1)60}}{32.2} = \frac{120 \pm \sqrt{10536}}{32.2} = \frac{120 \pm 102.64}{32.2} = 6.91, \ 0.54$$

After $t = 0.54$ s the object is 60 ft above the ground and rising. After $t = 6.91$ s, the object is 60 ft above the ground and falling.

(b) The object is at its highest point when

$$t = \frac{-b}{2a} = \frac{-(-120)}{2(16.1)} = 3.73 \, \text{s}. \text{ Its height is given by } 120t - 16.1t^2 = 120(3.73) - 16.1(3.73)^2 = 223.6 \, \text{ft}.$$

6.11. Solve $9x^2 - 10x^2 + 1 = 0$.

Factor: $(x^2 - 1)(9x^2 - 1) = 0$. Then $x^2 - 1 = 0$, $9x^2 - 1 = 0$; $x = \pm 1$, $x = \pm\tfrac{1}{3}$.

6.12. Solve $x^4 - 6x^3 + 12x^2 - 9x + 2 = 0$.

Complete the square on the first two terms: $(x^4 - 6x^3 + 9x^2) + 3x^2 - 9x + 2 = 0$

or $(x^2 - 3x)^2 + 3(x^2 - 3x) + 2 = 0$

Factor: $[(x^2 - 3x) + 2][(x^2 - 3x) + 1] = 0$

Then $x^2 - 3x + 2 = (x - 2)(x - 1) = 0$ and $x = 1, 2$

$$x^2 - 3x + 1 = 0 \quad \text{and} \quad x = \frac{3 \pm \sqrt{9 - 4}}{2} = \frac{3 \pm \sqrt{5}}{2}$$

6.13.   Solve $\sqrt{5x - 1} - \sqrt{x} = 1$.

Transpose one of the radicals: $\sqrt{5x - 1} = \sqrt{x} + 1$

Square: $5x - 1 = x + 2\sqrt{x} + 1$

Collect terms: $4x - 2 = 2\sqrt{x}$ or $2x - 1 = \sqrt{x}$

Square:

$$4x^2 - 4x + 1 = x, \quad 4x^2 - 5x + 1 = (4x - 1)(x - 1) = 0, \quad \text{and} \quad x = \tfrac{1}{4}, 1$$

For $x = \tfrac{1}{4}$: $\sqrt{5(\tfrac{1}{4}) - 1} - \sqrt{\tfrac{1}{4}} = 0 \neq 1$.

For $x = 1$: $\sqrt{5(1) - 1} - \sqrt{1} = 1$.   The root is $x = 1$.

6.14.   Solve $\sqrt{6x + 7} - \sqrt{3x + 3} = 1$.

Transpose one of the radicals: $\sqrt{6x + 7} = 1 + \sqrt{3x + 3}$

Square: $6x + 7 = 1 + 2\sqrt{3x + 3} + 3x + 3$

Collect terms: $3x + 3 = 2\sqrt{3x + 3}$

Square:

$$9x^2 + 18x + 9 = 4(3x + 3) = 12x + 12, \quad 9x^2 + 6x - 3 = 3(3x - 1)(x + 1) = 0, \quad \text{and} \quad x = \tfrac{1}{3}, -1$$

For $x = \tfrac{1}{3}$: $\sqrt{6(\tfrac{1}{3}) + 7} - \sqrt{3(\tfrac{1}{3}) + 3} = 3 - 2 = 1$.

For $x = -1$: $1 - 0 = 1$.   The roots are $x = \tfrac{1}{3}, -1$.

6.15.   Solve $\dfrac{\sqrt{x + 1} + \sqrt{x - 1}}{\sqrt{x + 1} - \sqrt{x - 1}} = 3$.

Multiply the numerator and denominator of the fraction by $(\sqrt{x + 1} + \sqrt{x - 1})$:

$$\frac{(\sqrt{x + 1} + \sqrt{x - 1})(\sqrt{x + 1} + \sqrt{x - 1})}{(\sqrt{x + 1} - \sqrt{x - 1})(\sqrt{x + 1} + \sqrt{x - 1})} = \frac{(x + 1) + 2\sqrt{x^2 - 1} + (x - 1)}{(x + 1) - (x - 1)} = x + \sqrt{x^2 - 1} = 3$$

Then $x - 3 = -\sqrt{x^2 - 1}, \quad x^2 - 6x + 9 = x^2 - 1, \quad \text{and} \quad x = \frac{5}{3}$

Check: $\dfrac{\sqrt{\tfrac{8}{3}} + \sqrt{\tfrac{2}{3}}}{\sqrt{\tfrac{8}{3}} - \sqrt{\tfrac{2}{3}}} = \dfrac{2\sqrt{\tfrac{2}{3}} + \sqrt{\tfrac{2}{3}}}{2\sqrt{\tfrac{2}{3}} - \sqrt{\tfrac{2}{3}}} = \dfrac{3\sqrt{\tfrac{2}{3}}}{\sqrt{\tfrac{2}{3}}} = 3$

6.16.   Solve $3x^2 - 5x + \sqrt{3x^2 - 5x + 4} = 16$.

Note that the unknown appears in the same polynomials in both the expressions free of radicals and under the radical. Add 4 to both sides:

$$3x^2 - 5x + 4 + \sqrt{3x^2 - 5x + 4} = 20$$

Let $y = \sqrt{3x^2 - 5x + 4}$. Then

$$y^2 + y - 20 = (y + 5)(y - 4) = 0 \qquad \text{and} \qquad y = 4, -5$$

Now $\sqrt{3x^2 - 5x + 4} = -5$ is impossible. From $\sqrt{3x^2 - 5x + 4} = 4$ we have

$$3x^2 - 5x + 4 = 16, \qquad 3x^2 - 5x - 12 = (3x + 4)(x - 3) = 0, \qquad \text{and} \qquad x = 3, -\tfrac{4}{3}$$

The reader will show that both $x = 3$ and $x = -\tfrac{4}{3}$ are solutions.

6.17. Without solving determine the character of the roots of

(a) $x^2 - 8x + 9 = 0$. Here $b^2 - 4ac = 28$; the roots are irrational and unequal.

(b) $3x^2 - 8x + 9 = 0$. Here $b^2 - 4ac = -44$; the roots are imaginary and unequal.

(c) $6x^2 - 5x - 6 = 0$. Here $b^2 - 4ac = 169$; the roots are rational and unequal.

(d) $4x^2 - 4\sqrt{3}x + 3 = 0$. Here $b^2 - 4ac = 0$; the roots are real and equal.

(NOTE: Although the discriminant is the square of a rational number, the roots $\tfrac{1}{2}\sqrt{3}, \tfrac{1}{2}\sqrt{3}$ are not rational. Why?)

6.18. Without sketching, state whether the graph of each of the following functions crosses the x axis, is tangent to it, or lies wholly above or below it.

(a) $3x^2 + 5x - 2$. $\quad b^2 - 4ac = 25 + 24 > 0$; the graph crosses the x axis.

(b) $2x^2 + 5x + 4$. $\quad b^2 - 4ac = 25 - 32 < 0$ and the graph is either wholly above the wholly below the x axis. Since $f(0) > 0$ (the value of the function for any other value of x would do equally well), the graph lies wholly above the x axis.

(c) $4x^2 - 20x + 25$. $\quad b^2 - 4ac = 400 - 400 = 0$; the graph is tangent to the x axis.

(d) $2x - 9 - 4x^2$. $\quad b^2 - 4ac = 4 - 144 < 0$ and $f(0) < 0$; the graph lies wholly below the x axis.

6.19. Find the sum and product of the roots of

(a) $x^2 + 5x - 8 = 0$. $\qquad\qquad$ *Ans.*   Sum $= -\dfrac{b}{a} = -5$, product $= \dfrac{c}{a} = -8$.

(b) $8x^2 - x - 2 = 0$ or $x^2 + \tfrac{1}{8}x - \tfrac{1}{4} = 0$. $\qquad$ *Ans.*   Sum $= \tfrac{1}{8}$, product $= -\tfrac{1}{4}$.

(c) $5 - 10x - 3x^2 = 0$ or $x^2 + \tfrac{10}{3}x - \tfrac{5}{3} = 0$. $\qquad$ *Ans.*   Sum $= -\tfrac{10}{3}$, product $= -\tfrac{5}{3}$.

6.20. Form the quadratic equation whose roots x_1 and x_2 are:

(a) $3, \tfrac{2}{3}$. Here $x_1 + x_2 = \tfrac{17}{5}$ and $x_1 \cdot x_2 = \tfrac{6}{5}$. The equation is $x^2 - \tfrac{17}{5}x + \tfrac{6}{5} = 0$ or $5x^2 - 17x + 6 = 0$.

(b) $-2 + 3\sqrt{5}, -2 - 3\sqrt{5}$. Here $x_1 + x_2 = -4$ and $x_1 \cdot x_2 = 4 - 45 = -41$. The equation is $x^2 + 4x - 41 = 0$.

(c) $\dfrac{3 - i\sqrt{2}}{2}, \dfrac{3 + i\sqrt{2}}{2}$. The sum of the roots is 3 and the product is $\tfrac{11}{4}$. The equation is $x^2 - 3x + \tfrac{11}{4} = 0$ or $4x^2 - 12x + 11 = 0$.

6.21. Determine k so that the given equation will have the stated property, and write the resulting equation.

(a) $x^2 + 4kx + k + 2 = 0$ has one root 0. Since the product of the roots is to be 0, $k + 2 = 0$ and $k = -2$. The equation is $x^2 - 8x = 0$.

(b) $4x^2 - 8kx - 9 = 0$ has one root the negative of the other. Since the sum of the roots is to be 0, $2k = 0$ and $k = 0$. The equation is $4x^2 - 9 = 0$.

(c) $4x^2 - 8kx + 9 = 0$ has roots whose difference is 4. Denote the roots by r and $r + 4$. Then $r + (r + 4) = 2r + 4 = 2k$ and $r(r + 4) = \tfrac{9}{4}$. Solving for $r = k - 2$ in the first and substituting in the second, we have

$(k-2)(k+2) = \frac{9}{4}$; then $4k^2 - 16 = 9$ and $k = \pm\frac{5}{2}$. The equation are $4x^2 + 20x + 9 = 0$ and $4x^2 - 20x + 9 = 0$.

(d) $2x^2 - 3kx + 5k = 0$ has one root twice the other. Let the roots be r and $2r$. Then $r + 2r = 3r = \frac{3}{2}k$, $r = \frac{1}{2}k$, and $r(2r) = 2r^2 = \frac{5}{2}k$. Thus $k = 0, 5$. The equations are $2x^2 = 0$ with roots $0, 0$ and $2x^2 - 15x + 25 = 0$ with roots $\frac{5}{2}, 5$.

Supplementary Problems

6.22. Locate the vertex of each of the parabolas of Problem 3.11. Compare the results with those of Problem 3.14.

6.23. Solve for x by factoring.

(a) $3x^2 + 4x = 0$ (c) $x^2 + 2x - 3 = 0$ (e) $10x^2 - 9x + 2 = 0$

(b) $16x^2 - 25 = 0$ (d) $2x^2 + 9x - 5 = 0$ (f) $2x^2 - (a + 4b)x + 2ab = 0$

Ans. (a) $0, -\frac{4}{3}$ (b) $\pm\frac{5}{4}$ (c) $1, -3$ (d) $\frac{1}{2}, -5$ (e) $\frac{1}{2}, \frac{2}{5}$ (f) $\frac{1}{2}a, 2b$

6.24. Solve for x by completing the square.

(a) $2x^2 + x - 5 = 0$ (c) $3x^2 + 2x - 2 = 0$ (e) $15x^2 - (16m - 14)x + 4m^2 - 8m + 3 = 0$

(b) $2x^2 - 4x - 3 = 0$ (d) $5x^2 - 4x + 2 = 0$

Ans. (a) $\frac{1}{4}(-1 \pm \sqrt{41})$ (b) $\frac{1}{2}(2 \pm \sqrt{10})$ (c) $\frac{1}{3}(-1 \pm \sqrt{7})$ (d) $\frac{1}{5}(2 \pm i\sqrt{6})$

(e) $\frac{1}{3}(2m - 1), \frac{1}{5}(2m - 3)$

6.25. Solve the equations of Problem 6.24 using the quadratic formula.

6.26. Solve $6x^2 + 5xy - 6y^2 + x + 8y - 2 = 0$ for (a) y in terms of x, (b) x in terms of y.

Ans. (a) $\frac{1}{2}(3x + 2), \frac{1}{3}(1 - 2x)$ (b) $\frac{1}{2}(1 - 3y), \frac{2}{3}(y - 1)$

6.27. Solve.

(a) $x^4 - 29x^2 + 100 = 0$ (c) $1 - \dfrac{2}{2x^2 - x} = \dfrac{3}{(2x^2 - x)^2}$ (e) $\sqrt{2x + 3} - \sqrt{4 - x} = 2$

(b) $\dfrac{21}{x + 2} - \dfrac{1}{x - 4} = 2$ (d) $\sqrt{4x + 1} - \sqrt{3x - 2} = 5$ (f) $\sqrt{3x - 2} - \sqrt{x - 2} = 2$

Ans. (a) $\pm 2, \pm 5$ (b) $5, 7$ (c) $-1, \frac{3}{2}, \frac{1}{4}(1 \pm i\sqrt{7})$ (d) 342 (e) 3 (f) $2, 6$

6.28. Form the quadratic equation whose roots are

(a) The negative of the roots of $3x^2 + 5x - 8 = 0$.

(b) Twice the roots of $2x^2 - 5x + 2 = 0$.

(c) One-half the roots of $2x^2 - 5x - 3 = 0$.

Ans. (a) $3x^2 - 5x - 8 = 0$ (b) $x^2 - 5x + 4 = 0$ (c) $8x^2 - 10x - 3 = 0$

6.29. The length of a rectangle is 7 cm more than its width; its area is 228 cm^2. What are its dimensions?

Ans. 12 cm × 19 cm

6.30. A rectangular garden plot 16 m × 24 m is to be bordered by a strip of uniform width x meters so as to double the area. Find x.

Ans. 4 m

6.31. The interior of a cubical box is lined with insulating material $\frac{1}{2}$ cm thick. Find the original interior dimensions if the volume is thereby decreased by 271 cm^3.

    Ans.   10 cm

6.32. What are the dimensions of the largest rectangular field which can be enclosed by 1200 m of fencing?

    Ans.   300 m × 300 m

(NOTE. See Chapter 48 for additional material involving extrema.)

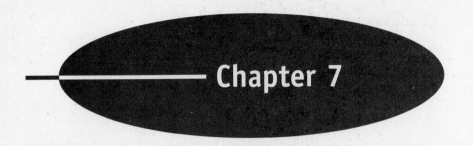

Inequalities

AN INEQUALITY is a statement that one (real) number is greater than or less than another; for example, $3 > -2, -10 < -5$.

Two inequalities are said to have the *same sense* if their signs of inequality point in the same direction. Thus, $3 > -2$ and $-5 > -10$ have the same sense; $3 > -2$ and $-10 < -5$ have opposite senses.

The sense of an equality is *not* changed:

    (*a*)   If the same number is added to or subtracted from both sides
    (*b*)   If both sides are multiplied or divided by the same *positive* number

The sense of an equality *is* changed if both sides are multiplied or divided by the same negative number. (See Problems 7.1–7.3.)

AN ABSOLUTE INEQUALITY is one which is true for all real values of the letters involved; for example, $x^2 + 1 > 0$ is an absolute inequality.

A CONDITIONAL INEQUALITY is one which is true for certain values of the letters involved; for example, $x + 2 > 5$ is a conditional inequality, since it is true for $x = 4$ but not for $x = 1$.

SOLUTION OF CONDITIONAL INEQUALITIES.   The solution of a conditional inequality in one letter, say x, consists of all values of x for which the inequality is true. These values lie on one or more intervals of the real number scale as illustrated in the examples below.

To solve a linear inequality, proceed as in solving a linear equality, keeping in mind the rules for keeping or reversing the sense.

EXAMPLE 1.   Solve the inequality $5x + 4 > 2x + 6$.
Subtract $2x$ from each member $\Big\}$ $3x > 2$
Subtract 4 from each member
Divide by 3: $x > \frac{2}{3}$
Graphical representation: (See Fig. 7-1.)

Fig. 7-1

(See Problems 7.5–7.6.)

To solve a quadratic inequality, $f(x) = ax^2 + bx + c > 0$, solve the equality $f(x) = 0$, locate the roots r_1 and r_2 on a number scale, and determine the sign of $f(x)$ on each of the resulting intervals.

EXAMPLE 2.   Solve the inequality $3x^2 - 8x + 7 > 2x^2 - 3x + 1$.

Subtract $2x - 3x + 1$ from each member:

$$x^2 - 5x + 6 > 0$$

Solve the equality $x^2 - 5x + 6 = 0$:

$$x = 2, \qquad x = 3$$

Locate the roots on a number scale (see Fig. 7-2).

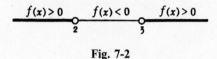

Fig. 7-2

Determine the sign of $f(x) = x^2 - 5x + 6$:

On the interval $x < 2$: $f(0) = 6 > 0$
On the interval $2 < x < 3$: $f(\frac{5}{2}) = \frac{25}{4} - \frac{25}{2} + 6 < 0$
On the interval $x > 3$: $f(4) = 16 - 20 + 6 > 0$

The given inequality is satisfied (see darkened portions of the scale) when $x < 2$ and $x > 3$. (See Problems 7.7 – 7.11.)

Solved Problems

7.1   Given the inequality $-3 < 4$, write the result when (a) 5 is added to both sides, (b) 2 is subtracted from both sides, (c) -3 is subtracted from both sides, (d) both sides are doubled, (e) both sides are divided by -2.

    Ans.   (a) $2 < 9$,   (b) $-5 < 2$,   (c) $0 < 7$,   (d) $-6 < 8$,   (e) $\frac{3}{2} > -2$

7.2   Square each of the inequalities: (a) $-3 < 4$,   (b) $-3 > -4$

    Ans.   (a) $9 < 16$,   (b) $9 < 16$

7.3   If $a > 0$, $b > 0$, prove that $a^2 > b^2$ if and only if $a > b$.

    Suppose $a > b$. Since $a > 0$, $a^2 > ab$ and, since $b > 0$, $ab > b^2$. Hence, $a^2 > ab > b^2$ and $a^2 > b^2$.
    Suppose $a^2 > b^2$. Then $a^2 - b^2 = (a-b)(a+b) > 0$. Dividing by $a + b > 0$, we have $a - b > 0$ and $a > b$.

7.4   Prove $\dfrac{a}{b^2} + \dfrac{b}{a^2} > \dfrac{1}{a} + \dfrac{1}{b}$ if $a > 0$, $b > 0$, and $a \neq b$.

    Suppose $a > b$; then $a^2 > b^2$ and $a - b > 0$. Now $a^2(a - b) > b^2(a - b)$ or $a^3 - a^2b > ab^2 - b^3$ and $a^3 + b^3 > ab^2 + a^2b$. Since $a^2b^2 > 0$,

$$\frac{a^3 + b^3}{a^2b^2} > \frac{ab^2 + a^2b}{a^2b^2} \qquad \text{and} \qquad \frac{a}{b^2} + \frac{b}{a^2} > \frac{1}{a} + \frac{1}{b}$$

Why is it necessary that $a > 0$ and $b > 0$?   **Hint:**   See Problem 7.3.

7.5   Solve $3x + 4 > 5x + 2$.

Subtract $5x + 4$ from each member:

$$-2x > -2$$

Divide by -2:

$$x < 1$$

See Fig. 7-3.

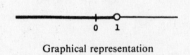

Graphical representation

Fig. 7-3

7.6   Solve $2x - 8 < 7x + 12$.

Subtract $7x - 8$ from each member:

$$-5x < 20$$

Divide by -5:

$$x > -4$$

See Fig. 7-4.

Graphical representation

Fig. 7-4

7.7   Solve $x^2 > 4x + 5$.

Subtract $4x + 5$ from each member:

$$x^2 - 4x - 5 > 0$$

Solve the equality $f(x) = x^2 - 4x - 5 = 0$:

$$x = -1, 5$$

Locate the roots on a number scale.

Determine the sign of $f(x)$

On the interval $x < -1$: $f(-2) = 4 + 8 - 5 > 0$
On the interval $-1 < x < 5$: $f(0) = -5 < 0$
On the interval $x > 5$: $f(6) = 36 - 24 - 5 > 0$

The inequality is satisfied when $x < -1$ and $x > 5$. See Fig. 7-5.

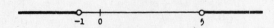

Fig. 7-5

7.8   Solve $3x^2 + 2x + 2 < 2x^2 + x + 4$.

Subtract $2x^2 + x + 4$ from each member:

$$x^2 + x - 2 < 0$$

Solve $f(x) = x^2 + x - 2 = 0$:

$$x = -2, 1$$

Locate the roots on a number scale.

Determine the sign of $f(x)$

On the interval $x < -2$: $f(-3) = 9 - 3 - 2 > 0$
On the interval $-2 < x < 1$: $f(0) = -2 < 0$
On the interval $x > 1$: $f(2) = 4 + 2 - 2 > 0$

The inequality is satisfied when $-2 < x < 1$. See Fig. 7-6.

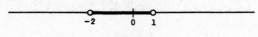

Fig. 7-6

7.9   Solve $(x + 5)(x - 1)(x - 2) < 0$.

Solve the equality $f(x) = (x + 5)(x - 1)(x - 2) = 0$. $x = 1, 2, -5$
Locate the roots on a number scale.
Determine the sign of $f(x)$

On the interval $x < -5$: $f(-6) = (-1)(-7)(-8) < 0$
On the interval $-5 < x < 1$: $f(0) = 5(-1)(-2) > 0$
On the interval $1 < x < 2$: $f(\frac{3}{2}) = (\frac{13}{2})(\frac{1}{2})(-\frac{1}{2}) < 0$
On the interval $x > 2$: $f(3) = 8 \cdot 2 \cdot 1 > 0$

The inequality is satisfied when $x < -5$ and $1 < x < 2$. See Fig. 7-7.

Fig. 7-7

7.10   Solve $(x - 2)^2(x - 5) > 0$.

Solve the equality $f(x) = (x - 2)^2(x - 5) = 0$: $x = 2, 2, 5$
Locate the roots on a number scale.
Determine the sign of $f(x)$

On the interval $x < 2$: $f(0) = (+)(-) < 0$
On the interval $2 < x < 5$: $f(3) = (+)(-) < 0$
On the interval $x > 5$: $f(6) = (+)(+) > 0$

The inequality is satisfied when $x > 5$. See Fig. 7-8.

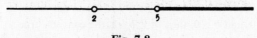

Fig. 7-8

(NOTE: The inequality $(x - 2)^2(x - 5) < 0$ is satisfied when $x < 2$ and $2 < x < 5$.
The inequality $(x - 2)^2(x - 5) \geq 0$ is satisfied when $x \geq 5$ and $x = 2$.)

7.11   Determine the values of k so that $3x^2 + kx + 4 = 0$ will have real roots.

The discriminant $b^2 - 4ac = k^2 - 48 = (k - 4\sqrt{3})(k + 4\sqrt{3}) \geq 0$. The roots will be real when $k \geq 4\sqrt{3}$ and when $k \leq -4\sqrt{3}$, that is, when $|k| \geq 4\sqrt{3}$.

Supplementary Problems

7.12   If $2y^2 + 4xy - 3x = 0$, determine the range of values of x for which the corresponding y roots are real.

Ans.   Here

$$y = \frac{-4x \pm \sqrt{16x^2 + 24x}}{4} = \frac{-2x \pm \sqrt{4x^2 + 6x}}{2}$$

will be real provided $4x^2 + 6x \geq 0$. Thus, y will be real for $x \leq -\frac{3}{2}$ and for $x \geq 0$.

7.13   Prove: If $a > b$ and $c > d$, then $a + c > b + d$.

   Hint: $(a - b) + (c - d) = (a + c) - (b + d) > 0$.

7.14   Prove: If $a \neq b$ are real numbers, then $a^2 + b^2 > 2ab$.

   Hint: $(a - b)^2 > 0$.

7.15   Prove: If $a \neq b \neq c$ are real numebrs, then $a^2 + b^2 + c^2 > ab + bc + ca$.

7.16   Prove: If $a > 0$, $b > 0$, and $a \neq b$, then $a/b + b/a > 2$.

7.17   Prove: If $a^2 + b^2 = 1$ and $c^2 + d^2 = 1$, then $ac + bd \leq 1$.

7.18   Solve: (a) $x - 4 > -2x + 5$ (c) $x^2 - 16 > 0$ (e) $x^2 - 6x > -5$

              (b) $4 + 3x < 2x + 24$ (d) $x^2 - 4x < 5$ (f) $5x^2 + 5x - 8 \leq 3x^2 + 4$

   Ans.   (a) $x > 3$ (b) $x < 20$ (c) $|x| > 4$ (d) $-1 < x < 5$

              (e) $x < 1, x > 5$ (f) $-4 \leq x \leq \frac{3}{2}$

7.19   Solve: (a) $(x + 1)(x - 2)(x + 4) > 0$ (b) $(x - 1)^3(x - 3)(x + 2) < 0$

              (c) $(x + 3)(x - 2)^2(x - 5)^3 < 0$

   Ans.   (a) $-4 < x < -1, \ x > 2$ (b) $x < -2, \ 1 < x < 3$ (c) $-3 < x < 2, \ 2 < x < 5$

7.20   In each of the following determine the domain of x for which y will be real:

   (a) $y = \sqrt{2x^2 - 7x + 3}$ (c) $y = \sqrt{61x^2 + 144x + 144}$ (e) $xy^2 + 3xy + 3x - 4y - 4 = 0$

   (b) $y = \sqrt{6 - 5x - 4x^2}$ (d) $y^2 + 2xy + 4y + x + 14 = 0$ (f) $6x^2 + 5xy - 6y^2 + x + 8y - 2 = 0$

   Ans.   (a) $x \leq \frac{1}{2}, x \geq 3$ (c)   all values of x (e) $-4 \leq x \leq \frac{4}{3}$

              (b) $-2 \leq x \leq \frac{3}{4}$ (d) $x \leq -5, x \geq 2$ (f)   all values of x

7.21   Prove that if $a < b$, $a > 0$, and $b > 0$, then $a^2 < b^2$. Is the converse true?

7.22   Solve Problem 7.12 using a graphing calculator.

7.23   Solve Problem 7.12 using a computer sofware package.

Chapter 8

The Locus of an Equation

WHAT IS A LOCUS?   *Locus*, in Latin, means location. The plural of locus is loci. A locus of points is the set of points, and only those points, that satisfy conditions.

For example, the locus of points that are 2 m from a given point P is the set of points 2 m from P. These points lie on the circle with P as center and radius 2 m.

To determine a locus:
(1) State what is given and the condition to be satisfied.
(2) Find several points satisfying the condition which indicate the shape of the locus.
(3) Connect the points and describe the locus fully. For example, the locus $x(y - x) = 0$ consists of the lines $x = 0$ and $y - x = 0$. The reader can easily sketch several points and the locus.

DEGENERATE LOCI.   The locus of an equation $f(x, y) = 0$ is called *degenerate* if $f(x, y)$ is the product of two or more real factors $g(x, y), h(x, y), \ldots$. The locus of $f(x, y) = 0$ then consists of the loci of $g(x, y) = 0, h(x, y) = 0, \ldots$ (See Problem 8.1.)

INTERCEPTS.   The intercepts on the coordinate axes of a locus are the directed distances from the origin to the points of intersection of the locus and the coordinate axes.

To find the x intercepts, set $y = 0$ in the equation of the locus and solve for x; to find the y intercepts, set $x = 0$ and solve for y. (See Problem 8.2.)

SYMMETRY.   Two points P and Q are said to be symmetric with respect to a point R if R is the midpoint of the segment PQ (see Fig. 8-1). Each of the points is called the symmetric point of the other with respect to the point R, the *center of symmetry*.

Two points P and Q are said to be symmetric with respect to a line l if l is the perpendicular bisector of the segment PQ (see Fig. 8-2). Each of the points P, Q is called the symmetric point of the other with respect to l, the *axis of symmetry*.

A locus is said to be symmetric with respect to a point R or to a line l if the symmetric point with respect to R or l of every point of the locus is also a point of the locus (see Figs. 8-3 and 8-4). Also see Chapter 22 for more on symmetry.

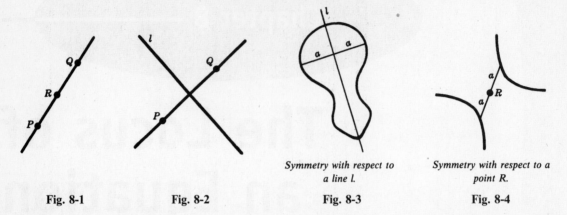

Symmetry with respect to   Symmetry with respect to a
a line l.   point R.

Fig. 8-1   Fig. 8-2   Fig. 8-3   Fig. 8-4

SYMMETRY OF A LOCUS.   The locus of a given equation $f(x, y) = 0$ is symmetric with respect to the x axis if an equivalent equation is obtained when y is replaced by $-y$, is symmetric with respect to the y axis if an equivalent equation is obtained when x is replaced by $-x$, and is symmetric with respect to the origin if an equivalent equation is obtained when x is replaced by $-x$ and y is replaced by $-y$ simultaneously.

An equation whose graph is symmetric with respect to the y axis is called *even*; one whose graph is symmetric with respect to the x axis is *odd*. See Chapter 22 for more on symmetry.

EXAMPLE 1.   Examine $x^2 + 2y^2 + x = 0$ for symmetry with respect to the coordinate axes and the origin.

When y is replaced by $-y$, we have $x^2 + 2y^2 + x = 0$; the locus is symmetric with respect to the x axis.

When x is replaced by $-x$, we have $x^2 + 2y^2 - x = 0$; the locus is not symmetric with respect to the y axis.

When x is replaced by $-x$ and y by $-y$, we have $x^2 + 2y^2 - x = 0$; the locus is not symmetric with respect to the origin. (See Problem 8.3.)

ASYMPTOTES.   The line $x = a$ is a vertical asymptote to the graph of an equation (locus) if, as x gets arbitrarily close to a, the locus gets arbitrarily large.

The line $y = b$ is a horizontal asymptote to a locus if the locus gets arbitrarily close to b as x gets arbitrarily large. See Fig. 8-5. (In calculus terms, if $\lim_{x \to \infty} f(x) = a$, then $y = a$ is a horizontal asymptote to $f(x)$. If $\lim_{x \to b} f(x) = \infty$, then $x = b$ is a vertical asymptote.)

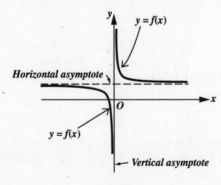

Fig. 8-5

EXAMPLE 2.   Show that the line $y - 2 = 0$ is a horizontal asymptote of the curve $xy - 2x - 1 = 0$.

The locus exists for all $x \neq 0$ and for all $y \neq 2$. Since $y > 0$ for $x > 0$, there is a branch of the locus in the first quadrant. On this branch choose a point, say, $A(1, 3)$, and let a point $P(x, y)$, moving to the right from A, trace the locus.

From Table 8.1, which lists a few of the positions assumed by P, it is clear that as P moves to the right, (a) its ordinate y remains greater than 2 and (b) the difference $y - 2$ may be made as small as we please by taking x sufficiently large. For example, if we wish $y - 2 < 1/10^{12}$, we take $x > 10^{12}$; if we wish $y - 2 < 1/10^{999}$, we take $x > 10^{999}$, and so on. The line $y - 2 = 0$ is therefore a horizontal asymptote.

Table 8.1

x	y
1	3
10	2.1
100	2.01
1000	2.001
10 000	2.0001

To find the horizontal asymptotes: Solve the equation of the locus for x and set each *real* linear factor of the denominator, if any, equal to zero.

To find the vertical asymptotes: Solve the equation of the locus for y and set each *real* linear factor of the denominator, if any, equal to zero.

EXAMPLE 3.   The locus of $x^2 + 4y^2 = 4$ has neither horizontal nor vertical asymptotes. The locus of $x = (y + 6)/(y + 3)$ has $y + 3 = 0$ as a horizontal asymptote and $x - 1 = 0$ as a vertical asymptote.

Solved Problems

8.1   (a)   The locus $xy + x^2 = x(y + x) = 0$ consists of the lines $x = 0$ and $y + x = 0$.

(b)   The locus $y^2 + xy^2 - xy - x^2 = (x + y)(y^2 - x) = 0$ consists of the line $x + y = 0$ and the parabola $y^2 - x = 0$.

(c)   The locus $y^4 + y^2 - x^2 - x = (y^2 - x)(y^2 + x + 1) = 0$ consists of the parabolas $y^2 - x = 0$ and $y^2 + x + 1 = 0$.

8.2   Examine for intercepts.

(a) $4x^2 - 9y^2 = 36$.

Set $y = 0$: $4x^2 = 36$, $x^2 = 9$; the x intercepts are ± 3.
Set $x = 0$: $-9y^2 = 36$, $y^2 = -4$; the y intercepts are imaginary.

(b) $x^2 - 2x = y^2 - 4y + 3$.

Set $y = 0$: $x^2 - 2x - 3 = (x - 3)(x + 1) = 0$; the x intercepts are -1 and 3.
Set $x = 0$: $y^2 - 4y + 3 = (y - 1)(y - 3) = 0$; the y intercepts are 1 and 3.

8.3   Examine for symmetry with respect to the coordinate axis and the origin.

(a) $4x^2 - 9y^2 = 36$.

Replacing y by $-y$, we have $4x^2 - 9y^2 = 36$; the locus is symmetric with respect to the x axis.
Replacing x by $-x$, we have $4x^2 - 9y^2 = 36$; the locus is symmetric with respect to the y axis.
Replacing x by $-x$ and y by $-y$, we have $4x^2 - 9y^2 = 36$; the locus is symmetric with respect to the origin.
Note that if a locus is symmetric with respect to the coordinate axes, it is automatically symmetric with respect to the origin. It will be shown in the next problem that the converse is not true.

(b) $x^3 - x^2y + y^3 = 0$.

Replacing y by $-y$, we have $x^3 + x^2y - y^3 = 0$; the locus is not symmetric with respect to the x axis.
Replacing x by $-x$, we have $-x^3 - x^2y + y^3 = 0$; the locus is not symmetric with respect to the y axis.

Replacing x by $-x$ and y by $-y$, we have $-x^3 + x^2y - y^3 = -(x^3 - x^2y + y^3) = 0$; the locus is symmetric with respect to the origin.

(c) $x^2 - 4y^2 - 2x + 8y = 0$.

Replacing y by $-y$, we have $x^2 - 4y^2 - 2x - 8y = 0$, replacing x by $-x$, we have $x^2 - 4y^2 + 2x + 8y = 0$, and replacing x by $-x$ and y by $-y$, we have $x^2 - 4y^2 + 2x - 8y = 0$; the locus is not symmetric with respect to either axis or the origin.

8.4 Investigate for horizontal and vertical asymptotes.

(a) $3x + 4y - 12 = 0$.

Solve for y: $y = \dfrac{12 - 3x}{4}$. Solve for x: $x = \dfrac{12 - 4y}{3}$.

Since the denominators do not involve the variables, there are neither horizontal nor vertical asymptotes.

(b) $xy = 8$.

Solve for y: $y = 8/x$. Solve for x: $x = 8/y$. Set each denominator equal to zero:

$x = 0$ is the vertical asymptote,
$y = 0$ is the horizontal asymptote.

(c) $xy - y - x - 2 = 0$.

Solve for y: $y = \dfrac{x + 2}{x - 1}$. Solve for x: $x = \dfrac{y + 2}{y - 1}$.

Set each denominator equal to zero:

$x = 1$ is the vertical asymptote.
$y = 1$ is the horizontal asymptote.

(d) $x^2y - x - 4y = 0$.

Solve for y: $y = \dfrac{x}{x^2 - 4}$. Solve for x: $x = \dfrac{1 \pm \sqrt{1 + 16y^2}}{2y}$.

Then $x = 2$ and $x = -2$ are vertical asymptotes and $y = 0$ is the horizontal asymptote.

(e) $x^2y - x^2 + 4y = 0$.

Solve for y: $y = \dfrac{x^2}{x^2 + 4}$. Solve for x: $x = \pm 2\sqrt{\dfrac{y}{1 - y}}$.

There are no vertical asymptotes since when $x^2 + 4 = 0$, x is imaginary. The horizontal asymptote is $y = 1$.

Discuss the following equations and sketch their loci.

8.5 $y^2 = -8x$.

Intercepts: When $y = 0$, $x = 0$ (x intercept); when $x = 0$, $y = 0$ (y intercept).

Symmetry: When y is replaced by $-y$, the equation is unchanged; the locus is symmetric with respect to the x axis.

The locus is a parabola with vertex at $(0, 0)$. It may be sketched after locating the following points: $(-1, \pm 2\sqrt{2}), (-2, \pm 4)$, and $(-3, \pm 2\sqrt{6})$. See Fig. 8-6.

8.6 $x^2 - 4x + 4y + 8 = 0$.

Intercepts: When $y = 0$, x is imaginary; when $x = 0$, $y = -2$ (y intercept).

Symmetry: The locus is not symmetric with respect to the coordinate axes or the origin.

The locus is a parabola with vertex at $(2, -1)$. Other points on the locus are: $(-2, -5), (4, -2)$, and $(6, -5)$. See Fig. 8-7.

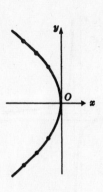
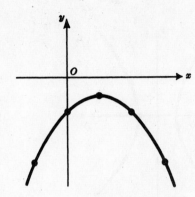

Fig. 8-6   Fig. 8-7

8.7 $x^2 + y^2 - 4x + 6y - 23 = 0$.

Intercepts: When $y = 0$, $x = \dfrac{4 \pm \sqrt{16 + 92}}{2} = 2 \pm 3\sqrt{3}$ (x intercepts); when $x = 0$, $y = \dfrac{-6 \pm \sqrt{36 + 92}}{2}$
$= -3 \pm 4\sqrt{2}$ (y intercepts).

Symmetry: There is no symmetry with respect to the coordinate axes or the origin.

Completing the squares, we have

$$(x^2 - 4x + 4) + (y^2 + 6y + 9) = 23 + 4 + 9 = 36 \quad \text{or} \quad (x - 2)^2 + (y + 3)^2 = 36$$

the equation of a circle having center at $C(2, -3)$ and radius 6. See Fig. 8-8.

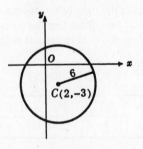
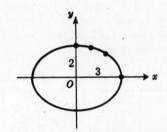

Fig. 8-8   Fig. 8-9

8.8 $4x^2 + 9y^2 = 36$.

Intercepts: When $y = 0$, $x = \pm 3$ (x intercepts); when $x = 0$, $y = \pm 2$ (y intercepts).
Symmetry: The locus is symmetric with respect to the coordinate axes and the origin.

Since the locus is symmetric with respect to both the axes, only sufficient points to sketch the portion of the locus in the first quadrant are needed. Two such points are $(1, 4\sqrt{2}/3)$ and $(2, 2\sqrt{5}/3)$. The locus is called an *ellipse*. See Fig. 8-9. Also see Chapter 36.

8.9 $9x^2 - 4y^2 = 36$.

Intercepts: When $y = 0$, $x = \pm 2$ (x intercepts); when $x = 0$, y is imaginary.
Symmetry: The locus is symmetric with respect to the coordinate axes and the origin.

The locus consists of two separate pieces and is not closed. The portion in the first quadrant has been sketched using the points $(3, 3\sqrt{5}/2)$, $(4, 3\sqrt{3})$, and $(5, 3\sqrt{21}/2)$. The locus is called a *hyperbola*. See Fig. 8-10. Also see Chapter 40.

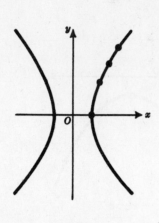

Fig. 8-10   **Fig. 8-11**

8.10 $xy - y - x - 2 = 0$.

Intercepts: The x intercept is -2; the y intercept is -2.
Symmetry: There is no symmetry with respect to the coordinate axes or the origin.
Asymptotes: $x = 1$, $y = 1$.

To sketch the locus, first draw in the asymptotes $x = 1$ and $y = 1$ (dotted lines). While the asymptotes are *not* a part of the locus, they serve as very convenient guide lines. Since the locus does not exist for $x = 1$ and $y = 1$, it does not cross the asymptotes. Since there is one value of y for each value of $x \neq 1$, that is, since y is single-valued, the locus appears in only two of the four regions into which the plane is separated by the asymptotes.

From Table 8.2 it is evident that the locus lies in the region to the right of the vertical asymptote and above the horizontal asymptote (see the portion of the table to the right of the double line) and in the region to the left of the vertical asymptote and below the horizontal asymptote (see the portion of the table to the left of the double line). The locus is shown in Fig. 8-11; note that it is symmetric with respect to $(1, 1)$, the point of intersection of the asymptotes.

Table 8.2

x	-10	-4	-3	-2	-1	0	$\frac{1}{2}$	$\frac{3}{4}$	$\frac{5}{4}$	$\frac{3}{2}$	2	10
y	$\frac{8}{11}$	$\frac{2}{5}$	$\frac{1}{4}$	0	$-\frac{1}{2}$	-2	-5	-11	13	7	4	$\frac{4}{3}$

8.11 $x^2 y - x^2 - 4y = 0$.

Intercepts: The x intercept is 0; the y intercept is 0.
Symmetry: The locus is symmetric with respect to the y axis.
Asymptotes: $x = \pm 2$, $y = 1$.

The asymptotes divide the plane into six regions. Since the locus does not exist when $x = \pm 2$ and when $y = 1$ (that is, does not cross an asymptote) and since y is single-valued, the locus appears in only three of these regions. By means of Table 8.3 the locus is sketched in Fig. 8-12. Note that only half of the table is necessary since the locus is symmetric with respect to the y axis.

Table 8.3

x	-10	-5	-4	-3	$-\frac{5}{2}$	$-\frac{7}{4}$	$-\frac{3}{2}$	-1	0	1	$\frac{3}{2}$	$\frac{7}{4}$
y	$\frac{25}{24}$	$\frac{25}{21}$	$\frac{4}{3}$	$\frac{9}{5}$	$\frac{25}{9}$	$-\frac{49}{15}$	$-\frac{9}{7}$	$-\frac{1}{3}$	0	$-\frac{1}{3}$	$-\frac{9}{7}$	$-\frac{49}{15}$

x	$\frac{5}{2}$	3	4	5	10
y	$\frac{25}{9}$	$\frac{9}{5}$	$\frac{4}{3}$	$\frac{25}{21}$	$\frac{25}{24}$

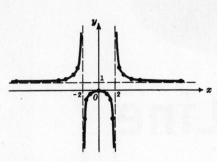

Fig. 8-12

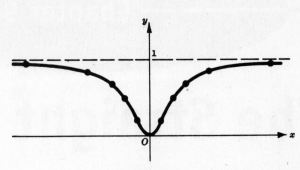

Fig. 8-13

8.12 $x^2y - x^2 + 4y = 0$.

Intercepts: The x intercept is 0; the y intercept is 0.
Symmetry: The locus is symmetric with respect to the y axis.

Since the locus exists only for $0 \le y < 1$, it lies entirely below its asymptote. The locus is sketeched in Fig. 8-13 using Table 8.4.

Table 8.4

x	-10	-5	-3	-2	-1	0	1	2	3	5	10
y	$\frac{25}{26}$	$\frac{25}{29}$	$\frac{9}{13}$	$\frac{1}{2}$	$\frac{1}{5}$	0	$\frac{1}{5}$	$\frac{1}{2}$	$\frac{9}{13}$	$\frac{25}{29}$	$\frac{25}{26}$

Only half of the table is necessary since the locus is symmetric with respect to the y axis.

Supplementary Problems

8.13   Discuss and sketch.

(a) $x^2 - 4y^2 = 0$
(b) $x^2 + 2xy + y^2 = 4$
(c) $y = 9x^2$
(d) $y^2 = 6x - 3$
(e) $y^2 = 4 - 2x$
(f) $x^2 + y^2 = 16$
(g) $x^2 + y^2 = 0$
(h) $4x^2 + 9y^2 = 36$

(i) $9x^2 - 4y^2 = 36$
(j) $9x^2 - 4y^2 + 36 = 0$
(k) $xy = 4$
(l) $xy = -4$
(m) $xy^2 = -9$
(n) $y^3 + xy^2 = 2xy - 2x^2$
(o) $y = x^3$
(p) $y = -x^3$

(q) $xy - 3x - y = 0$
(r) $x^2 + xy + y - 2 = 0$
(s) $x^2y - x - 4y = 0$
(t) $x^2y - 4xy + 3y - x - 2 = 0$
(u) $x^2y - x^2 + xy + 3x - 2 = 0$
(v) $x^3 + xy^2 - y^2 = 0$
(w) $x^2 + y^2 = -4$

8.14   Use a graphing calculator to sketch the following loci:

(a) $x^2 - 9y^2 = 0$
(b) $xy = 16$
(c) $9x^2 + 4y^2 = 36$

The Straight Line

THE EQUATION OF THE STRAIGHT LINE parallel to the y axis at a distance a from that axis is $x = a$.

The equation of the straight line having slope m and passing through the point (x_1, y_1) is

$$y - y_1 = m(x - x_1) \qquad (\textit{Point-slope form})$$

(See Problem 9.1.)

The equation of the line having slope m and y intercept b is

$$y = mx + b \qquad (\textit{Slope-intercept form})$$

(See Problem 9.2.)

The equation of the line passing through the points $(x_1, \ y_1)$ and $(x_2, \ y_2)$, where $x_1 \neq x_2$. is

$$y - y_1 = \frac{y_2 - y_1}{x_2 - x_1}(x - x_1) \qquad (\textit{Two-point form})$$

(See Problem 9.3.)

The equation of the line whose x intercept is a and whose y intercept is b, where $ab \neq 0$, is

$$\frac{x}{a} + \frac{y}{b} = 1 \qquad (\textit{Intercept form})$$

(See Problem 9.4.)

The length of the line segment with end points (x_1, y_1) and (x_2, y_2) is

$$d = \sqrt{(x_2 - x_1)^2 + (y_2 - y_1)^2}$$

THE GENERAL EQUATION of the straight line is $Ax + By + C = 0$, where A, B, C are arbitrary constants except that not both A and B are zero.

If $C = 0$, the line passes through the origin. If $B = 0$, the line is vertical; if $A = 0$, the line is horizontal. Otherwise, the line has slope $m = -A/B$ and y intercept $b = -C/B$.

If two nonvertical lines are parallel, their slopes are equal. Thus, the lines $Ax + By + C = 0$ and $Ax + By + D = 0$ are parallel.

If two lines are perpendicular, the slope of one is the negative reciprocal of the slope of the other. If m_1 and m_2 are the slopes of two perpendicular lines, the $m_1 = -1/m_2$ or $m_1 m_2 = -1$. Thus, $Ax + By + C = 0$ and $Bx - Ay + D = 0$, where $AB \neq 0$, are perpendicular lines. (See Problems 9.5–9.8.)

Solved Problems

9.1 Construct and find the equation of the straight line which passes though the point $(-1,-2)$ with slope (a) $\frac{3}{4}$ and (b) $-\frac{4}{5}$.

(a) A point tracing the line *rises* (slope is positive) 3 units as it moves a horizontal distance of 4 units to the right. Thus, after locating the point $A(-1,-2)$, move 4 units to the right and 3 up to the point $B(3,1)$. The required line is AB. [See Fig. 9-1(a).] Using $y - y_1 = m(x - x_1)$, the equation is

$$y + 2 = \tfrac{3}{4}(x + 1) \qquad \text{or} \qquad 3x - 4y - 5 = 0$$

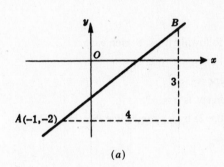

(a)   (b)

Fig. 9-1

(b) A point tracing the line falls (slope is negative) 4 units as it moves a horizontal distance of 5 units to the right. Thus, after locating the point $A(-1,-2)$, move 5 units to the right and 4 units down to the point $B(4,-6)$. The required line is AB as shown in Fig. 9-1(b). Its equation is

$$y + 2 = -\tfrac{4}{5}(x + 1) \qquad \text{or} \qquad 4x + 5y + 14 = 0$$

9.2 Determine the slope m and y intercept b of the following lines. Sketch each.

(a) $y = \frac{3}{2}x - 2$

Ans. $m = \frac{3}{2}$; $b = -2$

To sketch the locus, locate the point $(0,-2)$. Then move 2 units to the right and 3 units up to another point on the required line. See Fig. 9-2(a).

(b) $y = -3x + \frac{5}{2}$

Ans. $m = -3$; $b = \frac{5}{2}$

To sketch the locus, locate the point $(0,\frac{5}{2})$. Then move 1 unit to the right and 3 units down to another point on the line. See Fig. 9-2(b).

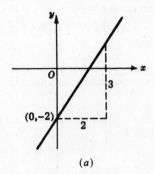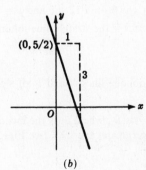

(a)   (b)

Fig. 9-2

9.3 Write the equation of the straight lines: (a) through (2, 3) and (−1, 4), (b) through (−7, −2) and (−2, −5), and (c) through (3, 3) and (3, 6).

We use $y - y_1 = \dfrac{y_2 - y_1}{x_2 - x_1}(x - x_1)$ and label each pair of points P_1 and P_2 in the order given.

(a) The equation is $y - 3 = \dfrac{4 - 3}{-1 - 2}(x - 2) = -\dfrac{1}{3}(x - 2)$ or $x + 3y - 11 = 0$.

(b) The equation is $y + 2 = \dfrac{-5 + 2}{-2 + 7}(x + 7) = -\dfrac{3}{5}(x + 7)$ or $3x + 5y + 31 = 0$.

(c) Here $x_1 = x_2 = 3$. The required equation is $x - 3 = 0$.

9.4 Determine the x intercept a and the y intercept b of the following lines. Sketch each.

(a) $3x - 2y - 4 = 0$

When $y = 0, 3x - 4 = 0$ and $x = \frac{4}{3}$; the x intercept is $a = \frac{4}{3}$.

When $x = 0, -2y - 4 = 0$ and $y = -2$; the y intercept is $b = -2$.

To obtain the locus, join the points $(\frac{4}{3}, 0)$ and $(0, -2)$ by a straight line. See Fig. 9-3(a).

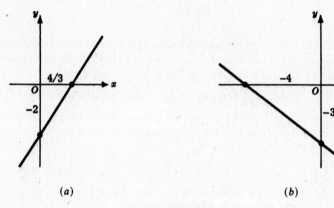

(a)   (b)

Fig. 9-3

(b) $3x + 4y + 12 = 0$

When $y = 0, 3x + 12 = 0$ and $x = -4$; the x intercept is $a = -4$.

When $x = 0, 4y + 12 = 0$ and $y = -3$; the y intercept is $b = -3$.

The locus is the straight line joining the point $(-4, 0)$ and $(0, -3)$. See Fig. 9-3(b).

9.5 Prove: If two oblique lines l_1 and l_2 of slope m_1 and m_2, respectively, are mutually perpendicular, then $m_1 = -1/m_2$.

Let $m_2 = \tan \theta_2$, where θ_2 is the inclination of l_2. The inclination of l_1 is $\theta_1 = \theta_2 \pm 90°$ according as θ_2 is less than or greater than 90° [see Figs. 9-4(a) and (b)]. Then

$$m_1 = \tan \theta_1 = \tan(\theta_2 \pm 90°) = -\cot \theta_2 - \frac{1}{\tan \theta_2} = -\frac{1}{m_2}$$

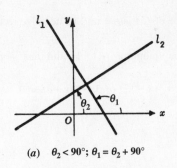 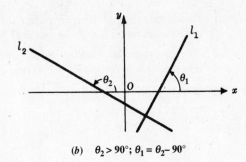

(a) $\theta_2 < 90°$; $\theta_1 = \theta_2 + 90°$ (b) $\theta_2 > 90°$; $\theta_1 = \theta_2 - 90°$

Fig. 9-4

9.6   Find the equation of the straight line (a) through (3, 1) and parallel to the line through (3, −2) and (−6, 5) and (b) through (−2, −4) and parallel to the line $8x - 2y + 3 = 0$.

Two lines are parallel provided their slopes are equal.

(a)   The slope of the line through (3, −2) and (−6, 5) is $m = \dfrac{y_2 - y_1}{x_2 - x_1} = \dfrac{5 + 2}{-6 - 3} = -\dfrac{7}{9}$. The equation of the line through (3, 1) with slope $-\frac{7}{9}$ is $y - 1 = -\frac{7}{9}(x - 3)$ or $7x + 9y - 30 = 0$.

(b)   **First Solution.**   From $y = 4x + \frac{3}{2}$, the slope of the given line is $m = 4$. The equation of the line through (−2, −4) with slope 4 is $y + 4 = 4(x + 2)$ or $4x - y + 4 = 0$.

   Second Solution.   The equation of the required line is of the form $8x - 2y + D = 0$. If (−2,−4) is on the line, then $8(-2) - 2(-4) + D = 0$ and $D = 8$. The required equation is $8x - 2y + 8 = 0$ or $4x - y + 4 = 0$.

9.7   Find the equation of the straight line (a) through (−1, −2) and perpendicular to the line through (−2, 3) and (−5, −6) and (b) through (2, −4) and perpendicular to the line $5x + 3y - 8 = 0$.

Two lines are perpendicular provided the slope of one is the negative reciprocal of the slope of the other.

(a)   The slope of the line through (−2, 3) and (−5, −6) is $m = 3$; the slope of the required line is $-1/m = -\frac{1}{3}$. The required equation is $y + 2 = -\frac{1}{3}(x + 1)$ or $x + 3y + 7 = 0$.

(b)   **First Solution.**   The slope of the given line is $-\frac{5}{3}$; the slope of the required line is $\frac{3}{5}$. The required equation is $y + 4 = \frac{3}{5}(x - 2)$ or $3x - 5y - 26 = 0$.

   Second Solution.   The equation of the required line is of the form $3x - 5y + D = 0$. If (2,−4) is on the line, then $3(2) - 5(-4) + D = 0$ and $D = -26$. The required equation is $3x - 5y - 26 = 0$.

9.8   Given the vertices $A(7, 9)$, $B(-5, -7)$, and $C(12, -3)$ of the triangle ABC (see Fig. 9-5), find

(a)   The equation of the side AB

(b)   The equation of the median through A

(c)   The equation of the altitude through B

(d)   The equation of the perpendicular bisector of the side AB

(e)   The equation of the line through C with slope that of AB

(f)   The equation of the line through C with slope the reciprocal of that of AB

(a) $y + 7 = \dfrac{9 + 7}{7 + 5}(x + 5) = \dfrac{4}{3}(x + 5)$ or $4x - 3y - 1 = 0$

(b)   The median through a vertex bisects the opposite side. The midpoint of BC is $L(\frac{7}{2}, -5)$. The equation of the median through A is

$$y - 9 = \dfrac{-5 - 9}{\frac{7}{2} - 7}(x - 7) = 4(x - 7) \qquad \text{or} \qquad 4x - y - 19 = 0$$

(c) The altitude through B is perpendicular to CA. The slope of CA is $-\frac{12}{5}$ and its negative reciprocal is $\frac{5}{12}$. The equation is $y + 7 = \frac{5}{12}(x + 5)$ or $5x - 12y - 59 = 0$.

(d) The perpendicular bisector of AB passes through the midpoint $N(1, 1)$ and has slope $-\frac{3}{4}$. The equation is $y - 1 = -\frac{3}{4}(x - 1)$ or $3x + 4y - 7 = 0$.

(e) $y + 3 = \frac{4}{3}(x - 12)$ or $4x - 3y - 57 = 0$.   (f) $y + 3 = \frac{3}{4}(x - 12)$ or $3x - 4y - 48 = 0$.

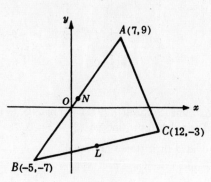

Fig. 9-5

Supplementary Problems

9.9   Determine the slope and the intercepts on the coordinate axes of each of the lines (a) $4x - 5y + 20 = 0$ and (b) $2x + 3y - 12 = 0$.

      Ans.   (a) $m = \frac{4}{5}$, $a = -5$, $b = 4$ (b) $m = -\frac{2}{3}$, $a = 6$, $b = 4$

9.10   Write the equation of each of the straight lines:

(a) With slope 3 and y intercept -2.   *Ans.* $y = 3x - 2$

(b) Through $(5, 4)$ and parallel to $2x + 3y - 12 = 0$.   *Ans.* $2x + 3y - 22 = 0$

(c) Through $(-3, 3)$ and with y intercept 6.   *Ans.* $x - y + 6 = 0$

(d) Through $(-3, 3)$ and with x intercept 4.   *Ans.* $3x + 7y - 12 = 0$

(e) Through $(5, 4)$ and $(-3, 3)$.   *Ans.* $x - 8y + 27 = 0$

(f) With $a = 3$ and $b = -5$.   *Ans.* $5x - 3y - 15 = 0$

(g) Through $(2, 3)$ and $(2, -5)$ *Ans.* $x - 2 = 0$

9.11   Find the value of k such that the line $(k - 1)x + (k + 1)y - 7 = 0$ is parallel to the line $3x + 5y + 7 = 0$.

      Ans. $k = 4$

9.12   Given the triangle whose vertices are $A(-3, 2)$, $B(5, 6)$, $C(1, -4)$. Find the equations of

(a) The sides.   *Ans.* $x - 2y + 7 = 0,\ 5x - 2y - 13 = 0,\ 3x + 2y + 5 = 0$

(b) The medians.   *Ans.* $x + 6y - 9 = 0,\ 7x - 6y + 1 = 0,\ x = 1$

(c) The altitudes.   *Ans.* $2x + 5y - 4 = 0,\ 2x - 3y + 8 = 0, 2x + y + 2 = 0$

(d) The perpendicular bisectors of the sides.   *Ans.* $2x + y - 6 = 0,\ 2x + 5y - 11 = 0, 2x - 3y - 1 = 0$

9.13   For the triangle of Problem 9.12: (*a*) Find the coordinates of the centroid *G* (intersection of the medians), the orthocenter *H* (intersection of the altitudes), and circumcenter *C* (intersection of the perpendicular bisectors of the sides). (*b*) Show that *G* lies on the line joining *H* and *C* and divides the segment *HC* in the ratio 2:1.

   Ans.   (*a*) $G(1, \frac{4}{3})$, $H(-\frac{7}{4}, \frac{3}{2})$, $C(\frac{19}{8}, \frac{5}{4})$

9.14   Show that the line passing through the points $(5, -4)$ and $(-2, 7)$ is the perpendicular bisector of the line segment whose end points are $(-4, -2)$ and $(7, 5)$.

9.15   Use the determinant form of the formula for the area of a triangle (Chapter 11) to show

   (*a*)   The equation of the straight line through $P_1(x_1, y_1)$ and $P_2(x_2, y_2)$ is given by $\begin{vmatrix} x & y & 1 \\ x_1 & y_1 & 1 \\ x_2 & y_2 & 1 \end{vmatrix} = 0$.

   (*b*)   Three points $P_1(x_1, y_1)$, $P_2(x_2, y_2)$, and $P_3(x_3, y_3)$ are on a straight line provided $\begin{vmatrix} x_1 & y_1 & 1 \\ x_2 & y_2 & 1 \\ x_3 & y_3 & 1 \end{vmatrix} = 0$.

   Is the converse of this statement true?

Chapter 10

Families of Straight Lines

THE EQUATION $y = 3x + b$ represents the set of all lines having slope 3. The quantity b is a variable over the set of lines, each value of b being associated with one and only one line of the set. To distinguish it from the variables x, y, which vary over the points of each line, b is called a *parameter*. Such sets of lines are also called *one-parameter systems* or *families* of lines. See Fig. 10-1. (See Chapter 45 for a fuller discussion of parametric equations.)

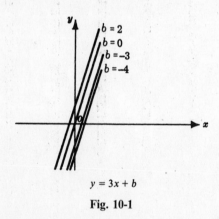

$$y = 3x + b$$

Fig. 10-1

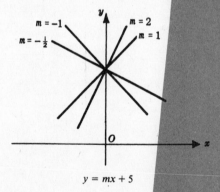

$$y = mx + 5$$

Fig. 10-2

Similarly, the equation $y = mx + 5$, in which m is the parameter, represents the one-parameter family of lines having y intercept $= 5$ or passing through the point $(0, 5)$. See Fig. 10-2. It is important to note that one line satisfying the geometric condition is not included since for no value of m does the equation $y = mx + 5$ yield the line $x = 0$. (See Problems 10.1–10.4.)

THE EQUATION

$$A_1 x + B_1 y + C_1 + k(A_2 x + B_2 y + C_2) = 0 \qquad (10.1)$$

represents the family of all lines, except for l_2, passing through the point of intersection of the lines

$$l_1: \quad A_1 x + B_1 y + C_1 = 0 \qquad \text{and} \qquad l_2: \quad A_2 y + C_2 = 0$$

(See Problem 10.5.)

60

Equation (*10.1*) provides a means of finding the equation of a line which passes through the point of intersection of two given lines and satisfies one other condition, without having first to compute the coordinate of the point of intersection.

EXAMPLE 1.   Write the equation of the line which passes through the point of intersection of the lines

$$l_1:\quad 3x+5y+29=0 \quad\text{and}\quad l_2:\quad 7x-11y-13=0$$

and through the point (2, 1).

Since the given point is not on l_2, the required line is among those given by

$$l_1+kl_2:\quad 3x+5y+29+k(7x-11y-13)=0$$

For the line which passes through $(2,1)$, $3(2)+5(1)+29+k[7(2)-11(1)-13]=0$ and $k=4$; hence, its equation is

$$l_1+4l_2:\quad 3x+5y+29+4(7x-11y-13)=31x-39y-23=0$$

(See Problem 10.6.)

Solved Problems

10.1   Write the equation of the family of lines satisfying the given condition. Name the parameter and list any lines satisfying the condition but not obtained by assigning a value to the parameter.

   (*a*) Parallel to the *x* axis.   (*d*) Perpendicular to $5x+3y-8=0$.

   (*b*) Through the point $(-3,2)$.   (*e*) The sum of whose intercepts is 10.

   (*c*) At a distance 5 from the origin.

   (*a*) The equation is $y=k$, k being the parameter.

   (*b*) The equation is $y-2=m(x+3)$, m being the parameter. The line $x+3=0$ is not included.

   (*c*) The equation is $x\cos\omega+y\sin\omega-5=0$, ω being the parameter.

   (*d*) The slope of the given lines is $-\frac{5}{3}$; hence, the slope of the perpendicular is $\frac{3}{5}$. The equation of the family is $y=\frac{3}{5}x+c$ or $3x-5y+5c=0$, c being the parameter. The latter is equivalent to $3x-5y+k=0$, with k as parameter.

   (*e*) Taking the *x* intercept as $a\neq0$, the *y* intercept is $10-a$ and the equation of the family is $\dfrac{x}{a}+\dfrac{y}{10-a}=1$. The parameter is a.

10.2   Describe each family of lines: (*a*) $x=k$, (*b*) $x\cos\omega+y\sin\omega-10=0$, (*c*) $x/\cos\theta+y/\sin\theta=1$, and (*d*) $kx+\sqrt{1-k^2}\,y-10=0$.

   (*a*) This is the family of all vertical lines.

   (*b*) This is the family of all tangents to the circle with center at the origin and radius $=10$.

   (*c*) This is the family of all lines the sum of the squares of whose intercepts is 1.

   (*d*) Since $\sqrt{1-k^2}$ is to be assumed real, the range of $k=\cos\omega$ is $-1\le k\le1$, while the range of $\sqrt{1-k^2}=\sin\omega$ is $0\le\sqrt{1-k^2}\le1$. The equation is that of the family of tangents to the upper half circle of (*b*).

10.3   In each of the following write the equation of the family of lines satisfying the first condition and then obtain the equation of the line of the family satisfying the second condition: (*a*) parallel to $3x-5y+12=0$, through $P(1,-2)$, (*b*) perpendicular to $3x-5y+12=0$, through $P(1,-2)$, and (*c*) through $P(-3,2)$, at a distance 3 from the origin.

   (*a*) The equation of the family is $3x-5y+k=0$. Substituting $x=1, y=-2$ in this equation and solving for k, we find $k=-13$. The required line has equation $3x-5y-13=0$.

(b) The equation of the family is $5x + 3y + k = 0$ [see Problem 10.1(d)]. Proceeding as in (a) above, we find $k = 1$; the required line has equation $5x + 3y + 1 = 0$.

(c) The equation of the family is $y - 2 = k(x + 3)$ or, in normal form, $\dfrac{kx - y + (3k + 2)}{\pm\sqrt{k^2 + 1}} = 0$. Setting the

undirected distance of a line of the family from the origin equal to 3, we have $\left|\dfrac{3k + 2}{\pm\sqrt{k^2 + 1}}\right| = 3$.

Then $\dfrac{(3k + 2)^2}{k^2 + 1} = 9$ and $k = \frac{5}{12}$. Thus, the required line has equation $y - 2 = \frac{5}{12}(x + 3)$ or $5x - 12y - 12y + 39 = 0$.

Now there is a second line, having equation $x + 3 = 0$, satisfying the conditions, but this line [see Problem 10.1(b)] was not included in the equation of the family.

10.4 Write the equation of the line which passes through the point of intersection of the lines $l_1: 3x + 5y + 26 = 0$ and $l_2: 7x - 11y - 13 = 0$ and satisfies the additional condition: (a) Passes through the origin. (b) Is perpendicular to the line $7x + 3y - 9 = 0$.

Each of the required lines is a member of the family

$$l_1 + kl_2: \quad 3x + 5y + 26 + k(7x - 11y - 13) = 0 \tag{A}$$

(a) Substitution $x = 0, y = 0$ in (A), we find $k = 2$; the required line has equation

$$l_1 + 2l_2: \quad 3x + 5y + 26 + 2(7x - 11y - 13) = 0 \quad\text{or}\quad x - y = 0$$

(b) The slope of the given line is $-\frac{7}{3}$ and the slope of a line of (A) is $-\dfrac{3 + 7k}{5 - 11k}$.

Setting one slope equal to the negative reciprocal of the other, we find $-\dfrac{7}{3} = \dfrac{5 - 11k}{3 + 7k}$ and $k = -\frac{9}{4}$. The required line has equation

$$l_1 - \tfrac{9}{4}l_2: \quad 3x + 5y + 26 - \tfrac{9}{4}(7x - 11y - 13) = 0 \quad\text{or}\quad 3x - 7y - 13 = 0$$

10.5 Write the equation of the line which passes through the point of intersection of the lines $l_1: x + 4y - 18 = 0$ and $l_2: x + 2y - 2 = 0$ and satisfies the additional condition: (a) Is parallel to the line $3x + 8y + 1 = 0$. (b) Whose distance from the origin is 2.

The equation of the family of lines through the point of intersection of the given lines is

$$l_1 + kl_2: \quad x + 4y - 18 + k(x + 2y - 2 = 0) \quad\text{or}\quad (1 + k)x + (4 + 2k)y - (18 + 2k) = 0 \tag{A}$$

(a) Since the required line is to have slope $-\frac{3}{8}$, we set $-\dfrac{1 + k}{4 + 2k} = \dfrac{3}{8}$. Then $k = 2$ and the line has equation

$$l_1 + 2l_2: \quad 3x + 8y - 22 = 0$$

(b) The normal form of (A) is $\dfrac{(1 + k)x + (4 + 2k)y - (18 - 2k)}{\pm\sqrt{17 + 18k + 5k^2}} = 0$. Setting $\left|\dfrac{1(18 + 2k)}{\pm\sqrt{17 + 18k + 5k^2}}\right| = 2$ and

squaring both members, we find $k = \pm 4$. The required lines have equations

$$l_1 + 4l_2: \quad 5x + 12y - 26 = 0 \quad\text{and}\quad l_1 - 4l_2: \quad 3x + 4y + 10 = 0$$

Supplementary Problems

10.6 Write the equation of the family of lines satisfying the given condition. List any lines satisfying the condition but not obtained by assigning a value to the parameter.

(a) Perpendicular to the x axis.

(b) Through the point $(3, -1)$.

(c) At the distance 6 from the origin.

(d) Parallel to $2x + 5y - 8 = 0$.

(e) The product of whose intercepts on the coordinate axes is 10.

Ans.   (a) $a = k$ (c) $x \cos \omega + y \sin \omega - 6 = 0$ (e) $10x + k^2 y - 10k = 0$
        (b) $y = kx - 3k - 1, x = 3$ (d) $2x + 5y + k = 0$

10.7 Write the equation of the family of lines satisfying the first condition and then obtain the equation of that line of the family satisfying the second condition.

(a) Parallel to $2x - 3y + 8 = 0$; passing through $P(2, -2)$.

(b) Perpendicular to $2x - 3y + 8 = 0$; passing through $P(2, -2)$.

(c) Sum of the intercepts on the coordinate axes is 2; forms with the coordinate axes a triangle of area 24.

(d) At a distance 3 from the origin; passes through $P(1, 3)$.

(e) Slope is $\frac{3}{2}$; product of the intercepts on the coordinate axes is -54.

(f) The y intercept is 6; makes an angle of $135°$ with the line $7x - y - 23 = 0$.

(g) Slope is $\frac{5}{12}$; 5 units from $(2, -3)$.

Ans.   (a) $2x - 3y - 10 = 0$ (e) $3x - 2y \pm 18 = 0$
        (b) $3x + 22y - 2 = 0$ (f) $3x - 4y + 24 = 0$
        (c) $3x - 4y - 24 = 0$, $4x - 3y + 24 = 0$ (g) $5x - 12y + 19 = 0$, $5x - 12y - 111 = 0$
        (d) $3x + 4y - 15 = 0$, $y = 3$

10.8 Find the equation of the line which passes through the point $P(3, -4)$ and has the additional property:

(a) The sum of its intercepts on the coordinate axes is -5.

(b) The product of its intercepts on the coordinate axes is the negative reciprocal of its slope.

(c) It forms with the coordinate axes a triangle of area 1.

Ans.   (a) $2x - y - 10 = 0$, $2x + 3y + 6 = 0$ (c) $2x + y - 2 = 0$, $8x + 9y + 12 = 0$
        (b) $x + y + 1 = 0$, $5x + 3y - 3 = 0$

10.9 Write the equation of the line which passes through the point of intersection of the lines l_1: $x - 2y - 4 = 0$ and l_2: $4x - y - 4 = 0$ and statisfies the additional condition:

(a) Passes through the origin.

(b) Passes through the point $P(4, -6)$.

(c) Is parallel to the line $16x - 11y + 3 = 0$.

(d) Is perpendicular to the line $9x + 22y - 8 = 0$.

(e) Is $\frac{12}{7}$ units from the origin.

(f) Makes an angle of $45°$ with the line $9x - 5y - 12 = 0$.

(g) Its y intercept is $-\frac{20}{11}$ its slope.

(h) Makes a triangle of area $\frac{12}{5}$ with the coordinate axes.

Ans.   (a) $3x + y = 0$ (e) $7y + 12 = 0$, $21x - 28y - 60 = 0$
        (b) $5x + 4y + 4 = 0$ (f) $49x + 14y - 4 = 0$
        (c) $16x - 11y - 28 = 0$ (g) $11x - 8y - 20 = 0$
        (d) $22x - 9y - 28 = 0$ (h) $6x - 5y - 12 = 0$, $15x - 2y - 12 = 0$,
                                                          $2(12 \mp \sqrt{89})x + (41 \pm 3\sqrt{89})y - 12(1 \mp \sqrt{89}) = 0$

Chapter 11

The Circle

A CIRCLE IS THE LOCUS of a point which moves in a plane so that it is always at a constant distance from a fixed point in the plane. The fixed point is called the *center* and the constant distance is the length of the radius of the circle. The circle is one of the "conic sections." That is, the circle also results from taking particular cross sections of the right-circular cone. See Chapter 36 for a discussion of three additional conic sections: the parabola, the hyperbola, and the ellipse. As you think about slicing a cone with a plane, what additional cross sections are generated?

THE STANDARD FORM of the equation of the circle whose center is at the point $C(h, k)$ and whose radius is the constant r is

$$(x-h)^2 + (y-k)^2 = r^2 \qquad (11.1)$$

See Fig. 11-1. (See Problem 11.1.)

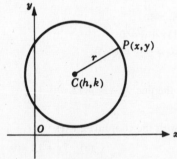

Fig. 11-1

THE GENERAL FORM of the equation of a circle is

$$x^2 + y^2 + 2Dx + 2Ey + F = 0 \qquad (11.2)$$

By completing the squares this may be put in the form

$$(x+D)^2 + (x+E)^2 = D^2 + E^2 - F$$

Thus, (*11.2*) represents a circle with center at $C(-D, -E)$ and radius $\sqrt{D^2 + E^2 - F}$ if $(D^2 + E^2 - F) > 0$, a point if $(D^2 + E^2 - F) = 0$, and an imaginary locus if $(D^2 + E^2 - F) = 0$. (See Problems 11.2–11.3.)

IN BOTH THE STANDARD AND GENERAL FORM the equation of a circle contains three independent arbitrary constants. It follows that a circle is uniquely determined by three independent conditions. (See Problems 11.4–11.6.)

THE EQUATION OF A TANGENT to a circle may be found by making use of the fact that a tangent and the radius drawn to the point of tangency are perpendicular. (See Problem 11.7.)

THE LENGTH OF A TANGENT to a circle from an external point P_1 is defined as the distance from the point P_1 to the point of tangency. The two tangents from an external point are of equal length. See Fig. 11-2, where $\overline{P_1T} \cong \overline{P_1T'}$.

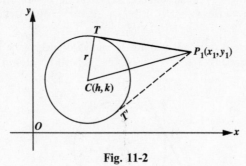

Fig. 11-2

The square of the length of a tangent from the external point $P_1(x_1, y_1)$ to a circle is obtained by substituting the coordinates of the point in the left member of the equation of the circle when written in the form $(x-h)^2 + (y-k)^2 - r^2 = 0$ or $x^2 + y^2 + 2Dx + 2Ey + F = 0$. (See Problems 11.8–11.9.)

THE EQUATION

$$x^2 + y^2 + 2D_1x + 2E_1y + F_1 + k(x^2 + y^2 + 2D_2x + 2E_2y + F_2) = 0 \qquad (11.3)$$

where

$$K_1: \quad x^2 + y^2 + 2D_1x + 2E_1y + F_1 = 0 \quad \text{and} \quad K_2: \quad x^2 + y^2 + 2D_2x + 2E_2y + F_2 = 0$$

are distinct circles and $k \neq -1$ is a parameter, represents a one-parameter family of circles.

If K_1 and K_2 are concentric, the circles of (11.3) are concentric with them.

If K_1 and K_2 are not concentric, the circles of (11.3) have a common line of centers with them and the centers of the circles of (11.3) divide the segment joining the centers of K_1 and K_2 in the ratio $k : 1$.

If K_1 and K_2 intersect in two distinct points P_1 and P_2, (11.3) consists of all circles except K_2 which pass through these points. If K_1 and K_2 are tangent to each other at the point P_1, (11.3) consists of all circles except K_2 which are tangent to each other at P_1. If K_1 and K_2 have no point in common, any two circles of the family (11.3) have no point in common with each other. See Fig. 11-3. (See Problems 11.10–11.11.)

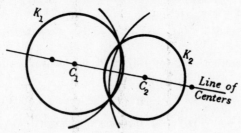

Fig. 11-3

Solved Problems

11.1   Write the equation of the circle satisfying the given conditions.

(a) $C(0, 0)$, $r = 5$ (d) $C(-5, 6)$ and tangent to the x axis

(b) $C(4, -2)$, $r = 8$ (e) $C(3, 4)$ and tangent to $2x - y + 5 = 0$

(c) $C(-4, -2)$ and passing through $P(1, 3)$ (f)   Center on $y = x$, tangent to both axes, $r = 4$

(a)   Using $(x - h)^2 + (y - k)^2 = r^2$, the equation is $(x - 0)^2 + (y - 0)^2 = 25$ or $x^2 + y^2 = 25$.

(b)   Using $(x - h)^2 + (y - k)^2 = r^2$, the equation is $(x - 4)^2 + (y + 2)^2 = 64$.

(c)   Since the center is at $C(-4, -2)$, the equation has the form $(x + 4)^2 + (y + 2)^2 = r^2$. The condition that $P(1, 3)$ lie on this circle is $(1 + 4)^2 + (3 + 2)^2 = r^2 = 50$. Hence the required equation is $(x + 4)^2 + (y + 2)^2 = 50$.

(d)   The tangent to a circle is perpendicular to the radius drawn to the point of tangency; hence $r = 6$. The equation of the circle is $(x + 5)^2 + (y - 6)^2 = 36$. See Fig. 11-4(a).

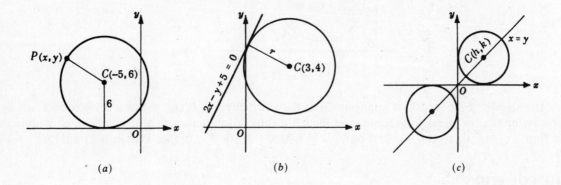

| (a) | (b) | (c) |

Fig. 11-4

(e)   The radius is the undirected distance of the point $C(3, 4)$ from the line $2x - y + 5 = 0$; thus $r = \left| \dfrac{2 \cdot 3 - 4 + 5}{-\sqrt{5}} \right| = \dfrac{7}{\sqrt{5}}$. The equation of the circle is $(x - 3)^2 + (y - 4)^2 = \frac{49}{5}$. See Fig. 11-4(b).

(f)   Since the center (h, k) lies on the line $x = y$, $h = k$; since the circle is tangent to both axes, $|h| = |k| = r$. Thus there are two circles satisfying the conditions, one with center $(4, 4)$ and equation $(x - 4)^2 + (y - 4)^2 = 16$, the other with center $(-4, -4)$ and equation $(x + 4)^2 + (y + 4)^2 = 16$. See Fig. 11-4(c).

11.2   Describe the locus represented by each of the following equations:

(a) $x^2 + y^2 - 10x + 8y + 5 = 0$ (c) $x^2 + y^2 + 4x - 6y + 24 = 0$

(b) $x^2 + y^2 - 6x - 8y + 25 = 0$ (d) $4x^2 + 4y^2 + 80x + 12y + 265 = 0$

(a)   From the standard form $(x - 5)^2 + (y + 4)^2 = 36$, the locus is a circle with center at $C(5, -4)$ and radius 6.

(b)   From the standard form $(x - 3)^2 + (y - 4)^2 = 0$, the locus is a point circle or the point $(3, 4)$.

(c)   Here we have $(x + 2)^2 + (y - 3)^2 = -11$; the locus is imaginary.

(d)   Dividing by 4, we have $x^2 + y^2 + 20x + 3y + \frac{265}{4} = 0$. From $(x + 10)^2 + (y + \frac{3}{2})^2 = 36$, the locus is a circle with center at $C(-10, -\frac{3}{2})$ and radius 6.

11.3   Show that the circles $x^2 + y^2 - 16x - 20y + 115 = 0$ and $x^2 + y^2 + 8x - 10y + 5 = 0$ are tangent and find the point of tangency.

The first circle has center $C_1(8, \ 10)$ and radius 7; the second has center $C_2(-4, 5)$ and radius 6. The two circles are tangent externally since the distance between their centers $C_1 C_2 = \sqrt{144 + 25} = 13$ is equal to the *sum* of the radii.

The point of tangency $P(x, y)$ divides the segment $C_2 C_1$ in the ratio $6 : 7$. Then

$$x = \frac{6 \cdot 8 + 7(-4)}{6 + 7} = \frac{20}{13}, \qquad y = \frac{6 \cdot 10 + 7 \cdot 5}{6 + 7} = \frac{95}{13}$$

and the point of tangency has coordinates $(\frac{20}{13}, \frac{95}{13})$.

11.4   Find the equation of the circle through the points $(5, 1)$, $(4, 6)$, and $(2, -2)$.

Take the equation in general form $x^2 + y^2 + 2Dx + 2Ey + F = 0$. Substituting successively the coordinates of the given points, we have

$$\begin{cases} 25 + \ 1 + 10D + \ 2E + F = 0 \\ 16 + 36 + \ 8D + 12E + F = 0 \\ 4 + \ 4 + \ 4D - \ 4E + F = 0 \end{cases} \quad \text{or} \quad \begin{cases} 10D + \ 2E + F = -26 \\ 8D + 12E + F = -52 \\ 4D - \ 4E + F = - \ 8 \end{cases}$$

with solution $D = -\frac{1}{3}$, $E = -\frac{8}{3}$, $F = -\frac{52}{3}$. Thus the required equation is

$$x^2 + y^2 - \tfrac{2}{3}x - \tfrac{16}{3}y - \tfrac{52}{3} = 0 \quad \text{or} \quad 3x^2 + 3y^2 - 2x - 16y - 52 = 0$$

11.5   Write the equations of the circle having radius $\sqrt{13}$ and tangent to the line $2x - 3y + 1 = 0$ at $(1, 1)$.

Let the equation of the circle be $(x - h)^2 + (y - k)^2 = 13$. Since the coordinates $(1, 1)$ satisfy this equation, we have

$$(1 - h)^2 + (1 - k)^2 = 13 \tag{1}$$

The undirected distance from the tangent to the center of the circle is equal to the radius; that is,

$$\left| \frac{2h - 3k + 1}{-\sqrt{3}} \right| = \sqrt{13} \quad \text{and} \quad \frac{2h - 3k + 1}{\sqrt{13}} = \pm\sqrt{13} \tag{2}$$

Finally, the radius through $(1, 1)$ is perpendicular to the tangent there; that is,

$$\text{Slope of radius through } (1, 1) = -\frac{1}{\text{slope of given line}} \quad \text{or} \quad \frac{k - 1}{h - 1} = -\frac{3}{2} \tag{3}$$

Since there are only two unknowns, we may solve simultaneously any two of the three equations. Using (2) and (3), noting that there are two equations in (2), we find $h = 3$, $k = -2$ and $h = -1, k = 4$. The equations of the circles are $(x - 3)^2 + (y + 2)^2 = 13$ and $(x + 1)^2 + (y - 4)^2 = 13$.

11.6   Write the equations of the circles satisfying the following sets of conditions:

(a)   Through $(2, 3)$ and $(-1, 6)$, with center on $2x + 5y + 1 = 0$.

(b)   Tangent to $5x - y - 17 = 0$ at $(4, 3)$ and also tangent to $x - 5y - 5 = 0$.

(c)   Tangent to $x - 2y + 2 = 0$ and to $2x - y - 17 = 0$, and passing through $(6, -1)$.

(d)   With center in the first quadrant and tangent to lines $y = 0$, $5x - 12y = 0$, $12x + 5y - 39 = 0$.

Take the equation of the circle in standard form $(x - h)^2 + (y - k)^2 = r^2$.
(a)   We obtain the following system of equations:

$$2h + 5k + 1 = 0 \quad [\text{center} \, (h, \ k) \, \text{on} \, 2x + 5y + 1 = 0] \tag{1}$$

$$(2-h)^2 + (3-k)^2 = r^2 \quad \text{[point } (2,3) \text{ on the circle]} \tag{2}$$

$$(-1-h)^2 + (6-k)^2 = r^2 \quad \text{[point } (-1,6) \text{ on the circle]} \tag{3}$$

The elimination of r between (2) and (3) yields $h - k + 4 = 0$ and when this is solved simultaneously with (1), we obtain $h = -3$, $k = 1$. By (2), $r^2 = (2+3)^2 + (3-1)^2 = 29$; the equation of the circle is $(x+3)^2 + (y-1)^2 = 29$.

(b) We obtain the following system of equations:

$$(4-h)^2 + (3-k)^2 = r^2 \quad \text{[point } (4,3) \text{ on the circle]} \tag{1}$$

$$\left(\frac{5h-k-17}{\sqrt{26}}\right)^2 = \left(\frac{h-5k-5}{\sqrt{26}}\right)^2 \quad \begin{array}{l}\text{(the square of the directed distance from} \\ \text{each tangent to the center is } r^2\text{)}\end{array} \tag{2}$$

$$\frac{k-3}{h-4} = -\frac{1}{5} \quad \begin{array}{l}\text{[the radius drawn to } (4,\ 3) \text{ is perpendicular} \\ \text{to the tangent there]}\end{array} \tag{3}$$

The elimination of h between (2) and (3) yields

$$\left(\frac{-26k+78}{\sqrt{26}}\right)^2 - \left(\frac{-10k+14}{\sqrt{26}}\right)^2$$

or $9k^2 - 59k + 92 = (k-4)(9k-23) = 0$; then $k = 4, \frac{23}{9}$.

When $k = 4$, $h = -5k + 19 = -1$; then $r^2 = (4+1)^2 + (3-4)^2 = 26$ and the equation of the circle is $(x+1)^2 + (y-4)^2 = 26$. When $k = \frac{23}{9}$, $h = \frac{56}{9}$ and $r^2 = \frac{416}{81}$; the equation of the circle is $(x - \frac{56}{9})^2 + (y - \frac{23}{9})^2 = \frac{416}{81}$.

(c) Observing the directions indicated in Fig. 11-5(a), we obtain for each circle the following system of equations:

$$-\frac{h-2k+2}{-\sqrt{5}} = r \quad (x - 2y + 2 = 0 \text{ is tangent to the circle}) \tag{1}$$

$$-\frac{2h-k-17}{\sqrt{5}} = r \quad (2x - y - 17 = 0 \text{ is tangent to the circle}) \tag{2}$$

$$(6-h)^2 + (-1-k)^2 = r^2 \quad \text{[point } (6, -1) \text{ is on the circle]} \tag{3}$$

The elimination of r between (1) and (2) yields

$$h - k - 5 = 0 \tag{4}$$

and the elimination of r between (1) and (3) yields

$$(6-h)^2 + (-1-k)^2 = \frac{(h-2k+2)^2}{5} \tag{5}$$

Eliminating h between (4) and (5), we have $(1-k)^2 + (-1-k)^2 = (-k+7)^2/5$ or $9k^2 + 14k - 39 = (k+3)(9k-13) = 0$ and $k = -3, \frac{13}{9}$.

When $k = -3$, $h = k + 5 = 2$, $r^2 = (6-2)^2 + [-1-(-3)]^2 = 20$ and the equation of the circle is $(x-2)^2 + (y+3)^2 = 20$. When $k = \frac{13}{9}$, $h = k + 5 = \frac{58}{9}$, $r^2 = (6 - \frac{58}{9})^2 + (-1 - \frac{13}{9})^2 = \frac{500}{81}$ and the circle has equation $(x - \frac{58}{9})^2 + (y - \frac{13}{9})^2 = \frac{500}{81}$.

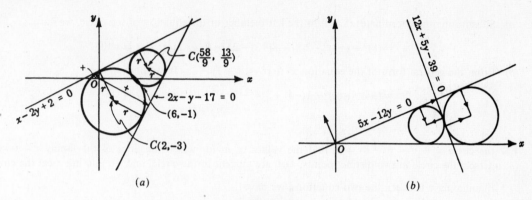

Fig. 11-5

(*d*) Observing the directions indicated in Fig. 11-5(*b*), we have for the inscribed circle the following system of equations:

$$k = r, \quad -\frac{5h - 12k}{-13} = r \quad \text{and} \quad -\frac{12h + 5k - 39}{13} = r$$

Then $h = \frac{5}{2}$, $k = r = \frac{1}{2}$, and the equation of the circle is $(x - \frac{5}{2})^2 + (y - \frac{1}{2})^2 = \frac{1}{4}$.
For the other circle, we have the following system of equations:

$$k = r, \quad -\frac{5h - 12k}{-13} = r \quad \text{and} \quad \frac{12h + 5k - 39}{13} = r$$

Here $h = \frac{15}{4}$, $k = r = \frac{3}{4}$, and the circle has equation $(x - \frac{15}{4})^2 + (y - \frac{3}{4})^2 = \frac{9}{16}$.

11.7 Find the equations of all tangents to the circle $x^2 + y^2 - 2x + 8y - 23 = 0$:

(*a*) at the point $(3, -10)$ on it, (*b*) having slope 3, and (*c*) through the external point $(8, -3)$.

(*a*) The center of the given circle is $C(1, -4)$. Since the tangent at any point P on the circle is perpendicular to the radius through P, the slope of the tangent is $\frac{1}{3}$. The equation of the tangent is then $y + 10 = \frac{1}{3}(x - 3)$ or $x - 3y - 33 = 0$.

(*b*) Each tangent belongs to the family of lines $y = 3x + b$ or, in normal form, $\dfrac{3x - y + b}{\pm\sqrt{10}} = 0$.

The undirected distance from any tangent to the center of the circle is equal to the radius of the circle; thus,

$$\left| \frac{3(1) - (-4) + b}{\pm\sqrt{10}} \right| = 2\sqrt{10} \quad \text{or} \quad \frac{7 + b}{\sqrt{10}} = \pm\sqrt{10} \quad \text{and} \quad b = 13, -27$$

There are then two tangents having equations $y = 3x + 13$ and $y = 3x - 27$.

(*c*) The tangents to a circle through the external point $(8, -3)$ belong to the family of lines $y + 3 = m(x - 8)$ or $y = mx - 8m - 3$. When this replacement for y is made in the equation of the circle, we have

$$x^2 + (mx - 8m - 3)^2 - 2x + 8(mx - 8m - 3) - 23 = (m^2 + 1)x^2$$
$$+ (-16m^2 + 2m - 2)x + (64m^2 - 16m - 38) = 0$$

Now this equation will have equal roots x provided the discriminant is zero, that is, provided $(-16m^2 + 2m - 2)^2 - 4(m^2 + 1)(64m^2 - 16m - 38) = -4(m - 3)(9m + 13) = 0$. Then $m = 3, -\frac{13}{9}$ and the equations of the tangents are $y + 3 = 3(x - 8)$ or $3x - y - 27 = 0$ and $y + 3 = -\frac{13}{9}(x - 8)$ or $13x + 9y - 77 = 0$.

11.8 Find the length of the tangent

(*a*) To the circle $x^2 + y^2 - 2x + 8y - 23 = 0$ from the point $(8, -3)$

(*b*) To the circle $4x^2 + 4y^2 - 2x + 5y - 8 = 0$ from the point $(-4, 4)$

Denote the required length by t.

(a) Substituting the coordinates $(8, -3)$ in the left member of the equation of the circle, we have

$$t^2 = (8)^2 + (-3)^2 - 2(8) + 8(-3) - 23 = 0 \quad \text{and} \quad t = \sqrt{10}$$

(b) From the general form of the equation $x^2 + y^2 - \frac{1}{2}x + \frac{5}{4}y - 2 = 0$, we find

$$t^2 = (-4)^2 + (4)^2 - \frac{1}{2}(-4) + \frac{5}{4}(4) - 2 = 37 \quad \text{and} \quad t = \sqrt{37}$$

11.9 For the circle $x^2 + y^2 + 6x - 8y = 0$, find the values of m for which the lines of the family $y = mx - \frac{1}{3}$ (a) intersect the circle in two distinct points, (b) are tangent to the circle, and (c) do not meet the circle.

Eliminating y between the two equations, we have

$$x^2 + \left(mx - \frac{1}{3}\right)^2 + 6x - 8\left(mx - \frac{1}{3}\right) = (m^2 + 1)x^2 + (6 - \frac{26}{3}m)x + \frac{25}{9} = 0$$

This equation will have real and distinct roots, two equal roots, or two imaginary roots according as its discriminant

$$(6 - \tfrac{26}{3}m)^2 - 4(m^2 + 1)\left(\tfrac{25}{9}\right) = \tfrac{8}{9}(72m^2 - 117m + 28) = \tfrac{8}{9}(3m - 4)(24m - 7) > , = , \text{ or } < 0$$

(a) The lines will intersect the circle in two distinct points when $m > \frac{4}{3}$ and $m < \frac{7}{24}$.

(b) The lines will be tangent to the circle when $m = \frac{4}{3}$ and $m = \frac{7}{24}$.

(c) The lines will not meet the circle when $\frac{7}{24} < m < \frac{4}{3}$.

11.10 Write the equation of the family of circles satisfying the given conditions: (a) having the common center $(-2, 3)$, (b) having radius $= 5$, and (c) with center on the x axis.

(a) The equation is $(x + 2)^2 + (y - 3)^2 = r^2$, r being a parameter.

(b) The equation is $(x - h)^2 + (y - k)^2 = 25$, h and k being parameters.

(c) Let the center have coordinates $(k_1, 0)$ and denote the radius by k_2. The equation of the family is $(x - k_1)^2 + y^2 = k_2^2$, k_1 and k_2 being parameters.

11.11 Write the equation of the family of circles which are tangent to the circles of Problem 11.3 at their common point and determine the circle of the family having the property (a) center on $x + 4y + 16 = 0$ (b) radius is $\frac{1}{2}$.

The equation of the family is

$$K_1 + kK_2: \quad x^2 + y^2 - 16x - 20y + 115 + k(x^2 + y^2 + 8x - 10y + 5) = 0$$

(a) The centers of K_1 and K_2 are $(8, 10)$ and $(-4, 5)$, respectively; the equation of their line of centers is $5x - 12y + 80 = 0$. This line meets $x + 4y + 16 = 0$ in the point $(-16, 0)$, the required center.

Now $(-16, 0)$ divides the segment joining $(8, 10)$ and $(-4, 5)$ in the ratio $k : 1$; thus,

$$\frac{5k + 10}{k + 1} = 0 \quad \text{and} \quad k = -2$$

The equation of the circle is $K_1 - 2K_2 : x^2 + y^2 + 32x - 105 = 0$.

(b) Setting the square of the radius of $K_1 + kK_2$ equal to $\left(\frac{1}{2}\right)^2$ and solving for k, we find

$$\left(\frac{4k - 8}{1 + k}\right)^2 + \left(-\frac{10 + 5k}{1 + k}\right)^2 - \frac{115 + 5k}{1 + k} = \frac{1}{4} \quad \text{and} \quad k = 1, \tfrac{15}{11}$$

The required circles have equations

$$K_1 + K_2 : x^2 + y^2 - 4x - 15y + 60 = 0 \quad \text{and} \quad K_1 + \tfrac{15}{11}K_2 : \quad 13x^2 + 13y^2 - 28x - 185y + 670 = 0$$

11.12 For each pair of circles

(a) $K_1 : x^2 + y^2 - 8x - 6y = 0$
       $K_2 : 4x^2 + 4y^2 - 10x - 10y - 13 = 0$

(b) $K_1 : x^2 + y^2 - 12x - 16y - 125 = 0$
       $K_2 : 3x^2 + 3y^2 - 60x - 16y + 113 = 0$

find the equation of the radical axis. Without finding the coordinates of their points of intersection, show that the circles (a) intersect in two distinct points while those of (b) are tangent internally.

(a) The equation of the radical axis is $K_2 - \frac{1}{4}K_2 : 22x + 14y - 13 = 0$. The undirected distance from the radical axis to the center $(4, 3)$ of K_1

$$\left| \frac{22(4) + 14(3) - 13}{2\sqrt{170}} \right| = \frac{117\sqrt{170}}{340}$$

is less than 5, the radius of K_1. Hence, the radical axis and the circle K_2 intersect K_1 in two distinct points

(b) The equation of the radical axis is $K_1 - \frac{1}{3}K_2 : 3x - 4y - 61 = 0$. The undirected distance from the radical axis to the center $(6, 8)$ of K_1

$$\left| \frac{3(6) - 4(8) - 61}{5} \right| = 15$$

is equal to the radius of K_1 and the circles are tangent to each other.

Since the distance between the centers of K_1 and $K_2 \sqrt{(10-6)^2 + (\frac{8}{3} - 8)^2} = \frac{20}{3}$ is equal to the difference between their radii, the circles are tangent internally.

Supplementary Problems

11.13 Write the equation of each of the following circles: (a) $C(0, 0)$, radius 7 (b) $C(-4, 8)$, radius 3 (c) $C(5, -4)$, through $(0, 0)$ (d) $C(-4, -3)$, through $(2, 1)$ (e) $C(-2, 5)$, tangent to x axis (f) $C(-2, -5)$, tangent to $2x - y + 3 = 0$ (g) tangent to both axes, radius 5 (h) circumscribed about the right triangle whose vertices are $(3, 4)$, $(-1, -4)$, $(5, -2)$ (i) circumscribed about the triangle of Problem 9.8.

    Ans. (a) $x^2 + y^2 = 49$

          (b) $x^2 + y^2 + 8x - 16y + 71 = 0$

          (c) $x^2 + y^2 - 10x + 8y = 0$

          (d) $x^2 + y^2 + 8x + 6y - 27 = 0$

          (e) $x^2 + y^2 + 4x - 10y + 4 = 0$

          (f) $5x^2 + 5y^2 + 20x + 50y + 129 = 0$

          (g) $x^2 + y^2 \pm 10x \pm 10y + 25 = 0,\ x^2 + y^2 \mp 10x \pm 10y + 25 = 0$

          (h) $x^2 + y^2 - 2x - 19 = 0$

          (i) $56x^2 + 56y^2 - 260x - y - 5451 = 0$

11.14 Find the center and radius of each of the circles.

    (a) $x^2 + y^2 - 6x + 8y - 11 = 0$ *Ans.* $C(3, -4),\ r = 6$

    (b) $x^2 + y^2 - 4x - 6y - \frac{10}{3} = 0$ *Ans.* $C(2, 3),\ r = 7\sqrt{3}/3$

    (c) $7x^2 + 7y^2 + 14x - 56y - 25 = 0$ *Ans.* $C(-1, 4),\ 4 = 12\sqrt{7}/7$

11.15 Explain why any line passing through $(4, -1)$ cannot be tangent to the circle $x^2 + y^2 - 4x + 6y - 12 = 0$.

11.16 (a) Show that the circles $x^2 + y^2 + 6x - 2y - 54 = 0$ and $x^2 + y^2 - 22x - 8y + 112 = 0$ do not intersect.

(b) Show that the circles $x^2 + y^2 + 2x - 6y + 9 = 0$ and $x^2 + y^2 + 8y - 6y + 9 = 0$ are tangent internally. Do this with and without a graphing calculator.

11.17 The equation of a given circle is $x^2 + y^2 = 36$. Find (a) the length of the chord which lies along the line $3x + 4y - 15 = 0$ (b) the equation of the chord whose midpoint is (3,2).

Hint: In (a) draw the normal to the given line.   *Ans.*   (a) $6\sqrt{3}$ (b) $3x + 2y - 13 = 0$

11.18 Find the equation of each circle satisfying the given conditions.

(a) Through (6,0) and (−2,−4), tangent to $4x + 3y - 25 = 0$.

(b) Tangent to $3x - 4y + 5 = 0$ at (1,2), radius 5.

(c) Tangent to $x - 2y - 4 = 0$ and $2x - y - 6 = 0$, passes through (−1, 2).

(d) Tangent to $2x - 3y - 7 = 0$ at (2,−1); passes through (4,1).

(e) Tangent to $3x + y + 3 = 0$ at (−3, 6), tangent to $x + 3y - 7 = 0$.

   Ans.   (a) $(x - 3)^2 + (y + 4)^2 = 25, (x - \frac{213}{121})^2 + (y + \frac{184}{121})^2 = 297\,025/14\,641$

   (b) $(x - 4)^2 + (y + 2)^2 = 25, (x + 2)^2 + (y - 6)^2 = 25$

   (c) $x^2 + y^2 - 2x - 2y - 3 = 0, x^2 + y^2 + 118x - 122y + 357 = 0$

   (d) $x^2 + y^2 + 4x - 10y - 23 = 0$

   (e) $x^2 + y^2 - 6x - 16y + 33 = 0, x^2 + y^2 + 9x - 11y + 48 = 0$

11.19 Repeat Problem 11.14 using a graphing calculator.

11.20 Find the equation of the tangent to the given circle at the given point on it.

(a) $x^2 + y^2 = 169, (5,-12)$ (b) $x^2 + y^2 - 4x + 6y - 37 = 0, (3, 4)$

   Ans.   (a) $5x - 12y - 169 = 0$ (b) $x + 7y - 31 = 0$

11.21 Find the equations of the tangent to each circle through the given external point.

(a) $x^2 + y^2 = 25, (7, 1)$ (b) $x^2 + y^2 - 4x + 2y - 31 = 0, (-1, 5)$

   Ans.   (a) $3x + 4y - 25 = 0, 4x - 3y - 25 = 0$ (b) $y - 5 = 0, 4x - 3y + 19 = 0$

11.22 Show that the circles $x^2 + y^2 + 4x - 6y = 0$ and $x^2 + y^2 + 6x + 4y = 0$ are *orthogonal*, that is, that the tangents to the two circles at a point of intersection are mutually perpendicular. Also, that the square of the distance between the centers of the circles is equal to the sum of the squares of the radii.

11.23 Determine the equation of the circle of the family of Problem 11.11: (a) which passes through the point (0, 3) (b) which is tangent to the line $4x + 3y - 25 = 0$.

   Ans.   (a) $5x^2 + 5y^2 + 16x - 60y + 135 = 0$

   (b) $x^2 + y^2 - 40x - 30y + 225 = 0, 8x^2 + 8y^2 - 20x - 115y + 425 = 0$

11.24 Repeat Problem 11.23 using a graphing calculator or a computer software package such as Maple.

PART II

TOPICS IN DISCRETE MATHEMATICS

Arithmetic and Geometric Progressions

A SEQUENCE IS A SET OF NUMBERS, called *terms*, arranged in a definite order; that is, there is a rule by which the terms are formed. Sequences may be finite or infinite. Only finite sequences will be treated in this chapter.

EXAMPLE 1

    (*a*) Sequence: 3, 7, 11, 15, 19, 23, 27.
               Type: Finite of 7 terms.
               Rule: Add 4 to a given term to produce the next. The first term is 3.
    (*b*) Sequence: 3, 6, 12, 24, 48, 96.
               Type: Finite of 6 terms.
               Rule: Multiply a given term by 2 to produce the next. The first term is 3.

AN ARITHMETIC PROGRESSION is a sequence in which each term after the first is formed by adding a fixed amount, called the *common difference*, to the preceding term. The sequence of Example 1(*a*) is an arithmetic progression whose common difference is 4. (See Problems 12.1–12.2.)

If a is the first term, d is the common difference, and n is the number of terms of an arithmetic progression, the successive terms are

$$a, a + d, a + 2d, a + 3d, \ldots, a + (n-1)d \qquad (12.1)$$

Thus, the *last* term (or *n*th term) l is given by

$$l = a + (n-1)d \qquad (12.2)$$

The *sum S* of the n terms of this progression is given by

$$S = \frac{n}{2}(a + l) \qquad \text{or} \qquad S = \frac{n}{2}[2a + (n-l)d] \qquad (12.3)$$

(For a proof, see Problem 12.3.)

EXAMPLE 2.   Find the twentieth term and the sum of the first 20 terms of the arithmetic progression $4, 9, 14, 19, \ldots$.

For this progression $a = 4$, $d = 5$, and $n = 20$; then the twentieth term is $l = a + (n-1)d = 4 + 19 \cdot 5 = 99$ and the sum of the first 20 terms is

$$S = \frac{n}{2}(a + l) = \frac{20}{2}(4 + 99) = 1030$$

(See Problems 12.4–12.8.)

THE TERMS BETWEEN THE FIRST AND LAST TERMS of an arithmetic progression are called *arithmetic means* between these two terms. Thus, to insert k arithmetic means between two numbers is to form an arithmetic progression of $(k + 2)$ terms having the two given numbers as first and last terms.

EXAMPLE 3.   Insert five arithmetic means between 4 and 22.

We have $a = 4$, $l = 22$, and $n = 5 + 2 = 7$. Then $22 = 4 + 6d$ and $d = 3$. The first mean is $4 + 3 = 7$, the second is $7 + 3 = 10$, and so on. The required means are 7, 10, 13, 16, 19, and the resulting progression is 4, 7, 10, 13, 16, 19, 22.

THE ARITHMETIC MEAN.   When just one mean is to be inserted between two numbers to form an arithmetic progression, it is called the *arithmetic mean* (also, the average) of the two numbers.

EXAMPLE 4.   Find the arithmetic mean of the two numbers a and l.

We seek the middle term of an arithmetic progression of three terms having a and l as the first and third terms, respectively. If d is the common difference, then $a + d = l - d$ and $d = \frac{1}{2}(l - a)$. The arithmetic mean is $a + d = a + \frac{1}{2}(l - a) = \frac{1}{2}(a + l)$. (See Problem 12.9.)

A GEOMETRIC PROGRESSION is a sequence in which each term after the first is formed by multiplying the preceding term by a fixed number, called the *common ratio*. The sequence $3, 6, 12, 24, 48, 96$ of Example 1(b) is a geometric progression whose common ratio is 2. (See Problems 12.10–12.11.)

If a is the first term, r is the common ratio, and n is the number of terms, the geometric progression is

$$a, \ ar, \ ar^2, \ldots, ar^{n-1} \tag{12.4}$$

Thus, the last (or nth) term l is given by

$$l = ar^{n-1} \tag{12.5}$$

The *sum S* of the first n terms of the geometric progression (12.4) is given by

$$S = \frac{a - rl}{1 - r} \qquad \text{or} \qquad S = \frac{a(1 - r^n)}{1 - r} \tag{12.6}$$

(For a proof, see Problem 12.12.)

EXAMPLE 5.   Find the ninth term and the sum of the first nine terms of the geometric progression $8, 4, 2, 1, \ldots$.

Here $a = 8, r = \frac{1}{2}$, and $n = 9$; the ninth term is $l = ar^{n-1} = 8\left(\frac{1}{2}\right)^8 = \left(\frac{1}{2}\right)^5 = \frac{1}{32}$, and the sum of the first nine terms is

$$S = \frac{a - rl}{1 - r} = \frac{8 - \frac{1}{2}\left(\frac{1}{32}\right)}{1 - \frac{1}{2}} = 16 - \frac{1}{32} = \frac{511}{32}$$

(See Problems 12.13–12.18.)

THE TERMS BETWEEN THE FIRST AND LAST TERMS of a geometric progression are called *geometric means* between the two terms. Thus, to insert k geometric means between two numbers is to form a geometric progression of $(k + 2)$ terms having the two given numbers as first and last terms.

EXAMPLE 6. Insert four geometric means between 25 and $\frac{1}{125}$.

We have $a = 25$, $l = \frac{1}{125}$, and $n = 4 + 2 = 6$. Using $l = ar^{n-1}$, $\frac{1}{125} = 25r^5$; then $r^5 = \left(\frac{1}{5}\right)^5$ and $r = \frac{1}{5}$. The first mean is $25\left(\frac{1}{5}\right) = 5$, the second is $5\left(\frac{1}{5}\right) = 1$, and so on. The required means are $5, 1, \frac{1}{5}, \frac{1}{25}$ and the geometric progression is $25, 5, 1, \frac{1}{5}, \frac{1}{25}, \frac{1}{125}$.

THE GEOMETRIC MEAN. When one mean is to be inserted between two numbers to form a geometric progression, it is called the *geometric mean* of the two numbers. The geometric mean of two numbers a and l, having like signs, is $(\pm)\sqrt{a \cdot l}$. The sign to be used is the common sign of a and l. (See Problem 12.19.)

Solved Problems

ARITHMETIC PROGRESSIONS

12.1 Determine which of the following sequences are arithmetic progressions (A.P.). In the case of an A.P., write three more terms.

(a) 3, 6, 9, 12, 15, 18.
   Since $6 - 3 = 9 - 6 = 12 - 9 = 15 - 12 = 18 - 15 = 3$, the sequence is an A.P. with common difference 3. The next three terms are $18 + 3 = 21$, $21 + 3 = 24$, and $24 + 3 = 27$.

(b) 25, 19, 13, 7, 1, −5.
   Since $19 - 25 = 13 - 19 = 7 - 13 = 1 - 7 = -5 - 1 = -6$, the sequence is an A.P. with $d = -6$. The next three terms are $-5 + (-6) = -11$, $-11 + (-6) = -17$, and $-17 + (-6) = -23$.

(c) 5, 10, 14, 20, 25.
   Since $10 - 5 \neq 14 - 10$, the sequence is not an A.P.

(d) $3a - 2b, 4a - b, 5a, 6a + b$.
   Since $(4a - b) - (3a - 2b) = 5a - (4a - b) = (6a + b) - 5a = a + b$, the sequence is an A.P. with $d = a + b$. The next three terms are $(6a + b) + (a + b) = 7a + 2b, 8a + 3b$, and $9a + 4b$.

12.2 Find the value of k such that each sequence is an A.P.

(a) $k - 1, k + 3, 3k - 1$.
   If the sequence is to form an A.P., $(k + 3) - (k - 1) = (3k - 1) - (k + 3)$. Then $k = 4$ and the A.P. is 3, 7, 11.

(b) $3k^2 + k + 1, 2k^2 + k; 4k^2 - 6k + 1$.
   Setting $(2k^2 + k) - (3k^2 + k + 1) = (4k^2 - 6k + 1) - (2k^2 + k)$, we have $3k^2 - 7k + 2 = 0$ and $k = 2, \frac{1}{3}$. The progressions are 15, 10, 5 when $k = 2$ and $\frac{5}{3}, \frac{5}{9}, -\frac{5}{9}$ when $k = \frac{1}{3}$.

12.3 Prove the formula $S = \frac{n}{2}[2a + (n - 1)d]$ for an arithmetic progression.

Write the indicated sum of the n terms in the order given by (*12.1*), then write this sum in reverse order, and sum term by term. Thus,

$$S = \{a \qquad\quad\} + \{a + d \qquad\} + \cdots + \{a + (n - 2)d\} + \{a + (n - 1)d\}$$
$$S = \{a + (n - 1)d\} + \{a + (n - 2)d\} + \cdots + \{a + d \qquad\} + \{a \qquad\quad\}$$
$$2S = \{2a + (n - 1)d\} + \{2a + (n - 1)d\} + \cdots + \{2a + (n - 1)d\} + \{2a + (n - 1)d\}$$
$$= n[2a + (n - 1)d] \qquad \text{and} \qquad S = \frac{n}{2}[2a + (n - 1)d]$$

12.4 (a) Find the eighteenth term and the sum of the first 18 terms of the A.P. $2, 6, 10, 14, \ldots$.
   Here $a = 2, d = 4, n = 18$, Then $l = a + (n - 1)d = 2 + 17 \cdot 4 = 70$ and $S = \frac{n}{2}(a + 1) = \frac{18}{2}(2 + 70) = 648$.

(b) Find the forty-ninth term and the sum of the first 49 terms of the A.P. $10, 4, -2, -8, \ldots$.
   Here $a = 10, d = -6, n = 49$. Then $l = 10 + 48(-6) = -278$ and $S = \frac{49}{2}(10 - 278) = -6566$.

(c) Find the twelfth term and the sum of the first 15 terms of the A.P. 8, $\frac{19}{3}$, $\frac{14}{3}$, 3,....

Since $a = 8$ and $d = -\frac{5}{3}$, the twelfth term is $l = 8 + 11(-\frac{5}{3}) = -\frac{31}{3}$ and the sum of the first 15 terms is

$$S = \frac{n}{2}[2a + (n-1)d] = \frac{15}{2}\left[16 + 14\left(-\frac{5}{3}\right)\right] = \frac{15}{2}\left(-\frac{22}{3}\right) = -55$$

(d) Find the tenth term, the sum of the first 10 terms, and the sum of the first 13 terms of the A.P. $2x + 3y, x + y, -y, \ldots$.

Here $a = 2x + 3y$ and $d = -x - 2y$. The tenth term is $l = (2x + 3y) + 9(-x - 2y) = -7x - 15y$.

Sum of first 10 terms is $S = 5[(2x + 3y) + (-7x - 15y)] = -25x - 60y$.

Sum of first 13 terms is $S = \frac{13}{2}[2(2x + 3y) + 12(-x - 2y)] = \frac{13}{2}(-8x - 18y) = -52x - 117y$.

12.5 The seventh term of an A.P. is 41 and the thirteenth term is 77. Find the twentieth term.

If a is the first term and d is the common difference, then for

the seventh term $a + 6d = 41$

and for the thirteenth term $a + 12d = 77$

Subtracting, $6d = 36$; then $d = 6$ and $a = 41 - 6 \cdot 6 = 5$. The twentieth term is $l = 5 + 19 \cdot 6 = 119$.

12.6 The sixth term of an A.P. is 21 and the sum of the first 17 terms is 0. Write the first three terms.

If a is the first term and d is the common difference,

$$a + 5d = 21 \quad \text{and} \quad 0 = \frac{17}{2}(2a + 16d) \quad \text{or} \quad a + 8d = 0$$

Then $d = -7$ and $a = -8d = 56$. The first three terms are 56, 49, 42.

12.7 Obtain formulas for (a) l in terms of a, n, S; (b) a in terms of d, n, S.

(a) From $S = \frac{n}{2}(a + l)$, $\quad a + l = \frac{2S}{n}$ and $l = \frac{2S}{n} - a$.

(b) From $S = \frac{n}{2}[2a + (n-1)d]$, $2a + (n-1)d = \frac{2S}{n}$, $2a = \frac{2S}{n} - (n-1)d$ and $a = \frac{S}{n} - \frac{1}{2}(n-1)d$.

12.8 If a body is dropped, the distance (s meters) through which it falls freely in t seconds is approximately $16t^2$. (a) Show that the distances through which it falls during the first, second, third,... seconds form an A.P. (b) How far will the body fall in the tenth second? (c) How far will it fall in the first 20 seconds?

(a) The distance through which the body falls during the first second is 16 m, during the second second is $6(2)^2 - 16 = 48$ m, during the third second is $16(3)^2 - 16(2)^2 = 80$ m, during the fourth second is $16(4)^2 - 16(3)^2 = 112$ m, and so on. These are the first four terms of an A.P. whose common difference is 32.

(b) When $n = 10, l = 16 + 9(32) = 304$ m.

(c) In the first 20 s, the body falls $16(20)^2 = 6400$ m.

12.9 (a) Insert six arithmetic means between 7 and 77.

For the A.P. having $a = 7$, $l = 77$ and $n = 6 + 2 = 8$, $77 = 7 + 7d$ and $d = 10$. The required means are 17, 27, 37, 47, 57, 67 and the A.P. is 7, 17, 27, 37, 47, 57, 67, 77.

(b) Find the arithmetic mean of 8 and −56.

From Example 4, the arithmetic means is $\frac{1}{2}(a + l) = \frac{1}{2}[8 + (-56)] = -24$.

GEOMETRIC PROGRESSIONS

12.10   Determine which of the following sequences are geometric progressions (G.P.). In the case of a G.P., write the next three terms.

(a)   4, 8, 16, 32, 64.

Since $\frac{8}{4} = \frac{16}{8} = \frac{32}{16} = \frac{64}{32} = 2$, the sequence is a G.P. with common ratio 2. The next three terms are 128, 256, 512.

(b) $1, \frac{1}{4}, \frac{1}{16}, \frac{1}{48}$.

Since $\frac{1}{16} / \frac{1}{4} \neq \frac{1}{48} / \frac{1}{16}$, the sequence is not a G.P.

(c) $12, -4, \frac{4}{3}, -\frac{4}{9}$.

Since $-\frac{4}{12} = \frac{4}{3} / -4 = -\frac{4}{9} / \frac{4}{3} = -\frac{1}{3}$, the sequence is a G.P. with common ratio $-\frac{1}{3}$. The next three terms are $\frac{4}{27}, -\frac{8}{81}, \frac{4}{243}$.

12.11   Find the value of k so that the sequence $2k - 5$, $k - 4$, $10 - 3k$ forms a G.P.

If the sequence is to form a G.P.,

$$\frac{k-4}{2k-5} = \frac{10-3k}{k-4} \qquad \text{or} \qquad k^2 - 8k + 16 = -6k^2 + 35k - 50$$

Then $7k^2 - 43k + 66 = (k-3)(7k-22) = 0$ and $k = 3, \frac{22}{7}$.

The sequences are $1, -1, 1$ when $k = 3$ and $\frac{9}{7}, -\frac{6}{7}, \frac{4}{7}$ when $k = \frac{22}{7}$.

12.12   Obtain the formula $S = \dfrac{a(1 - r^n)}{1 - r}$ for a geometric progression.

Write the indicated sum of the n terms given by (*12.4*), then multiply this sum by r, and subtract term by term. Thus,

$$S = a + ar + ar^2 + \cdots + ar^{n-1}$$

$$rS = \qquad ar + ar^2 + \cdots + ar^{n-1} + ar^n$$

$$S - rS = a \qquad\qquad\qquad\qquad\qquad - ar^n$$

Then $S(1 - r) = a - ar^n = a(1 - r^n)$ and $S = \dfrac{a(1 - r^n)}{1 - r}$.

12.13   (a)   Find the seventh term and the sum of the first seven terms of the G.P. $12, 16, \frac{64}{3}, \ldots$.

Here $a = 12, r = \frac{4}{3}, n = 7$. The seventh term is $l = ar^{n-1} = 12\left(\frac{4}{3}\right)^6 = 4^7/3^5 = 16\,384/243$ and the sum of the first seven terms is

$$S = \frac{a - rl}{1 - r} = \frac{12 - \frac{4}{3}(4^7/3^5)}{1 - \frac{4}{3}} = \frac{4^8}{3^5} - 36 = \frac{65\,536 - 8748}{243} = \frac{56\,788}{243}$$

(b)   Find the sixth term and the sum of the first nine terms of the G.P. $4, -6, 9, \ldots$.

Since $a = 4$ and $r = -\frac{3}{2}$, the sixth term is $l = 4\left(-\frac{3}{2}\right)^5 = -3^5/2^3 = -\frac{243}{8}$ and the sum of the first nine terms is

$$S = \frac{a(1 - r^n)}{1 - r} = \frac{4[1 - \left(-\frac{3}{2}\right)^9]}{1 - \left(-\frac{3}{2}\right)} = \frac{8(1 + 3^9/2^9)}{5} = \frac{2^9 + 3^9}{5 \cdot 2^6} = \frac{4039}{64}$$

(c)   Find the sum of the G.P. $8, -4, 2, \ldots, \frac{1}{128}$.

$$S = \frac{a - rl}{1 - r} = \frac{8 - \left(-\frac{1}{2}\right)\left(\frac{1}{128}\right)}{1 - \left(-\frac{1}{2}\right)} = \frac{2^4 + \left(\frac{1}{2}\right)^7}{3} = \frac{2^{11} + 1}{3 \cdot 2^7} = \frac{683}{128}$$

12.14 The fourth term of a G.P. is 1 and the eighth term is $\frac{1}{256}$. Find the tenth term.

Since the fourth term is $1, ar^3 = 1$; since the eighth term is $\frac{1}{256}, ar^7 = \frac{1}{256}$. Then $ar^7/ar^3 = \frac{1}{256}, r^4 = \frac{1}{256}$, and $r = \pm\frac{1}{4}$. From $ar^3 = 1$, we have $a = \pm 64$. In each case, the tenth term is $\frac{1}{4096}$.

12.15 Given $S = \frac{3367}{64}, r = \frac{3}{4}, l = \frac{243}{64}$. Find a and n.

Since $S = \dfrac{3367}{64} = \dfrac{a - (\frac{3}{4})(\frac{243}{64})}{1 - \frac{3}{4}} = 4a - \dfrac{729}{64},\quad 4a = \dfrac{4096}{64}\quad$ and $\quad a = 16.$

Now $l = \dfrac{243}{64} = 16\left(\dfrac{3}{4}\right)^{n-1},\quad \left(\dfrac{3}{4}\right)^{n-1} = \dfrac{243}{16 \cdot 64} = \left(\dfrac{3}{4}\right)^5,\quad n - 1 = 5\quad$ and $\quad n = 6.$

12.16 Given $a = 8, r = \frac{3}{2}, S = \frac{2059}{8}$. Find l and n.

Since $S = \dfrac{2059}{8} = \dfrac{8 - (\frac{3}{2})l}{1 - \frac{3}{2}} = 3l - 16,\quad 3l = \dfrac{2059}{8} + 16 = \dfrac{2187}{8}\quad$ and $\quad l = \dfrac{729}{8}.$

Now $l = \frac{729}{8} = 8\left(\frac{3}{2}\right)^{n-1},\quad \left(\frac{3}{2}\right)^{n-1} = \frac{729}{64} = \left(\frac{3}{2}\right)^6,\quad n - 1 = 6\quad$ and $\quad n = 7.$

12.17 If a boy undertakes to deposit 1¢ on Sept. 1, 2¢ on Sept. 2, 4¢ on Sept. 3, 8¢ on Sept. 4, and so on, (a) how much will he deposit from Sept. 1 to Sept. 15 inclusive, (b) how much would he deposit on Sept. 30?

Here, $a = 0.01$ and $r = 2$.
(a) When $n = 15$,

$$S = \frac{0.01(1 - 2^{15})}{1 - 2} = 0.01(2^{15} - 1) = \$327.67$$

(b) When $n = 30, l = 0.01(2)^{29} = \$5\,368\,709.12.$

12.18 A rubber ball is dropped from a height of 81 m. Each time it strikes the ground, it rebounds two-thirds of the distance through which it last fell. (a) Through what distance did the ball fall when it struck the ground for the sixth time? (b) Through what distance had it traveled from the time it was dropped until it struck the ground for the sixth time?

(a) The successive distances through which the ball falls form a G.P. in which $a = 81, r = \frac{2}{3}$. When $n = 6$,

$$l = 81\left(\tfrac{2}{3}\right)^5 = \tfrac{32}{3} \text{ m}$$

(b) The required distance is the sum of the distances for the first six falls and the first five rebounds.
For the falls: $a = 81, r = \frac{2}{3}, n = 6$, and

$$S = \frac{81[1 - (\frac{2}{3})^6]}{1 - \frac{2}{3}} = 81\left(3 - \frac{2^6}{3^5}\right) = \frac{3^6 - 2^6}{3} = \frac{665}{3} \text{ m}$$

For the rebounds: $a = 54, r = \frac{2}{3}, n = 5$, and

$$S = \frac{54[1 - (\frac{2}{3})^5]}{1 - \frac{2}{3}} = 54\left(3 - \frac{2^5}{3^4}\right) = \frac{2}{3}(3^5 - 2^5) = \frac{422}{3} \text{ m}$$

Thus, the total distance is $\frac{665}{3} + \frac{422}{3} = 362\frac{1}{3}$ m.

12.19 (a) Insert five geometric means between 8 and $\frac{1}{8}$.

We have $a = 8, l = \frac{1}{8}, n = 5 + 2 = 7$. Since $l = ar^{n-1}, \frac{1}{8} = 8r^6$ and $r = \pm\frac{1}{2}$.

When $r = \frac{1}{2}$, the first mean is $8(\frac{1}{2}) = 4$, the second is $4(\frac{1}{2}) = 2$, and so on. The required means are 4, 2, 1, $\frac{1}{2}$, $\frac{1}{4}$ and the G.P. is 8, 4, 2, 1, $\frac{1}{2}$, $\frac{1}{4}$, $\frac{1}{8}$.

When $r = -\frac{1}{2}$, the means are -4, 2, -1, $\frac{1}{2}$, $-\frac{1}{4}$ and the G.P. is 8, -4, 2, -1, $\frac{1}{2}$, $-\frac{1}{4}$, $\frac{1}{8}$.

(b) Insert four geometric means between $\frac{81}{2}$ and $-\frac{16}{3}$.

Here $a = \frac{81}{2}, l = -\frac{16}{3}, n = 6$. Then $-\frac{16}{3} = \frac{81}{2}r^5, r^5 = \frac{32}{243}$ and $r = -\frac{2}{3}$. The required means are $-27, 18, -12, 8$ and the G.P. is $\frac{81}{2}, -27, 18, -12, 8, -\frac{16}{3}$.

(c) Find the geometric mean of $\frac{1}{3}$ and 243.

The required mean is $\sqrt{\frac{1}{3}(243)} = \sqrt{81} = 9$.

(d) Find the geometric mean of $-\frac{2}{3}$ and $-\frac{32}{27}$.

The required mean is $-\sqrt{(-\frac{2}{3})(-\frac{32}{27})} = -\sqrt{\frac{64}{81}} = -\frac{8}{9}$.

Supplementary Problems

ARITHMETIC PROGRESSIONS

12.20 Find

(a) The fifteenth term and the sum of the first 15 terms of the A.P. 3, 8, 13, 18 . . . .
(b) The twelfth term and the sum of the first 20 terms of the A.P. 11, 8, 5, 2, ...
(c) The sum of the A.P. for which $a = 6\frac{3}{4}, l = -3\frac{1}{4}, n = 17$.
(d) The sum of all the integers from 1 to 200 which are divisible by 3.

Ans. (a) 73 570   (b) $-22, -350$ (c) $29\frac{3}{4}$ (d) 6633

12.21 The fourth term of an A.P. is 14 and the ninth term is 34. Find the thirteenth term.

Ans. 50

12.22 The sum of the first 7 terms of an A.P. is 98 and the sum of the first 12 terms is 288. Find the sum of the first 20 terms.

Ans. 800

12.23 Find the sum of (a) the first n positive integers, (b) the first n odd positive integers.

Ans. (a) $\frac{1}{2}n(n+1)$ (b) n^2

12.24 (a) Sum all the integers between 200 and 1000 that are divisible by 3.

Ans. 160 200

(b) Sum all the even positive integers less than 200 which are not divisible by 6.

Ans. 6534

12.25 In a potato race, 10 potatoes are placed 8 ft apart in a straight line. If the potatoes are to be picked up singly and returned to the basket, and if the first potato is 20 ft in front of the basket, find the total distance covered by a contestant who finishes the race.

Ans. 1120 ft

12.26 In a lottery, tickets are numbered consecutively from 1 to 100. Customers draw a ticket at random and pay an amount in cents corresponding to the number on the ticket except for those tickets with numbers divisible by 5, which are free. How much is realized if 100 tickets are sold?

  Ans. $40

12.27 Find the arithmetic mean between (*a*) 6 and 60, (*b*) $a - 2d$ and $a + 6d$.

  Ans. (*a*) 33 (*b*) $a + 2d$

12.28 Insert five arithmetic means between 12 and 42.

  Ans. 17, 22, 27, 32, 37

12.29 After inserting x arithmetic means between 2 and 38, the sum of the resulting progression is 200. Find x.

  Ans. 8

GEOMETRIC PROGRESSIONS

12.30 Find

  (*a*) The eight term and the sum of the first eight terms of the G.P. $4, 12, 36, \ldots$.
  (*b*) The tenth term and the sum of the first 12 terms of the G.P. $8, 4, 2, \ldots$.
  (*c*) The sum of the G.P. for which $a = 64, l = 729$, and $n = 7$.

  Ans. (*a*) 8748, 13 120 (*b*) 1/64, 15 255/256 (*c*) 2059, 463

12.31 The third term of a G.P. is 36 and the fifth term is 16. Find the tenth term.

  Ans. $\pm \frac{512}{243}$

12.32 The sum of the first three terms of a G.P. is 21 and the sum of the first six terms is $20\frac{2}{9}$. Find the sum of the first nine terms.

  Ans. $20\frac{61}{243}$

12.33 Given $S = \frac{255}{192}, l = -\frac{1}{64}, r = -\frac{1}{2}$; find a and n.

  Ans. $a = 2, n = 8$

12.34 Find three numbers in geometric progression such that their sum is 14 and the sum of their squares is 84.

  Ans. 2, 4, 8

12.35 Prove: $x^n - y^n = (x - y)(x^{n-1} + x^{n-2}y + \cdots + xy^{n-2} + y^{n-1})$, n being a positive integer.

12.36 In a certain colony of bacteria each divides into two every hour. How many will be produced from a single bacillus if the rate of division continues for 12 hr?

  Ans. 4096

12.37 Find the geometric mean between (*a*) 2 and 32, (*b*) −4 and −25.

  Ans. (*a*) 8 (*b*) −10

12.38   Insert five geometric means between 6 and 384.

       Ans.   12, 24, 48, 96, 192

12.39   Show that for $p > q$, positive integers, their arithmetic mean A is greater than their geometric mean G.

       Hint: Consider $A - G$.

12.40   The sum of three numbers in A.P. is 24. If the first is decreased by 1 and the second is decreased by 2, the three numbers are in G.P. Find the A.P.

       Ans.   4, 8, 12 or 13, 8, 3

Chapter 13

Infinite Geometric Series

THE INDICATED SUM of the terms of a finite or infinite sequence is called a finite or infinite *series*. The sums of arithmetic and geometric progressions in the preceding chapter are examples of finite series.

Of course, it is impossible to add up all the terms of an infinite series; that is, in the usual meaning of the word *sum*, there is no such thing as the sum of such a series. However, it is possible to associate with certain infinite series a well-defined number which, for convenience, will be called the sum of the series.

Infinite series will be treated in some detail in Part IV. For the study of the infinite geometric series here, we shall need only to examine the behavior of r^n, where $|r| < 1$, as n increases indefinitely.

EXAMPLE 1.   From the table of values of $\left(\frac{1}{2}\right)^n$ in Table 13.1, it appears that, as n increases indefinitely, $\left(\frac{1}{2}\right)^n$ decreases indefinitely while remaining positive. Moreover, it can be made to have a value as near 0 as we please by choosing n sufficiently large. We describe this state of affairs by saying: The limit of $\left(\frac{1}{2}\right)^n$, as n increases indefinitely, is 0.

Table 13.1

n	1	3	5	10
$\left(\frac{1}{2}\right)^n$	0.5	0.125	0.03125	0.0009765625

By examining the behavior of r^n for other values of r, it becomes tolerably clear that

The limit of r^n, as n increases indefinitely, is 0 when $|r| < 1$.

Using a calculator, one can easily see this for very small positive values of r and for negative values of r near 0, such as $r = -0.000001$.

THE SUM S of the infinite geometric series

$$a + ar + ar^2 + \cdots + ar^{n-1} + \cdots, \quad |r| < 1, \quad \text{is } S = \frac{a}{1-r}.$$

(For a proof, see Problem 13.1.) (Note carefully the $|r| < 1$ restriction.)

84

EXAMPLE 2. For the infinite geometric series $12 + 4 + \frac{4}{3} + \frac{4}{9} + \cdots$, $a = 12$ and $r = \frac{1}{3}$. The sum of the series is $S = \frac{a}{1-r} = \frac{12}{1 - \frac{1}{3}} = 18$. (See Problems 13.2–13.5.)

EVERY INFINITELY REPEATING DECIMAL represents a rational number. This rational number is also called the *limiting value of the decimal*.

EXAMPLE 3. Find the limiting value of the repeating decimal $0.727272\ldots$. We write $0.727272\ldots = 0.72 + 0.0072 + 0.000072 + \cdots$ and note that for this infinite geometric series $a = 0.72$ and $r = 0.01$.

Then
$$S = \frac{a}{1-r} = \frac{.72}{1 - .01} = \frac{.72}{.99} = \frac{72}{99} = \frac{8}{11}$$

(See Problem 13.6.)

SIGMA NOTATION is a convenient notation for expressing sums. $\sum_{i=1}^{n} a_i$ means add $+a_1 + a_2 + \cdots + a_{n-1} + a_n$. For example,

$$\sum_{i=1}^{4} 2i = 2(1) + 2(2)\ 2(3) + 2(4)$$
$$= 2 + 4 + 6 + 8 = 20$$

$$\sum_{j=2}^{6} j^2 = 2^2 + 3^2 + 4^2 + 5^2 + 6^2$$

Thus,
$$a + ar + ar^2 + \cdots + ar^{n-1} + \cdots = \frac{a}{1-r} = \sum_{i=0}^{\infty} ar^i, \qquad |r| < 1$$

Solved Problems

13.1 Prove: The sum of the infinite geometric series $a + ar + ar^2 + \cdots + ar^{n-1} + \cdots$, where $|r| < 1$, is $S = \frac{a}{1-r}$.

The sum of the first n terms of the series is
$$S_n = \frac{a(1 - r^n)}{1 - r} = \frac{a}{1-r} - \frac{a}{1-r} r^n$$

As n increases indefinitely, the first term $\frac{a}{1-r}$ remains fixed while r^n, and hence, $\frac{a}{1-r} r^n$, approaches zero in value. Thus, $S = \frac{a}{1-r}$.

13.2 Determine the sum of each of the following infinite geometric series:

(a) $18 + 12 + 8 + \cdots$

Here $a = 18$, $r = \frac{2}{3}$, and $S = \frac{a}{1-r} = \frac{18}{1 - \frac{2}{3}} = 54$.

(b) $25 - 20 + 16 - \cdots$

Here $a = 25, r = -\frac{4}{5}$, and $S = \dfrac{a}{1-r} = \dfrac{25}{1-(-\frac{4}{5})} = \dfrac{125}{9}$.

(c) $.6 + .06 + .006 + \cdots$

Here $a = .6$, $r = .1$, and $S = \dfrac{.6}{1-.1} = \dfrac{.6}{.9} = \dfrac{6}{9} = \dfrac{2}{3}$.

13.3   An equilateral triangle has a perimeter of 30 cm. Another triangle is formed by joining the midpoints of the sides of the given triangle, another is formed by joining the midpoints of the sides of the second triangle, and so on. Find the sum of the perimeters of the triangles thus formed.

Since the side of each new triangle is $\frac{1}{2}$ the side of the triangle from which it is formed, the perimeters of the triangles are 30, 15, $\frac{15}{2}$, ...

Then $30 + 15 + \dfrac{15}{2} + \cdots = \dfrac{30}{1-\frac{1}{2}} = 60$ cm.

13.4   A rubber ball is dropped from a height of 81 m. Each time it strikes the ground, it rebounds two-thirds of the distance through which it last fell. Find the total distance it travels in coming to rest.

For the falls: $a = 81$ and $r = \frac{2}{3}$; $S = \dfrac{a}{1-r} = \dfrac{81}{1-\frac{2}{3}} = 243$ m.

For the rebounds: $a = 54$ and $r = \frac{2}{3}$; $S = \dfrac{54}{1-\frac{2}{3}} = 162$ m.

Thus, the total distance traveled is $243 + 162 = 405$ m.

13.5   For what values of x does $\dfrac{1}{x+1} + \dfrac{1}{(x+1)^2} + \dfrac{1}{(x+1)^3} + \cdots$ have a sum? Find the sum.

There will be a sum provided $|r| = \left|\dfrac{1}{x+1}\right| < 1$.

When $\left|\dfrac{1}{x+1}\right| = 1, |x+1| = 1, (x+1)^2 = 1, x^2 + 2x = 0$, and $x = -2, 0$. By examining the intervals $x < -2, -2 < x < 0$, and $x > 0$, we find that $\left|\dfrac{1}{x+1}\right| < 1$ when $x < -2$ and $x > 0$.

Thus, the series has a sum $S = \dfrac{1/(x+1)}{1 - 1/(x+1)} = \dfrac{1}{x}$ when $x < -2$ and $x > 0$.

13.6   Find the limiting value of each of the repeating decimals.

(a) $.0123123123\ldots$

Since $.0123123123\ldots = .0123 + .0000123 + .0000000123 + \cdots$ in which $a = .0123$ and $r = .001$,

$$S = \frac{a}{1-r} = \frac{.0123}{1-.001} = \frac{.0123}{.999} = \frac{123}{9990} = \frac{41}{3330}$$

(b) $2.373737\ldots$

The given number may be written as $2 + [.37 + .0037 + .000037 + \cdots]$. For the infinite geometric series in the brackets, $a = .37$ and $r = .01$; hence, $S = \dfrac{.37}{1-.01} = \dfrac{.37}{.99} = \dfrac{37}{99}$.

The limiting value is $2 + \frac{37}{99} = \frac{235}{99}$.

(c) $23.1454545\ldots$

Write $23.1454545\ldots = 23.1 + [.045 + .00045 + .0000045 + \cdots]$

$$= 23.1 + \frac{.045}{1-.01} = 23.1 + \frac{45}{990} = \frac{231}{10} + \frac{1}{22} = \frac{1273}{55}$$

Supplementary Problems

13.7   Sum the following infinite geometric series:

(a) $36 + 12 + 4 + \cdots$ (d) $5.6 - 2.24 + 0.896 - \cdots$ (f) $3 - \dfrac{3}{\sqrt{2}-1} + \dfrac{3}{3-2\sqrt{2}} - \cdots$

(b) $18 - 12 + 8 - \cdots$ (e) $1 + \frac{1}{2}\sqrt{2} + \frac{1}{2} + \cdots$

(c) $5 + 3 + 1.8 + \cdots$

   Ans.   (a) 54   (b) $\frac{54}{5}$ (c) $12\frac{1}{2}$ (d) 4   (e) $2 + \sqrt{2}$ (f) $\frac{3}{2}(2 - \sqrt{2})$

13.8   A swinging pendulum bob traverses the following distances: $40, 30, 22\frac{1}{2}, \ldots$ cm. Find the distance which it travels before coming to rest.

   Ans.   160 cm.

13.9   An unlimited sequence of squares are inscribed one within another by joining the midpoints of the sides of each preceding square. If the initial square is 8 cm on a side, find the sum of the perimeters of these squares.

   Ans. $32(2 + \sqrt{2})$ cm.

13.10   Express each repeating decimal as a rational fraction.

(a) $0.272727\ldots$ (b) $1.702702\ldots$ (c) $2.4242\ldots$ (d) $0.076923076923\ldots$

   Ans.   (a) $\frac{3}{11}$ (b) $\frac{189}{111}$ (c) $\frac{80}{33}$ (d) $\frac{1}{13}$

13.11   Find the values of x for which each of the following geometric series may be *summed*:

(a) $3 + 3x + 3x^2 + \cdots$ **Hint:** $|r| = |x|$

(b) $1 + (x - 1) + (x - 1)^2 + \cdots$ **Hint:** $|r| = |x - 1|$

(c) $5 + 5(x - 3) + 5(x - 3)^2 + \cdots$

   Ans.   (a) $-1 < x < 1$ (b) $0 < x < 2$ (c) $2 < x < 4$

13.12   Find the exact error when $\frac{1}{6}$ is approximated as 0.1667.

   Ans. $0.000033\ldots = 1/30\,000$

13.13   Evaluate

(a) $\displaystyle\sum_{j=1}^{2} 3j^2$

   Ans.   15

(b) $\displaystyle\sum_{i=0}^{\infty} 3 \cdot (0.5)^i$

   Ans.   6

13.14   Do you think that $\displaystyle\sum_{i=1}^{\infty} \frac{1}{n}$ is finite or infinite? What is $\displaystyle\sum_{i=1}^{10} \frac{1}{n}$? $\displaystyle\sum_{i=1}^{100} \frac{1}{n}$? $\displaystyle\sum_{i=1}^{1000} \frac{1}{n}$, etc.?

Does the result surprise you?

13.15   Repeat Problem 13.2 using a scientific calculator.

Chapter 14

Mathematical Induction

EVERYONE IS FAMILIAR with the process of reasoning, called *ordinary induction*, in which a generalization is made on the basis of a number of simple observations.

EXAMPLE 1. We observe that $1 = 1^2, 1 + 3 = 4 = 2^2, 1 + 3 + 5 = 9 = 3^2, 1 + 3 + 5 + 7 = 16 = 4^2$, and conclude that

$$1 + 3 + 5 + \cdots + (2n - 1) = n^2$$

or, in words, the sum of the first n odd integers is n^2.

EXAMPLE 2. We observe that 2 points determine $1 = \frac{1}{2} \cdot 2(2 - 1)$ line; that 3 points, not on a line, determine $3 = \frac{1}{2} \cdot 3(3 - 1)$ lines; that 4 points, no 3 on a line, determine $6 = \frac{1}{2} \cdot 4(4 - 1)$ lines; that 5 points, no 3 on a line, determine $10 = \frac{1}{2} \cdot 5(5 - 1)$ lines; and conclude that n points, no 3 on a line, determine $\frac{1}{2}n(n - 1)$ lines.

EXAMPLE 3. We observe that for $n = 1, 2, 3, 4, 5$ the values of

$$f(n) = \frac{n^4}{8} - \frac{17n^3}{12} + \frac{47n^3}{8} - \frac{103n}{12} + 6$$

are 2, 3, 5, 7, 11, respectively, and conclude that $f(n)$ is a prime number for every positive integral value of n.

The conclusions in Examples 1 and 2 are valid as we shall prove later. The conclusion in Example 3 is false since $f(6) = 22$ is not a prime number.

MATHEMATICAL INDUCTION is a type of reasoning by which such conclusions as were drawn in the above examples may be proved or disproved.
The steps are

(*1*) The verfication of the proposed formula or theorem for some positive integral value of n, usually the smallest. (Of course, we would not attempt to prove an unknown theorem by mathematical induction without first verifying it for several values of n.)

(*2*) The proof that if the proposed formula or theorem is true for $n = k$, some positive integer, it is true also for $n = k + 1$.

(*3*) The conclusion that the proposed formula or theorem is true for all values of n greater than the one for which verification was made in Step 1.

The following analogy is helpful: If one can climb to the first step of a ladder, and if for every step one reaches, one can reach the next step, then one can reach every step of the ladder, no matter how many steps there are.

EXAMPLE 4.   Prove: $1 + 3 + 5 + \cdots + (2n - 1) = n^2$.

(1)   The formula is true for $n = 1$ since $1 = 1^2$ (the first step of the ladder).

(2)   Let us assume the formula true for $n = k$, any positive integer; that is, let us assume that

$$1 + 3 + 5 + \cdots + (2k - 1) = k^2 \quad \text{(for every step you reach} \cdots \text{)} \tag{14.1}$$

We wish to show that, when (14.1) is true, the proposed formula is then true for $n = k + 1$; that is, that

$$1 + 3 + 5 + \cdots + (2k - 1) + (2k + 1) = (k + 1)^2 \quad (\cdots \text{ you can reach the next.)} \tag{14.2}$$

[NOTE: Statements (14.1) and (14.2) are obtained by replacing n in the proposed formula by k and $k + 1$, respectively. Now it is clear that the left member of (14.2) can be obtained from the left member of (14.1) by adding $(2k + 1)$. At this point the proposed formula is true or false according as we do or do not obtain the right member of (14.2) when $(2k + 1)$ is added to the right member of (14.1).]

Adding $(2k + 1)$ to both members of (14.1), we have

$$1 + 3 + 5 + \cdots + (2k - 1) + (2k + 1) = k^2 + (2k + 1) = (k + 1)^2 \tag{14.3}$$

Now (14.3) is identical with (14.2); thus, if the proposed formula is true for any positive integer $n = k$, it is true for the next positive integer $n = k + 1$.

(3)   Since the formula is true for $n = k = 1$ (Step 1), it is true for $n = k + 1 = 2$; being true for $n = k = 2$, it is true for $n = k + 1 = 3$; and so on. Hence, the formula is true for all positive integral values of n.

Solved Problems

Prove by mathematical induction.

14.1 $1 + 7 + 13 + \cdots + (6n - 5) = n(3n - 2)$.

(1)   The proposed formula is true for $n = 1$, since $1 = 1(3 - 2)$.

(2)   Assume the formula to be true for $n = k$, a positive integer; that is, assume

$$1 + 7 + 13 + \cdots + (6k - 5) = k(3k - 2) \tag{1}$$

Under this assumption we wish to show that

$$1 + 7 + 13 + \cdots + (6k - 5) + (6k + 1) = (k + 1)(3k + 1) \tag{2}$$

When $(6k + 1)$ is added to both members of (1), we have on the right

$$k(3k - 2) + (6k + 1) = 3k^2 + 4k + 1 = (k + 1)(3k + 1)$$

Hence, if the formula is true for $n = k$ it is true for $n = k + 1$.

(3)   Since the formula is true for $n = k = 1$ (Step 1), it is true for $n = k + 1 = 2$; being true for $n = k = 2$ it is true for $n = k + 1 = 3$; and so on, for every positive integral value of n.

14.2 $1 + 5 + 5^2 + \cdots + 5^{n-1} = \frac{1}{4}(5^n - 1)$.

(1)   The proposed formula is true for $n = 1$, since $1 = \frac{1}{4}(5 - 1)$.

(2)   Assume the formula to be true for $n = k$, a positive integer; that is, assume

$$1 + 5 + 5^2 + \cdots + 5^{k-1} = \tfrac{1}{4}(5^k - 1) \tag{1}$$

Under this assumption we wish to show that

$$1 + 5 + 5^2 + \cdots + 5^{k-1} + 5^k = \tfrac{1}{4}(5^{k+1} - 1) \tag{2}$$

When 5^k is added to both members of (1), we have on the right

$$\tfrac{1}{4}(5^k - 1) + 5^k = \tfrac{5}{4}(5^k) - \tfrac{1}{4} = \tfrac{1}{4}(5 \cdot 5^k - 1) = \tfrac{1}{4}(5^{k+1} - 1)$$

Hence, if the formula is turue for $n = k$ it is true for $n = k + 1$.

(3)   Since the formula is true for $n = k = 1$ (Step 1), it is true for $n = k + 1 = 2$; being true for $n = k = 2$ it is true for $n = k + 1 = 3$; and so on, for every positive integral value of n.

14.3 $\dfrac{5}{1 \cdot 2 \cdot 3} + \dfrac{6}{2 \cdot 3 \cdot 4} + \dfrac{7}{3 \cdot 4 \cdot 5} + \cdots + \dfrac{n+4}{n(n+1)(n+2)} = \dfrac{n(3n+7)}{2(n+1)(n+2)}.$

(1)   The formula is true for $n = 1$, since $\dfrac{5}{1 \cdot 2 \cdot 3} = \dfrac{1(3+7)}{2 \cdot 2 \cdot 3} = \dfrac{5}{6}.$

(2)   Assume the formula to be true for $n = k$, a positive integer; that is, assume

$$\frac{5}{1 \cdot 2 \cdot 3} + \frac{6}{2 \cdot 3 \cdot 4} + \cdots + \frac{k+4}{k(k+1)(k+2)} + \frac{k(3k+7)}{2(k+1)(k+2)} \tag{1}$$

Under this assumption we wish to show that

$$\frac{5}{1 \cdot 2 \cdot 3} + \frac{6}{2 \cdot 3 \cdot 4} + \cdots + \frac{k+4}{k(k+1)(k+2)} + \frac{k+5}{(k+1)(k+2)(k+3)} = \frac{(k+1)(3k+10)}{2(k+2)(k+3)} \tag{2}$$

When $\dfrac{k+5}{(k+1)(k+2)(k+3)}$ is added to both members of (1), we have on the right

$$\frac{k(3k+7)}{2(k+1)(k+2)} + \frac{k+5}{(k+1)(k+2)(k+3)} = \frac{1}{(k+1)(k+2)}\left[\frac{k(3k+7)}{2} + \frac{k+5}{k+3}\right]$$

$$= \frac{1}{(k+1)(k+2)} \cdot \frac{k(3k+7)(k+3) + 2(k+5)}{2(k+3)}$$

$$= \frac{1}{(k+1)(k+2)} \cdot \frac{3k^3 + 16k^2 + 23k + 10}{2(k+3)}$$

$$= \frac{1}{(k+1)(k+2)} \cdot \frac{(k+1)^2(3k+10)}{2(k+3)}$$

$$= \frac{(k+1)(3k+10)}{2(k+2)(k+3)}$$

Hence, if the formula is true for $n = k$ it is true for $n = k + 1$.

(3)   Since the formula is true for $n = k = 1$ (Step 1), it is true for $n = k + 1 = 2$; being true for $n = k = 2$, it is true for $n = k + 1 = 3$; and so on, for all positive integral values of n.

14.4 $x^{2n} - y^{2n}$ is divisible by $x + y$.

(1)   The theorem is true for $n = 1$, since $x^2 - y^2 = (x - y)(x + y)$ is divisible by $x + y$.

(2)   Let us assume the theorem true for $n = k$, a positive integer; that is, let us assume

$$x^{2k} - y^{2k} \text{ is divisible by } x + y. \tag{1}$$

We wish to show that, when (1) is true.

$$x^{2k+2} - y^{2k+2} \text{ is divisible by } x + y. \tag{2}$$

Now $x^{2k+2} - y^{2k+2} = (x^{2k+2} - x^2 y^{2k}) + (x^2 y^{2k} - y^{2k+2}) = x^2(x^{2k} - y^{2k}) + y^{2k}(x^2 - y^2)$. In the first term $(x^{2k} - y^{2k})$ is divisible by $(x + y)$ by assumption, and in the second term $(x^2 - y^2)$ is divisible by $(x + y)$ by Step 1; hence, if the theorem is true for $n = k$, a positive integer, it is true for the next one, $n = k + 1$.

(3) Since the theorem is true for $n = k = 1$, it is true for $n = k + 1 = 2$; being true for $n = k = 2$, it is true for $n = k + 1 = 3$; and so on, for every positive integral value of n.

14.5 The number of straight lines determined by $n > 1$ points, no three on the same straight line, is $\frac{1}{2}n(n - 1)$.

(1) The theorem is true when $n = 2$, since $\frac{1}{2} \cdot 2(2 - 1) = 1$ and two points determine one line.

(2) Let us assume that k points, no three on the same straight line, determine $\frac{1}{2}k(k - 1)$ lines.

When an aditional point is added (not on any of the lines already determined) and is joined to each of the original k points, k new lines are determined. Thus, altogether we have $\frac{1}{2}k(k - 1) + k = \frac{1}{2}k(k - 1 + 2) = \frac{1}{2}k(k + 1)$ lines and this agrees with the theorem when $n = k + 1$.

Hence, if the theorem is true for $n = k$, a positive integer greater than 1, it is true for the next one, $n = k + 1$.

(3) Since the theorem is true for $n = k = 2$ (Step 1), it is true for $n = k + 1 = 3$; being true for $n = k = 3$, it is true for $n = k + 1 = 4$; and so on, for every possible integral value >1 of n.

Supplementary Problems

Prove by mathematical induction, n being a positive integer.

14.6 $1 + 2 + 3 + \cdots + n = \frac{1}{2}n(n + 1)$

14.7 $1 + 4 + 7 + \cdots + (3n - 2) = \frac{1}{2}n(3n - 1)$

14.8 $1^2 + 3^2 + 5^2 + \cdots + (2n - 1)^2 = \frac{1}{3}n(4n^2 - 1)$

14.9 $1^2 + 2^2 + 3^2 + \cdots + n^2 = \frac{1}{6}n(n + 1)(2n + 1)$

14.10 $1^3 + 2^3 + 3^3 + \cdots + n^3 = \frac{1}{4}n^2(n + 1)^2$

14.11 $1^4 + 2^4 + 3^4 + \cdots + n^4 = \frac{1}{30}n(n + 1)(2n + 1)(3n^2 + 3n - 1)$

14.12 $1 \cdot 2 + 2 \cdot 3 + 3 \cdot 4 + \cdots + n(n + 1) = \frac{1}{3}n(n + 1)(n + 2)$

14.13 $\dfrac{1}{1 \cdot 3} + \dfrac{1}{3 \cdot 5} + \dfrac{1}{5 \cdot 7} + \cdots + \dfrac{1}{(2n - 1)(2n + 1)} = \dfrac{n}{2n + 1}$

14.14 $1 \cdot 3 + 2 \cdot 3^2 + 3 \cdot 3^3 + \cdots + n \cdot 3^n = \frac{3}{4}[(2n - 1)3^n + 1]$

14.15 $\dfrac{3}{1 \cdot 2 \cdot 2} + \dfrac{4}{2 \cdot 3 \cdot 2^2} + \dfrac{5}{3 \cdot 4 \cdot 2^3} + \cdots + \dfrac{(n + 2)}{n(n + 1)2^n} = 1 - \dfrac{1}{(n + 1)2^n}$

14.16 A convex polygon of n sides has $\frac{1}{2}n(n - 3)$ diagonals.

14.17 The sum of the interior angles of a regular polygon of n sides is $(n - 2)180°$.

The Binomial Theorem

By actual multiplication

$$(a+b)^1 = a+b, \qquad (a+b)^2 = a^2 + 2ab + b^2, \qquad (a+b)^3 = a^3 + 3a^2b + 3ab^2 + b^3,$$
$$(a+b)^4 = a^4 + 4a^3b + 6a^2b^2 + 4ab^3 + b^4,$$
$$(a+b)^5 = a^5 + 5a^4b + 10a^3b^2 + 10a^2b^3 + 5ab^4 + b^5, \text{ etc.}$$

From these cases we conclude that, when n is a positive integer,

$$(a+b)^n = a^n + na^{n-1}b + \frac{n(n-1)}{1\cdot 2}a^{n-2}b^2 + \frac{n(n-1)(n-2)}{1\cdot 2\cdot 3}a^{n-3}b^3 + \cdots + nab^{n-1} + b^n \qquad (15.1)$$

and note the following properties:

 (1) The number of terms in the expansion is $(n+1)$.
 (2) The first term a of the binomial enters the first term of the expansion with exponent n, the second term with exponent $(n-1)$, the third term with exponent $(n-2)$, and so on.
 (3) The second term b of the binomial enters the second term of the expansion with exponent 1, the third term with exponent 2, the fourth term with exponent 3, and so on.
 (4) The sum of the exponents of a and b in any term is n.
 (5) The coefficient of the first term in the expansion is 1, of the second term is $n/1$, of the third term is $\frac{n(n-1)}{1\cdot 2}$, of the fourth term is $\frac{n(n-1)(n-2)}{1\cdot 2\cdot 3}$, etc.
 (6) The coefficients of terms equidistant from the ends of the expansion are the same. Note that the number of factors in the numerator and denominator of any coefficient except the first and last is then either the exponent of a or of b, whichever is the smaller.

The above properties may be proved by mathematical induction.

EXAMPLE 1.   Expand $(3x + 2y^2)^5$ and simplify term by term.
   We put the several powers of $(3x)$ in first, then the powers of $(2y^2)$, and finally the coefficients, recalling Property 6 and using (15.1).

$$(3x + 2y^2)^5 = (3x)^5 + \frac{5}{1}(3x)^4(2y^2) + \frac{5\cdot 4}{1\cdot 2}(3x)^3(2y^2)^2 + \frac{5\cdot 4}{1\cdot 2}(3x)^2(2y^2)^3 + \frac{5}{1}(3x)(2y^2)^4 + (2y^2)^5$$
$$= 3^5 x^5 + 5\cdot 3^4 x^4 \cdot 2y^2 + 10\cdot 3^3 x^3 \cdot 2^2 y^4 + 10\cdot 3^2 x^2 \cdot 2^3 y^6 + 5\cdot 3x\cdot 2^4 y^8 + 2^5 y^{10}$$
$$= 243x^5 + 810x^4 y^2 + 1080x^3 y^4 + 720x^2 y^6 + 240xy^8 + 32y^{10}$$

(See Problems 15.1–15.2.)

THE rth TERM $(r \le n + 1)$ in the expansion of $(a + b)^n$ is

$$\frac{n(n - 1)(n - 2) \cdots (n - r + 2)}{1 \cdot 2 \cdot 3 \cdots (r - 1)} a^{n-r+1} b^{r-1}.$$

(See Problem 15.3.)

WHEN THE LAWS ABOVE are used to expand $(a + b)^n$, where n is real but not a positive integer, an endless succession of terms is obtained. Such expansions are valid [see Problem 15.7(a) for a verification] when $|b| < |a|$.

EXAMPLE 2. Write the first five terms in the expansion of $(a + b)^{-3}$, $|b| < |a|$.

$$(a + b)^{-3} = a^{-3} + (-3)a^{-4}b + \frac{(-3)(-4)}{1 \cdot 2} a^{-5} b^2 + \frac{(-3)(-4)(-5)}{1 \cdot 2 \cdot 3} a^{-6} b^3$$

$$+ \frac{(-3)(-4)(-5)(-6)}{1 \cdot 2 \cdot 3 \cdot 4} a^{-7} b^4 + \cdots$$

$$= \frac{1}{a^3} - \frac{3b}{a^4} + \frac{6b^2}{a^5} - \frac{10b^3}{a^6} + \frac{15b^4}{a^7} - \cdots$$

(See Problems 15.5–15.8.)

Solved Problems

15.1   Expand and simplify term by term.

(a) $\left(x^2 + \frac{1}{2}y\right)^6 = (x^2)^6 + \frac{6}{1}(x^2)^5\left(\frac{1}{2}y\right) + \frac{6 \cdot 5}{1 \cdot 2}(x^2)^4\left(\frac{1}{2}y\right)^2 + \frac{6 \cdot 5 \cdot 4}{1 \cdot 2 \cdot 3}(x^2)^3\left(\frac{1}{2}y\right)^3$

$\qquad + \frac{6 \cdot 5}{1 \cdot 2}(x^2)^2\left(\frac{1}{2}y\right)^4 + \frac{6}{1}(x^2)\left(\frac{1}{2}y\right)^5 + \left(\frac{1}{2}y\right)^6$

$\qquad = x^{12} + 6(x^{10})\frac{1}{2}y + 15(x^8)\frac{1}{4}y^2 + 20(x^6)\frac{1}{8}y^3 + 15(x^4)\frac{1}{16}y^4 + 6(x^2)\frac{1}{32}y^5 + \frac{1}{64}y^6$

$\qquad = x^{12} + 3x^{10}y + \frac{15}{4}x^8y^2 + \frac{5}{2}x^6y^3 + \frac{15}{16}x^4y^4 + \frac{3}{16}x^2y^5 + \frac{1}{64}y^6$

(b) $(x^{1/2} + 2y^{1/3})^4 = (x^{1/2})^4 + \frac{4}{1}(x^{1/2})^3(2y^{1/3}) + \frac{4 \cdot 3}{1 \cdot 2}(x^{1/2})^2(2y^{1/3})^2 + \frac{4}{1}(x^{1/2})(2y^{1/3})^3 + (2y^{1/3})^4$

$\qquad = x^2 + 4(x^{3/2})2y^{1/3} + 6(x)4y^{2/3} + 4(x^{1/2})8y + 16y^{4/3}$

$\qquad = x^2 + 8x^{3/2}y^{1/3} + 24xy^{2/3} + 32x^{1/2}y + 16y^{4/3}$

(c) $\left(\frac{2}{3}x^{1/2} - \frac{1}{2x}\right)^6 = \left(\frac{2}{3}x^{1/2}\right)^4 + \frac{6}{1}\left(\frac{2}{3}x^{1/2}\right)^5\left(-\frac{1}{2x}\right)$

$\qquad + \frac{6 \cdot 5}{1 \cdot 2}\left(\frac{2}{3}x^{1/2}\right)^4\left(-\frac{1}{2x}\right)^2 + \frac{6 \cdot 5 \cdot 4}{1 \cdot 2 \cdot 3}\left(\frac{2}{3}x^{1/2}\right)^3\left(-\frac{1}{2x}\right)^3$

$\qquad + \frac{6 \cdot 5}{1 \cdot 2}\left(\frac{2}{3}x^{1/2}\right)^2\left(-\frac{1}{2x}\right)^4 + \frac{6}{1}\left(\frac{2}{3}x^{1/2}\right)\left(-\frac{1}{2x}\right)^5 + \left(-\frac{1}{2x}\right)^6$

$\qquad = \frac{64}{729}x^3 - \frac{32}{81}x^{3/2} + \frac{20}{27} - \frac{20x^{1/2}}{27x^2} + \frac{5}{12x^3} - \frac{x^{1/2}}{8x^5} + \frac{1}{64x^6}$

15.2 Write the first five terms in each expansion and simplify term by term.

(a) $\left(\dfrac{2}{3}m^{1/2}+\dfrac{3}{2m^{3/2}}\right)^{12}=\left(\dfrac{2}{3}m^{1/2}\right)^{12}+\dfrac{12}{1}\left(\dfrac{2}{3}m^{1/2}\right)^{11}\left(\dfrac{3}{2m^{3/2}}\right)+\dfrac{12\cdot 11}{1\cdot 2}\left(\dfrac{2}{3}m^{1/2}\right)^{10}\left(\dfrac{3}{2m^{3/2}}\right)^2$

$$+\dfrac{12\cdot 11\cdot 10}{1\cdot 2\cdot 3}\left(\dfrac{2}{3}m^{1/2}\right)^{9}\left(\dfrac{3}{2m^{3/2}}\right)^3+\dfrac{12\cdot 11\cdot 10\cdot 9}{1\cdot 2\cdot 3\cdot 4}\left(\dfrac{2}{3}m^{1/2}\right)^{8}\left(\dfrac{3}{2m^{3/2}}\right)^4+\cdots$$

$$=\dfrac{2^{12}}{3^{12}}m^6+12\left(\dfrac{2^{11}}{3^{11}}m^{11/2}\right)\dfrac{3}{2m^{3/2}}+66\left(\dfrac{2^{10}}{3^{10}}m^5\right)\dfrac{3^2}{2^2m^3}$$

$$+220\left(\dfrac{2^9}{3^9}m^{9/2}\right)\dfrac{3^3}{2^3m^{9/2}}+495\left(\dfrac{2^8}{3^8}m^4\right)\dfrac{3^4}{2^4m^6}+\cdots$$

$$=\dfrac{2^{12}}{3^{12}}m^6+\dfrac{2^{12}}{3^9}m^4+11\dfrac{2^9}{3^7}m^2+55\dfrac{2^8}{3^6}+55\dfrac{2^4}{3^2m^2}+\cdots$$

(b) $\left(\dfrac{x^{1/2}}{y^{2/3}z}-\dfrac{yz^2}{2x}\right)^{11}=\left(\dfrac{x^{1/2}}{y^{2/3}z}\right)^{11}+\dfrac{11}{1}\left(\dfrac{x^{1/2}}{y^{2/3}z}\right)^{10}\left(-\dfrac{yz^2}{2x}\right)+\dfrac{11\cdot 10}{1\cdot 2}\left(\dfrac{x^{1/2}}{y^{2/3}z}\right)^{9}\left(-\dfrac{yz^2}{2x}\right)^2$

$$+\dfrac{11\cdot 10\cdot 9}{1\cdot 2\cdot 3}\left(\dfrac{x^{1/2}}{y^{2/3}z}\right)^{8}\left(-\dfrac{yz^2}{2x}\right)^3+\dfrac{11\cdot 10\cdot 9\cdot 8}{1\cdot 2\cdot 3\cdot 4}\left(\dfrac{x^{1/2}}{y^{2/3}z}\right)^{7}\left(-\dfrac{yz^2}{2x}\right)^4+\cdots$$

$$=\dfrac{x^{11/2}}{y^{22/3}z^{11}}-11\left(\dfrac{x^5}{y^{20/3}z^{10}}\right)\dfrac{yz^2}{2x}+55\left(\dfrac{x^{9/2}}{y^6z^9}\right)\dfrac{y^2z^4}{2^2x^2}$$

$$-165\left(\dfrac{x^4}{y^{16/3}z^8}\right)\dfrac{y^3z^6}{2^3x^3}+330\left(\dfrac{x^{7/2}}{y^{14/3}z^7}\right)\dfrac{y^4z^8}{2^4x^4}-\cdots$$

$$=\dfrac{x^{11/2}y^{2/3}}{y^8z^{11}}-\dfrac{11x^4y^{1/3}}{2y^6z^8}+\dfrac{55x^{5/2}}{4y^4z^5}-\dfrac{165xy^{2/3}}{8y^3z^2}+\dfrac{165x^{1/2}y^{1/3}z}{8xy}-\cdots$$

15.3 Find the indicated term and simplify.

(a) The seventh term of $(a+b)^{15}$.

In the seventh term the exponent of b is $7-1=6$, the exponent of a is $15-6=9$, and the coefficient has six factors in the numerator and denominator. Hence, the term is

$$\dfrac{15\cdot 14\cdot 13\cdot 12\cdot 11\cdot 10}{1\cdot 2\cdot 3\cdot 4\cdot 5\cdot 6}a^9b^6=5005a^9b^6$$

(b) The ninth term of $\left(x-\dfrac{1}{x^{1/2}}\right)^{12}$.

In the ninth term the exponent of $b=-\dfrac{1}{x^{1/2}}$ is $9-1=8$, the exponent of $a=x$ is $12-8=4$, and the coefficient has four factors in numerator and denominator. Hence, the required term is

$$\dfrac{12\cdot 11\cdot 10\cdot 9}{1\cdot 2\cdot 3\cdot 4}(x)^4\left(-\dfrac{1}{x^{1/2}}\right)^8=495(x^4)\dfrac{1}{x^4}=495$$

(c) The twelfth term of $\left(\dfrac{x^{1/2}}{4}-\dfrac{2y}{x^{3/2}}\right)^{18}$.

The required term is

$$\dfrac{18\cdot 17\cdot 16\cdot 15\cdot 14\cdot 13\cdot 12}{1\cdot 2\cdot 3\cdot 4\cdot 5\cdot 6\cdot 7}\left(\dfrac{x^{1/2}}{4}\right)^7\left(-\dfrac{2y}{x^{3/2}}\right)^{11}=-9\cdot 17\cdot 16\cdot 13\left(\dfrac{x^{7/2}}{2^{14}}\right)\dfrac{2^{11}y^{11}}{x^{33/2}}=-3978\dfrac{y^{11}}{x^{13}}$$

(d) The middle term in the expansion of $\left(x^{2/3} + \dfrac{1}{x^{1/2}}\right)^{10}$.

Since there are 11 terms in all, the middle term is the sixth. This term is

$$\frac{10 \cdot 9 \cdot 8 \cdot 7 \cdot 6}{1 \cdot 2 \cdot 3 \cdot 4 \cdot 5}(x^{2/3})^5 \left(\frac{1}{x^{1/2}}\right)^5 = 252 \frac{x^{10/3}}{x^{5/2}} = 252 x^{5/6}$$

(e) The term involving y^{12} in the expansion of $\left(y^3 - \dfrac{x}{3}\right)^9$.

The first term of the binomial must be raised to the fourth power to produce y^{12}; hence, the second term must be raised to the fifth power and we are to write the sixth term. This term is

$$\frac{9 \cdot 8 \cdot 7 \cdot 6}{1 \cdot 2 \cdot 3 \cdot 4}(y^3)^4 \left(-\frac{x}{3}\right)^5 = -3^2 \cdot 14 y^{12} \frac{x^5}{3^5} = -\frac{14}{27} x^5 y^{12}$$

(f) The term involving x^4 in the expansion of $\left(\dfrac{2}{x} + \dfrac{x^2}{4}\right)^{14}$.

Let p and q be positive integers so that $p + q = 14$. We are required to determine p and q so that $\left(\dfrac{2}{x}\right)^p \left(\dfrac{x^2}{4}\right)^q$ yields a term in x^4. Then $2q - p = 4$ or $2q - (14 - q) = 3q - 14 = 4$ and $q = 6$. The required term, the seventh in the expansion, is

$$\frac{14 \cdot 13 \cdot 12 \cdot 11 \cdot 10 \cdot 9}{1 \cdot 2 \cdot 3 \cdot 4 \cdot 5 \cdot 6}\left(\frac{2}{x}\right)^8 \left(\frac{x^2}{4}\right)^6 = 3003 \frac{2^8}{x^8} \frac{x^{12}}{2^{12}} = \frac{3003}{16} x^4$$

15.4 Evaluate $(1.02)^{12}$ correct to four decimal places.

$$(1.02)^{12} = (1 + .02)^{12} = 1 + 12(.02) + \frac{12 \cdot 11}{1 \cdot 2}(.02)^2 + \frac{12 \cdot 11 \cdot 10}{1 \cdot 2 \cdot 3}(.02)^3 + \frac{12 \cdot 11 \cdot 10 \cdot 9}{1 \cdot 2 \cdot 3 \cdot 4}(.02)^4 + \cdots$$

$$= 1 + .24 + .0264 + .00176 + .00008 + \cdots = 1.26824 \text{ (approximately)}$$

Thus, $(1.02)^{12} = 1.2682$ correct to four decimal places.

15.5 Write the first five terms and simplify term by term.

(a) $\left(x^2 - \dfrac{2}{x^4}\right)^{1/2} = (x^2)^{1/2} + \dfrac{1}{2}(x^2)^{-1/2}\left(-\dfrac{2}{x^4}\right) + \dfrac{(\frac{1}{2})(-\frac{1}{2})}{1 \cdot 2}(x^2)^{-3/2}\left(-\dfrac{2}{x^4}\right)^2$

$\qquad + \dfrac{(\frac{1}{2})(-\frac{1}{2})(-\frac{3}{2})}{1 \cdot 2 \cdot 3}(x^2)^{-5/2}\left(-\dfrac{2}{x^4}\right)^3 + \dfrac{(\frac{1}{2})(-\frac{1}{2})(-\frac{3}{2})(-\frac{5}{2})}{1 \cdot 2 \cdot 3 \cdot 4}(x^2)^{-7/2}\left(-\dfrac{2}{x^4}\right)^4 + \cdots$

$\qquad = x - \dfrac{1}{2} \cdot \dfrac{1}{x} \cdot \dfrac{2}{x^4} - \dfrac{1}{2^3} \cdot \dfrac{1}{x^3} \cdot \dfrac{2^2}{x^8} - \dfrac{1}{2^4} \cdot \dfrac{1}{x^5} \cdot \dfrac{2^3}{x^{12}} - \dfrac{5}{2^7} \cdot \dfrac{1}{x^7} \cdot \dfrac{2^4}{x^{16}} - \cdots$

$\qquad = x - \dfrac{1}{x^5} - \dfrac{1}{2x^{11}} - \dfrac{1}{2x^{17}} - \dfrac{5}{8x^{23}} - \cdots, \qquad (|x| > \sqrt[6]{2})$

(b) $(1 - x^3)^{-2/3} = 1^{-2/3} + \left(-\dfrac{2}{3}\right)1^{-5/3}(-x^3) + \dfrac{(-\frac{2}{3})(-\frac{5}{3})}{1 \cdot 2}1^{-8/3}(-x^3)^2$

$\qquad + \dfrac{(-\frac{2}{3})(-\frac{5}{3})(-\frac{8}{3})}{1 \cdot 2 \cdot 3}1^{-11/3}(-x^3)^3 + \dfrac{(-\frac{2}{3})(-\frac{5}{3})(-\frac{8}{3})(-\frac{11}{3})}{1 \cdot 2 \cdot 3 \cdot 4}1^{-14/3}(-x^3)^4 + \cdots$

$\qquad = 1 + \tfrac{2}{3}x^3 + \tfrac{5}{9}x^6 + \tfrac{40}{81}x^9 + \tfrac{110}{243}x^{12} + \cdots, \qquad (|x| < 1)$

15.6 Find and simplify the sixth term in the expansion of $\left(x^2 - \dfrac{3}{2x}\right)^{5/3}$, $|x| > \sqrt[3]{\tfrac{3}{2}}$.

The required term is

$$\frac{(\frac{5}{3})(\frac{2}{3})(-\frac{1}{3})(-\frac{4}{3})(-\frac{7}{3})}{1 \cdot 2 \cdot 3 \cdot 4 \cdot 5}(x^2)^{-10/3}\left(-\frac{3}{2x}\right)^5 = \frac{7}{3^6}x^{-20/3}\frac{3^5}{2^5 x^5} = \frac{7x^{1/3}}{96x^{12}}$$

15.7   (a) Evaluate $\sqrt{26}$ correct to four decimal places.

$$\sqrt{26} = (5^2 + 1)^{1/2} = (5^2)^{1/2} + \frac{1}{2}(5^2)^{-1/2}(1) + \frac{(\frac{1}{2})(-\frac{1}{2})}{1 \cdot 2}(5^2)^{-3/2}(1)^2$$

$$+ \frac{(\frac{1}{2})(-\frac{1}{2})(-\frac{3}{2})}{1 \cdot 2 \cdot 3}(5^2)^{-5/2}(1)^3 + \frac{(\frac{1}{2})(-\frac{1}{2})(-\frac{3}{2})(-\frac{5}{2})}{1 \cdot 2 \cdot 3 \cdot 4}(5^2)^{-7/2}(1)^4 + \cdots$$

$$= 5 + \frac{1}{2} \cdot \frac{1}{5} - \frac{1}{2^3} \cdot \frac{1}{5^3} + \frac{1}{2^4} \cdot \frac{1}{5^5} - \frac{1}{2^7} \cdot \frac{1}{5^6} + \cdots$$

$$= 5.00000 + .10000 - .00100 + .00002 - \cdots$$

$$= 5.09902 \quad \text{(approximately)}$$

Thus, $\sqrt{26} = 5.0990$, correct to four decimal places.

[NOTE: If we write $\sqrt{26} = (1 + 5^2)^{1/2} = 1 + \frac{1}{2}(5^2) - \frac{1}{8}(5^2)^2 + \frac{1}{16}(5^2)^3 - \cdots$

$$= 1 + 12.5 - 78.125 + 976.5625 - \cdots$$

it is clear that the expansion is not valid. The condition $|b| < |a|$ in $(a + b)^n$ in essential when n is not a positive integer.]

(b) Evaluate $\sqrt{23}$ correct to four decimal places.

$$\sqrt{23} = (5^2 - 2)^{1/2} = (5^2)^{1/2} + \frac{1}{2}(5^2)^{-1/2}(-2) + \frac{(\frac{1}{2})(-\frac{1}{2})}{1 \cdot 2}(5^2)^{-3/2}(-2)^2$$

$$+ \frac{(\frac{1}{2})(-\frac{1}{2})(-\frac{3}{2})}{1 \cdot 2 \cdot 3}(5^2)^{-5/2}(-2)^3 + \frac{(\frac{1}{2})(-\frac{1}{2})(-\frac{3}{2})(-\frac{5}{2})}{1 \cdot 2 \cdot 3 \cdot 4}(5^2)^{-7/2}(-2)^4 + \cdots$$

$$= 5 - \frac{1}{5} - \frac{1}{2} \cdot \frac{1}{5^3} - \frac{1}{2} \cdot \frac{1}{5^5} - \frac{1}{8} \cdot \frac{1}{5^6} - \cdots$$

$$= 4.79583 \quad \text{(approximately)}$$

Thus, $\sqrt{23} = 4.7958$, correct to four decimal places.

15.8   (a) Show that the sum of the coefficients in the expansion of $(a + b)^n$, n a positive integer, is 2^n.
     (b) Show that the sum of the coefficients in the expansion of $(a - b)^n$, n a positive integer, is 0.

(a) In $(a + b)^n = a^n + na^{n-1}b + \frac{n(n-1)}{1 \cdot 2}a^{n-2}b^2 + \cdots + \frac{n(n-1)}{1 \cdot 2}a^2b^{n-2} + nab^{n-1} + b^n$, let $a = b = 1$;
     then $(1 + 1)^n = 2^n = 1 + n + \frac{1}{2}n(n-1) + \cdots + \frac{1}{2}n(n-1) + n + 1$, as was to be proved.

(b) Similarly, let $a = b = 1$ in the expansion of $(a - b)^n$ and obtain

$$1 - n + \tfrac{1}{2}n(n-1) - \cdots + (-1)^{n-2}[\tfrac{1}{2}n(n-1)] + (-1)^{n-1}n + (-1)^n = (1 - 1)^n = 0$$

Supplementary Problems

15.9   Expand by the binomial theorem and simplify term by term.

(a) $(a + \frac{1}{2}b)^6 = a^6 + 3a^5b + \frac{15}{4}a^4b^2 + \frac{5}{2}a^3b^3 + \frac{15}{16}a^2b^4 + \frac{3}{16}ab^5 + \frac{1}{64}b^6$

(b) $(4x + \frac{1}{4}y)^5 = 1024x^5 + 320x^4y + 40x^3y^2 + \frac{5}{2}x^2y^3 + \frac{5}{64}xy^4 + \frac{1}{1024}y^5$

(c) $\left(\frac{x}{4y^3} - \frac{2y}{x^2}\right)^5 = \frac{x^5}{1024y^{15}} - \frac{5x^2}{128y^{11}} + \frac{5}{8xy^7} - \frac{5}{x^4y^3} + \frac{20y}{x^7} - \frac{32y^5}{x^{10}}$

15.10 Find the indicated term and simplify.

(a) Fifth term of $(\frac{1}{2} + x)^{10}$

(d) Seventh term of $\left(x^{1/3} - \frac{1}{2x^{2/3}}\right)^{10}$

(b) Sixth term of $\left(\frac{2}{x^{1/2}} - \frac{x^{1/4}}{4}\right)^9$

(e) Middle term of $\left(\frac{1}{x} - x^2\right)^{12}$

(c) Tenth term of $\left(\frac{27a^2}{b^3} + \frac{b^2}{6a^4}\right)^{12}$

(f) Middle term of $\left(a^{1/2}b^{1/2} - \frac{a}{2b^{3/2}}\right)^9$

(g) The term involving x^{14} in the expansion of $(2/x - x^2)^{10}$

(h) The term free of y in the expansion of $(xy^{1/6} - y^{-2/3})^{15}$

Ans. (a) $\frac{105}{32}x^4$ (b) $-\frac{63x^{1/4}}{32x}$ (c) $\frac{55b^9}{128a^{30}}$ (d) $\frac{105x^{1/3}}{32x^3}$ (e) $924x^6$

(f) $\frac{63a^{13/2}b^{1/2}}{8b^4}, -\frac{63a^7b^{1/2}}{16b^6}$ (g) $180x^{14}$ (h) $-455x^{12}$

15.11 Expand $(a + b - c)^3$. **Hint:** Write $(a + b - c)^3 = [(a + b) - c]^3$.

Ans. $a^3 + b^3 - c^3 + 3a^2b - 3a^2c + 3ab^2 - 3b^2c + 3ac^2 + 3bc^2 - 6abc$

15.12 Find the value of n if the coefficients of the sixth and sixteenth terms in the expansion of $(a + b)^n$ are equal.

Ans. $n = 20$

15.13 Find the first five terms in the expansion and simplify term by term.

(a) $\left(\frac{x}{a} + \frac{a^3}{x^2}\right)^{-3} = \frac{a^3}{x^3} - \frac{3a^7}{x^6} + \frac{6a^{11}}{x^9} - \frac{10a^{15}}{x^{12}} + \frac{15a^{19}}{x^{15}} - \cdots, (|x| > a^{4/3})$

(b) $\left(x^2 - \frac{2y^2}{x}\right)^{3/2} = x^3 - 3y^2 + \frac{3y^4}{2x^3} + \frac{y^6}{2x^6} + \frac{3y^8}{8x^9} + \cdots, (|x| > \sqrt[3]{2y^2})$

(c) $\left(1 - \frac{3y^2}{x}\right)^{1/3} = 1 - \frac{y^2}{x} - \frac{y^4}{x^2} - \frac{5y^6}{3x^3} - \frac{10y^8}{3x^4} - \cdots, (|x| > 3y^2)$

15.14 Find and simplify the indicated term.

(a) The sixth term in $(a^2 - 4b^2)^{1/2}$, $(|a| > 2|b|)$.

Ans. $-28\frac{b^{10}}{a^9}$

(b) The seventh term in $\left(x^{1/4} - \frac{3}{x^{1/2}}\right)^{-1}$, $(|x| > 3^{4/3})$.

Ans. $729\frac{x^{1/4}}{x^5}$

(c) The term involving x^{20} in $\left(\frac{2}{x} + x^2\right)^{-2}$, $(|x| < \sqrt[3]{2})$.

Ans. $\frac{7x^{20}}{256}$

15.15 Prove that the number of terms in the binomial theorem expansion of $(a + b)^n$ is $n + 1$.

(Hint: Use induction.)

15.16 Prove that the sum of the exponents of a and b in any term of the expansion of $(a + b)^n$ is n.

Chapter 16

Permutations

ANY ARRANGEMENT OF A SET OF OBJECTS in a definite order is called a *permutation* of the set taken all at a time. For example, *abcd*, *acbd*, *bdca* are permutations of a set of letters, *a*, *b*, *c*, *d* taken all at a time.

If a set contains *n* objects, any ordered arrangement of any $r \le n$ of the objects is called a permutation of the *n* object taken *r* at a time. For example, *ab*, *ba*, *ca*, *db* are permutations of the $n = 4$ letters *a*, *b*, *c*, *d* taken $r = 2$ at a time, while *abc*, *adb*, *bad*, *cad* are permutations of the $n = 4$ letters taken $r = 3$ at a time. The number of permutations of *n* objects taken *r* at a time is denoted by *nPr*, where $r \le n$.

THE NUMBER OF PERMUTATIONS which may be formed in each situation can be found by means of the

FUNDAMENTAL PRINCIPLE: If one thing can be done in *u* different ways, if after it has been done in any one of these, a second thing can be done in *v* different ways, if after it has been done in any one of these, a third thing can be done in *w* different ways, ..., the several things can be done in the order stated in $u \cdot v \cdot w \cdots$ different ways.

EXAMPLE 1. In how many ways can 6 students be assigned to (*a*) row of 6 seats, (*b*) a row of 8 seats?

(*a*) Let the seats be denoted xxxxxx. The seat on the left may be assigned to any one of the 6 students; that is, it may be assigned in 6 different ways. After the assignment has been made, the next seat may be assigned to any one of the 5 remaining students. After the assignment has been made, the next seat may be assigned to any one of the 4 remaining students, and so on. Placing the number of ways in which each seat may be assigned under the x marking the seat, we have

$$\begin{array}{cccccc} x & x & x & x & x & x \\ 6 & 5 & 4 & 3 & 2 & 1 \end{array}$$

By the fundamental principle, the seats may be assigned in

$$6 \cdot 5 \cdot 4 \cdot 3 \cdot 2 \cdot 1 = 720 \text{ ways}$$

The reader should assure him- or herself that the seats might have been assigned to the students with the same result.

(*b*) Here each student must be assigned a seat. The first student may be assigned any one of the 8 seats, the second student any one of the 7 remaining seats, and so on. Letting x represent a student, we have

$$\begin{array}{cccccc} x & x & x & x & x & x \\ 8 & 7 & 6 & 5 & 4 & 3 \end{array}$$

and the assignment may be made in $8 \cdot 7 \cdot 6 \cdot 5 \cdot 4 \cdot 3 = 20\,160$ ways.

(NOTE: If we attempt to assign students to seats, we must first select the six seats to be used. The problem of selections will be considered under Combinations in Chapter 17.) (See Problem 16.1–16.5.)

Define $n!$ (n factorial) to be

$$n \cdot (n-1) \cdot (n-2) \cdot \cdots \cdot (2)(1) \text{ for positive integers } n,$$

where $0! = 1! = 1$. Then, if we define $\binom{n}{k}$ to be $\dfrac{n!}{k!(n-k)!}$, we call $\binom{n}{k}$ a binomial coefficient and note that

$$(a+b)^n = \sum_{k=0}^{n} \binom{n}{k} a^{n-k} b^k, \qquad n \geq 1.$$

(See Problems 16.6–16.7.) (See Chapter 15 for a discussion of the binomial theorem.)

PERMUTATIONS OF OBJECTS NOT ALL DIFFERENT.

If there are n objects of which k are alike while the remaining $(n-k)$ objects are different from them and from each other, it is clear that the number of different permutations of the n objects taken all together is not $n!$.

EXAMPLE 2. How many different permutations of four letters can be formed using the letters of the word *bass*?

For the moment, think of the given letters as b, a, s_1, s_2 so that they are all different. Then

$$bas_1s_2 \quad as_1bs_2 \quad s_2s_1ba \quad s_1as_2b \quad bs_1s_2a$$

$$bas_2s_1 \quad as_2bs_1 \quad s_1s_2ba \quad s_2as_1b \quad bs_2s_1a$$

are 10 of the 24 permutations of the four letters taken all together. However, when the subscripts are removed, it is seen that the two permutations in each column are alike.

Thus, there are $\dfrac{1 \cdot 2 \cdot 3 \cdot 4}{1 \cdot 2} = 12$ different permutations.

In general, given n objects of which k_1 are one sort, k_2 of another, k_3 of another,..., then the number of different permutations that can be made from the n objects taken all together is

$$\frac{n!}{k_1! k_2! k_3! \ldots}$$

(See Problem 16.8.)

IN GENERAL: The number of permutations, nPr, of different objects taken $r < n$ at a time is $\dfrac{n!}{(n-r)!}$. The number of permutation of n objects taken n at a time, nPn, is $n!$.

Solved Problems

16.1   Using the letters of the word MARKING and calling any arrangement a word, (a) how many different 7-letter words can be formed, (b) how many different 3-letter words can be formed?

(a)   We must fill each of the positions xxxxxxx with a different letter. The first position may be filled in 7 ways, the second in 6 ways, and so on.

Thus, we have $\begin{matrix} x & x & x & x & x & x & x \\ 7 & 6 & 5 & 4 & 3 & 2 & 1 \end{matrix}$ and there are $7 \cdot 6 \cdot 5 \cdot 4 \cdot 3 \cdot 2 \cdot 1 = 5040$ words.

(b)   We must fill of the positions xxx with a different letter. The first position can be filled in 7 ways, the second in 6 ways, and the third in 5 ways. Thus, there are $7 \cdot 6 \cdot 5 = 210$ words.

16.2   In forming 5-letter words using the letters of the word EQUATIONS, (*a*) how many consist only of vowels, (*b*) how many contain all of the consonants, (*c*) how many begin with E and end in S, (*d*) how many begin with a consonant, (*e*) how many contain N, (*f*) how many in which the vowels and consonants alternate, (*g*) how many in which Q is immediately followed by U?

There are 9 letters, consisting of 5 vowels and 4 consonants.

(*a*)   There are five places to be filled and 5 vowels at our disposal. Hence, we can form $5 \cdot 4 \cdot 3 \cdot 2 \cdot 1 = 120$ words.

(*b*)   Each word is to contain the 4 consonants and one of the 5 vowels. There are now six things to do: first pick the vowel to be used (in 5 ways), next place the vowel (in 5 ways), and fill the remaining four positions with consonants.

We have $\begin{smallmatrix} x & x & x & x & x & x \\ 5 & 5 & 4 & 3 & 2 & 1 \end{smallmatrix}$; hence there are $5 \cdot 5 \cdot 4 \cdot 3 \cdot 2 \cdot 1 = 600$ words.

(*c*)   Indicate the fact that the position of certain letters is fixed writing $\begin{smallmatrix} & E & & & S \\ x & x & x & x & x \end{smallmatrix}$.

Now there are just three positions to be filled and 7 letters at our disposal. Thus, there are $7 \cdot 6 \cdot 5 = 210$ words.

(*d*)   Here we have $\begin{smallmatrix} c & & & & \\ x & x & x & x & x \\ 4 & 8 & 7 & 6 & 5 \end{smallmatrix}$ since, after filling the first position with any one of the 4 consonants, there are 8 letters remaining. Hence, there are $4 \cdot 8 \cdot 7 \cdot 6 \cdot 5 = 6720$ words.

(*e*)   There are five things to do: first, place the letter N in any one of the five positions and then fill the other four positions from among the 8 letters remaining.

We have $\begin{smallmatrix} x & x & x & x & x \\ 5 & 8 & 7 & 6 & 5 \end{smallmatrix}$; hence, there are $5 \cdot 8 \cdot 7 \cdot 6 \cdot 5 = 8400$ words.

(*f*)   We may have $\begin{smallmatrix} v & c & v & c & v \\ x & x & x & x & x \\ 5 & 4 & 4 & 3 & 3 \end{smallmatrix}$ or $\begin{smallmatrix} c & v & c & v & c \\ x & x & x & x & x \\ 4 & 5 & 3 & 4 & 2 \end{smallmatrix}$. Hence, there are $5 \cdot 4 \cdot 4 \cdot 3 \cdot 3 + 4 \cdot 5 \cdot 3 \cdot 4 \cdot 2 = 1200$ words.

(*g*)   First we place Q so that U may follow it (Q may occupy any of the first four positions but not the last), next we place U (in only 1 way), and then we fill the three other positions from among the 7 letters remaining.

Thus, we have $\begin{smallmatrix} x & x & x & x & x \\ 4 & 1 & 7 & 6 & 5 \end{smallmatrix}$ and there $4 \cdot 1 \cdot 7 \cdot 6 \cdot 5 = 840$ words.

16.3   If repetitions are not allowed, (*a*) how many three-digit numbers can be formed with the digits $0, 1, 2, 3, 4, 5, 6, 7, 8, 9$? (*b*) How many of these are odd numbers? (*c*) How many are even numbers? (*d*) How many are divisble by 5? (*e*) How many are greater than 600?

In each case we have $\begin{smallmatrix} \neq 0 & & \\ x & x & x \end{smallmatrix}$.

(*a*)   The position on the left can be filled in 9 ways (0 cannot be used), the middle position can be filled in 9 ways (0 can be used), and the position on the right can be filled in 8 ways. Thus, there are $9 \cdot 9 \cdot 8 = 648$ numbers.

(*b*)   We have $\begin{smallmatrix} \neq 0 & & odd \\ x & x & x \end{smallmatrix}$. Care must be exercised here in choosing the order in which to fill the positions. If the position on the left is filled first (in 9 ways), we cannot determine the number of ways in which the position on the right can be filled since, if the former is filled with an odd digit there are 4 ways of filling the latter but if the former is filled with an even digit there are 5 ways of filling the latter.

We fill first the position on the right (in 5 ways), then the position on the left (in 8 ways, since one odd digit and 0 are excluded), and the middle position (in 8 ways, since two digits are now excluded). Thus, there are $8 \cdot 8 \cdot 5 = 320$ numbers.

(c)   We have $\overset{\neq 0}{\underset{x}{}}\ \overset{}{\underset{x}{}}\ \overset{even}{\underset{x}{}}$. We note that the argument above excludes the possibility of first filling the position on the left. But if we fill the first position on the right (in 5 ways), were unable to determine the number of ways the position on the left can be filled (9 ways if 0 was used on the right, 8 ways if 2, 4, 6, or 8 was used). Thus, we must separate the two cases.

First, we form all numbers ending in 0; there are $9 \cdot 8 \cdot 1$ of them. Next, we form all numbers ending in 2, 4, 6, or 8; there $8 \cdot 8 \cdot 4$ of them. Thus, in all, there are $9 \cdot 8 \cdot 1 + 8 \cdot 8 \cdot 4 = 328$ numbers.

As a check, we have 320 odd and 328 even numbers for a total of 648 as found in (a) above.

(d)   A number is divisible by 5 if and only if it ends in 0 or 5. There are $9 \cdot 8 \cdot 1$ numbers ending in 0 and $8 \cdot 8 \cdot 1$ numbers ending in 5. Hence, in all, there are $9 \cdot 8 \cdot 1 + 8 \cdot 8 \cdot 1 = 136$ numbers divisible by 5.

(e)   The position on the left can be filled in 4 ways (with 6, 7, 8, or 9) and the remaining positions in $9 \cdot 8$ ways. Thus, there are $4 \cdot 9 \cdot 8 = 288$ numbers.

16.4   Solve Problem 16.3 (a), (b), (c), (d) if any digit may be used once, twice, or three times in forming the three-digit number.

(a)   The position on the left can be filled in 9 ways and each of other position can be filled in 10 ways. Thus, there are $9 \cdot 10 \cdot 10 = 900$ numbers.

(b)   The position on the right can be filled in 5 ways, the middle position in 10 ways, and the position on the left in 9 ways. Thus, there are $9 \cdot 10 \cdot 5 = 450$ numbers.

(c)   There are $9 \cdot 10 \cdot 5 = 450$ even numbers.

(d)   There are $9 \cdot 10 \cdot 1 = 90$ numbers ending in 0 and the same number ending in 5. Thus, there are 180 numbers divisible by 5.

16.5   In how many ways can 10 boys be arranged (a) in a straight line, (b) in a circle?

(a)   The boys may be arranged in a straight line in $10 \cdot 9 \cdot 8 \cdot 7 \cdot 6 \cdot 5 \cdot 4 \cdot 3 \cdot 2 \cdot 1$ ways.

(b)   We first place a boy at any point on the circle. The other 9 boys may then be arranged in $9 \cdot 8 \cdot 7 \cdot 6 \cdot 5 \cdot 4 \cdot 3 \cdot 2 \cdot 1$ ways.

This is an example of a *circular permutation*. In general, n objects may be arranged in a circle in $(n-1)(n-2) \cdots 2 \cdot 1$ ways.

16.6   Evaluate.

(a) $\dfrac{8!}{3!} = \dfrac{1 \cdot 2 \cdot 3 \cdot 4 \cdot 5 \cdot 6 \cdot 7 \cdot 8}{1 \cdot 2 \cdot 3} = 6720$ (c) $\dfrac{10!}{3!3!4!} = \dfrac{1 \cdot 2 \cdot 3 \cdot 4 \cdot 5 \cdot 6 \cdot 7 \cdot 8 \cdot 9 \cdot 10}{1 \cdot 2 \cdot 3 \cdot 1 \cdot 2 \cdot 3 \cdot 1 \cdot 2 \cdot 3 \cdot 4} = 4200$

(b) $\dfrac{7!}{6!} = \dfrac{1 \cdot 2 \cdot 3 \cdot 4 \cdot 5 \cdot 6 \cdot 7}{1 \cdot 2 \cdot 3 \cdot 4 \cdot 5 \cdot 6} = 7$ (d) $\dfrac{(r+1)!}{(r-1)!} = \dfrac{1 \cdot 2 \cdots (r-1)r(r+1)}{1 \cdot 2 \cdots (r-1)} = r(r+1)$

16.7   Solve for n, given (a) $_nP_2 = 110$, (b) $_nP_4 = 30\,_nP_2$.

(a) $_nP_2 = n(n-1) = n^2 - n - 110$. Then $n^2 - n = 110 = (n-11)(n+10) = 0$ and, since n is positive, $n = 11$.

(b)   We have $n(n-1)(n-2)(n-3) = 30n(n-1)$ or $n(n-1)(n-2)(n-3) - 30n(n-1) = 0$.
Then $n(n-1)[(n-2)(n-3) - 30] = n(n-1)(n^2 - 5n - 24) = n(n-1)(n-8)(n+3) = 0$.
Since $n \geq 4$, the required solution is $n = 8$.

16.8   (a)   How many permutations can be made of the letters, taken all together, of the "word" MASSESS?
      (b)   In how many ways will the four S's be together? (c) How many will end in SS?

(a)   There are seven letters of which four are S's. The number of permutations is $7!/4! = 210$.

(b)   First, permute the non-S's in $1 \cdot 2 \cdot 3 = 6$ ways and then place the four S's at the ends or between any two letters in each of the six permutations. Thus, there will be $4 \cdot 6 = 24$ permutations.

(c) After filling the last two places with S, we have to fill five places with 5 letters of which 2 are S's. Thus, there are $5!/2! = 60$ permutations.

Supplementary Problems

16.9 In how many different ways can 5 persons be seated on a bench?

   Ans. 120

16.10 In how many ways can the offices of chairman, vice-chairman, secretary, and treasurer be filled from a committee of seven?

   Ans. 840

16.11 How many 3-digit numbers can be formed with the digits $1, 2, \ldots, 9$, if no digit is repeated in any number?

   Ans. 504

16.12 How many 3-digit odd numbers can be formed with the digits $1, 2, 3, \ldots, 9$, if no digit is repeated in any number?

   Ans. 280

16.13 How many 3-digit number > 300 can be formed with the digits $1, 2, 3, 4, 5, 6$, if no digit is repeated in any number?

   Ans. 80

16.14 How many 4-digit numbers > 3000 can be formed with the digits $2, 3, 4, 5$ if repetitions of digits (a) are not allowed, (b) are allowed?

   Ans. (a) 18   (b) 192

16.15 In how many ways can 3 girls and 3 boys be seated in a row if boys and girls alternate?

   Ans. 72

16.16 In how many ways can 2 letters be mailed if 5 letter boxes are available?

   Ans. 25

16.17 Seven-letter words are formed using the letters of the word BLACKER. (a) How many can be formed? (b) How many which end in R? (c) How many in which E immediately follows K? (d) How many do not begin with B? (e) How many in which the vowels are separated by exactly two letters? (f) How many in which the vowels are separated by two or more letters?

   Ans. (a) 5040   (b) 720   (c) 720   (d) 4320   (e) 960   (f) 2400

16.18 Eight books are to be arranged on a shelf. (a) In how many ways can this be done? (b) In how many ways if two of the books are to be placed together? (c) In how many ways if five of the books have red binding and three have blue binding, and the books of the same color are to be kept together? (d) In how many ways if four of the books belong to a numbered set and are to be kept together and in order?

   Ans. (a) 40 320   (b) 10 080   (c) 1440   (d) 120

16.19 How many six-letter words can be formed using the letters of the word ASSIST (*a*) in which the S's alternate with other letters? (*b*) in which the three S's are together? (*c*) which begin and end with S? (*d*) which neither begin nor end with S?

   Ans. (*a*) 12   (*b*) 24   (*c*) 24   (*d*) 24

16.20 (*a*) In how many ways can 8 persons be seated about a round table? (*b*) With 8 beads of different colors, how many bracelets can be formed by stringing them all together?

   Ans. (*a*) 5040   (*b*) 2520

16.21 How many signals can be made with 3 white, 3 green, and 2 blue flags by arranging them on a mast (*a*) all at a time? (*b*) three at a time? (*c*) five at a time?

   Ans. (*a*) 560   (*b*) 26   (*c*) 170

16.22 A car will hold 2 in the front seat and 1 in the rear seat. If among 6 persons only 2 can drive, in how many ways can the car be filled?

   Ans. 40

16.23 A chorus consists of 6 boys and 6 girls. How many arrangements of them can be made (*a*) in a row, facing front, the boys and girls alternating? (*b*) in two rows, facing front, with a boy behind each girl? (*c*) in a ring with the boys facing the center and the girls facing away from the center? (*d*) in two concentric rings, both facing the center, with a boy behind each girl?

   Ans. (*a*) 1 036 800   (*b*) 518 400   (*c*) 39 916 800   (*d*) 86 400

16.24 (*a*) In how many ways can 10 boys take positions in a straight line if two particular boys must not stand side by side? (*b*) In how many ways can 10 boys take positions about a round table if two particular boys must not be seated side by side?

   Ans. (*a*) $8 \cdot 9!$ (*b*) $7 \cdot 8!$

16.25 A man has 5 large books, 7 medium-sized books, and 3 small books. In how many different ways can they be arranged on a shelf if all books of the same size are to be kept together?

   Ans. 21 772 800

16.26 (*a*) How many words can be made from the letters of the word MASSACHUSETTS taken all together? (*b*) Of the words in (*a*), how many begin and end with SS? (*c*) Of the words in (*a*), how many begin and end with S? (*d*) Show that there are as many words having H as middle letter as there are circular permutations, using all letters.

   Ans. (*a*) 64 864 800   (*b*) 90 720   (*c*) 4 989 600

Chapter 17

Combinations

THE COMBINATIONS of n objects taken r at a time consist of all possible sets of r of the objects, without regard to the order of arrangement. The number of combinations of n objects taken r at a time will be denoted by $_nC_r$.

For example, the combinations of the $n = 4$ letters a, b, c, d taken $r = 3$ at a time, are

$$abc, \qquad abd, \qquad acd, \qquad bcd$$

Thus, $_4C_3 = 4$. When the letters of each combination are rearranged (in 3! ways), we obtain the $_4P_3$ permutations of the 4 letters taken 3 at a time. Hence, $_4P_3 = 3!(_4C_3)$ and $_4C_3 = {_4P_3}/3!$. See Chapter 16 for a discussion of permutations.

The number of combinations of n different objects taken r at a time is equal to the number of permutations of the n objects taken r at a time divided by factorial r, or

$$_nC_r = \frac{_nP_r}{r!} = \frac{n(n-1)\cdots(n-r+1)}{1\cdot 2 \cdots r}$$

(For a proof, see Problem 17.1.)

EXAMPLE. From a shelf containing 12 different toys, a child is permitted to select 3. In how many ways can this be done?

The required number is

$$_{12}C_3 = \frac{_{12}P_3}{3!} = \frac{12\cdot 11 \cdot 10}{1\cdot 2\cdot 3} = 220.$$

Notice that $_nC_r$ is the rth term's coefficient in the binomial theorem. See Chapter 15.

Solved Problems

17.1 Derive the formula $_nC_r = \dfrac{_nP_r}{r!}$.

From each of the $_nC_r$ combinations of n objects taken r at a time, $r!$ permutations can be formed. Since two combinations differ at least in one element, the $_nC_r \cdot r!$ permutations thus formed are precisely the number of permutations $_nP_r$ of n objects taken r at a time. Thus,

$$_nC_r \cdot r! = {_nP_r} \qquad \text{and} \qquad _nC_r = \frac{_nP_r}{r!}$$

17.2 Show (a) $_nC_r = \dfrac{n!}{r!(n-r)!}$, (b) $_nC_r = {}_nC_{n-r}$.

(a) $_nC_r = \dfrac{n(n-1)\cdots(n-r+1)}{1\cdot 2\cdots r} = \dfrac{n(n-1)\cdots(n-r+1)}{1\cdot 2\cdots r}\cdot\dfrac{(n-r)\cdots 2\cdot 1}{1\cdot 2\cdots(n-r)} = \dfrac{n!}{r!(n-r)!}$

(b) $_nC_r = \dfrac{n!}{r!(n-r)!} = \dfrac{n(n-1)\cdots(r+1)r(r-1)\cdots 2\cdot 1}{1\cdot 2\cdots r\cdot(n-r)!} = \dfrac{n(n-1)\cdots(r+1)}{(n-r)!} = \dfrac{{}_nP_{n-r}}{(n-r)!} = {}_nC_{n-r}$

17.3 Compute (a) $_{10}C_2$, (b) $_{12}C_5$, (c) $_{15}C_{12}$, (d) $_{25}C_{21}$.

(a) $_{10}C_2 = \dfrac{10\cdot 9}{1\cdot 2} = 45$

(c) $_{15}C_{12} = {}_{15}C_3 = \dfrac{15\cdot 14\cdot 13}{1\cdot 2\cdot 3} = 455$

(b) $_{12}C_5 = \dfrac{12\cdot 11\cdot 10\cdot 9\cdot 8}{1\cdot 2\cdot 3\cdot 4\cdot 5} = 792$

(d) $_{25}C_{21} = {}_{25}C_4 = \dfrac{25\cdot 24\cdot 23\cdot 22}{1\cdot 2\cdot 3\cdot 4} = 12\,650$

17.4 A lady gives a dinner party for six guests. (a) In how many ways may they be selected from among 10 friends? (b) In how many ways if two of the friends will not attend the party together?

(a) The six guests may be selected in $_{10}C_6 = {}_{10}C_4 = \dfrac{10\cdot 9\cdot 8\cdot 7}{1\cdot 2\cdot 3\cdot 4} = 210$ ways.

(b) Let A and B denote the two who will not attend together. If neither A nor B is included, the guests may be selected in $_8C_6 = {}_8C_2 = \dfrac{8\cdot 7}{1\cdot 2} = 28$ ways. If one of A and B is included, the guests may be selected in $2\cdot {}_8C_5 = 2\cdot {}_8C_3 = 2\left(\dfrac{8\cdot 7\cdot 6}{1\cdot 2\cdot 3}\right) = 112$ ways. Thus, the six guests may be selected in $28 + 112 = 140$ ways.

17.5 A committee of 5 is to be selected from 12 seniors and 8 juniors. In how many ways can this be done (a) if the committee is to consist of 3 seniors and 2 juniors, (b) if the committee is to contain at least 3 seniors and 1 junior?

(a) With each of the $_{12}C_3$ selections of 3 seniors, we may associate any one of the $_8C_2$ selections of 2 juniors. Thus, a committee can be selected in $_{12}C_3 \cdot {}_8C_2 = \dfrac{12\cdot 11\cdot 10}{1\cdot 2\cdot 3}\cdot\dfrac{8\cdot 7}{1\cdot 2} = 6160$ ways.

(b) The committee may consist of 3 seniors and 2 juniors or of 4 seniors and 1 junior. A committee of 3 seniors and 2 juniors can be selected in 6160 ways, and a committee of 4 seniors and 1 junior can be selected in $_{12}C_4 \cdot {}_8C_1 = 3960$ ways. In all, a committee may be selected in $6160 + 3960 = 10\,120$ ways.

17.6 There are ten points $A, B, \ldots,$ in a plane, no three on the same straight line. (a) How many lines are determined by the points? (b) How many of the lines pass through A? (c) How many triangles are determined by the points? (d) How many of the triangles have A as a vertex? (e) How many of the triangles have AB as a side?

(a) Since any two points determine a line, there are $_{10}C_2 = 45$ lines.

(b) To determine a line through A, one other point must be selected. Thus, there are nine lines through A.

(c) Since any three of the points determine a triangle, there are $_{10}C_3 = 120$ triangles.

(d) Two additional points are needed to form a triangle. These points may be selected from the nine points in $_9C_2 = 36$ ways.

(e) One additional point is needed; there are eight triangles having AB as a side.

17.7 In how many ways may 12 persons be divided into three groups (a) of 2, 4, and 6 persons, (b) of 4 persons each?

(a) The groups of two can be selected in $_{12}C_2$ ways, then the group of four in $_{10}C_4$ ways, and the group of six in $_6C_6 = 1$ way. Thus, the division may be made in $_{12}C_2 \cdot {}_{10}C_4 \cdot 1 = 13\,860$ ways.

(b) One group of four can be selected in $_{12}C_4$ ways, then another in $_8C_4$ ways, and the third in 1 way. Since the order in which the groups are formed is now immaterial, the division may be made in $_{12}C_4 \cdot _8C_4 \cdot 1 \div 3! = 5775$ ways.

17.8 The English alphabet consists of 21 consonants and 5 vowels.

(a) In how many ways can 4 consonants and 2 vowels be selected?
(b) How many words consisting of 4 consonants and 2 vowels can be formed?
(c) How many of the words in (b) begin with R?
(d) How many of the words in (c) contain E?

(a) The 4 consonants can be selected in $_{21}C_4$ ways and the 2 vowels can be selected in $_5C_2$ ways. Thus, the selections may be made in $_{21}C_4 \cdot _5C_2 = 59\,850$ ways.

(b) From each of the selections in (a), 6! words may be formed by permuting the letters. Therefore, $59850 \cdot 6! = 43\,092\,000$ words can be formed.

(c) Since the position of the consonant R is fixed, we must select 3 other consonants (in $_{20}C_3$ ways) and 2 vowels (in $_5C_2$ ways), and arrange each selection of 5 letters in all possible ways. Thus, there are $_{20}C_3 \cdot _5C_2 \cdot 5! = 1\,368\,000$ words.

(d) Since the position of the consonant R is fixed but the position of the vowel E is not, we must select 3 other consonants (in $_{20}C_3$ ways) and 1 other vowel (in four ways), and arrange each set of 5 letters in all possible ways. Thus, there are $_{20}C_3 \cdot 4 \cdot 5! = 547\,200$ words.

17.9 From an ordinary deck of playing cards, in how many different ways can five cards be dealt (a) consisting of spades only, (b) consisting of black cards only, (c) containing the four aces, (d) consisting of three cards of one suit and two of another, (e) consisting of three kings and a pair, (f) consisting of three of one kind and two of another?

(a) From the 13 spades, 5 can be selected in $_{13}C_5 = 1287$ ways.

(b) From the 26 black cards, 5 can be selected in $_{26}C_5 = 65\,780$ ways.

(c) One card must be selected from the 48 remaining cards. This can be done in 48 different ways.

(d) A suit can be selected in four ways and three cards from the suit can be selected in $_{13}C_3$ ways; a second suit can now be selected in three ways and two cards of this suit in $_{13}C_2$ ways. Thus, three cards of one suit and two of another can be selected in $4 \cdot _{13}C_3 \cdot 3 \cdot _{13}C_2 = 267\,696$ ways.

(e) Three kings can be selected from the four kings in $_4C_3$ ways, another kind can be selected in 12 ways, and two cards of this kind can be selected in $_4C_2$ ways. Thus, three kings and another pair can be dealt in $_4C_3 \cdot 12 \cdot _4C_2 = 288$ ways.

(f) A kind can be selected in 13 ways and three of this kind in $_4C_3$ ways; another kind can be selected in 12 ways and two of this kind in $_4C_2$ ways. Thus, 3 of one kind and 2 of another can be dealt in $13 \cdot _4C_3 \cdot 12 \cdot _4C_2 = 3744$ ways.

17.10 (a) Prove: The total number of combinations of n objects taken successively $1, 2, 3, \ldots, n$ at a time is $2^n - 1$.
(b) In how many different ways can one invite one or more of five friends to the movies.

(a) The total number of combinations is

$$_nC_1 + _nC_2 + _nC_3 + \cdots + _nC_n = 2^n - 1$$

since, from Problem 42.8(a),

$$_nC_0 + _nC_1 + _nC_2 + \cdots + _nC_n = 2^n$$

(b) The number is $_5C_1 + _5C_2 + _5C_3 + _5C_4 + _5C_5 = 2^5 - 1 = 31$.

Supplementary Problems

17.11   Evaluate   (a) $_6C_2$ (b) $_8C_6$ (c) $_nC_3$.

   Ans.   (a) 15   (b) 28   (c) $\dfrac{n(n-1)(n-2)}{1 \cdot 2 \cdot 3}$

17.12   Find n if (a) $_nC_2 = 55$,   (b) $_nC_3 = 84$,   (c) $_{2n}C_3 = 11 \cdot {_nC_3}$.

   Ans.   (a) 11   (b) 9   (c) 6

17.13   Two dice can be tossed in 36 ways. In how many of these is the sum equal to (a) 4, (b) 7, (c) 11?

   Ans.   (a) 3   (b) 6   (c) 2

17.14   Four delegates are to be chosen from eight members of a club. (a) How many choices are possible? (b) How many contain member A? (c) How many contain A or B but not both?

   Ans.   (a) 70   (b) 35   (c) 40

17.15   A party of 8 boys and 8 girls are going on a picnic. Six of the party go in one automobile, four go in another, and the rest walk. (a) In how many ways can the party be distributed for the trip? (b) In how many ways if no girl walks?

   Ans.   (a) 1 681 680   (b) 5880

17.16   Solve Problem 17.15 if the owner of each car (a boy) drives his own car.

   Ans.   (a) 168 168   (b) 56

17.17   How many selection of five letters each can be made from the letters of the word CANADIANS?

   Ans.   41

17.18   A bag contains nine balls numbered $1, 2, \ldots, 9$. In how many ways can two balls be drawn so that (a) both are odd? (b) their sum is odd?

   Ans.   (a) 10   (b) 20

17.19   How many diagonals has (a) a hexagon, (b) an octagon, (c) an n-gon?

   Ans.   (a) 9   (b) 20   (c) $\frac{1}{2}n(n-3)$

17.20   (a) How many words consisting of 3 consonants and 2 vowels can be formed from 10 consonants and 5 vowels? (b) In how many of these will the consonants occupy the odd places?

   Ans.   (a) 144 000   (b) 14 400

17.21 Three balls are drawn from a bag containing five red, four white, and three black balls. In how many ways can this can be done if (*a*) each is of a different color? (*b*) they are of the same color? (*c*) exacly two are red? (*d*) at least two are red?

   Ans. (*a*) 60   (*b*) 15   (*c*) 70   (*d*) 80

17.22 A squad is made up of 10 privates and 5 privates first class. (*a*) In how many ways can a detail of 4 privates and 2 privates first class be formed? (*b*) On how many of these details will private X serve? (*c*) On how many will private X but not private first class Y serve?

   Ans. (*a*) 2100   (*b*) 840   (*c*) 504

17.23 A civic club has 60 members including 2 bankers, 4 lawyers, and 5 doctors. In how many ways can a committee of 10 be formed to contain 1 banker, 2 lawyers, and 2 doctors?

   Ans.   228 826 080

17.24 How many committees of two or more can be selected from 10 people?

   Ans. $2^{10} - 11$

17.25 Hands consisting of three cards are dealt from an ordinary deck. Show that a hand consisting of three different kinds should show 352 times as often as a hand consisting of three cards of the same kind.

17.26 Prove (*a*) $_nC_r + _nC_{r+1} = _{n+1}C_{r+1}$ (*b*) $_{2n}C_n = 2 \cdot _{2n-1}C_{n-1}$.

17.27 Prove that $_nC_n = 1$.

17.28 Prove that $_{n+1}C_n = 2(n+1)$.

17.29 Derive a formula for $_{n+2}C_n$.

Chapter 18

Probability

IN ESTIMATING THE PROBABILITY that a given event will or will not happen, we may, as in the case of drawing a face card from an ordinary deck, count the different ways in which this event may or may not happen. On the other hand, in the case of estimating the probability that a person who is now 25 years old will live to receive a bequest at age 30, we are forced to depend upon such knowledge of what has happened on similar occasions in the past as is available. In the first case, the result in called *mathematical* or *theoretical probability*; in the latter case, the result is called *statistical* or *empirical probability*.

MATHEMATICAL PROBABILITY. If an event must result in some one of n, $(n \neq 0)$ different but *equally likely ways* and if a certain s of these ways are considered successes and the other $f = n - s$ ways are considered failures, then the probability of success in a given trial is $p = s/n$ and the probability of failure is $q = f/n$. Since $p + q = \dfrac{s+f}{n} = \dfrac{n}{n} = 1$, $p = 1 - q$ and $q = 1 - p$.

EXAMPLE 1. One card is drawn from an ordinary deck. What is the probability (a) that it is a red card, (b) that it is a spade, (c) that it is a king, (d) that it is not the ace of hearts?

One card can be drawn from the desk in $n = 52$ different ways.

(a) A red card can be drawn from the deck in $s = 26$ different ways. Thus, the probability of drawing a red card is $s/n = \frac{26}{52} = \frac{1}{2}$.
(b) A spade can be drawn from the deck in 13 different ways. The probability of drawing a spade is $\frac{13}{52} = \frac{1}{4}$.
(c) A king can be drawn in 4 ways. The required probability is $\frac{4}{52} = \frac{1}{13}$.
(d) The ace of hearts can be drawn in 1 way; the probability of drawing the ace of hearts is $\frac{1}{52}$. Thus, the probability of *not* drawing the ace of hearts is $1 - \frac{1}{52} = \frac{51}{52}$.

(See Problems 18.1–18.4.)

[NOTE: We write $P(A)$ to denote the "probability that A occurs."]

Two or more events are called *mutually exclusive* if not more than one of them can occur in a single trial. Thus, the drawing of a jack and the drawing of a queen on a single draw from an ordinary deck are mutually exclusive events; however, the drawing of a jack and the drawing of a spade are not mutually exclusive.

EXAMPLE 2. Find the probability of drawing a jack or a queen from an ordinary deck of cards.

Since there are four jacks and four queens, $s = 8$ and $p = \frac{8}{52} = \frac{2}{13}$. Now the probability of drawing a jack is $\frac{1}{13}$, the probability of drawing a queen is $\frac{1}{13}$, and the required probability is $\frac{2}{13} = \frac{1}{13} + \frac{1}{13}$.

We have verified

THEOREM A.   The probability that some one of a set of mutually exclusive events will happen at a single trial is the sum of their separate probabilities of happening.

THEOREM A'.   The probability that A will occur or B will occur, $P(A \cup B) = P(A) + P(B) - P(A \cap B)$. (NOTE: "$\cup$" refers to set union and "$\cap$" to set intersection.)

$A \cup B$ is the set of all elements belonging either to A or to B or to both

$$= \{x | x \text{ is in } A \text{ or } B \text{ or both}\}.$$

$A \cap B$ is the set of all elements common to A and B

$$= \{x | x \text{ is in } A \text{ and } x \text{ is in } B\}.$$

(See Fig. 18-1.)

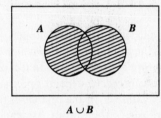
$A \cup B$

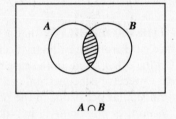
$A \cap B$

Fig. 18-1

For example, if $A = \{1, 2, 3\}$ and $B = \{2, 3, 7\}$, then $A \cup B = \{1, 2, 3, 7\}$ and $A \cap B = \{2, 3\}$.

Two events A and B are called *independent* if the happening of one does not affect the happening of the other. Thus, in a toss of two dice, the fall of either does not affect the fall of the other. However, in drawing two cards from a deck, the probability of obtaining a red card on the second draw depends upon whether or not a red card was obtained on the draw of the first card. Two such events are called *dependent*. More explicitly, if $P(A \cap B) = P(A) \cdot P(B)$, then A and B are independent.

EXAMPLE 3.   One bag contains 4 white and 4 black balls, a second bag contains 3 white and 6 black balls, and a third contains 1 white and 5 black balls. If one ball is drawn from each bag, find the probability that all are white.
   A ball can be drawn from the first bag in any one of 8 ways, from the second in any one of 9 ways, and from the third in any one of 6 ways; hence, three balls can be drawn from each bag in $8 \cdot 9 \cdot 6$ ways. A white ball can be drawn from the first bag in 4 ways, from the second in 3 ways, and from the third in 1 way; hence, three white balls can be drawn one from each bag in $4 \cdot 3 \cdot 1$ ways. Thus the required probability is

$$\frac{4 \cdot 3 \cdot 1}{8 \cdot 9 \cdot 6} = \frac{1}{36}$$

Now drawing a white ball from one bag does not affect the drawing of a white ball from another, so that here we are concerned with three independent events. The probability of drawing a white ball from the first bag is $\frac{4}{8}$, from the second is $\frac{1}{9}$, and from the third bag is $\frac{1}{6}$.

Since the probability of drawing three white balls, one from each bag, is $\frac{4}{8} \cdot \frac{3}{9} \cdot \frac{1}{6}$, we have verified

THEOREM B.   The probability that all of a set of independent events will happen in a single trial is the product of their separate probabilities.

THEOREM C.   (concerning dependent events). If the probability that an event will happen is p_1, and if after it has happened the probability that a second event will happen is p_2, the probability that the two events will happen in that order is $p_1 p_2$.

EXAMPLE 4.   Two cards are drawn from an ordinary deck. Find the probability that both are face cards (king, queen, jack) if (*a*) the first card drawn is replaced before the second is drawn, (*b*) the first card drawn is not replaced before the second is drawn.

(*a*)   Since each drawing is made from a complete deck, we have the case of two independent events. The probability of drawing a face card in a single draw is $\frac{12}{52}$; thus the probability of drawing two face cards, under the conditions imposed, is $\frac{12}{52} \cdot \frac{12}{52} = \frac{9}{169}$.

(*b*)   Here the two events are dependent. The probability that the first drawing results in a face card is $\frac{12}{52}$. Now, of the 51 cards remaining in the deck, there are 11 face cards; the probability that the second drawing results in a face card is $\frac{11}{51}$. Hence, the probability of drawing two face cards is $\frac{12}{52} \cdot \frac{11}{51} = \frac{11}{221}$.

(See Problems 18.5–18.10.)

EXAMPLE 5.   Two dice are tossed six times. Find the probability (*a*) that 7 will show on the first four tosses and will not show on the other two, (*b*) that 7 will show on exactly four of the tosses.

The probability that 7 will show on a single toss is $p = \frac{1}{6}$ and the probability that 7 will not show is $q = 1 - p = \frac{5}{6}$.

(*a*)   The probability that 7 will show on the first four tosses and will not show on the other two is

$$\frac{1}{6} \cdot \frac{1}{6} \cdot \frac{1}{6} \cdot \frac{1}{6} \cdot \frac{5}{6} \cdot \frac{5}{6} = \frac{25}{46\,656}$$

(*b*)   The four tosses on which 7 is to show may be selected in $_6C_4 = 15$ ways. Since these 15 ways constitue mutually exclusive events and the probability of any one of them is $\left(\frac{1}{6}\right)^4 \left(\frac{5}{6}\right)^2$, the probability that 7 will show exactly four times in six tosses is $_6C_4 \left(\frac{1}{6}\right)^4 \left(\frac{5}{6}\right)^2 = \frac{125}{15\,552}$.

We have verified

THEOREM D.   If p is the probability that an event will happen and q is the probability that it will fail to happen at a given trial, the probability that it will happen exactly r times in n trials is $_nC_r p^r q^{n-r}$. (See Problems 18.11–18.12.)

EMPIRICAL PROBABILITY. If an event has been observed to happen s times in n trials, the ratio $p = s/n$ is defined as the *empirical probability* that the event will happen at any future trial. The confidence which can be placed in such probabilities depends in a large measure on the number of observations used. Life insurance companies, for example, base their preminum rate on empirical probabilities. For this purpose they use a mortality table based on an enormous number of observations over the years.

The American Experience Table of Mortality begins with 100 000 persons all of age 10 years and indicates the number of the group who die each year thereafter. In using this table, it will be assumed that the laws stated above for mathematical probability hold also for empirical probability.

EXAMPLE 6.   Find the probability that a person 20 years old (*a*) will die during the year, (*b*) will die during the next 10 years, (*c*) will reach age 75.

(*a*)   Of the 100 000 persons alive at age 10 years, 92 637 are alive at age 20 years. Of these 92 637 a total of 723 will die during the year. The probability that a person 20 years of age will die during the year is $\frac{723}{92\,637} = 0.0078$.

(b) Of the 92 637 who reach age 20 years, 85 441 reach age 30 years; thus, $92\,637 - 85\,441 = 7196$ die during the 10-year period. The required probability is $\dfrac{7196}{92\,637} = 0.0777$.

(c) Of the 92 637 alive at age 20 years, 26 237 will reach age 75 years. The required probability is $\dfrac{26\,237}{92\,637} = 0.2832$.

(See Problem 18.13.)

Solved Problems

18.1 One ball is drawn from a bag containing 3 white, 4 red, and 5 black balls. What is the probability (a) that it is white, (b) that it is white or red, (c) that it is not red?

   A ball can be drawn from the bag in $n = 3 + 4 + 5 = 12$ different ways.

(a) A white ball can be drawn in $s = 3$ different ways. Thus, the probability of drawing a white ball is $p = s/n = \frac{3}{12} = \frac{1}{4}$.

(b) Here success consists in drawing either a white or a red ball; hence, $s = 3 + 4 = 7$ and the required probability is $\frac{7}{12}$.

(c) The probability of drawing a red ball is $p = \frac{4}{12} = \frac{1}{3}$. The probability that the ball drawn is *not* red is $1 - p = 1 - \frac{1}{3} = \frac{2}{3}$. This problem may also be solved as in (b).

18.2 If two dice are tossed, what is the probability (a) of throwing a total of 7, (b) of throwing a total of 8, (c) of throwing a total of 10 or more, (d) of both dice showing the same number?

   Two dice may turn up in $6 \times 6 = 36$ ways.

(a) A total of 7 may result in 6 ways. (1, 6; 6, 1; 2, 5; 5, 2; 3, 4; 4, 3). The probability of a throw of 7 is then $\frac{6}{36} = \frac{1}{6}$.

(b) A total of 8 may result in 5 ways (2, 6; 6, 2; 3, 5; 5, 3; 4, 4). The probability of a throw of 8 is $\frac{5}{36}$.

(c) Here, success consists in throwing a total of 10, 11, or 12. Since a total of 10 may result in 3 ways, a total of 11 in 2 ways, and a total of 12 in 1 way, the probability is $(3 + 2 + 1)/36 = \frac{1}{6}$.

(d) The probability that the second die will show the same number as the first is $\frac{1}{6}$. This problem may also be solved by counting the number of successes 1, 1; 2, 2; etc.

18.3 If five coins are tossed, what is the probability (a) that all will show heads, (b) that exactly three will show heads, (c) that at least three will show heads?

   Each coin can turn up in 2 ways; hence, the five coins can turn up in $2^5 = 32$ ways. (The assumption here is that HHHTT and THHHT are different results.)

(a) Five heads can turn up in only 1 way; hence, the probability of a toss of five heads is $\frac{1}{32}$.

(b) Exactly three heads can turn up in $_5C_3 = 10$ ways; thus, the probability of a toss of exactly three heads is $\frac{10}{32} = \frac{5}{16}$.

(c) Success here consists of a throw of exactly three, exactly four, or all heads. Exactly three heads can turn up in 10 ways, exactly four heads in 5 ways, and all heads in 1 way. Thus, the probability of throwing at least three heads is $(10 + 5 + 1)/32 = \frac{1}{2}$.

18.4   If three cards are drawn from an ordinary deck, find the probability (*a*) that all are red cards, (*b*) that all are of the same suit, (*c*) that all are aces.

Three cards may be drawn from a deck in $_{52}C_3$ ways.

(*a*)   Three red cards may be drawn from the 26 red cards in $_{26}C_3$ ways. Hence, the probability of drawing three red cards in a single draw is $\dfrac{_{26}C_3}{_{52}C_3} = \dfrac{26 \cdot 25 \cdot 24}{1 \cdot 2 \cdot 3} \cdot \dfrac{1 \cdot 2 \cdot 3}{52 \cdot 51 \cdot 50} = \dfrac{2}{17}$.

(*b*)   There are 4 ways of choosing a suit and $_{13}C_3$ ways of selecting three cards of that suit. Thus, the probability of drawing three cards of the same suit is $4 \cdot _{13}C_3 / _{52}C_3 = \frac{22}{425}$.
     This problem may also be solved as follows: The first card drawn determines a suit. The deck now contains 51 cards of which 12 are of that suit; hence, the probability that the next two cards drawn will be of that suit is $_{12}C_2 / _{51}C_2 = \frac{22}{425}$, as before.

(*c*)   Three aces may be selected from the four aces in 4 ways; hence, the required probability is $\dfrac{4}{_{52}C_3} = \dfrac{1}{5525}$.

18.5   One bag contains 8 black balls and a second bag contains 1 white and 6 black balls. One of the bags is selected and then a ball is drawn from that bag. What is the probability that it is the white ball?

The probability that the second bag is chosen is $\frac{1}{2}$ and the probability that the white ball is drawn from this bag is $\frac{1}{7}$. Thus, the required probability is $\frac{1}{2}(\frac{1}{7}) = \frac{1}{14}$.

18.6   Two cards are drawn in succession from an ordinary deck. What is the probability (*a*) that the first will be the jack of diamonds and the second will be the queen of spades, (*b*) that the first will be a diamond and the second a spade, (*c*) that both cards are diamonds or both are spades?

(*a*)   The probability that the first card is the jack of diamonds is $\frac{1}{52}$ and the probability that the second is the queen of spades is $\frac{1}{51}$. The required probability is $(\frac{1}{52})(\frac{1}{51}) = \frac{1}{2652}$.

(*b*)   The probability that the first card is a diamond is $\frac{13}{52}$ and the probability that the second card is a spade is $\frac{13}{51}$. The probability of the required draw is $\frac{13}{52} \cdot \frac{13}{51} = \frac{13}{204}$.

(*c*)   The probability that both cards are of a specified suit is $\frac{13}{52} \cdot \frac{12}{51}$. Thus, the probability that both are diamonds or both are spades is $\frac{13}{52} \cdot \frac{12}{51} + \frac{13}{52} \cdot \frac{12}{51} = \frac{2}{17}$.

18.7   *A*, *B*, and *C* work independently on a problem. If the respective probabilities that they will solve it are $\frac{1}{2}, \frac{1}{3}, \frac{2}{5}$, find the probability that the problem will be solved.

The problem will be solved unless all three fail; the probability that this will happen is $\frac{1}{2} \cdot \frac{2}{3} \cdot \frac{3}{5} = \frac{1}{5}$. Thus, the probability that the problem will be solved is $1 - \frac{1}{5} = \frac{4}{5}$.

18.8   *A* tosses a coin and if a head appears he wins the game; if a tail appears, *B* tosses the coin under the same conditions, and so on. If the stakes are $15, find the expectation of each.

We first compute the probability that *A* will win. The probability that he will win on the first toss is $\frac{1}{2}$; the probability that he will win on his second toss (that is, that *A* first tosses a tail, *B* tosses a tail, and *A* then tosses a head) is $\frac{1}{2} \cdot \frac{1}{2} \cdot \frac{1}{2} = \left(\frac{1}{2}\right)^3$; the probability that he will win on his third toss (that is, that *A* first tosses a tail, *B* tosses a tail, *A* tosses a tail, *B* tosses a tail, and *A* then tosses a head) is $\frac{1}{2} \cdot \frac{1}{2} \cdot \frac{1}{2} \cdot \frac{1}{2} \cdot \frac{1}{2} = \left(\frac{1}{2}\right)^5$, and so on. Thus, the probability that *A* will win is

$$\frac{1}{2} + \frac{1}{2^3} + \frac{1}{2^5} + \cdots = \frac{\frac{1}{2}}{1 - 1/2^2} = \frac{2}{3}$$

and his expectation is $\frac{2}{3}(\$15) = \$10$. Then *B*'s expectation is $5.

18.9 On a toss of two dice, X throws a total of 5. Find the probability that he will throw another 5 before he throws 7.

X will succeed should he throw a total of 5 on the next toss, or should he not throw 5 or 7 on this toss but throw 5 on the next, or should he not throw 5 or 7 on either of these tosses but throw 5 on the next, and so on. The respective probabilities are $\frac{4}{36}, \frac{26}{36} \cdot \frac{4}{36}, \frac{26}{36} \cdot \frac{26}{36} \cdot \frac{4}{36}$, and so on. Thus, the probability that he throws 5 before 7 is

$$\frac{4}{36} + \frac{26}{36} \cdot \frac{4}{36} + \frac{26}{36} \cdot \frac{26}{36} \cdot \frac{4}{36} + \cdots = \frac{\frac{4}{36}}{1 - \frac{26}{36}} = \frac{2}{5}$$

18.10 A bag contains 2 white and 3 black balls. A ball is drawn 5 times, each being replaced before another is drawn. Find the probability that (a) the first 4 balls drawn are white and the last is black, (b) exactly 4 of the balls drawn are white, (c) at least 4 of the balls drawn are white, (d) at least 1 ball is white.

The probability of drawing a white ball is $p = \frac{2}{5}$ and the probability of drawing a black ball is $\frac{3}{5}$.

(a) The probability that the first 4 are white and the last black is $\frac{2}{5} \cdot \frac{2}{5} \cdot \frac{2}{5} \cdot \frac{2}{5} \cdot \frac{3}{5} = \frac{48}{3125}$.

(b) Here $n = 5, r = 4$; the probability that exactly 4 of the balls drawn are white is

$$nC_r p^r q^{n-r} = {}_5C_4\left(\tfrac{2}{5}\right)^4\left(\tfrac{3}{5}\right) = \frac{48}{625}$$

(c) Since success consists of drawing either 4 white and 1 black ball or 5 white balls, the probability is

$$_5C_4\left(\tfrac{2}{5}\right)^4\left(\tfrac{3}{5}\right) + {}_5C_5\left(\tfrac{2}{5}\right)^5 = \frac{272}{3125}$$

(d) Here failure consists of drawing 5 black balls. Since the probability of failure is $\left(\tfrac{3}{5}\right)^5 = \frac{243}{3125}$, the probability of success is $1 - \frac{243}{3125} = \frac{2882}{3125}$. The problem may also be solved as in (c).

18.11 One bag contains 2 white balls and 2 black balls, and another contains 3 white balls and 5 black balls. At five different trials, a bag is chosen at random and 1 ball is drawn from that bag and replaced. Find the probability (a) that exactly 3 white balls are drawn, (b) that at least 3 white balls are drawn.

At any trial the probability that a white ball is drawn from the first bag is $\frac{1}{2} \cdot \frac{1}{2}$ and the probability that a white ball is drawn from the second bag is $\frac{1}{2}\left(\frac{3}{8}\right)$. Thus, the probability that a white ball is drawn at any trial is $p = \frac{1}{2} \cdot \frac{1}{2} + \frac{1}{2}\left(\frac{3}{8}\right) = \frac{7}{16}$, and the probability that a black ball is drawn is $q = \frac{9}{16}$.

(a) The probability of drawing exactly 3 white balls in 5 trials is $_5C_3\left(\dfrac{7}{16}\right)^3\left(\dfrac{9}{16}\right)^2 = \dfrac{138\,915}{524\,288}$.

(b) The probability of drawing at least 3 white balls in 5 trials is

$$_5C_3\left(\frac{7}{16}\right)^3\left(\frac{9}{16}\right)^2 + {}_5C_4\left(\frac{7}{16}\right)^4\left(\frac{9}{16}\right) + {}_5C_5\left(\frac{7}{16}\right)^5 = \frac{201\,341}{524\,288}$$

18.12 A husband is 35 years old and his wife is 28. Find the probability that at the end of 20 years (a) the husband will be alive, (b) the wife will be alive, (c) both will be alive, (d) both will be dead, (e) the wife will be alive and the husband will not, (f) one will be alive but not the other.

(a) Of the 81 822 alive at age 35, 64 563 will reach age 55. The probability that the husband will be alive at the end of 20 years is $\dfrac{64\,563}{81\,822} = 0.7890$.

(b) The probability that the wife will be alive at the end of 20 years is $\dfrac{71\,627}{86\,878} = 0.8245$.

(c) Since the survival of the husband and of the wife are independent events, the probability that both are alive after 20 years is $\dfrac{64\,563}{81\,822} \cdot \dfrac{71\,627}{86\,878} = 0.6506$.

(d) From (a) 17 259 of the 81 822 alive at age 35 will not reach age 55; thus, the probability that the husband will not live for 20 years is $\dfrac{17\,259}{81\,822}$. Similarly, the probability that the wife will not live for 20 years is $\dfrac{15\,251}{86\,878}$. Hence, the probability that after 20 years both will be dead is $\dfrac{17\,259}{81\,822} \cdot \dfrac{15\,251}{86\,878} = 0.0370$.

(e) The probability that the husband will be dead and the wife will be alive after 20 years is $\dfrac{17\,259}{81\,822} \cdot \dfrac{71\,627}{86\,878} = 0.1739$.

(f) The probability that the wife will survive but the husband will not is found in (e). The probability that the husband will survive but the wife will not is $\dfrac{64\,563}{81\,822} \cdot \dfrac{15\,251}{86\,878}$. Thus, the probability that just one will survive is $\dfrac{17\,259}{81\,822} \cdot \dfrac{71\,627}{86\,878} + \dfrac{64\,563}{81\,822} \cdot \dfrac{15\,251}{86\,878} = 0.3116$.

Supplementary Problems

18.13 One ball is drawn from a bag containing 4 white and 6 black balls. Find the probability that it is (a) white, (b) black.

Ans. (a) $\frac{2}{5}$ (b) $\frac{3}{5}$

18.14 Three balls are drawn together from a bag containing 8 white and 12 black balls. Find the probability that (a) all are white, (b) just two are white, (c) just one is white, (d) all are black.

Ans. (a) $\frac{14}{285}$ (b) $\frac{28}{95}$ (c) $\frac{44}{95}$ (d) $\frac{11}{57}$

18.15 Ten students are seated at random in a row. Find the probability that two particular students are not seated side by side.

Ans. $\frac{4}{5}$

18.16 If a die is cast three times, find the probability (a) that an even number will be thrown each time, (b) that an odd number will appear just once, (c) that the sum of the three numbers will be even.

Ans. (a) $\frac{1}{8}$ (b) $\frac{3}{8}$ (c) $\frac{1}{2}$

18.17 From a box containing 10 cards numbered 1, 2, 3, ..., 10, four cards are drawn. Find the probability that their sum will be even (a) if the cards are drawn together, (b) if each card drawn is replaced before the next is drawn.

Ans. (a) $\frac{11}{21}$ (b) $\frac{1}{2}$

18.18 A and B, having equal skill, are playing a game of three points. After A has won 2 points and B has won 1 point, what is the probability that A will win the game?

Ans. $\frac{3}{4}$

18.19 One bag contains 3 white and 2 black balls, and another contains 2 white and 3 black balls. A ball is drawn from the second bag and placed in the first; then a ball is drawn from the first bag and placed in the second. When the pair of operations is repeated, what is the probability that the first bag will contain 5 white balls?

   Ans. $\frac{1}{225}$

18.20 Three bags contain respectively 2 white and 1 back ball, 3 white and 3 black balls, 6 white and 2 black balls. Two bags are selected and a ball is drawn from each. Find the probability (*a*) that both balls are white, (*b*) that both balls are of the same color.

   Ans. (*a*) $\frac{29}{72}$ (*b*) $\frac{19}{36}$

18.21 If four trials are made in Problem 18.20, find the probability (*a*) that the first two will result in pairs of white balls and the other two in pairs of black balls, (*b*) that a pair of black balls will be obtained at least three times.

   Ans. (*a*) $\frac{841}{331\,776}$ (*b*) $\frac{125}{36\,864}$

18.22 Five cards numbered 1, 2, 3, 4, 5 respectively are placed in a revolving box. If the cards are drawn one at a time from the box, what is the probability that they will be drwn in their natural order?

   Ans. $\frac{1}{120}$

18.23 Brown, Jones, and Smith shoot at a target in alphabetical order with probabilities $\frac{1}{4}, \frac{1}{3}, \frac{1}{2}$ respectively of hitting the bull's eye. (*a*) Find the probability that on his first shot each will be the first to hit the bull's-eye. (*b*) Find the probability that the bull's-eye is not hit on the first round. (*c*) Find the probability that the first to hit the bull's-eye is Jones on his second shot.

   Ans. (*a*) $\frac{1}{4}, \frac{1}{4}, \frac{1}{4}$ (*b*) $\frac{1}{4}$ (*c*) $\frac{1}{16}$

18.24 The probability that X will win a game of checkers is $\frac{2}{5}$. In a five-game match, what is the probability (*a*) that X will win the first, third, and fifth games, and lose the others? (*b*) that he will win exactly three games? (*c*) that he will win at least three games?

   Ans. (*a*) $\frac{72}{3125}$ (*b*) $\frac{144}{625}$ (*c*) $\frac{992}{3125}$

18.25 Three pennies are tossed at the same time. Find the probability that two are heads and one is a tail.

   Ans. $\frac{3}{8}$

Chapter 19

Determinants of Orders Two and Three

DETERMINANTS OF ORDER TWO. The symbol $\begin{vmatrix} a_1 & b_1 \\ a_2 & b_2 \end{vmatrix}$, consisting of 2^2 numbers called *elements* arranged in two rows and two columns, is called a *determinant of order two*. The elements a_1 and b_2 are said to lie along the *principal diagonal*; the elements a_2 and b_1 are said to lie along the *secondary diagonal*. Row 1 consists of a_1 and b_1. Row 2 consists of a_2 and b_2. Column 1 consists of a_1 and a_2, and column 2 consists of b_1 and b_2.

The *value* of the determinant is obtained by forming the product of the elements along the principal diagonal and subtracting from it the product of the elements along the secondary diagonal; thus,

$$\begin{vmatrix} a_1 & b_1 \\ a_2 & b_2 \end{vmatrix} = a_1 b_2 - a_2 b_1$$

(See Problem 19.1.)

THE SOLUTION of the consistent and independent equations

$$\begin{cases} a_1 x + b_1 y = c_1 \\ a_2 x + b_2 y = c_2 \end{cases} \tag{19.1}$$

may be expressed as quotients of determinants of order two:

$$x = \frac{c_1 b_2 - c_2 b_1}{a_1 b_2 - a_2 b_1} = \frac{\begin{vmatrix} c_1 & b_1 \\ c_2 & b_2 \end{vmatrix}}{\begin{vmatrix} a_1 & b_1 \\ a_2 & b_2 \end{vmatrix}}, \qquad y = \frac{a_1 c_2 - a_2 c_1}{a_1 b_2 - a_2 b_1} = \frac{\begin{vmatrix} a_1 & c_1 \\ a_2 & c_2 \end{vmatrix}}{\begin{vmatrix} a_1 & b_1 \\ a_2 & b_2 \end{vmatrix}}$$

These equations are consistent and independent if and only if $\begin{vmatrix} a_1 & b_1 \\ a_2 & b_2 \end{vmatrix} \neq 0$. See Chapter 5.

EXAMPLE 1. Solve $\begin{cases} y = 3x + 1 \\ 4x + 2y - 7 = 0 \end{cases}$ using determinants.

Arrange the equations in the form *(19.1)*: $\begin{cases} 3x - y = -1 \\ 4x + 2y = 7 \end{cases}$. The solution requires the values of three determinants:

117

The denominator, D, formed by writing the coefficients of x and y in order

$$D = \begin{vmatrix} 3 & -1 \\ 4 & 2 \end{vmatrix} = 3 \cdot 2 - 4(-1) = 6 + 4 = 10$$

The numerator of x, N_x, formed from D by replacing the coefficients of x by the constant terms

$$N_x = \begin{vmatrix} -1 & -1 \\ 7 & 2 \end{vmatrix} = -1 \cdot 2 - 7(-1) = -2 + 7 = 5$$

The numerator of y, N_y, formed from D by replacing the coefficients of y by the constant terms

$$N_y = \begin{vmatrix} 3 & -1 \\ 4 & 7 \end{vmatrix} = 3 \cdot 7 - 4(-1) = 21 + 4 = 25$$

Then $x = \dfrac{N_x}{D} = \dfrac{5}{10} = \dfrac{1}{2}$ and $y = \dfrac{N_y}{D} = \dfrac{25}{10} = \dfrac{5}{2}$. (See Problem 19.2.)

DETERMINANTS OF ORDER THREE.

The symbol

$$\begin{vmatrix} a_1 & b_1 & c_1 \\ a_2 & b_1 & c_2 \\ a_3 & b_3 & c_3 \end{vmatrix},$$

consisting of 3^2 elements arranged in three rows and three columns, is called a *determinant of order three*. Its value is

$$a_1 b_2 c_3 + a_2 b_3 c_1 + a_3 b_1 c_2 - a_1 b_3 c_2 - a_2 b_1 c_3 - a_3 b_2 c_1$$

This may be written as

$$a_1(b_2 c_3 - b_3 c_2) - b_1(a_2 c_3 - a_3 c_2) + c_1(a_2 b_3 - a_3 b_2)$$

or

$$a_1 \begin{vmatrix} b_2 & c_2 \\ b_3 & c_3 \end{vmatrix} - b_1 \begin{vmatrix} a_2 & c_2 \\ a_3 & c_3 \end{vmatrix} + c_1 \begin{vmatrix} a_2 & b_2 \\ a_3 & b_3 \end{vmatrix} \qquad (19.2)$$

to involve three determinants of order two. Note that the elements which multiply the determinants of order two are the elements of the first row of the given determinant. (See Problem 19.3.)

THE SOLUTION

of the system of consistent and independent equations

$$\begin{cases} a_1 x + b_1 y + c_1 z = d_1 \\ a_2 x + b_2 y + c_2 z = d_2 \\ a_3 x + b_3 y + c_3 z = d_3 \end{cases}$$

in determinant form is given by

$$x = \frac{N_x}{D} = \frac{\begin{vmatrix} d_1 & b_1 & c_1 \\ d_2 & b_2 & c_2 \\ d_3 & b_3 & c_3 \end{vmatrix}}{\begin{vmatrix} a_1 & b_1 & c_1 \\ a_2 & b_2 & c_2 \\ a_3 & b_3 & c_3 \end{vmatrix}}, \qquad y = \frac{N_y}{D} = \frac{\begin{vmatrix} a_1 & d_1 & c_1 \\ a_2 & d_2 & c_2 \\ a_3 & d_3 & c_3 \end{vmatrix}}{D}, \qquad z = \frac{N_z}{D} = \frac{\begin{vmatrix} a_1 & b_1 & d_1 \\ a_2 & b_2 & d_2 \\ a_3 & b_3 & d_3 \end{vmatrix}}{D}.$$

The determinant D is formed by writing the coefficients of x, y, z in order, while the determinant appearing in the numerator for any unknown is obtained from D by replacing the column of coefficients of that unknown by the column of constants.

The system is consistent and independent if and only if $D \neq 0$.

EXAMPLE 2. Solve, using determinants: $\begin{cases} x + 3y + 2z = -13 \\ 2x - 6y + 3z = 32 \\ 3x - 4y - z = 12 \end{cases}$

The solution requires the values of four determinants:
The denominator,

$$D = \begin{vmatrix} 1 & 3 & 2 \\ 2 & -6 & 3 \\ 3 & -4 & -1 \end{vmatrix} = 1(6 + 12) - 3(-2 - 9) + 2(-8 + 18)$$

$$= 18 + 33 + 20 = 71$$

The numerator of x,

$$N_x = \begin{vmatrix} -13 & 3 & 2 \\ 32 & -6 & 3 \\ 12 & -4 & -1 \end{vmatrix} = -13(6 + 12) - 3(-32 - 36) + 2(-128 + 72)$$

$$= -234 + 204 - 112 = -142$$

The numerator of y,

$$N_y = \begin{vmatrix} 1 & -13 & 2 \\ 2 & 32 & 3 \\ 3 & 12 & -1 \end{vmatrix} = 1(-32 - 36) - (-13)(-2 - 9) + 2(24 - 96)$$

$$= -68 - 143 - 144 = -355$$

The numerator of z,

$$N_z = \begin{vmatrix} 1 & 3 & -13 \\ 2 & -6 & 32 \\ 3 & -4 & 12 \end{vmatrix} = 1(-72 + 128) - 3(24 - 96) + (-13)(-8 + 18)$$

$$= 56 + 216 - 130 = 142$$

Then $\qquad x = \dfrac{N_x}{D} = \dfrac{-142}{71} = -2, \qquad y = \dfrac{N_y}{D} = \dfrac{-355}{71} = -5, \qquad z = \dfrac{N_z}{D} = \dfrac{142}{71} = 2$

(See Problems 19.4–19.5.)

Solved Problems

19.1 Evaluate each of the following determinants:

(a) $\begin{vmatrix} 2 & 3 \\ 4 & 5 \end{vmatrix} = 2 \cdot 5 - 4 \cdot 3 = -2$ (b) $\begin{vmatrix} 5 & -2 \\ 3 & 1 \end{vmatrix} = 5 \cdot 1 - 3(-2) = 11$

19.2 Solve for x and y, using determinants: (a) $\begin{cases} x + 2y = -4 \\ 5x + 3y = 1 \end{cases}$,   (b) $\begin{cases} ax - 2by = c \\ 2ax - 3by = 4c \end{cases}$

(a) $\quad D = \begin{vmatrix} 1 & 2 \\ 5 & 3 \end{vmatrix} = 3 - 10 = -7, \qquad N_x = \begin{vmatrix} -4 & 2 \\ 1 & 3 \end{vmatrix} = -12 - 2 = -14, \qquad N_y = \begin{vmatrix} 1 & -4 \\ 5 & 1 \end{vmatrix} = 1 + 20 = 21$

$$x = \frac{N_x}{D} = \frac{-14}{-7} = 2, \quad y = \frac{N_y}{D} = \frac{21}{-7} = -3$$

(b) $D = \begin{vmatrix} a & -2b \\ 2a & -3b \end{vmatrix} = ab, \qquad N_x = \begin{vmatrix} c & -2b \\ 4c & -3b \end{vmatrix} = 5bc, \qquad N_y = \begin{vmatrix} a & c \\ 2a & 4c \end{vmatrix} = 2ac$

$$x = \frac{N_x}{D} = \frac{5bc}{ab} = \frac{5c}{a}, \qquad y = \frac{N_y}{D} = \frac{2ac}{ab} = \frac{2c}{b}$$

19.3   Evaluate the following determinants:

(a) $\begin{vmatrix} 2 & -2 & -1 \\ 6 & 1 & -1 \\ 4 & 3 & 5 \end{vmatrix} = 2\begin{vmatrix} 1 & -1 \\ 3 & 5 \end{vmatrix} - (-2)\begin{vmatrix} 6 & -1 \\ 4 & 5 \end{vmatrix} + (-1)\begin{vmatrix} 6 & 1 \\ 4 & 3 \end{vmatrix}$

$$= 2(5+3) + 2(30+4) - (18-4) = 2 \cdot 8 + 2 \cdot 34 - 14 = 70$$

(b) $\begin{vmatrix} 2 & 5 & 0 \\ 0 & 3 & 4 \\ -5 & 3 & 6 \end{vmatrix} = 2\begin{vmatrix} 3 & 4 \\ 3 & 6 \end{vmatrix} - 5\begin{vmatrix} 0 & 4 \\ -5 & 6 \end{vmatrix} + 0\begin{vmatrix} 0 & 3 \\ -5 & 3 \end{vmatrix} = 2(18-12) - 5(0+20) + 0 = -88$

(c) $\begin{vmatrix} 3 & 2 & 1 \\ 1 & -2 & 4 \\ 4 & 2 & 3 \end{vmatrix} = 3(-6-8) - 2(3-16) + 1(2+8) = -6$

(d) $\begin{vmatrix} 4 & -3 & 2 \\ 5 & 9 & -7 \\ 4 & -1 & 7 \end{vmatrix} = 4(36-7) + 3(20+28) + 2(-5-36) = 178$

19.4   Solve using determinants: $\begin{cases} 2x - 3y + 2z = 6 \\ x + 8y + 3z = -31 \\ 3x - 2y + z = -5 \end{cases}$

We evaluate $D = \begin{vmatrix} 2 & -3 & 2 \\ 1 & 8 & 3 \\ 3 & -2 & 1 \end{vmatrix} = 2(8+6) + 3(1-9) + 2(-2-24) = -48$

$$N_x = \begin{vmatrix} 6 & -3 & 2 \\ -31 & 8 & 3 \\ -5 & -2 & 1 \end{vmatrix} = 68(8+6) + 3(-31+15) + 2(62+40) = 240$$

$$N_y = \begin{vmatrix} 2 & 6 & 2 \\ 1 & -31 & 3 \\ 3 & -5 & 1 \end{vmatrix} = 2(-31+15) - 6(1-9) + 2(-5+93) = 192$$

$$N_z = \begin{vmatrix} 2 & -3 & 6 \\ 1 & 8 & -31 \\ 3 & -2 & -5 \end{vmatrix} = 2(-40-62) + 3(-5+93) + 6(-2-24) = -96$$

Then $\qquad x = \dfrac{N_x}{D} = \dfrac{240}{-48} = -5, \quad y = \dfrac{N_y}{D} = \dfrac{192}{-48} = -4, \quad z = \dfrac{N_z}{D} = \dfrac{-96}{-48} = -2$

19.5   Solve, using determinants: $\begin{cases} 2x + y = 2 \\ z - 4y = 0 \\ 4x + z = 6 \end{cases}$

$D = \begin{vmatrix} 2 & 1 & 0 \\ 0 & -4 & 1 \\ 4 & 0 & 1 \end{vmatrix} = -4, \quad N_x = \begin{vmatrix} 2 & 1 & 0 \\ 0 & -4 & 1 \\ 6 & 0 & 1 \end{vmatrix} = -2, \quad N_y = \begin{vmatrix} 2 & 2 & 0 \\ 0 & 0 & 1 \\ 4 & 6 & 1 \end{vmatrix} = -4, \quad N_z = \begin{vmatrix} 2 & 1 & 2 \\ 0 & -4 & 0 \\ 4 & 0 & 6 \end{vmatrix} = -16$

Then $\qquad x = \dfrac{-2}{-4} = \dfrac{1}{2}, \quad y = \dfrac{-4}{-4} = 1, \text{ and } z = \dfrac{-16}{-4} = 4.$

Supplementary Problems

19.6 Evaluate. (a) $\begin{vmatrix} 1 & 2 \\ -3 & 1 \end{vmatrix}$ (b) $\begin{vmatrix} 2 & 0 \\ -5 & 1 \end{vmatrix}$ (c) $\begin{vmatrix} 3 & 2 \\ 1 & -4 \end{vmatrix}$ (d) $\begin{vmatrix} 2 & -10 \\ 3 & -15 \end{vmatrix}$

      Ans. (a) 7 (b) 2 (c) −14 (d) 0

19.7 Solve, using determinants. (a) $\begin{cases} 2x + y = 4 \\ 3x + 4y = 1 \end{cases}$ (b) $\begin{cases} 5x + 2y = 2 \\ 3x - 5y = 26 \end{cases}$ (c) $\begin{cases} 5x + 3y = -6 \\ 3x + 5y = -18 \end{cases}$

      Ans. (a) $x = 3, \ y = -2$ (b) $x = 2, \ y = -4$ (c) $x = \frac{3}{2}, \ y = -\frac{9}{2}$

19.8 Evaluate. (a) $\begin{vmatrix} 1 & 2 & 3 \\ 2 & -3 & 4 \\ -3 & 4 & 5 \end{vmatrix}$ (b) $\begin{vmatrix} 2 & -1 & 3 \\ 2 & 1 & 5 \\ -2 & 3 & -2 \end{vmatrix}$ (c) $\begin{vmatrix} -2 & 3 & 1 \\ 3 & 5 & 1 \\ 4 & -2 & 1 \end{vmatrix}$ (d) $\begin{vmatrix} 2 & 6 & 1 \\ 4 & -4 & 1 \\ 3 & 1 & 1 \end{vmatrix}$

      Ans. (a) −78 (b) −4 (c) −37 (d) 0

19.9 Solve, using determinants.

(a) $\begin{cases} x + 2y + 2z = 4 \\ 3x - y + 4z = 25 \\ 3x + 2y - z = -4 \end{cases}$ (b) $\begin{cases} 2x - 3y + 5z = 4 \\ 3x - 2y + 2z = 3 \\ 4x + y - 4z = -6 \end{cases}$ (c) $\begin{cases} \dfrac{1}{x} + \dfrac{2}{y} + \dfrac{1}{z} = 2 \\ \dfrac{3}{x} - \dfrac{4}{y} - \dfrac{2}{z} = 1 \\ \dfrac{2}{x} + \dfrac{5}{y} - \dfrac{2}{z} = 3 \end{cases}$

      Hint: In (c) solve first for $1/x, \ 1/y, \ 1/z$.

      Ans. (a) $x = 2, \ y = -3, \ z = 4$ (b) $x = -\frac{1}{3}, \ y = \frac{2}{3}, \ z = \frac{4}{3}$ (c) $x = 1, \ y = z = 3$

19.10 Verify, by evaluating the determinants.

      (a) $\begin{vmatrix} 1 & 2 & 3 \\ a & 2a & 3a \\ 8 & 9 & 10 \end{vmatrix} = 0$ (b) $\begin{vmatrix} 2 & 1 & -1 \\ 3 & 4 & 2 \\ -2 & -5 & 3 \end{vmatrix} = \begin{vmatrix} 0 & 1 & 0 \\ -5 & 4 & 6 \\ 8 & -5 & -2 \end{vmatrix}$

      (c) $\begin{vmatrix} 3 & 4 & 5 \\ 7 & -2 & 3 \\ 2 & 5 & -1 \end{vmatrix} = - \begin{vmatrix} 4 & -2 & 5 \\ 3 & 7 & 2 \\ 5 & 3 & -1 \end{vmatrix}$ (d) $\begin{vmatrix} 4 & 2 & 5 \\ -7 & -3 & 1 \\ 9 & 4 & 8 \end{vmatrix} + \begin{vmatrix} -1 & 2 & 5 \\ 5 & -3 & 1 \\ -3 & 4 & 8 \end{vmatrix} = \begin{vmatrix} 3 & 2 & 5 \\ -2 & -3 & 1 \\ 6 & 4 & 8 \end{vmatrix}$

19.11 Solve, using determinants. $\begin{vmatrix} x & y & 1 \\ 1 & -1 & 1 \\ 13 & 2 & 1 \end{vmatrix} = 0, \quad \begin{vmatrix} x & y & 1 \\ 3 & 2 & 1 \\ -6 & -4 & 1 \end{vmatrix} = 0.$

      Ans. $x = -3, \ y = -2$

Chapter 20

Determinants of Order n

A DETERMINANT of order n consists of n^2 numbers called elements arranged in n rows and n columns, and enclosed by two vertical lines. For example,

$$D_1 = |a_1| \qquad D_2 = \begin{vmatrix} a_1 & b_1 \\ a_2 & b_2 \end{vmatrix} \qquad D_3 = \begin{vmatrix} a_1 & b_1 & c_1 \\ a_2 & b_2 & c_2 \\ a_3 & b_3 & c_3 \end{vmatrix} \qquad D_4 = \begin{vmatrix} a_1 & b_1 & c_1 & d_1 \\ a_2 & b_2 & c_2 & d_2 \\ a_3 & b_3 & c_3 & d_3 \\ a_4 & b_4 & c_4 & d_4 \end{vmatrix}$$

are determinants of orders one, two, three, and four, respectively. In this notation the letters designate columns and the subscripts designate rows. Thus, all elements with letter c are in the third column and all elements with subscript 2 are in the second row.

THE MINOR OF A GIVEN ELEMENT of a determinant is the determinant of the elements which remain after deleting the row and the column in which the given element stands. For example, the minor of a_1 in D_4 is

$$\begin{vmatrix} b_2 & c_2 & d_2 \\ b_3 & c_3 & d_3 \\ b_4 & c_4 & d_4 \end{vmatrix}$$

and the minor of b_3 is

$$\begin{vmatrix} a_1 & c_1 & d_1 \\ a_2 & c_2 & d_2 \\ a_4 & c_4 & d_4 \end{vmatrix}$$

Note that the minor of a given element contains no element having either the letter or the subscript of the given element. (See Problem 20.1.)

THE VALUE OF A DETERMINANT of order one is the single element of the determinant. A determinant of order $n > 1$ may be expressed as the sum of n products formed by multiplying each element of any chosen row (column) by its minor and prefixing a proper sign. The proper sign associated with each product is $(-1)^{i+j}$, where i is the number of the row and j is the number of the column in which the element stands. For example, for D_3 above,

$$D_3 = -a_2 \begin{vmatrix} b_1 & c_1 \\ b_3 & c_3 \end{vmatrix} + b_2 \begin{vmatrix} a_1 & c_1 \\ a_3 & c_3 \end{vmatrix} - c_2 \begin{vmatrix} a_1 & b_1 \\ a_3 & b_3 \end{vmatrix}$$

is the expansion of D_3, along the second row. The sign given to the first product is $-$, since a_2 stands in the second row and first column, and $(-1)^{2+1} = -1$. In all, there are six expansions of D_3 along its rows and columns yielding identical results when the minors are evaluated.

There are eight expansions of D_4 along its rows and columns, of which

$$D_4 = +a_1 \begin{vmatrix} b_2 & c_2 & d_2 \\ b_3 & c_3 & d_3 \\ b_4 & c_4 & d_4 \end{vmatrix} - b_1 \begin{vmatrix} a_2 & c_2 & d_2 \\ a_3 & c_3 & d_3 \\ a_4 & c_4 & d_4 \end{vmatrix} + c_1 \begin{vmatrix} a_2 & b_2 & d_2 \\ a_3 & b_3 & d_3 \\ a_4 & b_4 & d_4 \end{vmatrix} - d_1 \begin{vmatrix} a_2 & b_2 & c_2 \\ a_3 & b_3 & c_3 \\ a_4 & b_4 & c_4 \end{vmatrix}$$

(along the first row)

$$D_4 = +a_1 \begin{vmatrix} b_2 & c_2 & d_2 \\ b_3 & c_3 & d_3 \\ b_4 & c_4 & d_4 \end{vmatrix} - a_2 \begin{vmatrix} b_1 & c_1 & d_1 \\ b_3 & c_3 & d_3 \\ b_4 & c_4 & d_4 \end{vmatrix} + a_3 \begin{vmatrix} b_1 & c_1 & d_1 \\ b_2 & c_2 & d_2 \\ b_4 & c_4 & d_4 \end{vmatrix} - a_4 \begin{vmatrix} b_1 & c_1 & d_1 \\ b_2 & c_2 & d_2 \\ b_3 & c_3 & d_3 \end{vmatrix}$$

(along the first column)

$$D_4 = -a_4 \begin{vmatrix} b_1 & c_1 & d_1 \\ b_2 & c_2 & d_2 \\ b_3 & c_3 & d_3 \end{vmatrix} + b_4 \begin{vmatrix} a_1 & c_1 & d_1 \\ a_2 & c_2 & d_2 \\ a_3 & c_3 & d_3 \end{vmatrix} - c_4 \begin{vmatrix} a_1 & b_1 & d_1 \\ a_2 & b_2 & d_2 \\ a_3 & b_3 & d_3 \end{vmatrix} + d_4 \begin{vmatrix} a_1 & b_1 & c_1 \\ a_2 & b_2 & c_2 \\ a_3 & b_3 & c_3 \end{vmatrix}$$

(along the fourth row)

$$D_4 = -b_1 \begin{vmatrix} a_2 & c_2 & d_2 \\ a_3 & c_3 & d_3 \\ a_4 & c_4 & d_4 \end{vmatrix} + b_2 \begin{vmatrix} a_1 & c_1 & d_1 \\ a_3 & c_3 & d_3 \\ a_4 & c_4 & d_4 \end{vmatrix} - b_3 \begin{vmatrix} a_1 & c_1 & d_1 \\ a_2 & c_2 & d_2 \\ a_4 & c_4 & d_4 \end{vmatrix} + b_4 \begin{vmatrix} a_1 & c_1 & d_1 \\ a_2 & c_2 & d_2 \\ a_3 & c_3 & d_3 \end{vmatrix}$$

(along the second column)

are examples. (See Problem 20.2.)

All computer mathematics packages of software give you the capability to easily evaluate determinants. The reader should try at least one of these (Maple, or others) to gain some familiarity with these kinds of computer computations.

THE COFACTOR OF AN ELEMENT of a determinant is the minor of that element together with the sign associated with the product of that element and its minor in the expansion of the determinant. The cofactors of the elements $a_1, a_2, b_1, b_3, c_1, \ldots$ will be denoted by $A_1, A_2, B_1, B_3, C_1, \ldots$. Thus, the cofactor of c_1 in D_3 is $C_1 = + \begin{vmatrix} a_2 & b_2 \\ a_3 & b_3 \end{vmatrix}$ and the cofactor of b_3 is $B_3 = - \begin{vmatrix} a_1 & c_1 \\ a_2 & c_2 \end{vmatrix}$.

When cofactors are used, the expansions of D_4 given above take the more compact form

$$\begin{aligned} D_4 &= a_1 A_1 + b_1 B_1 + c_1 C_1 + d_1 D_1 && \text{(along the first row)} \\ &= a_1 A_1 + a_2 A_2 + a_3 A_3 + a_4 A_4 && \text{(along the first column)} \\ &= a_4 A_4 + b_4 B_4 + c_4 C_4 + d_4 D_4 && \text{(along the fourth row)} \\ &= b_1 B_1 + b_2 B_2 + b_3 B_3 + b_4 B_4 && \text{(along the second column)} \end{aligned}$$

(See Problems 20.3–20.4.)

PROPERTIES OF DETERMINANTS. Subject always to our assumption of equivalent expansions of a determinant along any of its rows or columns, the following theorems may be proved by mathematical induction.

THEOREM I. If two rows (or two columns) of a determinant are identical, the value of the determinant is zero. For example,

$$\begin{vmatrix} 2 & 3 & 2 \\ 3 & 1 & 3 \\ 1 & 4 & 1 \end{vmatrix} = 0$$

COROLLARY I. If each of the elements of a row (or a column) of a determinant is multiplied by the cofactor of the corresponding element of another row (column), the sum of the products is zero.

THEOREM II. If the elements of a row (or a column) of a determinant are multiplied by any number *m*, the determinant is multiplied by *m*. For example,

$$5\begin{vmatrix} 2 & 3 & 4 \\ 3 & -1 & 2 \\ 1 & 4 & -3 \end{vmatrix} = \begin{vmatrix} 10 & 3 & 4 \\ 15 & -1 & 2 \\ 5 & 4 & -3 \end{vmatrix} = \begin{vmatrix} 2 & 3 & 4 \\ 15 & -5 & 10 \\ 1 & 4 & -3 \end{vmatrix}$$

THEOREM III. If each of the elements of a row (or a column) of a determinant is expressed as the sum of two or more numbers, the determinant may be written as the sum of two or more determinants. For example,

$$\begin{vmatrix} 2 & 5 & 4 \\ 4 & -2 & 3 \\ 1 & -4 & 3 \end{vmatrix} = \begin{vmatrix} -2+4 & 5 & 4 \\ 3+1 & -2 & 3 \\ 1+0 & -4 & 3 \end{vmatrix} = \begin{vmatrix} -2 & 5 & 4 \\ 3 & -2 & 3 \\ 1 & -4 & 3 \end{vmatrix} + \begin{vmatrix} 4 & 5 & 4 \\ 1 & -2 & 3 \\ 0 & -4 & 3 \end{vmatrix}$$

THEOREM IV. If to the elements of any row (or any column) of a determinant there is added *m* times the corresponding elements of another row (another column), the value of the determinants is unchanged. For example,

$$\begin{vmatrix} -2 & 5 & 4 \\ 3 & -2 & 2 \\ 1 & -4 & 3 \end{vmatrix} = \begin{vmatrix} -2 & 5+4(-2) & 4 \\ 3 & -2+4(3) & 2 \\ 1 & -4+4(1) & 3 \end{vmatrix} = \begin{vmatrix} -2 & -3 & 4 \\ 3 & 10 & 2 \\ 1 & 0 & 3 \end{vmatrix}$$

(See Problem 20.5.)

EVALUATION OF DETERMINANTS. A determinant of any order may be evaluated by expanding it and all subsequent determinants (minors) thus obtained along a row or column. This procedure may be greatly simplified by the use of Theorem IV. In Problem 20.6, (*a*) and (*b*), a row (column) containing an element +1 or −1 is used to obtain an equivalent determinant having an element 0 in another row (column). In (*c*) and (*d*), the same theorem has been used to obtain an element +1 or −1; this procedure is to be followed when the given determinant is lacking in these elements.

The revised procedure consists in first obtaining an equivalent determinant in which all the elements, save one, in some row (column) are zeros and then expanding along that row (column).

EXAMPLE 1. Evaluate $\begin{vmatrix} 1 & 4 & 3 & 1 \\ 2 & 8 & 2 & 5 \\ 4 & -4 & -1 & -3 \\ 2 & 5 & 3 & 3 \end{vmatrix}$

Using the first column since it contains the element 1 in the first row, we obtain an equivalent determinant all of whose elements, save the first, in the first row are zeros. We have

$$\begin{vmatrix} 1 & 4 & 3 & 1 \\ 2 & 8 & 2 & 5 \\ 4 & -4 & -1 & -3 \\ 2 & 5 & 3 & 3 \end{vmatrix} = \begin{vmatrix} 1 & 4+(-4)1 & 3+(-3)1 & 1+(-1)1 \\ 2 & 8+(-4)2 & 2+(-3)2 & 5+(-1)2 \\ 4 & -4+(-4)4 & -1+(-3)4 & -3+(-1)4 \\ 2 & 5+(-4)2 & 3+(-3)2 & 3+(-1)2 \end{vmatrix} = \begin{vmatrix} 1 & 0 & 0 & 0 \\ 2 & 0 & -4 & 3 \\ 4 & -20 & -13 & -7 \\ 2 & -3 & -3 & 1 \end{vmatrix}$$

$$= \begin{vmatrix} 0 & -4 & 3 \\ -20 & -13 & -7 \\ -3 & -3 & 1 \end{vmatrix} \quad \text{(by expanding along the first row)}$$

Expanding the resulting determinant along the first row to take full advantage of the element 0, we have

$$\begin{vmatrix} 0 & -4 & 3 \\ -20 & -13 & -7 \\ -3 & -3 & 1 \end{vmatrix} = 4(-20-21)+3(60-39) = -101$$

Solved Problems

20.1 Write the minors of the elements a_1, b_3, c_2 of D_3.

The minor of a_1 is $\begin{vmatrix} b_2 & c_2 \\ b_3 & c_3 \end{vmatrix}$, the minor of b_3 is $\begin{vmatrix} a_1 & c_1 \\ a_2 & c_2 \end{vmatrix}$, the minor of c_1 is $\begin{vmatrix} a_2 & b_2 \\ a_3 & b_3 \end{vmatrix}$.

20.2 Evaluate: (a) $\begin{vmatrix} -1 & 3 & -4 \\ 0 & 2 & 0 \\ 2 & -3 & 5 \end{vmatrix}$ by expanding along the second row

(b) $\begin{vmatrix} 8 & 0 & 2 & 0 \\ 5 & 1 & -3 & 0 \\ -4 & 3 & 7 & -3 \\ 4 & 0 & 6 & 0 \end{vmatrix}$ by expanding along the fourth column

(a) $\begin{vmatrix} -1 & 3 & -4 \\ 0 & 2 & 0 \\ 2 & -3 & 5 \end{vmatrix} = -0\begin{vmatrix} 3 & -4 \\ -3 & 5 \end{vmatrix} + 2\begin{vmatrix} -1 & -4 \\ 2 & 5 \end{vmatrix} - 0\begin{vmatrix} -1 & 3 \\ 2 & -3 \end{vmatrix} = 2\begin{vmatrix} -1 & -4 \\ 2 & 5 \end{vmatrix} = 2 \cdot 3 = 6$

(b) $\begin{vmatrix} 8 & 0 & 2 & 0 \\ 5 & 1 & -3 & 0 \\ -4 & 3 & 7 & -3 \\ 4 & 0 & 6 & 0 \end{vmatrix} = -(-3)\begin{vmatrix} 8 & 0 & 2 \\ 5 & 1 & -3 \\ 4 & 0 & 6 \end{vmatrix} = 3\left\{+1\begin{vmatrix} 8 & 2 \\ 4 & 6 \end{vmatrix}\right\} = 3(48-8) = 120$

20.3 (a) Write the cofactors of the elements a_1, b_3, c_2, d_4 of D_4.

The cofactor of a_1 is $A_1 = +\begin{vmatrix} b_2 & c_2 & d_2 \\ b_3 & c_3 & d_3 \\ b_4 & c_4 & d_4 \end{vmatrix}$, of b_3 is $B_3 = -\begin{vmatrix} a_1 & c_1 & d_1 \\ a_2 & c_2 & d_2 \\ a_4 & c_4 & d_4 \end{vmatrix}$

of c_2 is $C_2 = -\begin{vmatrix} a_1 & b_1 & d_1 \\ a_3 & b_3 & d_3 \\ a_4 & b_4 & d_4 \end{vmatrix}$, of d_4 is $D_4 = +\begin{vmatrix} a_1 & b_1 & c_1 \\ a_2 & b_2 & c_2 \\ a_3 & b_3 & c_3 \end{vmatrix}$

(b) Write the expansion of D_4 along (1) the second row, (2) the third column, using cofactors.
 (1) $D_4 = a_2A_2 + b_2B_2 + c_2C_2 + d_2D_2$
 (2) $D_4 = c_1C_1 + c_2C_2 + c_3C_3 + c_4C_4$

20.4 Express $g_1C_1 + g_2C_2 + g_3C_3 + g_4C_4$, where the C_i are cofactors of the elements c_i of D_4, as a determinant.

Since the cofactors C_i contain no elements with basal letter c, we replace c_1, c_2, c_3, c_4 by g_1, g_2, g_3, g_4 respectively, in Problem 20.3(b) and obtain

$$g_1 C_1 + g_2 C_2 + g_3 C_3 + g_4 C_4 = \begin{vmatrix} a_1 & b_1 & g_1 & d_1 \\ a_2 & b_2 & g_2 & d_2 \\ a_3 & b_3 & g_3 & d_3 \\ a_4 & b_4 & g_4 & d_4 \end{vmatrix}$$

20.5   Prove by induction: If two rows (or two columns) of a determinant are identical, the value of the determinant is zero.

The theorem is true for determinants of order two since $\begin{vmatrix} a_1 & a_1 \\ a_2 & a_2 \end{vmatrix} = a_1 a_2 - a_1 a_2 = 0$.

Let us assume the theorem true for determinants of order k and consider a determinant D of order $(k + 1)$ in which two columns are identical. When D is expanded along any column, other than the two with identical elements, each cofactor involved is a determinant of order k with two columns identical and, by assumption, is equal to zero. Thus, D is equal to zero and the theorem is proved by induction.

20.6   From the determinant $\begin{vmatrix} 1 & 4 & 3 & 1 \\ 2 & 8 & 2 & 5 \\ 4 & -4 & -1 & -3 \\ 2 & 5 & 3 & 3 \end{vmatrix}$ obtain an equivalent determinant.

(a)   By adding -4 times the elements of the first column to the corresponding elements of the second column

(b)   By adding 3 times the elements of the third row to the corresponding elements of the fourth row

(c)   By adding -1 times the elements of the third column to the corresponding elements of the second column

(d)   By adding -2 times the elements of the fourth row to the corresponding elements of the second row

$$(a) \quad \begin{vmatrix} 1 & 4 & 3 & 1 \\ 2 & 8 & 2 & 5 \\ 4 & -4 & -1 & -3 \\ 2 & 5 & 3 & 3 \end{vmatrix} = \begin{vmatrix} 1 & 4+(-4)1 & 3 & 1 \\ 2 & 8+(-4)2 & 2 & 5 \\ 4 & -4+(-4)4 & -1 & -3 \\ 2 & 5+(-4)2 & 3 & 3 \end{vmatrix} = \begin{vmatrix} 1 & 0 & 3 & 1 \\ 2 & 0 & 2 & 5 \\ 4 & -20 & -1 & -3 \\ 2 & -3 & 3 & 3 \end{vmatrix}$$

$$(b) \quad \begin{vmatrix} 1 & 4 & 3 & 1 \\ 2 & 8 & 2 & 5 \\ 4 & -4 & -1 & -3 \\ 2 & 5 & 3 & 3 \end{vmatrix} = \begin{vmatrix} 1 & 4 & 3 & 1 \\ 2 & 8 & 2 & 5 \\ 4 & -4 & -1 & -3 \\ 2+(3)4 & 5+(3)(-4) & 3+(3)(-1) & 3+(3)(-3) \end{vmatrix} = \begin{vmatrix} 1 & 4 & 3 & 1 \\ 2 & 8 & 2 & 5 \\ 4 & -4 & -1 & -3 \\ 14 & -7 & 0 & -6 \end{vmatrix}$$

$$(c) \quad \begin{vmatrix} 1 & 4 & 3 & 1 \\ 2 & 8 & 2 & 5 \\ 4 & -4 & -1 & -3 \\ 2 & 5 & 3 & 3 \end{vmatrix} = \begin{vmatrix} 1 & 4+(-1)3 & 3 & 1 \\ 2 & 8+(-1)2 & 2 & 5 \\ 4 & -4+(-1)(-1) & -1 & -3 \\ 2 & 5+(-1)3 & 3 & 3 \end{vmatrix} = \begin{vmatrix} 1 & 1 & 3 & 1 \\ 2 & 6 & 2 & 5 \\ 4 & -3 & -1 & -3 \\ 2 & 2 & 3 & 3 \end{vmatrix}$$

$$(d) \quad \begin{vmatrix} 1 & 4 & 3 & 1 \\ 2 & 8 & 2 & 5 \\ 4 & -4 & -1 & -3 \\ 2 & 5 & 3 & 3 \end{vmatrix} = \begin{vmatrix} 1 & 4 & 3 & 1 \\ 2+(-2)2 & 8+(-2)5 & 2+(-2)3 & 5+(-2)3 \\ 4 & -4 & -1 & -3 \\ 2 & 5 & 3 & 3 \end{vmatrix} = \begin{vmatrix} 1 & 4 & 3 & 1 \\ -2 & -2 & -4 & -1 \\ 4 & -4 & -1 & -3 \\ 2 & 5 & 3 & 3 \end{vmatrix}$$

20.7   Evaluate:

$$(a) \quad \begin{vmatrix} 50 & 2 & -9 \\ 250 & -10 & 45 \\ -150 & 6 & 27 \end{vmatrix} = 50 \cdot 2 \cdot 9 \begin{vmatrix} 1 & 1 & -1 \\ 5 & -5 & 5 \\ -3 & 3 & 3 \end{vmatrix} = 900 \cdot 5 \cdot 3 \begin{vmatrix} 1 & 1 & -1 \\ 1 & -1 & 1 \\ -1 & 1 & 1 \end{vmatrix} = 13\,500 \begin{vmatrix} 2 & 0 & 0 \\ 1 & -1 & 1 \\ -1 & 1 & 1 \end{vmatrix}$$

$$= 27\,000 \begin{vmatrix} -1 & 1 \\ 1 & 1 \end{vmatrix} = -54\,000$$

$$(b)\begin{vmatrix} 3 & 2 & -3 & 1 \\ -1 & -3 & 5 & 2 \\ 2 & 1 & 6 & -3 \\ 5 & 4 & -3 & 4 \end{vmatrix} = \begin{vmatrix} 3+(-3)1 & 2+(-2)1 & -3+3(1) & 1 \\ -1+(-3)2 & -3+(-2)2 & 5+(3)2 & 2 \\ 2+(-3)(-3) & 1+(-2)(-3) & 6+(3)(-3) & -3 \\ 5+(-3)4 & 4+(-2)4 & -3+(3)4 & 4 \end{vmatrix} = \begin{vmatrix} 0 & 0 & 0 & 1 \\ -7 & -7 & 11 & 2 \\ 11 & 7 & -3 & -3 \\ -7 & -4 & 9 & 4 \end{vmatrix}$$

$$= -\begin{vmatrix} -7 & -7 & 11 \\ 11 & 7 & -3 \\ -7 & -4 & 9 \end{vmatrix} = -\begin{vmatrix} 0 & -7 & 11 \\ 4 & 7 & -3 \\ -3 & -4 & 9 \end{vmatrix} = -[7(36-9)+11(-16+21)] = -244$$

$$(c)\begin{vmatrix} 2 & -1 & -2 & 3 \\ 3 & 2 & 4 & -1 \\ 2 & 4 & 1 & -5 \\ 4 & -3 & 2 & 1 \end{vmatrix} = \begin{vmatrix} 0 & -1 & 0 & 0 \\ 7 & 2 & 0 & 5 \\ 10 & 4 & -7 & 7 \\ -2 & -3 & 8 & -8 \end{vmatrix} = \begin{vmatrix} 7 & 0 & 5 \\ 10 & -7 & 7 \\ -2 & 8 & -8 \end{vmatrix} = 5 \cdot 66 = 330$$

Supplementary Problems

20.8 Verify: The value of a determinant is unchanged if the rows are written as columns or if its columns are written as rows.

Hint: Show that $\begin{vmatrix} a_1 & b_1 & c_1 \\ a_2 & b_2 & c_2 \\ a_3 & b_3 & c_3 \end{vmatrix} = \begin{vmatrix} a_1 & a_2 & a_3 \\ b_1 & b_2 & b_3 \\ c_1 & c_2 & c_3 \end{vmatrix}$.

20.9 Prove by induction:

(a) The expansion of a determinant of order n contains $n!$ terms.

(b) If two rows (or two columns) of a determinant are interchanged, the sign of the determinant is changed.

Hint: Show that $\begin{vmatrix} a_1 & b_1 \\ a_2 & b_2 \end{vmatrix} = -\begin{vmatrix} b_1 & a_1 \\ b_2 & a_2 \end{vmatrix}$ and proceed as in Problem 20.5.

20.10 Show, without expanding the determinants, that

$$(a)\begin{vmatrix} 3 & 2 & 1 & 4 \\ -1 & 5 & 2 & 6 \\ 2 & -4 & 7 & -5 \\ -2 & 1 & 3 & 5 \end{vmatrix} = -\begin{vmatrix} 3 & 2 & 1 & 4 \\ 2 & -4 & 7 & -5 \\ -1 & 5 & 2 & 6 \\ -2 & 1 & 3 & 5 \end{vmatrix} = -\begin{vmatrix} 3 & 2 & 1 & 4 \\ -2 & 1 & 3 & 5 \\ 2 & -4 & 7 & -5 \\ -1 & 5 & 2 & 6 \end{vmatrix} = -\begin{vmatrix} 3 & -2 & 2 & -1 \\ 2 & 1 & -4 & 5 \\ 1 & 3 & 7 & 2 \\ 4 & 5 & -5 & 6 \end{vmatrix}$$

$$(b)\begin{vmatrix} 10 & 0 & -2 \\ -10 & 3 & -4 \\ -5 & -2 & 3 \end{vmatrix} = -5\begin{vmatrix} -2 & 0 & -2 \\ 2 & 3 & -4 \\ 1 & -2 & 3 \end{vmatrix} = 10\begin{vmatrix} 1 & 0 & 1 \\ 2 & 3 & -4 \\ 1 & -2 & 3 \end{vmatrix}$$

$$(c)\begin{vmatrix} -3 & -1 & 2 \\ 1 & -3 & 3 \\ 4 & -2 & 1 \end{vmatrix} = 0 \quad \textbf{Hint:} \text{ Subtract the first row from the second.}$$

$$(d)\begin{vmatrix} a-3b & a+b & a+5b & e \\ a-2b & a+2b & a+6b & f \\ a-b & a+3b & a+7b & g \\ a & a+4b & a+8b & h \end{vmatrix} = 0 \quad (\text{NOTE: The elements of the first three columns form an arithmetic progression.})$$

$$(e)\begin{vmatrix} 0 & a_1 & a_2 \\ -a_1 & 0 & a_3 \\ -a_2 & -a_3 & 0 \end{vmatrix} = 0 \quad \textbf{Hint:} \text{ Write the rows as columns and factor } -1 \text{ from each row.}$$

20.11 Verify: If the corresponding elements of two rows (or two columns) of a determinant are proportional, the value of the determinant is zero.

20.12 Evaluate each of the following determinants:

(a) $\begin{vmatrix} -3 & 6 & -1 & 1 \\ -4 & -3 & -2 & 4 \\ 5 & -4 & 1 & 3 \\ -1 & -5 & 0 & -1 \end{vmatrix}$ (b) $\begin{vmatrix} 1 & -1 & 1 & 2 \\ 3 & -2 & 4 & -3 \\ 5 & 4 & 1 & 2 \\ -3 & 0 & 3 & 1 \end{vmatrix}$ (c) $\begin{vmatrix} 3 & -1 & -1 & -4 \\ 2 & 3 & -6 & 1 \\ 4 & -1 & 3 & 1 \\ 3 & -1 & 5 & 2 \end{vmatrix}$

(d) $\begin{vmatrix} 2 & 4 & 4 & -3 & 4 \\ 1 & 3 & 1 & 0 & -1 \\ -1 & 3 & 1 & 2 & -1 \\ 4 & 8 & 11 & -10 & 9 \\ 2 & 6 & 9 & -12 & 5 \end{vmatrix}$

   Ans. (a) -50 (b) -397 (c) -78 (d) -316

20.13 Show that $\begin{vmatrix} 1 & 1 & 1 & 1 \\ a & b & c & d \\ a^2 & b^2 & c^2 & d^2 \\ a^3 & b^3 & c^3 & d^3 \end{vmatrix} = (b-a)(c-a)(d-a)(c-b)(d-b)(d-c)$

20.14 Write a determinant that is equal to $x^2 - 1$. (See Problem 20.13.)

20.15 Repeat Problem 20.7 above using a computer software package to evaluate the given determinants.

Chapter 21

Systems of Linear Equations

SYSTEMS OF n LINEAR EQUATIONS IN n UNKNOWNS.   Consider, for the sake of brevity, the system of four linear equations in four unknowns

$$\begin{cases} a_1x + b_1y + c_1z + d_1w = k_1 \\ a_2x + b_2y + c_2z + d_2w = k_2 \\ a_3x + b_3y + c_3z + d_3w = k_3 \\ a_4x + b_4y + c_4z + d_4w = k_4 \end{cases} \qquad (21.1)$$

in which each equation is written with the unknowns x, y, z, w in that order on the left side and the constant term on the right side. Form

$$D = \begin{vmatrix} a_1 & b_1 & c_1 & d_1 \\ a_2 & b_2 & c_2 & d_2 \\ a_3 & b_3 & c_3 & d_3 \\ a_4 & b_4 & c_4 & d_4 \end{vmatrix}, \qquad \text{the determinant of the coefficients of the unknowns,}$$

and from it the determinants

$$N_x = \begin{vmatrix} k_1 & b_1 & c_1 & d_1 \\ k_2 & b_2 & c_2 & d_2 \\ k_3 & b_3 & c_3 & d_3 \\ k_4 & b_4 & c_4 & d_4 \end{vmatrix}, \qquad N_y = \begin{vmatrix} a_1 & k_1 & c_1 & d_1 \\ a_2 & k_2 & c_2 & d_2 \\ a_3 & k_3 & c_3 & d_3 \\ a_4 & k_4 & c_4 & d_4 \end{vmatrix}, \qquad N_z = \begin{vmatrix} a_1 & b_1 & k_1 & d_1 \\ a_2 & b_2 & k_2 & d_2 \\ a_3 & b_3 & k_3 & d_3 \\ a_4 & b_4 & k_4 & d_4 \end{vmatrix}$$

$$N_w = \begin{vmatrix} a_1 & b_1 & c_1 & k_1 \\ a_2 & b_2 & c_2 & k_2 \\ a_3 & b_3 & c_3 & k_3 \\ a_4 & b_4 & c_4 & k_4 \end{vmatrix}$$

by replacing the column of coefficients of the indicated unknown by the column of constants.

CRAMER'S RULE STATES THAT:

(a) If $D \neq 0$, the system (21.1) has the unique solution

$$x = N_x/D, \quad y = N_y/D, \quad z = N_z/D, \quad w = N_w/D$$

(See Problems 21.1–21.2.)

(b) If $D = 0$ and at least one of $N_x, N_y, N_z, N_w \neq 0$, the system has no solution. For, if $D = 0$ and $N_x \neq 0$, then $x \cdot D = N_x$ leads to a contradiction. Such systems are called *inconsistent*. (See Problem 21.3.)

(c) If $D = 0$ and $N_x = N_y = N_z = N_w = 0$, the system may or may not have a solutions. A system having an infinite number of solutions is called *dependent*.

For systems of three or four equations, the simplest procedure is to evaluate D. If $D \neq 0$, proceed as in (A); if $D = 0$, proceed as in Chapter 5. (See Problem 21.4.)

SYSTEMS OF m LINEAR EQUATIONS IN $n > m$ UNKNOWNS.   Ordinarily if there are fewer equations than unknowns, the system will have an infinite number of solutions.

To solve a consistent system of m equations, solve for m of the unknowns (in certain cases for $p < m$ of the unknowns) in terms of the others. (See Problem 21.5.)

SYSTEMS OF n EQUATIONS IN $m < n$ UNKNOWNS.   Ordinarily if there are more equations than unknowns the system is inconsistent. However, if $p \leq m$ of the equations have a solution and if this solution satisfies each of the remaining equations, the system is consistent. (See Problem 21.6.)

A HOMOGENEOUS EQUATION is one in which all terms are of the same degree; otherwise, the equation is called *nonhomogeneous*. For example, the linear equation $2x + 3y - 4z = 5$ is nonhomogeneous, while $2x + 3y - 4z = 0$ is homogeneous. (The term "5" in the first equation has degree 0, while all other terms are of degree 1.)

Every system of homogeneous linear equations

$$a_1 x + b_1 y + c_1 z + \cdots = 0$$
$$a_2 x + b_2 y + c_2 z + \cdots = 0$$
$$\vdots$$
$$a_n x + b_n y + c_n z + \cdots = 0$$

always has the *trivial solution* $x = 0, y = 0, z = 0, \ldots$.

A system of n homogeneous linear equations in n unknowns has *only* the trivial solution if D, the determinant of the coefficients, is not equal to zero. If $D = 0$, the system has nontrivial solutions as well. (See Problem 21.7.)

Solved Problems

21.1   Solve the system $\begin{cases} 3x - 2y - z - 4w = 7 & (1) \\ x \quad\quad + 3z + 2w = -10 & (2) \\ x + 4y + 2z + w = 0 & (3) \\ 2x + 3y \quad\quad + 3w = 1 & (4) \end{cases}$

We find

$$D = \begin{vmatrix} 3 & -2 & -1 & -4 \\ 1 & 0 & 3 & 2 \\ 1 & 4 & 2 & 1 \\ 2 & 3 & 0 & 3 \end{vmatrix} = \begin{vmatrix} 3 & -2 & -10 & -10 \\ 1 & 0 & 0 & 0 \\ 1 & 4 & -1 & -1 \\ 2 & 3 & -6 & -1 \end{vmatrix} = - \begin{vmatrix} -2 & -10 & -10 \\ 4 & -1 & -1 \\ 3 & -6 & -1 \end{vmatrix} = 2 \begin{vmatrix} 1 & 5 & 5 \\ 4 & -1 & -1 \\ 3 & -6 & -1 \end{vmatrix} = -210$$

$$N_x = \begin{vmatrix} 7 & -2 & -1 & -4 \\ -10 & 0 & 3 & 2 \\ 0 & 4 & 2 & 1 \\ 1 & 3 & 0 & 3 \end{vmatrix} = \begin{vmatrix} 0 & -23 & -1 & -25 \\ 0 & 30 & 3 & 32 \\ 0 & 4 & 2 & 1 \\ 1 & 3 & 0 & 3 \end{vmatrix} = - \begin{vmatrix} -23 & -1 & -25 \\ 30 & 3 & 32 \\ 4 & 2 & 1 \end{vmatrix} = - \begin{vmatrix} -23 & -1 & -2 \\ 30 & 3 & 2 \\ 4 & 2 & -3 \end{vmatrix} = -105$$

$$N_y = \begin{vmatrix} 3 & 7 & -1 & -4 \\ 1 & -10 & 3 & 2 \\ 1 & 0 & 2 & 1 \\ 2 & 1 & 0 & 3 \end{vmatrix} = \begin{vmatrix} 3 & 7 & -7 & -7 \\ 1 & -10 & 1 & 1 \\ 1 & 0 & 0 & 0 \\ 2 & 1 & -4 & 1 \end{vmatrix} = \begin{vmatrix} 7 & -7 & -7 \\ -10 & 1 & 1 \\ 1 & -4 & 1 \end{vmatrix} = 7 \begin{vmatrix} 1 & -1 & -1 \\ -10 & 1 & 1 \\ 1 & -4 & 1 \end{vmatrix} = -315$$

$$N_z = \begin{vmatrix} 3 & -2 & 7 & -4 \\ 1 & 0 & -10 & 2 \\ 1 & 4 & 0 & 1 \\ 2 & 3 & 1 & 3 \end{vmatrix} = \begin{vmatrix} 3 & -14 & 7 & -7 \\ 1 & -4 & -10 & 1 \\ 1 & 0 & 0 & 0 \\ 2 & -5 & 1 & 1 \end{vmatrix} = \begin{vmatrix} -14 & 7 & -7 \\ -4 & -10 & 1 \\ -5 & 1 & 1 \end{vmatrix} = -7 \begin{vmatrix} 2 & -1 & 1 \\ -4 & -10 & 1 \\ -5 & 1 & 1 \end{vmatrix} = 525$$

$$N_w = \begin{vmatrix} 3 & -2 & -1 & 7 \\ 1 & 0 & 3 & -10 \\ 1 & 4 & 2 & 0 \\ 2 & 3 & 0 & 1 \end{vmatrix} = \begin{vmatrix} 3 & -14 & -7 & 7 \\ 1 & -4 & 1 & -10 \\ 1 & 0 & 0 & 0 \\ 2 & -5 & -4 & 1 \end{vmatrix} = \begin{vmatrix} -14 & -7 & 7 \\ -4 & 1 & -10 \\ -5 & -4 & 1 \end{vmatrix} = -7 \begin{vmatrix} 2 & 1 & -1 \\ -4 & 1 & -10 \\ -5 & -4 & 1 \end{vmatrix} = 315$$

Then

$$x = \frac{N_x}{D} = \frac{-105}{-210} = \frac{1}{2}, \quad y = \frac{N_y}{D} = \frac{-315}{-210} = \frac{3}{2}, \quad z = \frac{N_z}{D} = \frac{525}{-210} = -\frac{5}{2}, \quad w = \frac{N_w}{D} = \frac{315}{-210} = -\frac{3}{2}.$$

Check. Using (1), $3\left(\frac{1}{2}\right) - 2\left(\frac{3}{2}\right) - \left(-\frac{5}{2}\right) - 4\left(-\frac{3}{2}\right) = \frac{3 - 6 + 5 + 12}{2} = 7.$

[NOTE: The above system permits some variation in procedure. For example, having found $x = \frac{1}{2}$
and $y = \frac{3}{2}$ using determinants, the value of w may be obtained by substituting in (4)

$$2\left(\tfrac{1}{2}\right) + 3\left(\tfrac{3}{2}\right) + 3w = 1, \qquad 3w = -\tfrac{9}{2}, \qquad w = -\tfrac{3}{2}$$

and the value of z may then be obtained by substituting in (2)

$$\left(\tfrac{1}{2}\right) + 3z + 2\left(-\tfrac{3}{2}\right) = -10, \qquad 3z = -\tfrac{15}{2}, \qquad z = -\tfrac{5}{2}.$$

The solution may be checked by substituting in (1) or (3).]

21.2   Solve the system $\begin{cases} 2x + y + 5z + w = 5 & (1) \\ x + y - 3z - 4w = -1 & (2) \\ 3x + 6y - 2z + w = 8 & (3) \\ 2x + 2y + 2z - 3w = 2 & (4) \end{cases}$

We have $D = \begin{vmatrix} 2 & 1 & 5 & 1 \\ 1 & 1 & -3 & -4 \\ 3 & 6 & -2 & 1 \\ 2 & 2 & 2 & -3 \end{vmatrix} = -120, \qquad N_x = \begin{vmatrix} 5 & 1 & 5 & 1 \\ -1 & 1 & -3 & -4 \\ 8 & 6 & -2 & 1 \\ 2 & 2 & 2 & -3 \end{vmatrix} = -240,$

$$N_y = \begin{vmatrix} 2 & 5 & 5 & 1 \\ 1 & -1 & -3 & -4 \\ 3 & 8 & -2 & 1 \\ 2 & 2 & 2 & -3 \end{vmatrix} = -24, \qquad N_z = \begin{vmatrix} 2 & 1 & 5 & 1 \\ 1 & 1 & -1 & -4 \\ 3 & 6 & 8 & 1 \\ 2 & 2 & 2 & -3 \end{vmatrix} = 0.$$

Then $\quad x = \dfrac{N_x}{D} = \dfrac{-240}{-120} = 2, \qquad y = \dfrac{N_y}{D} = \dfrac{-24}{-120} = \dfrac{1}{5} \quad$ and $\quad z = \dfrac{Nz}{D} = \dfrac{0}{-120} = 0.$

Substituting in (1), $2(2) + (\frac{1}{5}) + 5(0) + w = 5$ and $w = \frac{4}{5}$.
Check. Using (2), $(2) + (\frac{1}{5}) - 3(0) - 4(\frac{4}{5}) = -1$.

21.3   Show that the system $\begin{cases} 2x - y + 5z + w = 2 \\ x + y - z - 4w = 1 \\ 3x + 6y + 8z + w = 3 \\ 2x + 2y + 2z - 3w = 1 \end{cases}$ is inconsistent.

Since $D = \begin{vmatrix} 2 & 1 & 5 & 1 \\ 1 & 1 & -1 & -4 \\ 3 & 6 & 8 & 1 \\ 2 & 2 & 2 & -3 \end{vmatrix} = 0$ while $N_x = \begin{vmatrix} 2 & 1 & 5 & 1 \\ 1 & 1 & -1 & -4 \\ 3 & 6 & 8 & 1 \\ 1 & 2 & 2 & -3 \end{vmatrix} = -80 \neq 0$, the system is inconsistent.

21.4   Solve when possible: (a) $\begin{cases} 2x - 3y + z = 0 & (1) \\ x + 5y - 3z = 3 & (2) \\ 5x + 12y - 8z = 9 & (3) \end{cases}$ (c) $\begin{cases} 6x - 2y + z = 1 & (1) \\ x - 4y + 2z = 0 & (2) \\ 4x + 6y - 3z = 0 & (3) \end{cases}$

(b) $\begin{cases} x + 2y + 3z = 2 & (1) \\ 2x + 4y + z = -1 & (2) \\ 3x + 6y + 5z = 2 & (3) \end{cases}$ (d) $\begin{cases} x + 2y - 3z + 5w = 11 & (1) \\ 4x - y + z - 2w = 0 & (2) \\ 2x + 4y - 6z + 10w = 22 & (3) \\ 5x + y - 2z + 3w = 11 & (4) \end{cases}$

(a) Here $D = 0$; we shall eliminate the variable x.

$$(1) - 2(2): \quad -13y + 7z = -6$$
$$(3) - 5(2): \quad -13y + 7z = -6$$

Then $y = \dfrac{7z + 6}{13}$ and from (2), $x = 3 - 5y + 3z = \dfrac{4z + 9}{13}$.

The solutions may be written as $x = \dfrac{4a + 9}{13}$, $y = \dfrac{7a + 6}{13}$, $z = a$, where a is arbitrary.

(b) Here $D = 0$; we shall eliminate x.

$$(2) - 2(1): \quad -5z = -5$$
$$(3) - 3(1): \quad -4z = -4$$

Then $z = 1$ and each of the given equations reduces to $x + 2y = -1$. Note that the same situation arises when y is eliminated.

The solution may be written $x = -1 - 2y$, $z = 1$ or as $x = -1 - 2a$, $y = a$, $z = 1$, where a is arbitrary.

(c) Here $D = 0$; we shall eliminate z.

$$(2) - 2(1): \quad -11x = -2$$
$$(3) + 3(1): \quad 22x = 3$$

The system is inconsistent.

(d) Here $D = 0$; we shall eliminate x.

$$(2) - 4(1): \quad -9y + 13z - 22w = -44$$
$$(3) - 2(1): \quad 0 = 0$$
$$(4) - 5(1): \quad -9y + 13z - 22w = -44$$

Then $y = \dfrac{44 + 13z - 22w}{9}$ and, from (1), $x = \dfrac{11 + z - w}{9}$. The solutions are $x = \dfrac{11 + a - b}{9}$, $y = \dfrac{44 + 13a - 22b}{9}$, $z = a$, $w = b$, where a and b are arbitrary.

21.5 (a) The system of two equations in four unknowns

$$\begin{cases} x + 2y + 3z - 4w = 5 \\ 3x - y - 5z - 5w = 1 \end{cases}$$

may be solved for any two of the unknowns in terms of the others; for example, $x = 1 + z + 2w$, $y = 2 - 2z + w$.

(b) The system of three equations in four unknowns

$$\begin{cases} x + 2y + 3z - 4w = 5 \\ 3x - y - 5z - 5w = 1 \\ 2x + 3y + z - w = 8 \end{cases}$$

may be solved for any three of the unknowns in terms of the fourth; for example, $x = 1 + 4w$, $y = 2 - 3w$, $z = 2w$.

(c) The system of three equations in four unknowns

$$\begin{cases} x + 2y + 3z - 4w = 5 \\ 3x - y - 5z - 5w = 1 \\ 2x + 3y + z - w = 8 \end{cases}$$

may be solved for any two of the unknowns in terms of the others. Note that the third equation is the same as the second minus the first. We solve any two of these equations, say the first and second, and obtain the solution given in (a) above.

21.6 Solve when possible.

(a) $\begin{cases} 3x - 2y = 1 \\ 4x + 3y = 41 \\ 6x + 2y = 23 \end{cases}$ (b) $\begin{cases} 3x + y = 1 \\ 5x - 2y = 20 \\ 4x + 5y = -17 \end{cases}$ (c) $\begin{cases} x + y + z = 2 \\ 4x + 5y - 3z = -15 \\ 5x - 3y + 4z = 23 \\ 7x - y + 6z = 27 \end{cases}$ (d) $\begin{cases} x + y = 5 \\ y + z = 8 \\ x + z = 7 \\ 5x - 5y + z = 1 \end{cases}$

(a) The system $\begin{cases} 3x - 2y = 1 \\ 4x + 3y = 41 \end{cases}$ has solution $x = 5$, $y = 7$.

Since $6x + 2y = 6(5) + 2(7) \neq 23$, the given system is inconsistent.

(b) The system $\begin{cases} 3x + y = 1 \\ 5x - 2y = 20 \end{cases}$ has solution $x = 2$, $y = -5$.

Since $4x + 5y = 4(2) + 5(-5) = -17$, the given system is consistent with solution $x = 2$, $y = -5$.

(c) The system $\begin{cases} x + y + z = 2 \\ 4x + 5y - 3z = -15 \\ 5x - 3y + 4z = 23 \end{cases}$ has solution $x = 1$, $y = -2$, $z = 3$.

Since $7x - y + 6z = 7(1) - (-2) + 6(3) = 27$, the given system is consistent with solution $x = 1$, $y = -2$, $z = 3$.

(d) The system $\begin{cases} x + y = 5 \\ y + z = 8 \\ x + z = 7 \end{cases}$ has solution $x = 2$, $y = 3$, $z = 5$.

Since $5x - 5y + z = 5(2) - 5(3) + (5) \neq 1$, the given system is inconsistent.

(NOTE: If the constant of the fourth equation of the system were changed from 1 to 0, the resulting system would be consistent.)

21.7 Examine the following systems for nontrivial solutions:

(a) $\begin{cases} 2x - 3y + 3z = 0 \\ 3x - 4y + 5z = 0 \\ 5x + y + 2z = 0 \end{cases}$ (b) $\begin{cases} 4x + y - 2z = 0 \\ x - 2y + z = 0 \\ 11x - 4y - z = 0 \end{cases}$ (c) $\begin{cases} x + 2y + z = 0 \\ 3x + 6y + 3z = 0 \\ 5x + 10y + 5z = 0 \end{cases}$ (d) $\begin{cases} 2x + y = 0 \\ 3y - 2z = 0 \\ 2y + w = 0 \\ 4x - w = 0 \end{cases}$

(a) Since $D = \begin{vmatrix} 2 & -3 & 3 \\ 3 & -4 & 5 \\ 5 & 1 & 2 \end{vmatrix} \neq 0$, the system has only the trivial solution.

(b) Since $D = \begin{vmatrix} 4 & 1 & -2 \\ 1 & -2 & 1 \\ 11 & -4 & -1 \end{vmatrix} = 0$, there are nontrivial solutions.

The system $\begin{cases} 4x + y = 2z \\ x - 2y = -z \end{cases}$, for which $D = \begin{vmatrix} 4 & 1 \\ 1 & -2 \end{vmatrix} \neq 0$, has the solution $x = \frac{2}{3}$, $y = \frac{2z}{3}$. This solution may be written as $x = \frac{a}{3}$, $y = \frac{2a}{3}$, $z = a$ or $x = a$, $y = 2a$, $z = 3a$, where a is arbitrary, or as $x : y : z = 1 : 2 : 3$.

(c) Here $D = \begin{vmatrix} 1 & 2 & 1 \\ 3 & 6 & 3 \\ 5 & 10 & 5 \end{vmatrix} = 0$ and there are nontrivial solutions.

Since the minor of every element of D is zero, we cannot proceed as in (b). We solve the first equation for $x = -2y - z$ and write the solution as $x = -2a - b$, $y = a$, $z = b$, where a and b are arbitrary.

(d) Here $D = \begin{vmatrix} 2 & 1 & 0 & 0 \\ 0 & 3 & -2 & 0 \\ 0 & 2 & 0 & 1 \\ 4 & 0 & 0 & -1 \end{vmatrix} = 0$ and there are nontrivial solutions.

Take $x = a$, where a is arbitrary. From the first equation, $y = -2a$; from the second, $2z = 3y = -6a$ and $z = -3a$; and from the fourth equation, $w = 4a$.

Thus the solution is $x = a$, $y = -2a$, $z = -3a$, $w = 4a$, or $x : y : z : w = 1 : -2 : -3 : 4$.

Supplementary Problems

21.8 Solve, using determinants.

(a) $\begin{cases} x + y + z = 6 \\ y + z + w = 9 \\ z + w + x = 8 \\ w + x + y = 7 \end{cases}$ (b) $\begin{cases} 3x - 2y + 2z + w = 5 \\ 2x + 4y - z - 2w = 3 \\ 3x + 7y - z + 3w = 23 \\ x - 3y + 2z - 3w = -12 \end{cases}$ (c) $\begin{cases} x + y + z + w = 2 \\ 2x + 3y - 2z - w = 5 \\ 3x - 2y + z + 3w = 4 \\ 5x + 2y + 3z - 2w = -4 \end{cases}$

Ans. (a) $x = 1$, $y = 2$, $z = 3$, $w = 4$ (b) $x = 2$, $y = 1$, $z = -1$, $w = 3$
(c) $x = \frac{1}{2}$, $y = 1$, $z = -\frac{3}{2}$, $w = 2$

21.9 Test for consistency, and solve when possible.

(a) $\begin{cases} 2x + 3y - 4z = 1 \\ 3x - y + 2z = -2 \\ 5x - 9y + 14z = 3 \end{cases}$ (b) $\begin{cases} x + 7y + 5z = -22 \\ x - 9y - 11z = 26 \\ x - y - 3z = -22 \end{cases}$ (c) $\begin{cases} x + y + z = 4 \\ 2x - 4y + 11z = -7 \\ 4x + 6y + z = 21 \end{cases}$

Ans. (a) Inconsistent   (b) $x = 2z - 1$, $y = -z - 3$ (c) $x = \frac{1}{2}(3 - 5z)$, $y = \frac{1}{2}(5 + 3z)$

21.10   Solve, when possible.

(a) $\begin{cases} x - 3y + 11 = 0 \\ 3x + 2y - 33 = 0 \\ 2x - 3y + 4 = 0 \end{cases}$ (b) $\begin{cases} x - 2y - 8 = 0 \\ 3x + y - 3 = 0 \\ x - 10y + 32 = 0 \end{cases}$ (c) $\begin{cases} 2x - 3y - 7 = 0 \\ 5x + 4y + 17 = 0 \\ 4x - y + 1 = 0 \end{cases}$

   Ans.   (a) $x = 7, y = 6$ (b)   No solution   (c) $x = -1, y = -3$

21.11   Solve the systems.

(a) $\begin{cases} 2x - y + 3z = 8 \\ x + 3y - 2z = -3 \end{cases}$ (b) $\begin{cases} 4x + 2y + z = 13 \\ 2x + y - 2z = -6 \end{cases}$ (c) $\begin{cases} 4x + 2y - 6z + w = 10 \\ 3x - y - 9z - w = 7 \\ 7x + y - 11z - w = 13 \end{cases}$

   Ans.   (a) $x = 3 - z, y = -2 + z$ (b) $y = 4 - 2x, z = 5$
           (c) $x = 7w/10, y = 2 - 23w/20, z = -1 + w/4$

21.12   Examine for nontrivial solutions.

(a) $\begin{cases} 3x + y - 9z = 0 \\ 4x - 3y + z = 0 \\ 6x - 11y + 21z = 0 \end{cases}$ (b) $\begin{cases} 2x - 3y - 5z = 0 \\ x + 2y - 13z = 0 \\ 9x - 10y - 30z = 0 \end{cases}$ (c) $\begin{cases} 2x - 3y + 2z - 9w = 0 \\ x + 4y - z + 3w = 0 \\ 3x - 2y - 2z - 6w = 0 \\ 7x + 11y + 3z - 6w = 0 \end{cases}$

   Ans.   (a) $x = 2z, y = 3z$ (b) $x = y = z = 0$ (c) $x = 2w, y = -w, z = w$

21.13   Is it the case that the converse of the statement "A system of n homogeneous linear equations in n unknowns has only the trivial solution if D, the determinant of the coefficients, is not equal to zero is true?" In other words, is this an "if and only if" statement?

21.14   Solve Problem 21.12 using a computer software package.

Chapter 22

Introduction to Transformational Geometry

INTRODUCTION TO TRANSFORMATIONS. If you look back at previous chapters that included topics in geometry, you will notice that, while we have concentrated on different topics from chapter to chapter, all the material had one very important thing in common: The *positions* of all the geometric figures were *fixed*. In other words, when we considered a triangle such as $\triangle ABC$ in Fig. 22-1, we did not move it. In this chapter, we consider objects in geometry as they change position. These objects (such as triangles, lines, points, and circles) will move as a result of transformations of the plane.

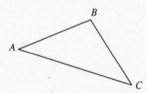

Fig. 22-1

DEFINITION: By a *transformation of the plane*, we mean a rule that assigns to each point in the plane a different point or the point itself.

Note that *each* point in the plane is assigned to *exactly one point*. Points that are assigned to themselves are called *fixed* points. If point P is assigned to point Q, then we say that the *image* of P is Q, and the image of Q is P.

REFLECTIONS. Imagine that a mirror is placed along line m in Fig. 22-2. What would be the image of point S in the mirror? How would you describe S', the image of S? If we actually placed a mirror along m, we would see that the image of S lies on l, on the other side of m, and that the distance from S to O is equal to the distance from O to S' (see Fig. 22-3). We say that S' is the image of S under a reflection in line m. Notice that, under this reflection, O is the image of O.

136

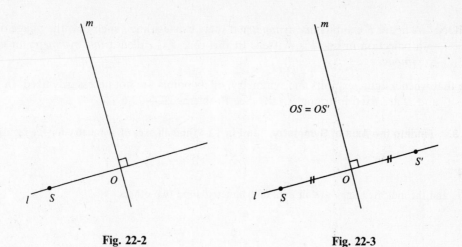

Fig. 22-2   Fig. 22-3

DEFINITION: A *reflection in line m* is a transformation of the plane having the property that the image of any point S not on m is S', where m is the perpendicular bisector of SS'; the image of any point O on m is O itself.

We write $R_m(S) = S'$ to mean S' is the image of S under the reflection in line m.

EXAMPLE 1.   Image of a Point.   Find the image of (a) A, (b) B, (c) C, (d) $\overline{AC}$, and (e) $\angle DAC$ under the reflection in line t indicated in Fig. 22-4.

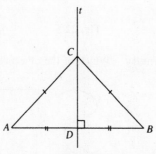

Fig. 22-4

SOLUTIONS

(a)   B, because t is the perpendicular bisector of $\overline{AB}$

(b)   A

(c)   C, because C is on t

(d) $\overline{BC}$ (Why?)

(e) $\angle DBC$, because D and C are fixed, and $R_t(B) = A$

EXAMPLE 2.   Image of a Triangle.   What is the image of $\triangle ABC$ in Fig. 22-4 under a reflection in line t?

SOLUTION

We saw that $R_t(A) = B$, $R_t(B) = A$, and $R_t(C) = C$; thus, $\triangle ABC$ is its own image. See Problems 22.1–22.3.

LINE SYMMETRY.   Notice that the images of angles are angles and the images of segments are segments under a reflection in a line. When a figure is its own image under a reflection in a line (like $\triangle ABC$ in Fig. 22-4), we say the figure has *line symmetry*.

DEFINITION: A figure *F* exhibits *line symmetry* if there exists a line *l* such that the image of *F* under a reflection in line *l* is *F* itself. In this case, *l* is called a *line symmetry* or an *axis of symmetry*.

Notice that when a figure exhibits line symmetry, all its points are not necessarily fixed. In Fig. 22-4, only points *C* and *D* are fixed in triangle *ABC*. See Problems 22.4–22.6.

EXAMPLE 3.   Finding the Axis of Symmetry.   In Fig. 22-5, find all axes of symmetry for the regular hexagon *ABCDEF*.

SOLUTION

$\overline{AD}$, $\overline{FC}$, $\overline{BE}$, and the indicated line *l* are all axes of symmetry. Find two others.

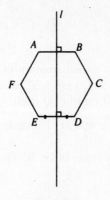

Fig. 22-5

EXAMPLE 4.   Discovering Line Symmetry.   Which of the objects in Fig. 22-6 exhibit line symmetry?

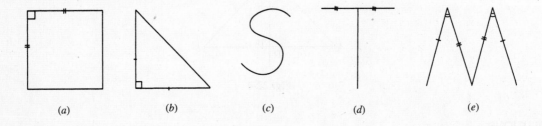

    (*a*)   (*b*)   (*c*)   (*d*)   (*e*)

Fig. 22-6

SOLUTION

All except (*c*).

POINT SYMMETRY.   Not only can we transform the plane by reflections in a line, but we can also reflect in a point *P*. In Fig. 22-7, for example, we can reflect *Q* in the point *P* by finding the point *Q'* such that $QP = PQ'$.

DEFINITION: A *reflection in the point P* is a transformation of the plane such that the image of any point *Q* except *P* is *Q'*, where $QP = PQ'$, *Q*, *P*, and *Q'* are collinear, and the image of *P* is *P* (i.e., *P* is fixed). If point *F* is its own image under such a transformation, then we say *F* exhibits *point symmetry*. (Note that $\overline{QP} = \overline{PQ'}$ means that they are equal in length. The segments $\overline{QP}$ and $\overline{PQ'}$ are congruent.

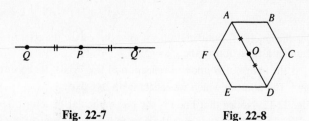

Fig. 22-7   Fig. 22-8

Figure 22-8 shows a regular hexagon $ABCDEF$, with $AO = OD$. Notice that A is the image of D under the reflection in O. We use the notation $R_O(A) = D$ and $R_O(D) = A$ to indicate that A and D are each other's images under a reflection in point O. See Problem 22.7.

EXAMPLE 5.   Finding Images under a Reflection in a Point. Referring to Fig. 22-7, find
(a) $R_O(B)$, (b) $R_O(C)$, (c) $R_O(\overline{AD})$, (d) $R_O(\angle AOB)$, and (e) $R_O(ABCDEF)$.

SOLUTIONS

(a) E (b) F (c) $\overline{AD}$ (d) $\overline{DOE}$.
(e)   Hexagon $ABCDEF$. (Thus, $ABCDEF$ exhibits point symmetry.)

EXAMPLE 6.   Finding Point Symmetry. Which of the following exhibit point symmetry?
   (a)   Squares   (b)   Rhombuses   (c)   Scalene triangles   (d)   S

SOLUTION

All except (c)

REFLECTIONS AND ANALYTIC GEOMETRY.   Since points can change position in transformational geometry, analytic geometry is a particularly useful tool for these transformations. Recall that in analytic geometry, we deal extensively with the positions of points; being able to locate points and determine distances is of great help in exploring the properties of transformations.

EXAMPLE 7.   Images under Reflections (Fig. 22-9).
   (a)   What is the image of point A under a reflection in the x axis? The y axis?
   (b)   What is the image of B under a reflection in the y axis?
   (c)   What is the image of O under a reflection in the point O?
   (d)   What is the image of B under a reflection in the line $y = x$?
   (e)   What is the image of A under a reflection in the line $x = -1$?
   (f)   What is the image of $\triangle AOB$ under a reflection in the y axis? Under a reflection in O?

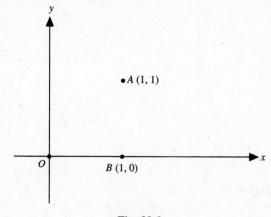

Fig. 22-9

SOLUTIONS

(a)   Point A' in Fig. 22-10 is the image of A under a reflection in the x axis; the coordinates of A' are $(1, -1)$. Point A'' is the image of A under a reflection in the y axis; $A'' = (-1, 1)$.

(b)   Point B' in Fig. 22-11 is the image of B under a reflection in the y axis. Its coordinates are $(-1, 0)$.

(c)   Point O is a fixed point. The point in which we reflect is always fixed.

(d) $R_l(B) = B'(0, 1)$ in Fig. 22-12. Notice that line l is the perpendicular bisector of $\overline{BB'}$.

(e) $R_m(A) = A'(-3, 1)$ in Fig. 22-13. Note that m is the perpendicular bisector of AA'.

(f)   The image of $\triangle AOB$ under a reflection in the y axis is $\triangle A'B'O$ in Fig. 22-14(a), where $A' = (-1, 1)$, $B' = (-1, 0)$, and $O = (0, 0)$. The image under a reflection in the origin is $\triangle A''B''O$ in Fig. 22-14(b), where $A'' = (-1, -1)$, $B'' = (-1, 0)$, and $O = (0, 0)$.

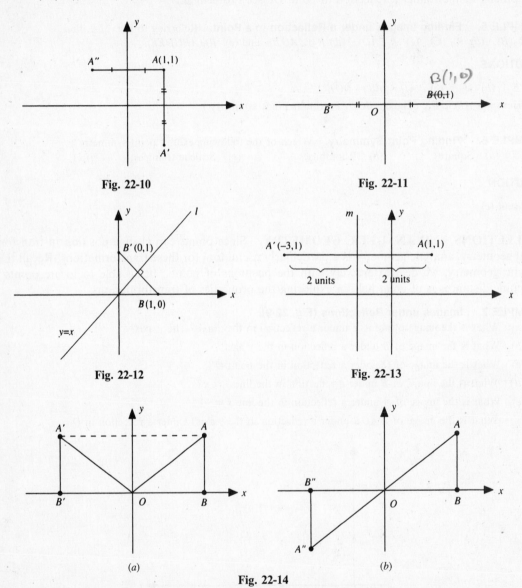

Fig. 22-10   Fig. 22-11

Fig. 22-12   Fig. 22-13

(a)   (b)

Fig. 22-14

PATTERNS IN REFLECTIONS. We can observe several patterns in the results of Example 7(f):

1.   The distance from A' to B' in Fig. 22-14(a) equals the distance from A to B. In other words, distance is *preserved* under a reflection. Observe that measures of angles are also preserved.

In other words, m∠BAO = m∠B'A'O in Fig. 22-14(a), and that property appears to hold for other reflections. As you will see, other properties are preserved as well.

2. Under a reflection in the x axis, the point (a, b) moves to $(a, -b)$; under a reflection in the y axis, (a, b) moves to $(-a, b)$; and under a reflection in the origin, (a, b) moves to $(-a, -b)$. These patterns hold only for these reflections.

EXAMPLE 8.   More Images under Reflections.   In Fig. 22-15, find

(a)   The reflection of C in the y axis

(b)   The reflection of B in the origin.

(c)   The reflection of △CAB in the x axis

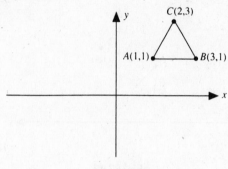

Fig. 22-15

SOLUTIONS

(a) $(-2, 3)$ (b) $(-3, -1)$ (c)   △C'A'B', where $C' = (2, -3), A' = (1, -1)$, and $B' = (3, -1)$. Sketch the triangle △A'B'C'.

TRANSLATIONS.   Let us transform △ABC in Fig. 22-16(a) by adding 1 to each x coordinate and 2 to each y coordinate. The result is shown in Fig. 22-16(b). Notice that △ABC does not change shape, but it does move in the plane, in the direction of ray $\overline{OD}$, where $D = (1, 2)$. The x coordinate of D is the "amount" by which the x coordinates of the triangle are shifted, and the y coordinate of D is the "amount" by which the y coordinates are shifted. We call this kind of transformation a *translation*.

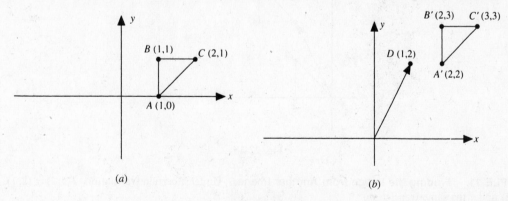

(a)   (b)

Fig. 22-16

DEFINITION: A *translation* is a transformation of the plane such that the image of every point (a, b) is the point $(a + h, b + k)$, where h and k are given values.

A translation has the effect of moving every point the *same distance* in the *same direction*. We use the notation $T_{(h,k)}(a,b)$ to mean the image of (a,b) under a translation of h units in the x direction and k units in the y direction.

As in a reflection, distance and angle measure are preserved in a translation.

EXAMPLE 9.   Finding the Image of a Point. Find $T_{(-1,1)}(1,4)$ and $T_{(-1,1)}(-1,2)$.

SOLUTION

$$T_{(-1,1)}(1,4) = (1+(-1), 4+1) = (0,5) \qquad T_{(-1,1)}(-1,2) = (-1+(-1), 2+1) = (-2,3)$$

Notice in Fig. 22-17 that $(1,4)$ and $(-1,2)$ are translated the same number of units in the same direction by the same translation T.

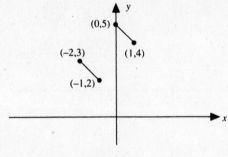

Fig. 22-17

EXAMPLE 10.   Finding the Image of a Triangle. Find the image of $\triangle ABC$ under the translation $T_{(1,2)}$, where $A = (0,0)$, $B = (1,1)$, and $C = (1,0)$.

SOLUTION

$T_{(1,2)}(0,0) = (1,2)$, $T_{(1,2)}(1,1) = (2,3)$, and $T_{(1,2)}(1,0) = (2,2)$. Hence, the image of $\triangle ABC$ is $\triangle A'B'C'$ in Fig. 22-18. All points are translated along ray $\overrightarrow{AA'} = \overrightarrow{OA'}$, where $A'(1,2)$ has the coordinates of the translation.

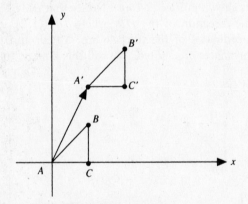

Fig. 22-18

EXAMPLE 11.   Finding the Image from Another Image. Under a certain translation, $T(5,2) = (7,1)$. Find $T(-3,6)$ under the same translation.

SOLUTION

We have $T_{(h,k)}(5,2) = (7,1)$. Thus, $5+h = 7$, or $h = 2$; and $2+k = 1$, or $k = -1$. Then $T_{(2,-1)}(-3,6) = (2+(-3), -1+6) = (-1,5)$.

EXAMPLE 12.   Finding Various Images Under Translation.

(a) Find $T_{(-1, 0)}(6, 2)$.

(b) Find h and k if $T_{(h, k)}(1, 7) = (0, 0)$.

(c) Find the image of square $ABCD$ under the translation $T_{(1, 1)}$, where $A = (0, 0), B = (1, 0), C = (0, 1)$, and $D = (1, 1)$.

(d) Find $T_{(h, k)}(1, 6)$ if $T_{(h, k)}(4, 1) = (0, -7)$.

(e) Find all fixed points under $T_{(-1, 4)}$.

SOLUTIONS

(a) $T(6, 2) = (6 + (-1), 2 + 0) = (5, 2)$ (b) $h = 0 - 1 = -1; k = 0 - 7 = -7$

(c) $A'B'C'D'$, where $A' = (1, 1), B' = (2, 1), C' = (1, 2)$, and $D' = (2, 2)$

(d) $h = 0 - 4 = -4$ and $k = -7 - 1 = -8$, so $T(1, 6) = (-3, -2)$

(e) Only $T_{(0,0)}$ has fixed points. Any other translation, including $T_{(-1,4)}$, has none.

EXAMPLE 13.   Finding Images of Figures.   Let $A = (1, 1)$, $B = (2, 2)$, and $C = (3, 1)$. Find the image under $T_{(2, -1)}$ of (a) $\overline{AB}$, (b) $\triangle ABC$, and (c) $\angle CBA$.

SOLUTIONS

(a) $A'B'$, where $A' = (3, 0)$ and $B' = (4, 1)$ (b) $A'B'C'$ with $C' = (5, 0)$ (c) $\angle C'B'A'$

ROTATIONS.   Consider square $ABCD$ in Fig. 22-19(a). Suppose we were to rotate that square counterclockwise 90° about P, as shown by the arrow. (Imagine that the square is separate from the page, but held to it by a pin through point P.) Then

The image of B would be A.
The image of D would be C.
The image of C would be B.
The image of A would be D.

Now consider point S in Fig. 22-19(b). We can rotate it counterclockwise by, say, 50° about P, as if it were one end of a ruler that was nailed to the page at P. The image of S' is in the diagram.

In both these rotations, the segment from P to the point being rotated is congruent to the segment from P to the image of that point.

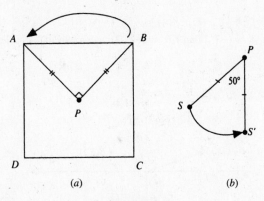

(a)   (b)

Fig. 22-19

DEFINITION: A rotation through an angle of measure θ degrees about a point P is a transformation of the plane such that the image of P is P and, for any point $B \neq P$, the image of B is B', where m$\angle BPB' = \theta$ and $BP = B'P$.

Figure 22-20 shows P, B, B', and θ. If $\theta > 0$, the rotation is counterclockwise. If $\theta < 0$, the rotation is clockwise.

We use the notation $\text{Rot}_{(P,\theta)}(B)$ to mean the rotation of point B about point P through $\theta°$. Segment lengths and measures of angles are preserved under rotations.

Let us rotate point B in Fig. 22-21 through $180°$ about O. The image of B is $B'(-2,-2)$. Notice that this is also the image of B under the reflection in O.

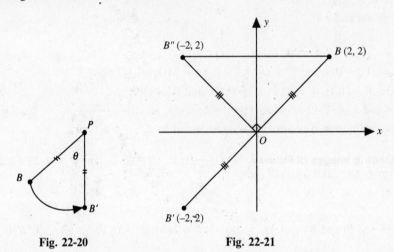

Fig. 22-20   Fig. 22-21

Now let us rotate point B through $90°$ about O. The image here is $B''(-2, 2)$. Notice that this is the image of B under the reflection in the y axis.

Be Careful.   These two similarities of transformations are coincidences. They come about because we are rotating $90°$ and $180°$ *about the origin*. Do not generalize beyond these cases! Note, though, that these coincidences yield the following formulas:

$$\text{Rot}_{(P,90°)}(a,b) = (-b, a) \qquad \text{and} \qquad \text{Rot}_{(O,180°)}(a,b) = (-a, -b)$$

EXAMPLE 14.   Finding the Rotation of a Point.   Let $A = (1, 3)$ and $B = (2, 1)$, and find (a) $\text{Rot}_{(O,\ 90°)}(A)$, (b) $\text{Rot}_{(O,\ 90°)}(B)$, and (c) the image of AB under a rotation of $90°$ about O.

SOLUTIONS

(a) $Rot_{(O,\ 90°)}(a, b) = (-b, a) = (-3, 1)$ (b) $Rot_{(O,\ 90°)}(2, 1) = (-1, 2)$

(c)   The image of $\overline{A'B'}$, where $A' = (-3, 1)$ and $B' = (-1, 2)$, as shown in Fig. 22-22.

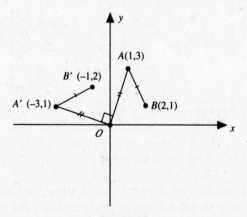

Fig. 22-22

EXAMPLE 15.   Finding the Image of a Triangle

(a)   Find the image of $\triangle ABC$ under a rotation of 180° about O, if $A = (1, 3)$, $B = (2, 1)$, and $C = (1, 1)$.

(b)   Find the image of $\angle BAC$.

SOLUTIONS

(a) $\text{Rot}_{(O, 180°)}(A) = (-1, -3) = A'$ (b)   The image of $\angle BAC$ is $\angle B'A'C'$.

  $\text{Rot}_{(O, 180°)}(B) = (-2, -1) = B'$

  $\text{Rot}_{(O, 180°)}(C) = (-1, -1) = C'$

  The image of $\triangle ABC$ is $\triangle A'B'C'$.

SYMMETRY OF ROTATION.   The image of the square in Fig. 22-19 under a rotation of 90° is the square itself. This is also true for a rotation of −90°, or 180°, and so on.

EXAMPLE 16.   Determining Rotational Symmetry (Fig. 22-23)

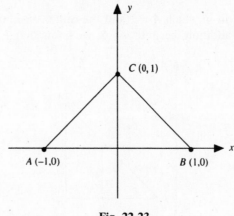

Fig. 22-23

(a)   Find $\text{Rot}_{(O, 45°)}(O)$.

(b)   Find $\text{Rot}_{(O, 90°)}$ of $A, B,$ and C.

(c)   Find $\text{Rot}_{(O, -90°)}$ of $A, B,$ and C.

(d)   Find the image of $\triangle ABC$ under $\text{Rot}_{(O, -90°)}$.

(e)   Does $\triangle ABC$ exhibit rotational symmetry?

SOLUTIONS

(a) $\text{Rot}(O) = 0$

(b) $\text{Rot}(A) = A'(0, 1); \text{Rot}(B) = B'(0, 1); \text{Rot}(C) = C'(-1, 0)$

(c) $\text{Rot}(A) = A''(0, 1); \text{Rot}(B) = B''(0, -1); \text{Rot}(C) = C''(1, 0)$

(d) $\text{Rot}(\triangle ABC) = A''B''C''$

(e)   No, because it is not its own image for any rotation except one of 360°.

DILATIONS.   Suppose we blew a balloon up most of the way and traced its outline, and then blew it up all the way and traced it again. The outlines might look like those in Fig. 22-24. Although the balloon has changed size from (a) to (b), its shape has not changed. Notice that if C is on AB, then its image C' is on $\overline{A'B'}$. Such a transformation in the plane is called a *dilation* (or dilatation). The "reverse" transformation is also a dilation: The balloon could be reduced in size in a transformation from that in (b) to that in (a).

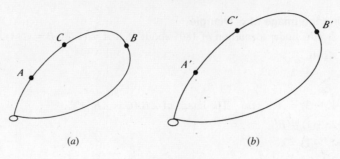

Fig. 22-24

DEFINITION: Given a point P in the plane and a positive number n, a transformation of the plane having the following properties is called a *dilation* of n, and P is called the *center of dilation*: Point P is fixed, and for any point Q, the image of Q is the point Q' such that $PQ' = (n)(PQ)$ and $\overline{PQ}$ and $\overline{PQ'}$ are identical rays. The point Q' is usually denoted $D_n(Q)$.

Figure 22-25 shows a dilation in which $n = 2$ and the center of dilation is P. Hence, $D_2(A) = A'$, $D_2(B) = B'$, and $D_2(P) = P$. In addition, because $n = 2$, we know that $PA' = 2PA$ and $PB' = 2PB$.

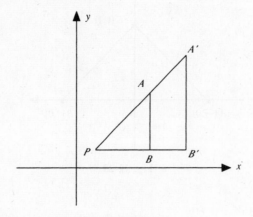

Fig. 22-25

Several properties of dilations are evident in Fig. 22-25:

(1)   Dilations *do not* preserve distance.
(2)   The image of a figure is similar to the figure under a dilation. In Fig. 22-25, $\triangle PAB \sim PA'B'$.
(3)   Angles *are* preserved under dilations (because of item 2 above).

When the center of a dilation is $O = (0,0)$, we can find the images of points very easily: $D_n(x, y) = (nx, ny)$.

EXAMPLE 17.   Finding the Dilation of a Triangle.   Find the image of $\triangle ABC$ in Fig. 22-26 under a dilation of $n = \frac{1}{2}$ with center of dilation at $(0, 0)$.

SOLUTION

$D_{1/2}(1, 1) = (\frac{1}{2}, \frac{1}{2}) = B'$ as shown in Fig. 22-26. Also, $D_{1/2}(1, 0) = (\frac{1}{2}, 0) = A'$ and $D_{1/2}(2, 1) = (1, \frac{1}{2}) = C'$. Then $\triangle A'B'C'$ is the image of $\triangle BAC$, and $\triangle B'A'C' \sim \triangle BAC$. Note that the image here is smaller than the original triangle because $n = \frac{1}{2}$.

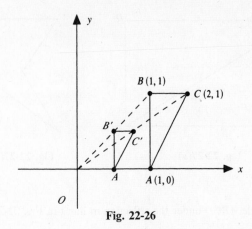

Fig. 22-26

EXAMPLE 18.   Finding an Unknown.   Given that $D_n(8,0) = (1,0)$, find n for a dilation in which $(0,0)$ is the center of dilation.

SOLUTION

Since the origin is the center of dilation, $(1,0) = (n8, n0)$. Therefore, $8n = 1$ and $n = \frac{1}{8}$.

EXAMPLE 19.   Dilating a Square.   Draw a square $ABCD$ in the coordinate plane such that $A = (1,1), B = (1,2), C = (2,1)$, and $D = (2,2)$. Then,

(a)   With O as the center of dilation, find the image of $ABCD$ under a dilation with $n = \frac{1}{3}$.

(b)   Find the midpoint M of $\overline{AB}$ and the midpoint M' of $\overline{A'B'}$.

(c)   Find $D_{1/3}(M)$.

SOLUTIONS

(a)   For a dilation with center $(0,0)$ and $n = \frac{1}{3}$, we have $D(x, y) = \left(\frac{1}{3}x, \frac{1}{3}y\right)$. The image of $ABCD$ is $A'B'C'D'$, where $A' = \left(\frac{1}{3}, \frac{1}{3}\right), B' = \left(\frac{1}{3}, \frac{2}{3}\right), C' = \left(\frac{2}{3}, \frac{1}{3}\right)$, and $D' = \left(\frac{2}{3}, \frac{2}{3}\right)$.

(b) $M = \left(\frac{1}{2}(1+1), \frac{1}{2}(1+2)\right) = \left(1, \frac{3}{2}\right)$ and $M' = \left(\frac{1}{2}(1), \frac{1}{2}\left(\frac{2}{3}\right)\right) = \left(\frac{1}{3}, \frac{1}{3}\right)$.

(c) $D(M) = M'$

PROPERTIES OF TRANSFORMATIONS.   We are now in a position to summarize the properties of transformations. In particular, we are interested in what is preserved under each kind of transformation.

(1)   Reflections preserve (a) distance, (b) angle measure, (c) midpoints, (d) parallelism, and (e) collinearity.

(2)   Translations preserve these same five properties, (a) through (e).

(3)   Rotations preserve all five properties as well.

(4)   Dilations preserve all except distance, that is, (b) through (e).

Solved Problems

22.1   Find the image of each of the following under the reflection in line t in Fig. 22-27(a): (a) point D, (b) point C, (c) point B, (d) $\overline{AC}$.

   Ans.   (a) C (b) D (c) B (d) $\overline{AD}$

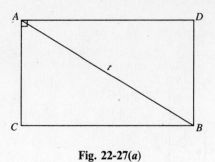

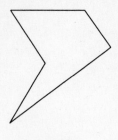

Fig. 22-27(a)   Fig. 22-27(b)

22.2   Find the image of rectangle *ABCD* under the reflection in line *t* in Fig. 22-27(a).

   Ans.   Rectangle *ABCD*

22.3   Is it true or false that every circle is its own image under a reflection in a diameter?

   Ans.   True

22.4   Find all axes of symmetry for the rectangle in Fig. 22-27(a).

   Ans.   *t* is one such axis. Are there any which are not diagonals?

22.5   Give (or draw) an example of a five-sided polygon that does not exhibit line symmetry.

   Ans.   There are many; Fig. 22-27(b) is one.

22.6   Explain why each figure in Fig. 22-28 exhibits line symmetry.

   Hint: For each, find an axis of symmetry.

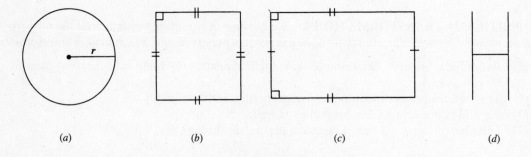

(a)   (b)   (c)   (d)

Fig. 22-28

22.7   In Fig. 22-29, find (a) $R_O(B)$,   (b) $R_O(A)$,   (c) $R_O(O)$,   (d) $R_O(\triangle AOB)$.

   Ans. (a) $(-2,0)$ (b) $(-2,-2)$ (c) O'

   (d) $\triangle A'OB'$ where $A' = (-2,-2)$ and $B' = (-2,0)$

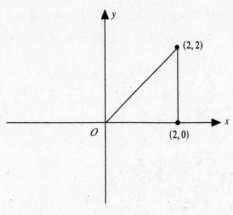

Fig. 22-29

Supplementary Problems

22.8   In Fig. 22-30, find:

    (*a*) The image of *E* under a reflection in the *y* axis   *Ans. E*

    (*b*) The image of *B* under a reflection in the *y* axis   *Ans. C*

    (*c*) The image of $\overline{AB}$ under a reflection in the *y* axis   *Ans. BA*

    (*d*) The image of *BC* under a reflection in the *x* axis   *Ans.* $B'C'$ where $B' = (-1, -1)$; $C' = (1, -1)$

    (*e*) The image of *ABCD* under a reflection in the *x* axis   *Ans.* $A'B'C'D'$ where $A' = A$, $D' = D$

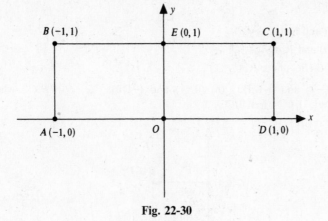

Fig. 22-30

22.9   Find (*a*) $T_{(1,3)}(2,8)$ (*b*) $T_{(1,3)}(6,5)$ (*c*) $T_{(1,3)}(0,0)$ (*d*) $T_{(1,3)}(1,1)$.

    Ans. (*a*) (3, 11)   (*b*) (7, 8)   (*c*) (1, 3)   (*d*) (2, 4)

22.10   Find the image of rectangle *ABCD* in Fig. 22-30 under the translation $T_{(3,6)}$.

    Ans. $A'B'C'D'$ where $A' = (2, 6)$, etc.

22.11   Under a particular translation, $T(3, 4) = (0, 0)$. Find $T(-8, -6)$ under that same translation.

    Ans. Then *T*, here, $= T_{(-3,-4)}$; $T(-8, -6) = (-11, -10)$

22.12 Find (a) $T_{(4,3)}(0,-6)$; (b) $T_{(h,k)}(3,7)$; (c) $T_{(h,k)}(e,f)$; (d) $T_{(h,k)}(4,1)$ if $T_{(h,k)}(1,1) = (2,2)$.

   Ans. (a) $(4,-3)$;   (b) $(4+h, 7+k)$;   (c) $(e+h, f+k)$;   (d) $(5,2)$

22.13 In Fig. 22-31, find (a) $T_{(0,0)}(EF)$; (b) $T_{(1,0)}(EF)$; (c) $T_{(0,1)}(EF)$; (d) $T_{(0,0)}(\triangle OEF)$.

   Ans. (a) EF;   (b) $E' = (1,1)$; $F' = (2,0)$;   (c) $E' = (0,2)$; $F' = (1,1)$;   (d) $O' = O$, etc.

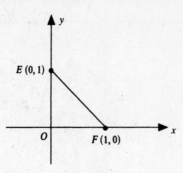

Fig. 22-31

22.14 Let $x = (4,1)$ and $y = (0,3)$, and find (a) $\text{Rot}_{(O,\,90°)}(x)$; (b) $\text{Rot}_{(O,\,90°)}(y)$; (c) the image of yx under a rotation of 90° about O.

   Ans. (a) $(-1,4)$;   (b) $(-3,0)$;   (c) $y'x'$, where $y' = (-3,0)$, $x' = (-1,4)$

22.15 Find the image of $\triangle EOF$ in Fig. 22-31 under a rotation of 180° about O.

   Ans. $O' = 0$, $E' = (0,-1)$, $F' = (-1,0)$; $\triangle E'O'F'$ is the image

24.16 In Fig. 22-32, find

   (a) $\text{Rot}_{(O,\,90°)}(A)$ and $\text{Rot}_{(O,\,90°)}(B)$
   (b) $\text{Rot}_{(O,\,-90°)}(A)$ and $\text{Rot}_{(O,\,90°)}(C)$
   (c) the image of $OABC$ under $\text{Rot}_{(O,\,-90°)}$.

   Ans. (a) $(0,-1)$ and $(-1,1)$ (b) $(0,1)$ and $(-1,0)$ (c) $O'A'B'C'$; where $O' = 0$, $A' = (0,1)$, $B' = (-1,1)$, $C' = (-1,0)$

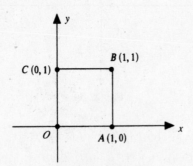

Fig. 22-32

22.17 Find (a) $D_{1/3}(-1,3)$, (b) $D_{1/2}(5,-3)$, (c) $D_4(0,0)$, and (d) $D_5(1,6)$, where the center of dilation in each case is the origin.

   Ans. (a) $\left(-\frac{1}{3}, 1\right)$;   (b) $\left(\frac{5}{2}, -\frac{3}{2}\right)$;   (c) $(0,0)$;   (d) $(5,30)$

22.18 If the center of a dilation D is $(0,0)$ and $D_n(3,6) = (5,10)$, find n and $D_n(0,-7)$.

      Ans. $n = \frac{5}{3}$; $\left(0, -\frac{35}{5}\right)$

22.19 For A and B as given in Fig. 22-33 and dilations with centers at $(0, 0)$, find

    (*a*) The image of $\triangle OAB$ under a dilation with $n = \frac{1}{2}$

    (*b*) The image of the midpoint of $\overline{AB}$ under a dilation with $n = 3$

      Ans. (*a*) $O' = 0, A' = \left(0, \frac{1}{2}\right)$, $B' = \left(\frac{1}{2}, 0\right)$; image is $\triangle OA'B'$ (*b*) midpoint: $\left(\frac{1}{2}, \frac{1}{2}\right)$; image is $\left(\frac{3}{2}, \frac{3}{2}\right)$

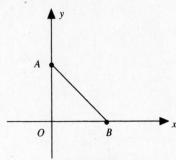

Fig. 22-33

22.20 Prove that dilations preserve "shapes" for triangles. In other words, prove that if the image of $\triangle ABC$ under a dilation is $\triangle A'B'C'$, then $\triangle A'B'C' \sim \triangle ABC$.

22.21 Prove that translations preserve angle measure.

PART III

TOPICS IN PRECALCULUS

Chapter 23

Angles and Arc Length

TRIGONOMETRY, as the word implies, is concerned with the measurement of the parts of a triangle. Plane trigonometry, considered in the next several chapters, is restricted to triangles lying in planes. Spherical trigonometry deals with certain triangles which lie on spheres.

The science of trigonometry is based on certain ratios, called trigonometric functions, to be defined in the next chapter. The early applications of the trigonometric functions were to surveying, navigation, and engineering. These functions also play an important role in the study of all sorts of vibratory phenomena—sound, light, electricity, etc. As a consequence, a considerable portion of the subject matter is concerned properly with a study of the properties of and relations among the trigonometric functions.

THE PLANE ANGLE XOP is formed by the two intersecting half lines OX and OP. The point O is called the *vertex* and the half lines are called the sides of the angle. See Fig. 23-1.

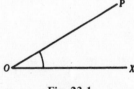

Fig. 23-1

More often, a plane angle is to be thought of as generated by revolving (in plane) a half line from the initial position OX to a terminal position OP. Then O is again the vertex, OX is called the *initial side*, and OP is called the *terminal side* of the angle.

An angle, so generated, is called *positive* if the direction of rotation (indicated by a curved arrow) is counterclockwise and *negative* if the direction of rotation is clockwise. The angle is positive in Figs. 23-2(*a*) and (*c*), and negative in Fig. 23-2(*b*).

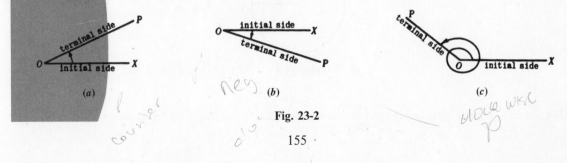

(a)   (b)   (c)

Fig. 23-2

MEASURES OF ANGLES

A. A *degree* (°) is defined as the measure of the central angle subtended by an arc of a circle equal to $\frac{1}{360}$ of the circumference of the circle.

A *minute* ($'$) is $\frac{1}{60}$ of a degree; a second ($''$) is $\frac{1}{60}$ of a minute.

EXAMPLE 1

(*a*) $\frac{1}{4}(36°24') = 9°6'$ (*b*) $\frac{1}{2}(127°24') = \frac{1}{2}(126°84') = 63°42'$

(*c*) $\frac{1}{2}(81°15') = \frac{1}{2}(80°75') = 40°37.5'$ or $40°37'30''$

(*d*) $\frac{1}{4}(74°29'20'') = \frac{1}{4}(72°149'20'') = \frac{1}{4}(72°128'80'') = 18°37'20''$

We write m $\angle A$ or m $\measuredangle A$ to denote the "measure of angle *A*."

B. A *radian* (rad) is defined as the measure of the central angle subtended by an arc of a circle equal to the radius of the circle. See Fig. 23-3.

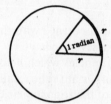

Fig. 23-3

The circumference of a circle = 2π (radius) and subtends an angle of 360°. Then 2π radians = 360°, from which we obtain

$$1 \text{ radian} = \frac{180°}{\pi} = 57.296° = 57°17'45'', \text{ approximately}$$

and

$$1 \text{ degree} = \frac{\pi}{180} \text{ radian} = 0.017453 \text{ rad, approximately}$$

where $\pi = 3.14159$ approximately.

EXAMPLE 2   (*a*) $\frac{7}{12}\pi \text{ rad} = \frac{7\pi}{12} \cdot \frac{180°}{\pi} = 105°$,   (*b*) $50° = 50 \cdot \frac{\pi}{180} \text{ rad} = \frac{5\pi}{18} \text{ rad}$. (See Problems 23.1–23.3.)

ARC LENGTH.   On a circle of radius *r*, a central angle of θ radians intercepts an arc of length $s = r\theta$; that is, arc length = radius × the measure of the central angle in radians. See Fig. 23-4.

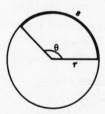

Fig. 23-4

(NOTE: *s* and *r* may be measured in any convenient unit of length but they must be expressed in the same unit.)

EXAMPLE 3

(a) On a circle of radius 30 in., the length of arc intercepted by a central angle of $\frac{1}{3}$ radian is

$$s = r\theta = 30(\tfrac{1}{3}) = 10 \text{ in.}$$

(b) On the same circle a central angle of 50° intercepts an arc length

$$s = r\theta = 30\left(\frac{5\pi}{18}\right) = \frac{25\pi}{3} \text{ in.}$$

(c) On the same circle an arc length $1\frac{1}{2}$ ft subtends a central angle

$$m \measuredangle \theta = \frac{s}{r} = \frac{18}{30} = \frac{3}{5} \text{ rad} \qquad \text{when } s \text{ and } r \text{ are expressed in inches}$$

or

$$m \measuredangle \theta = \frac{s}{r} = \frac{\frac{3}{2}}{\frac{5}{2}} = \frac{3}{5} \text{ rad} \quad \text{when } s \text{ and } r \text{ are expressed in feet}$$

(See Problems 23.4–23.5.)

(NOTE: Throughout the remainder of this book, degree measure and radian measure will be used. The reader should make certain that he or she knows how to use a calculator in both of these modes of angle measure.)

Solved Problems

23.1   Express each angle in radian measure: (a) 30°,   (b) 135°,   (c) 25°30′,   (d) 42°24′35″.

Since $1° = \frac{\pi}{180}$ radian = 0.017453 rad,

(a) $30° = 30 \times \frac{\pi}{180}$ rad $= \frac{\pi}{6}$ rad   or   0.5236 rad

(b) $135° = 135 \times \frac{\pi}{180}$ rad $= \frac{3\pi}{4}$ rad   or   2.3562 rad

(c) $25°30′ = 25.5° = 25.5 \times \frac{\pi}{180}$ rad $= 0.4451$ rad

(d) $42°24′35″ = 42° + \left(\frac{24 \times 60 + 35}{3600}\right)° = 42.41° = 42.41 + \frac{\pi}{180}$ rad $= 0.7402$ rad

23.2   Express each angle in degree measure: (a) $\frac{\pi}{3}$ rad,   (b) $\frac{5\pi}{9}$ rad,   (c) $\frac{2}{5}$ rad,   (d) $\frac{4}{3}$ rad.

Since 1 rad $= \frac{180°}{\pi} = 57°17′45″$,

(a) $\frac{\pi}{3}$ rad $= \frac{\pi}{3} \times \frac{180°}{\pi} = 60°$ (b) $\frac{5\pi}{9}$ rad $= \frac{5\pi}{9} \times \frac{180°}{\pi} = 100°$

(c) $\frac{2}{5}$ rad $= \frac{2}{5} \times \frac{180°}{\pi} = \frac{72°}{\pi}$ or $\frac{2}{5}(57°17′45″) = 22°55′6″$

(d) $\frac{4}{3}$ rad $= \frac{4}{3} \times \frac{180°}{\pi} = \frac{240°}{\pi}$ or $\frac{4}{3}(57°17′45″) = 76°23′40″$

23.3   A wheel is turning at the rate of 48 rpm (revolutions per minute or rev/min). Express this angular speed in (a) rev/s,   (b) rad/min,   (c) rad/s.

(a) 48 rev/min $= \frac{48}{60}$ rev/s $= \frac{4}{5}$ rev/s

(b) Since 1 rev $= 2 = 2\pi$ rad, 48 rev/min $= 48(2\pi)$ rad/min $= 301.6$ rad/min
(c) 48 rev/min $= \frac{4}{5}$ rev/s $= \frac{4}{5}(2\pi)$ rad/s $= 5.03$ rad/s   or

$$48 \text{ rev/min} = 96\pi \text{ rad/min} = \frac{96\pi}{60} \text{ rad/s} = 5.03 \text{ rad/s}$$

23.4   The minute hand of a clock is 12 in. long. How far does the tip of the hand move during 20 min?

During 20 min the hand moves through an angle $\theta = 120° = 2\pi/3$ rad and the tip of the hand moves over a distance $s = r\theta = 12(2\pi/3) = 8\pi$ in. $= 25.1$ in.

23.5   A central angle of a circle of radius 30 in. intercepts an arc of 6 in. Express the measure of the central angle θ in radians and in degrees.

$$\text{m} \angle \theta = \frac{s}{r} = \frac{6}{30} = \frac{1}{5} \text{ rad} = 11°27'33''$$

23.6   A railroad curve is to be laid out on a circle. What radius should be used if the track is to change direction by 25° in a distance of 120 ft?

We are required to find the radius of a circle on which a central angle θ, with measure 25° or $5\pi/36$ rad, intercepts an arc of 120 ft. Then

$$r = \frac{s}{\theta} = \frac{120}{5\pi/36} = \frac{864}{\pi} \text{ ft} = 275 \text{ ft}$$

23.7   Assuming the earth to be a sphere of radius 3960 miles, find the distance of a point in latitude 36°N from the equator.

Since $36° = \frac{\pi}{5}$ radian, $s = r\theta = 3960\left(\frac{\pi}{5}\right) = 2488$ miles.

23.8   Two cities 270 miles apart lie on the same meridian. Find their difference in latitude.

$$\text{m} \angle \theta = \frac{s}{r} = \frac{270}{3960} = \frac{3}{44} \text{ rad}   \text{or}   3°54.4'$$

23.9   A wheel 4 ft in diameter is rotating at 80 rpm. Find the distance (in ft) traveled by a point on the rim in 1 s, that is, the linear speed of the point (in ft/s).

$$80 \text{ rpm} = 80\left(\frac{2\pi}{60}\right) \text{ rad/s} = \frac{8\pi}{3} \text{ rad/s}$$

Then in 1 s the wheel turns through an angle θ measuring $8\pi/3$ rad and a point on the wheel will travel a distance $s = r\theta = 2(8\pi/3)$ ft $= 16.8$ ft. The linear velocity is 16.8 ft/s.

23.10   Find the diameter of a pulley which is driven at 360 rpm by a belt moving at 40 ft/s.

$$360 \text{ rev/min} = 360\left(\frac{2\pi}{60}\right) \text{ rad/s} = 12\pi \text{ rad/s}$$

Then in 1 s the pulley turns through an angle θ measuring 12π rad and a point on the rim travels a distance $s = 40$ ft.

$$d = 2r = 2\left(\frac{s}{\theta}\right) = 2\left(\frac{40}{12\pi}\right) \text{ ft} = \frac{20}{3\pi} \text{ ft} = 2.12 \text{ ft}$$

23.11   A point on the rim of a turbine wheel of diameter 10 ft moves with a linear speed 45 ft/s. Find the rate at which the wheel turns (angular speed) in rad/s and in rev/s.

In 1s a point on the rim travels a distance $s = 45$ ft. Then in 1 s the wheel turns through an angle $\theta = s/r = \frac{45}{5} = 9$ radians and its angular speed is 9 rad/s.

Since 1 rev $= 2\pi$ rad or 1 rad $= \dfrac{1}{2\pi}$ rev, 9 rad/s $= 9\left(\dfrac{1}{2\pi}\right)$ rev/s $= 1.43$ rev/s.

Supplementary Problems

23.12 Express in radian measure: (*a*) 25°, (*b*) 160°, (*c*) 75°30′, (*d*) 112°40′, (*e*) 12°12′20″.

   Ans. (*a*) 5π/36 rad or 0.4363 rad   (*c*) 151π/360 rad or 1.3177 rad   (*e*) 0.2130 rad
   (*b*) 8π/9 rad or 2.7925 rad   (*d*) 169π/270 rad or 1.9664 rad

23.13 Express in degree measure: (*a*) π/4 rad, (*b*) 7π/10, (*c*) 5π/6 rad, (*d*) $\frac{1}{4}$ rad, (*e*) $\frac{7}{5}$ rad.

   Ans. (*a*) 45°   (*b*) 126°   (*c*) 150°   (*d*) 14°19′26″   (*e*) 80°12′51″

23.14 On a circle of radius 24 in., find the length of arc subtended by a central angle (*a*) of $\frac{2}{3}$ rad, (*b*) of 3π/5 rad, (*c*) of 75°, (*d*) of 130°.

   Ans. (*a*) 16 in.   (*b*) 14.4π or 45.2 in.   (*c*) 10π or 31.4 in.   (*d*) 52π/3 or 54.5 in.

23.15 A circle has a radius of 30 in. How many radians are there in an angle at the center subtended by an arc (*a*) of 30 in.,   (*b*) of 20 in.,   (*c*) of 50 in.?

   Ans. (*a*) 1 rad   (*b*) $\frac{2}{3}$ rad   (*c*) $\frac{5}{3}$ rad

23.16 Find the radius of the circle for which an arc 15 in. long subtends an angle (*a*) of 1 rad, (*b*) of $\frac{2}{3}$ rad, (*c*) of 3 rad, (*d*) of 20°, (*e*) of 50°.

   Ans. (*a*) 15 in. (*b*) 22.5 in. (*c*) 5 in. (*d*) 43.0 in. (*e*) 17.2 in.

23.17 The end of a 40-in. pendulum describes an arc of 5 in. Through what angle does the pendulum swing?

   Ans. $\frac{1}{8}$ rad or 7°9′43″

23.18 A train is traveling at the rate of 12 mi/hr on a curve of radius 3000 ft. Through what angle has it turned in 1 min?

   Ans. 0.352 rad or 20°10′

23.19 A reversed curve on a railroad track consists of two circular arcs. The central angle of one measures 20° with radius 2500 ft and the central angle of the other measures 25° with radius 3000 ft. Find the total length of the two arcs.

   Ans. 6250π/9 ft or 2182 ft

23.20 A flywheel or radius 10 in. is turning at the rate of 900 rpm. How fast does a point on the rim travel in ft/s?

   Ans. 78.5 ft/s

23.21 An automobile tire is 30 in. in diameter. How fast (rpm) does the wheel turn on the axle when the automobile maintains a speed of 45 mph?

   Ans.   504 rpm

23.22 In grinding certain tools the linear velocity of the grinding surface should not exceed 6000 ft/s. Find the maximum number of revolutions per second (*a*) of a 12-in. (diameter) emery wheel, (*b*) of an 8-in. wheel.

   Ans.   (*a*) $6000/\pi$ rev/s or 1910 rev/s   (*b*)   2865 rev/s

23.23 If an automobile wheel 32 in. in diameter rotates at 800 rpm, what is the speed of the car in mph?

   Ans.   76.2 mph

Chapter 24

Trigonometric Functions of a General Angle

ANGLES IN STANDARD POSITION. With respect to a rectangular coordinate system, an angle is said to be *in standard position* when its vertex is at the origin and its initial side coincides with the positive *x* axis.

An angle is said to be a *first quadrant angle* or to be *in the first quadrant* if, when in standard position, its terminal side falls in the quadrant. Similar definitions hold for the other quadrants. For example, the angles 30°, 59°, and −330° are first quadrant angles; 119° is a second quadrant angle; −119° is a third quadrant angle; −10° and 710° are fourth quadrant angles. See Figs. 24-1 and 24-2.

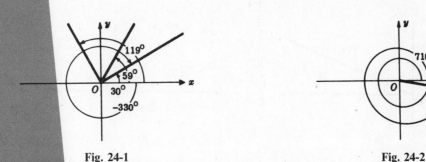

Fig. 24-1   Fig. 24-2

Two angles which, when placed in standard position, have coincident terminal sides are called *coterminal angles*. For example, 30° and −330°, −10° and 710° are pairs of coterminal angles. There are an unlimited number of angles coterminal with a given angle. (See Problem 24.1.)

The angles 0°, 90°, 180°, 270°, and all angles conterminal with them are called *quadrantal angles*.

TRIGONOMETRIC FUNCTIONS OF A GENERAL ANGLE. Let θ be an angle (not quadrantal) in standard position and let $P(x, y)$ be any point, distinct from the origin, on the terminal side of the angle.

161

The six trigonometric functions of θ are defined, in terms of the abscissa, ordinate, and distance of P from the origin O, as follows:

$$\sin \theta = \sin \theta = \frac{\text{ordinate}}{\text{distance}} = \frac{y}{r} \qquad\qquad \text{cotangent } \theta = \cot \theta = \frac{\text{abscissa}}{\text{ordinate}} = \frac{x}{y}$$

$$\text{cosine } \theta = \cos \theta = \frac{\text{abscissa}}{\text{distance}} = \frac{x}{r} \qquad\qquad \text{secant } \theta = \sec \theta = \frac{\text{distance}}{\text{abscissa}} = \frac{r}{x}$$

$$\text{tangent } \theta = \tan \theta = \frac{\text{ordinate}}{\text{abscissa}} = \frac{y}{x} \qquad\qquad \text{cosecant } \theta = \csc \theta = \frac{\text{distance}}{\text{ordinate}} = \frac{r}{y}$$

Note that $r = \sqrt{x^2 + y^2}$ (see Fig. 24-3).

As an immediate consequence of these definitions, we have the so-called *reciprocal relations*:

$$\sin \theta = \frac{1}{\csc \theta} \qquad\qquad \tan \theta = \frac{1}{\cot \theta} \qquad\qquad \sec \theta = \frac{1}{\cos \theta}$$

$$\cos \theta = \frac{1}{\sec \theta} \qquad\qquad \cot \theta = \frac{1}{\tan \theta} \qquad\qquad \csc \theta = \frac{1}{\sin \phi}$$

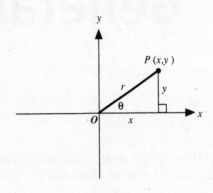

Fig. 24-3

It is evident from Figs. 24-4(a)–(d) that the values of the trigonometric functions of θ change as θ changes. The values of the functions of a given angle θ are, however, independent of the choice of the point P on its terminal side.

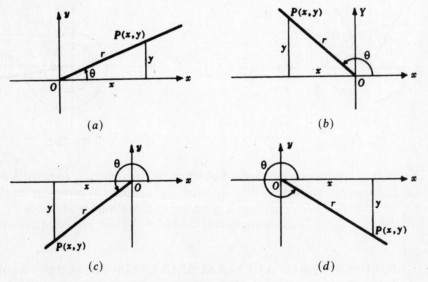

(a)

(b)

(c)

(d)

Fig. 24-4

ALGEBRAIC SIGNS OF THE FUNCTIONS.   Since r is always positive, the signs of the functions in the various quadrants depend upon the signs of x and y. To determine these signs one may visualize the angle in standard position or use some device as shown in Fig. 24-5 in which only the functions having signs are listed.

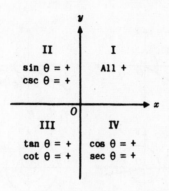

Fig. 24-5

When an angle is given, its trigonometric functions are uniquely determined. When, however, the value of one function of an angle is given, the angle is not uniquely determined. For example, if $\sin\theta = \frac{1}{2}$, then $\theta = 30°, 150°, 390°, 510°, \dots$. In general, two possible positions of the terminal side are found—for example, the terminal sides of $30°$ and $150°$ in Fig. 24-4(a), (b). The exceptions to this rule occur when the angle is quadrantal. (See Problems 24.2–24.10.)

TRIGONOMETRIC FUNCTIONS OF QUADRANTAL ANGLES.   For a quadrantal angle, the terminal side coincides with one of the axes. A point, P, distinct from the origin, on the terminal side has either $x = 0, y \neq 0$, or $x \neq 0, y = 0$. In either case, two of the six functions will not be defined. For example, the terminal side of the angle $0°$ coincides with the positive x axis and the ordinate of P is 0. Since the ordinate occurs in the denominator of the ratio defining the contangent and cosecant, these functions are not defined. Certain authors indicate this by writing $\cot 0° = \infty$ and others write $\cot 0° = \pm\infty$. The trigonometric functions of the quadrantal angles are given in Table 24.1.

Table 24.1

angle θ	$\sin\theta$	$\cos\theta$	$\tan\theta$	$\cot\theta$	$\sec\theta$	$\csc\theta$
0°	0	1	0		1	undefined
90°	1	0	undefined	0	undefined	1
180°	0	−1	0		−1	undefined
270°	−1	0	undefined	0	undefined	−1

Solved Problems

24.1   (a) Construct the following angles in standard position and determine those which are coterminal: $125°, 210°, -150°, 385°, 930°, -370°, -955°, -870°$.

(b) Give five other angles coterminal with $125°$.

(a) The angles 125° and −955° = 125° − 3 · 360° are coterminal. The angles 210°, −150° = 210° − 360°, 930° = 210° + 2 · 360°, and −870° = 210° − 3 · 360° are coterminal. See Fig. 24-6.

(b) 485° = 125° + 360°, 1205° = 125° + 3 · 360°, 1925° = 125° + 5 · 360°, −235° = 125° − 360°, −1315° = 125° − 4 · 360° are coterminal with 125°.

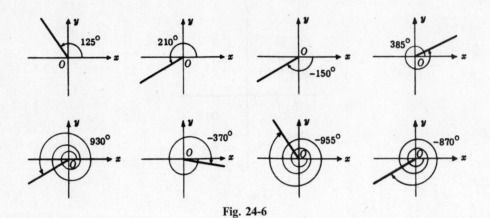

Fig. 24-6

24.2 Determine the values of the trigonometric functions of angle θ (smallest positive angle in standard position) if P is a point on the terminal side of θ and the coordinates of P are (a) $P\,(3,4)$, (b) $P(-3,4)$, (c) $P(-1,-3)$. See Fig. 24-7.

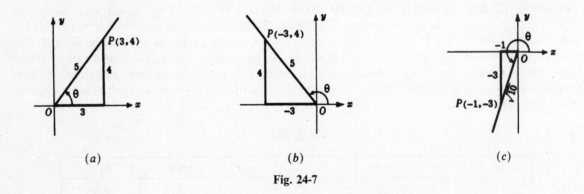

(a)   (b)   (c)

Fig. 24-7

(a) $r = \sqrt{3^2 + 4^2} = 5$ (b) $r = \sqrt{(-3)^2 + 4^2} = 5$ (c) $r = \sqrt{(-1)^2 + (-3)^2} = \sqrt{10}$

$\sin\theta = y/r = \frac{4}{5}$	$\sin\theta = \frac{4}{5}$	$\sin\theta = -3\sqrt{10} = -3\sqrt{10}/10$
$\cos\theta = x/r = \frac{3}{5}$	$\cos\theta = -\frac{3}{5}$	$\cos\theta = -1/\sqrt{10} = -\sqrt{10}/10$
$\tan\theta = y/x = \frac{4}{3}$	$\tan\theta = \frac{4}{-3} = -\frac{4}{3}$	$\tan\theta = -\frac{3}{1} = 3$
$\cot\theta = x/y = \frac{3}{4}$	$\cot\theta = -\frac{3}{4}$	$\cot\theta = -\frac{1}{3} = \frac{1}{3}$
$\sec\theta = r/x = \frac{5}{3}$	$\sec\theta = \frac{5}{-3} = -\frac{5}{3}$	$\sec\theta\sqrt{10}/(-1) = -\sqrt{10}$
$\csc\theta = r/y = \frac{5}{4}$	$\csc\theta = \frac{5}{4}$	$\csc\theta\sqrt{10}(-3) = -\sqrt{10}/3$

Note the reciprocal relationships. For example, in (b), $\sin\theta = 1/\csc\theta = \frac{4}{5}, \cos\theta = 1/\sec\theta = -\frac{3}{5}$, $\tan\theta = 1/\cot\theta = -\frac{4}{3}$, etc.

24.3 In what quadrant will θ terminate, if
   (a) $\sin\theta$ and $\cos\theta$ are both negative?   (c) $\sin\theta$ is positive and secant θ is negative?
   (b) $\sin\theta$ and $\tan\theta$ are both positive?   (d) $\sec\theta$ is negative and $\tan\theta$ is negative?

   (a) Since $\sin\theta = y/r$ and $\cos\theta = x/r$, both x and y are negative. (Recall that r is always positive.) Thus, θ is a third quadrant angle.

   (b) Since $\sin\theta$ is positive, y is positive; since $\tan\theta = y/x$ is positive, x is also positive. Thus, θ is a first quadrant angle.

   (c) Since $\sin\theta$ is positive, y is positive; since $\sec\theta$ is negative, x is negative. Thus, θ is a second quadrant angle.

   (d) Since $\sec\theta$ is negative, x is negative; since $\tan\theta$ is negative, y is then positive. Thus, θ is a second quadrant angle.

24.4 In what quadrants may θ terminate, if   (a) $\sin\theta$ is positive?   (b) $\cos\theta$ is negative?   (c) $\tan\theta$ is negative?
   (d) $\sin\theta$ is positive?

   (a) Since $\sin\theta$ is positive, y is positive. Then x may be positive or negative and θ is a first or second quadrant angle.

   (b) Since $\cos\theta$ is negative, x is negative. Then y may be positive or negative and θ is a second or third quadrant angle.

   (c) Since $\tan\theta$ is negative, either y is positive and x is negative or y is negative and x is positive. Thus, θ may be a second or fourth quadrant angle.

   (d) Since $\sec\theta$ is positive, x is positive. Thus, θ may be a first or fourth quadrant angle.

24.5 Find the values of $\cos\theta$ and $\tan\theta$, given $\sin\theta = \frac{8}{17}$ and θ in quadrant I.

   Let P be a point on the terminal line of θ. Since $\sin\theta = y/r = \frac{8}{17}$, we take $y = 8$ and $r = 17$. Since θ is in quadrant I, x is positive; thus $x = \sqrt{r^2 - y^2} = \sqrt{(17)^2 - (8)^2} = 15$.

   To draw Fig. 24-8, locate the point $P(15, 8)$, join it to the origin, and indicate the angle θ. Then $\cos\theta = x/y = \frac{15}{17}$ and $\tan\theta = y/x = \frac{8}{15}$.

   The choice of $y = 8, r = 17$ is one of convenience. Note that $\frac{8}{17} = \frac{16}{34}$ and we might have taken $y = 16$, $r = 34$. Then $x = 30, \cos\theta = \frac{30}{34} = \frac{15}{17}$ and $\tan\theta = \frac{16}{30} = \frac{8}{15}$.

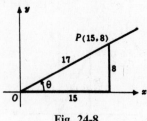

Fig. 24-8

24.6 Find the possible values of $\sin\theta$ and $\tan\theta$, given $\cos\theta = \frac{5}{6}$.

   Since $\cos\theta$ is positive, θ is in quadrant I or IV. Since $\cos\theta = x/r = \frac{5}{6}$, we take $x = 5, r = 6$; $y = \pm\sqrt{(6)^2 - (5)^2} = \pm\sqrt{11}$.

   (a) For θ in quadrant I [Fig. 24-9(a)] we have $x = 5, y = \sqrt{11}, r = 6$; then $\sin\theta = y/r = \sqrt{11}/6$ and $\tan\theta = y/x = \sqrt{11}/5$.

   (b) For θ in quadrant IV [Fig. 24-9(b)] we have $x = 5, y = -\sqrt{11}, r = 6$; then $\sin\theta = y/r = -\sqrt{11}/6$ and $\tan\theta = y/x = -\sqrt{11}/5$.

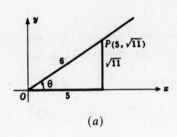

(a)   (b)

Fig. 24-9

24.7 Find the possible values of $\sin\theta$ and $\cos\theta$, given $\tan\theta = -\frac{3}{4}$.

Since $\tan\theta = y/x$ is negative, θ is in quadrant II (take $x = -4$, $y = 3$) or in quadrant IV (take $x = 4, y = -3$). In either case $r = \sqrt{16+9} = 5$.

(a) For θ in quadrant II [Fig. 24-10(a)], $\sin\theta = y/r = \frac{3}{5}$ and $\cos\theta = x/r = -\frac{4}{5}$.

(b) For θ in quadrant IV [Fig. 24-10(b)], $\sin\theta = y/r = -\frac{3}{5}$ and $\cos\theta = x/r = \frac{4}{5}$.

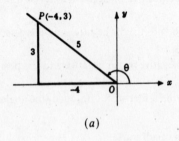

(a)   (b)

Fig. 24-10

24.8 Find the values of the remaining functions of θ, given $\sin\theta = \sqrt{3}/2$ and $\theta = -\frac{1}{2}$.

Since $\sin\theta = y/r$ is positive, y is positive. Since $\cos\theta = x/r$ is negative, x is negative. Thus, θ is in quadrant II.

Taking $x = -1, y = \sqrt{3}, r = \sqrt{(-1)^2 + (\sqrt{3})^2} = 2$ (Fig. 24-11), we have

$$\tan\theta = y/x = \sqrt{3}/-1 = -\sqrt{3} \qquad \cot\theta = 1/\tan\theta = -1\sqrt{3} = -\sqrt{3}/3$$
$$\sec\theta = 1/\cos\theta = -2 \qquad \csc\theta = 1/\sin\theta = 2/\sqrt{3} = 2\sqrt{3}/3$$

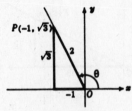

Fig. 24-11

24.9 Determine the possible values of $\cos\theta$ and $\tan\theta$ if $\sin\theta = m/n$, a negative fraction.

Since $\sin\theta$ is negative, θ is in quadrant III or IV.

(a) In quadrant III: Take $y = m, r = n, x = -\sqrt{n^2 - m^2}$; then $\cos\theta = x/r = -\sqrt{n^2 - m^2}/n$ and $\tan\theta = y/x = -m/\sqrt{n^2 - m^2}$.

(b) In quadrant IV: Take $y = m, r = n, x = +\sqrt{n^2 - m^2}$; then $\cos\theta = x/r = \sqrt{n^2 - m^2}/n$ and $\tan\theta = y/x = m\sqrt{n^2} = m^2$.

24.10 Evaluate:

(a) $\sin 0° + 2\cos 0° + 3\sin 90° + 4\cos 90° + 5\sec 0° + 6\csc 90°$

(b) $\sin 180° + 2\cos 180° + 3\sin 270° + 4\cos 270° - 5\sec 180° - 6\csc 270°$

(a) $0 + 2(1) + 3(1) + 4(0) + 5(1) + 6(1) = 16$

(b) $0 + 2(-1) + 3(-1) + 4(0) - 5(-1) - 6(-1) = 6$

Supplementary Problems

24.11 State the quadrant in which each angle terminates and the signs of the sine, cosine, and tangent of each angle.

(a) 125°   (b) 75°   (c) 320°   (d) 212°   (e) 460°   (f) 750°   (g) −250°   (h) −1000°

Ans.   (a) II; +, −, −   (b) I; +, +, +   (c) IV; −, +, −   (d) III; −, −, +   (e) II
(f) I   (g) II   (h) I

24.12 In what quadrant will θ terminate if

(a) $\sin\theta$ and $\cos\theta$ are both positive?

(b) $\cos\theta$ and $\tan\theta$ are both positive?

(c) $\sin\theta$ and $\sec\theta$ are both negative?

(d) $\cos\theta$ and $\cot\theta$ are both negative?

(e) $\tan\theta$ is positive and $\sec\theta$ is negative?

(f) $\tan\theta$ is negative and $\sec\theta$ is positive?

(g) $\sin\theta$ is positive and $\cos\theta$ is negative?

(h) $\sec\theta$ is positive and $\csc\theta$ is negative?

Ans.   (a) I   (b) I   (c) III   (d) II   (e) III   (f) IV   (g) II   (h) IV

24.13 Denote by θ the smallest positive angle whose terminal side passes through the given point and find the trigonometric functions of θ:

(a) $P(-5, 12)$ (b) $P(7, -24)$ (c) $P(2, 3)$ (d) $P(-3, -5)$

Ans.   (a) $\frac{12}{13}, -\frac{5}{13}, -\frac{12}{5}, -\frac{5}{12}, -\frac{13}{5}, \frac{13}{12}$

(b) $-\frac{24}{25}, \frac{7}{25}, -\frac{24}{7}, -\frac{7}{24}, \frac{25}{7}, -\frac{25}{24}$

(c) $3/\sqrt{13}, 2/\sqrt{13}, \frac{3}{2}, \frac{2}{3}, \sqrt{13}/2, \sqrt{13}/3$

(d) $-5/\sqrt{34}, -3/\sqrt{34}, \frac{5}{3}, \frac{3}{5}, -\sqrt{34}/3, -\sqrt{34}/5$

24.14 Find the possible values of the trigonometric functions of θ, given

(a) $\sin\theta = \frac{7}{25}$ (d) $\cot\theta = \frac{24}{7}$ (g) $\tan\theta = \frac{3}{5}$ (j) $\csc\theta = -2/\sqrt{3}$

(b) $\cos\theta = -\frac{4}{5}$ (e) $\sin\theta = -\frac{2}{3}$ (h) $\cot\theta = \sqrt{6}/2$

(c) $\tan\theta = -\frac{5}{12}$ (f) $\cos\theta = \frac{5}{6}$ (i) $\sec\theta = -\sqrt{5}$

Ans.   (a) I: $\frac{7}{25}, \frac{24}{25}, \frac{7}{24}, \frac{24}{7}, \frac{25}{24}, \frac{25}{7}$;   II: $\frac{7}{25}, -\frac{24}{25}, -\frac{7}{24}, -\frac{24}{7}, -\frac{25}{24}, \frac{25}{7}$

(b) II: $\frac{3}{5}, -\frac{4}{5}, -\frac{3}{4}, -\frac{4}{3}, -\frac{5}{4}, \frac{5}{3}$;   III: $-\frac{3}{5}, -\frac{4}{5}, \frac{3}{4}, \frac{4}{3}, -\frac{5}{4}, -\frac{5}{3}$

(c) II: $\frac{5}{13}, -\frac{12}{13}, -\frac{5}{12}, -\frac{12}{5}, -\frac{13}{12}, \frac{13}{5}$;   IV: $-\frac{5}{13}, \frac{12}{13}, -\frac{5}{12}, -\frac{12}{5}, \frac{13}{12}, -\frac{13}{5}$

(d) I: $\frac{7}{25}, \frac{24}{25}, \frac{7}{24}, \frac{24}{7}, \frac{25}{24}, \frac{25}{7}$;   III: $-\frac{7}{25}, -\frac{24}{25}, \frac{7}{24}, \frac{24}{7}, -\frac{25}{24}, -\frac{25}{7}$

(e) III: $-\frac{2}{3}, -\sqrt{5}/3, 2/\sqrt{5}, \sqrt{5}/2, -3/\sqrt{5}, -\frac{3}{2}$;   IV: $-\frac{2}{3}, \sqrt{5}/3, -2\sqrt{5}, -\sqrt{5}/2, 3/\sqrt{5}, -\frac{3}{2}$

(f) I: $\sqrt{11}/6, \frac{5}{6}, \sqrt{11}/5, 5/\sqrt{11}, \frac{6}{5}, 6/\sqrt{11}$;   IV: $-\sqrt{11}/6, \frac{5}{6}, -\sqrt{11}/5, -5/\sqrt{11}, \frac{6}{5}, -6\sqrt{/11}$

(g) I: $3/\sqrt{34}, 5\sqrt{34}, \frac{3}{5}, \frac{5}{3}, \sqrt{34}/5, \sqrt{34}/3$;   III: $-3/\sqrt{34}, -5/\sqrt{34}, \frac{3}{5}, \frac{5}{3}, -\sqrt{34}/5, -\sqrt{34}/3$

(h) I: $2/\sqrt{10}, \sqrt{3}/\sqrt{5}, 2/\sqrt{6}, \sqrt{6}/2, \sqrt{5}/\sqrt{3}, \sqrt{10}/2$;   III: $-2/\sqrt{10}, -\sqrt{3}/\sqrt{5}, 2/\sqrt{6}, \sqrt{6}/2, -\sqrt{5}/\sqrt{3}, -\sqrt{10}/2$

(i)   II: $2/\sqrt{5}, -1\sqrt{5}, -2, -\frac{1}{2}, -\sqrt{5}, \sqrt{5}/2$;   III: $-2/\sqrt{5}, -1\sqrt{5}, 2, \frac{1}{2}, -\sqrt{5}, -\sqrt{5}/2$

(j)   III: $-\sqrt{3}/2, -\frac{1}{2}, \sqrt{3}, 1/\sqrt{3}, -2, -2/\sqrt{3}$;   IV: $-\sqrt{3}/2, \frac{1}{2}, -\sqrt{3}, -1/\sqrt{3}, 2, -2\sqrt{3}$

24.15 Evaluate each of the following:

(a) $\tan 180° - 2\cos 180° + 3\csc 270° + \sin 90°$

(b) $\sin 0° + 3\cot 90° + 5\sec 180° - 4\cos 270°$

    Ans.   (a) 0
           (b) −5

Chapter 25

Trigonometric Functions of an Acute Angle

TRIGONOMETRIC FUNCTIONS OF AN ACUTE ANGLE. In dealing with any right triangle, it will be convenient (see Fig. 25-1) to denote the vertices as A, B, C such that C is the vertex of the right triangle; to denote the angles of the triangle as A, B, C, such that m $\angle C = 90°$; and to denote the sides opposite the angles as a, b, c, respectively. With respect to angle A, a will be called the *opposite side* and b will be called the *adjacent side*; with respect to angle B, a will be called the *adjacent side* and b the *opposite side*. Side c will always be called the *hypotenuse*.

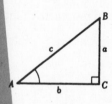

Fig. 25-1

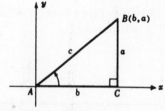

Fig. 25-2

If now the right triangle is placed in a coordinate system (Fig. 25-2) so that angle A is in standard position, point B on the terminal side of angle A has coordinates (b, a) and distance $c = \sqrt{a^2 + b^2}$. Then the trigonometric functions of angle A may be defined in terms of the sides of the right triangle, as follows:

$$\sin A = \frac{a}{c} = \frac{\text{length of opposite side}}{\text{length of hypotenuse}} \qquad \cot A = \frac{b}{a} = \frac{\text{length of adjacent side}}{\text{length of opposite side}}$$

$$\cos A = \frac{b}{c} = \frac{\text{length of adjacent side}}{\text{length of hypotenuse}} \qquad \sec A = \frac{c}{b} = \frac{\text{length of hypotenuse}}{\text{length of adjacent side}}$$

$$\tan A = \frac{a}{b} = \frac{\text{length of opposite side}}{\text{length of adjacent side}} \qquad \csc A = \frac{c}{a} = \frac{\text{length of hypotenuse}}{\text{length of opposite side}}$$

169

TRIGONOMETRIC FUNCTIONS OF COMPLEMENTARY ANGLES.   The acute angles A and B of the right triangle ABC are complementary; that is, m $\measuredangle\ A +$ m $\measuredangle\ B = 90°$. From Fig. 25-1, we have

$$\sin B = \frac{b}{c} = \cos A \qquad \cot B = \frac{a}{b} = \tan A$$

$$\cos B = \frac{a}{c} = \sin A \qquad \sec B = \frac{c}{a} = \csc A$$

$$\tan B = \frac{b}{a} = \cot A \qquad \csc B = \frac{c}{b} = \sec A$$

These relations associate the functions in pairs—sine and cosine, tangent and cotangent, secant and cosecant—each function of a pair being called the *cofunction* of the other. Thus, any function of an acute angle is equal to the corresponding cofunction of the complementary angle.

TRIGONOMETRIC FUNCTIONS OF 30°, 45°, 60°.   The results in Table 25.1 are obtained in Problems 25.8–25.9.

Table 25.1

Angle θ	$\sin \theta$	$\cos \theta$	$\tan \theta$	$\cot \theta$	$\sec \theta$	$\csc \theta$
30°	$\frac{1}{2}$	$\frac{1}{2}\sqrt{3}$	$\frac{1}{3}\sqrt{3}$	$\sqrt{3}$	$\frac{2}{3}\sqrt{3}$	2
45°	$\frac{1}{2}\sqrt{2}$	$\frac{1}{2}\sqrt{2}$	1	1	$\sqrt{2}$	$\sqrt{2}$
60°	$\frac{1}{2}\sqrt{3}$	$\frac{1}{2}$	$\sqrt{3}$	$\frac{1}{3}\sqrt{3}$	2	$\frac{2}{3}\sqrt{3}$

Problems 25.10–25.16 illustrate a number of simple applications of the trigonometric functions. For this purpose, Table 25.2 will be used.

Table 25.2

Angle θ	$\sin \theta$	$\cos \theta$	$\tan \theta$	$\cot \theta$	$\sec \theta$	$\csc \theta$
15°	0.26	0.97	0.27	3.7	1.0	3.9
20°	0.34	0.94	0.36	2.7	1.1	2.9
30°	0.50	0.87	0.58	1.7	1.2	2.0
40°	0.64	0.77	0.84	1.2	1.3	1.6
45°	0.71	0.71	1.0	1.0	1.4	1.4
50°	0.77	0.64	1.0	0.84	1.6	1.3
60°	0.87	0.50	1.7	0.58	2.0	1.2
70°	0.94	0.34	2.7	0.36	2.9	1.1
75°	0.97	0.26	3.7	0.27	3.9	1.0

Solved Problems

[NOTATION: We will write AB (or c) to denote the length of AB, and $\overline{AB}$ to denote "the segment AB."
$\overleftrightarrow{AB}$ denotes "the line AB."]

25.1   Find the values of the trigonometric functions of the acute angles of the right triangle ABC given $b = 24$ and $c = 25$.

Since $a^2 = c^2 - b^2 = (25)^2 - (24)^2 = 49$, $a = 7$. See Fig. 25-3. Then

$$\sin A = \frac{\text{opposite side}}{\text{hypotenuse}} = \frac{7}{25} \qquad \cos A = \frac{\text{adjacent side}}{\text{opposite side}} = \frac{24}{7}$$

$$\cos A = \frac{\text{adjacent side}}{\text{hypotenuse}} = \frac{24}{25} \qquad \sec A = \frac{\text{hypotenuse}}{\text{adjacent side}} = \frac{25}{24}$$

$$\tan A = \frac{\text{opposite side}}{\text{adjacent side}} = \frac{7}{24} \qquad \csc A = \frac{\text{hypotenuse}}{\text{opposite side}} = \frac{25}{7}$$

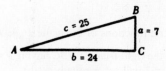

Fig. 25-3

and

$$\sin B = \tfrac{24}{25} \qquad \cot B = \tfrac{7}{24}$$

$$\cos B = \tfrac{7}{25} \qquad \sec B = \tfrac{25}{7}$$

$$\tan B = \tfrac{24}{7} \qquad \csc B = \tfrac{25}{24}$$

25.2   Find the values of the trigonometric functions of the acute angles of the right triangle ABC, given $a = 2$, $c = 2\sqrt{5}$.

Since $b^2 = c^2 - a^2 = (2\sqrt{5})^2 - 2^2 = 20 - 4 = 16$, $b = 4$. See Fig. 25-4. Then

$$\sin A = \frac{2}{2\sqrt{5}} = \frac{\sqrt{5}}{5} = \cos B \qquad \cot A = \tfrac{4}{2} = 2 = \tan B$$

$$\cos A = \frac{4}{2\sqrt{5}} = \frac{2\sqrt{5}}{5} = \sin B \qquad \sec A = \frac{2\sqrt{5}}{4} = \frac{\sqrt{5}}{2} = \csc B$$

$$\tan A = \tfrac{2}{4} = \tfrac{1}{2} = \cot B \qquad \csc A = \frac{2\sqrt{5}}{2} = \sqrt{5} = \sec B$$

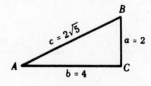

Fig. 25-4

25.3   Find the values of the trigonometric functions of the acute angle A, given $\sin A = \frac{3}{7}$.

Construct the right triangle ABC having $a = 3$, $c = 7$, and $b = \sqrt{7^2 - 3^2} = 2\sqrt{10}$ units. See Fig. 25-5. Then

$$\sin A = \tfrac{3}{7} \qquad\qquad\qquad \cot A = \frac{2\sqrt{10}}{3}$$

$$\cos A = \frac{2\sqrt{10}}{7} \qquad\qquad \sec A = \frac{7}{2\sqrt{10}} = \frac{7\sqrt{10}}{20}$$

$$\tan A = \frac{3}{2\sqrt{10}} = \frac{3\sqrt{10}}{20} \qquad \csc A = \tfrac{7}{3}$$

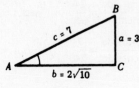

Fig. 25-5

25.4   Find the values of the trigonometric functions of the acute angle B, given $\tan B = 1.5$.

Refer to Fig. 25-6. Construct the right triangle ABC having $b = 15$ and $a = 10$ units. (Note that $1.5 = \frac{3}{2}$ and a right triangle with $b = 3$, $a = 2$ will serve equally well.)

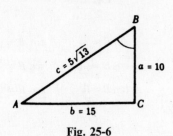

Fig. 25-6

Then $c = \sqrt{a^2 + b^2} = \sqrt{10^2 + 15^2} = 5\sqrt{13}$ and

$$\sin B = \frac{15}{5\sqrt{13}} = \frac{3\sqrt{13}}{13} \qquad\qquad \cot B = \tfrac{2}{3}$$

$$\cos B = \frac{10}{5\sqrt{13}} = \frac{2\sqrt{13}}{13} \qquad\qquad \sec B = \frac{5\sqrt{13}}{10} = \frac{\sqrt{13}}{2}$$

$$\tan B = \tfrac{15}{10} = \tfrac{3}{2} \qquad\qquad\qquad \csc B = \frac{5\sqrt{13}}{15} = \frac{\sqrt{13}}{3}$$

25.5   If A is acute and $\sin A = 2x/3$, determine the values of the remaining functions.

Construct the right triangle ABC having $a = 2x < 3$ and $c = 3$, as in Fig. 25-7. Then $b = \sqrt{c^2 - a^2} = \sqrt{9 - 4x^2}$ and

$$\sin A = \frac{2x}{3}, \qquad \cos A = \frac{\sqrt{9 - 4x^2}}{3}, \qquad \tan A = \frac{2x}{\sqrt{9 - 4x^2}}, \qquad \cot A = \frac{\sqrt{9 - 4x^2}}{2x},$$

$$\sec A = \frac{3}{\sqrt{9 - 4x^2}}, \qquad \csc A = \frac{3}{2x}.$$

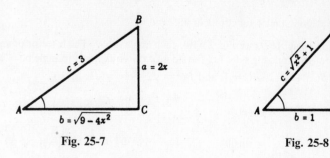

Fig. 25-7   Fig. 25-8

25.6   If A is acute and $\tan A = x = x/1$, determine the values of the remaining functions.

Construct the right triangle ABC having $a = x$ and $b = 1$, as in Fig. 25-8. Then $c = \sqrt{x^2 + 1}$ and

$$\sin A = \frac{x}{\sqrt{x^2 + 1}}, \quad \cos A = \frac{1}{\sqrt{x^2 + 1}}, \quad \tan A = x, \quad \cot A = \frac{1}{x}, \quad \sec A = \sqrt{x^2 + 1}, \quad \csc A = \frac{\sqrt{x^2 + 1}}{x}.$$

25.7   If A is an acute angle:

(*a*) Why is $\sin A < 1$?   (*d*) Why is $\sin A < \tan A$?

(*b*) When is $\sin A = \cos A$?   (*e*) When is $\sin A < \cos A$?

(*c*) Why is $\sin A < \csc A$?   (*f*) When is $\tan A > 1$?

In any right triangle ABC:

(*a*) Side $a <$ side c; therefore $\sin A = a/c < 1$.

(*b*) $\sin A = \cos A$ when $a/c = b/c$; then $a = b$, $A = B$, and $A = 45°$.

(*c*) $\sin A < 1$ (above) and $\csc A = 1/\sin A > 1$.

(*d*) $\sin A = a/c$, $\tan A = a/b$, and $b < c$; therefore $a/c < a/b$ or $\sin A < \tan A$.

(*e*) $\sin A < \cos A$ when $a < b$; then $A < B$ or $A < 90° - A$, and $A < 45°$.

(*f*) $\tan A = a/b > 1$ when $a > b$; then $A > B$ and $A > 45°$.

25.8   Find the value of the trigonometric functions of 45°.

In any isosceles right triangle ABC, $A = B = 45°$ and $a = b$. See Fig. 25-9. Let $a = b = 1$; then $c = \sqrt{1 + 1} = \sqrt{2}$ and

$$\sin 45° = \frac{1}{\sqrt{2}} = \frac{1}{2}\sqrt{2} \qquad \cot 45° = 1$$

$$\cos 45° = \frac{1}{\sqrt{2}} = \frac{1}{2}\sqrt{2} \qquad \sec 45° = \sqrt{2}$$

$$\tan 45° = \frac{1}{1} = 1 \qquad \csc 45° = \sqrt{2}$$

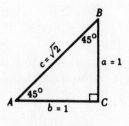

Fig. 25-9

25.9   Find the values of the trigonometric functions of 30° and 60°.

In any equilateral triangle ABD (see Fig. 25-10), each angle is 60°. The bisector of any angle, as B, is the perpendicular bisector of the opposite side. Let the sides of the equilateral triangle be of length 2 units. Then in the right triangle ABC, $AB = 2$, $AC = 1$, and $BC = \sqrt{2^2 - 1^2} = \sqrt{3}$.

$$\sin 30° = \tfrac{1}{2} = \cos 60° \qquad\qquad \cot 30° = \sqrt{3} = \tan 60°$$

$$\cos 30° = \frac{\sqrt{3}}{2} = \sin 60° \qquad\qquad \sec 30° = \frac{2}{\sqrt{3}} = \frac{2\sqrt{3}}{3} = \csc 60°$$

$$\tan 30° = \frac{1}{\sqrt{3}} = \frac{\sqrt{3}}{3} = \cot 60° \qquad\qquad \csc 30° = 2 = \sec 60°$$

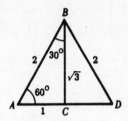

Fig. 25-10

25.10   When the sun is 20° above the horizon, how long is the shadow cast by a building 150 ft high?

In Fig. 25-11, $A = 20°$ and $CB = 150$. Then $\cot A = AC/CB$ and $AC = CB \cot A = 150 \cot 20° = 150(2.7) = 405$ ft.

25.11   A tree 100 ft tall casts a shadow 120 ft long. Find the measure of the angle of elevation of the sun.

In Fig. 25-12, $CB = 100$ and $AC = 120$. Then $\tan A = CB/AC = \frac{100}{120} = 0.83$ and m ∡ $A = 40°$.

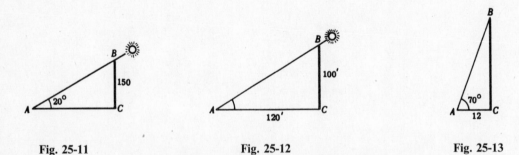

Fig. 25-11   Fig. 25-12   Fig. 25-13

25.12   A ladder leans against the side of a building with its foot 12 m from the building. How far from the ground is the top of the ladder and how long is the ladder if it makes an angle of 70° with the ground?

From Fig. 25-13, $\tan A = CB/AC$; then $CB = AC \tan A = 12 \tan 70° = 12(2.7) = 32.4$. The top of ladder is 32 m above the ground.

Sec $A = AB/AC$; then $AB = AC \sec A = 12 \sec 70° = 12(2.9) = 34.8$. The ladder is 35 m long.

25.13 From the top of a lighthouse, 120 ft above the sea, the angle of depression of a boat is 15°. How far is the boat from the lighthouse?

In Fig. 25-14, the right triangle ABC has A measuring 15° and $CB = 120$; then $\cot A = AC/CB$ and $AC = CB \cot A = 120 \cot 15° = 120(3.7) = 444$ ft.

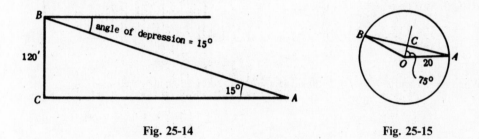

Fig. 25-14   **Fig. 25-15**

25.14 Find the length of the chord of a circle of radius 20 cm subtended by a central angle of 150°.

In Fig. 25-15, OC bisects $\angle AOB$. Then $BC = AC$ and OAC is a right triangle. In $\triangle OAC$,

$$\sin \angle COA = \frac{AC}{OA} \text{ and } AC = OA \sin \angle COA = 20 \sin 75° = 20(0.97) = 19.4$$

Then $BA = 38.8$ and the length of the chord is 39 cm.

25.15 Find the height of a tree if the angle of elevation of its top changes from 20° to 40° as the observer advances 75 ft toward its base. See Fig. 25-16.

In the right triangle ABC, $\cot A = AC/CB$; then $AC = CB \cot A$ or $DC + 75 = CB \cot 20°$. In the right triangle DBC, $\cot D = DC/CB$; then $DC = CB \cot 40°$.

Then $DC = CB \cot 20° - 75 = CB \cot 40°,$ $CB(\cot 20° - \cot 40°) = 75,$

$$CB(2.7 - 1.2) = 75, \quad \text{and} \quad CB = \frac{75}{1.5} = 50 \text{ ft.}$$

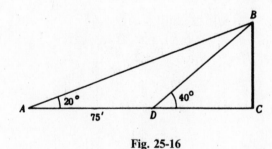

Fig. 25-16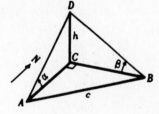

Fig. 25-17

25.16 A tower standing on level ground is due north of point A and due west of point B, a distance c ft from A. If the angles of elevation of the top of the tower as measured from A and B are α and β, respectively, find the height h of the tower.

In the right triangle ACD of Fig. 25-17, $\cot \alpha = AC/h$; in the right triangle BCD, $\cot \beta = BC/h$. Then $AC = h \cot \alpha$ and $BC = h \cot \beta$.

Since ABC is a right triangle, $(AC)^2 + (BC)^2 = c^2 = h^2(\cot \alpha)^2 + h^2(\cot \beta)^2$ and

$$h = \frac{c}{\sqrt{(\cot \alpha)^2 + (\cot \beta)^2}}$$

25.17 If holes are to be spaced regularly on a circle, show that the distance d between the centers of two successive holes is given by $d = 2r\sin(180°/n)$, where r = radius of the circle and n = number of holes. Find d when $r = 20$ in. and $n = 4$.

Let A and B be the centers of two consecutive holes on the circle of radius r and center O. See Fig. 25-18. Let the bisector of the angle O of the triangle AOB meet AB at C. In right triangle AOC, $\sin\angle AOC = AC/r = \frac{1}{2}d/r = d/2r$. Then

$$d = 2r\sin\angle AOC = 2r\sin\tfrac{1}{2}\angle AOB = 2r\sin\frac{1}{2}\left(\frac{360°}{n}\right) = 2r\sin\frac{180°}{n}$$

When $r = 20$ and $n = 4$, $d = 2 \cdot 20 \sin 45° = 2 \cdot 20 \cdot \frac{\sqrt{2}}{2} = 20\sqrt{2}$ in.

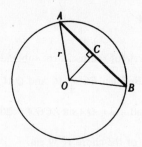

Fig. 25-18

Supplementary Problems

25.18 Find the values of the trigonometric functions of the acute angles of the right triangle ABC, given
 (a) $a = 3, b = 1$ (b) $a = 2, c = 5$ (c) $b = \sqrt{7}, c = 4$

    Ans. (a) A: $3/\sqrt{10}, 1/\sqrt{10}, 3, \frac{1}{3}, \sqrt{10}, \sqrt{10}/3$; B: $1/\sqrt{10}, 3/\sqrt{10}, \frac{1}{3}, 3, \sqrt{10}/3, \sqrt{10}$

            (b) A: $\frac{2}{5}, \sqrt{21}/5, 2/\sqrt{21}, \sqrt{21}/2, 5/\sqrt{21}, \frac{5}{2}$; B: $\sqrt{21}/5, \frac{2}{5}, \sqrt{21}/2, 2/\sqrt{21}, \frac{5}{2}, 5/\sqrt{21}$

            (c) A: $\frac{3}{4}, \sqrt{7}/4, 3/\sqrt{7}, \sqrt{7}/3, 4/\sqrt{7}, \frac{4}{3}$; B: $\sqrt{7}/4, \frac{3}{4}, \sqrt{7}/3, 3/\sqrt{7}, \frac{4}{3}, 4/\sqrt{7}$

25.19 Which is the greater and why:

(a) $\sin 55°$ or $\cos 55°$?   (c) $\tan 15°$ or $\cot 15°$?

(b) $\sin 40°$ or $\cos 40°$?   (d) $\sec 55°$ or $\csc 55°$?

Hint: Consider a right triangle having as acute angle the given angle.
    Ans. (a) $\sin 55°$ (b) $\cos 40°$ (c) $\cot 15°$ (d) $\sec 55°$

25.20 Find the value of each of the following:

(a) $\sin 30° + \tan 45°$ (e) $\dfrac{\tan 60° - \tan 30°}{1 + \tan 60° \tan 30°}$

(b) $\cot 45° + \cos 60°$

(c) $\sin 30° \cos 60° + \cos 30° \sin 60°$ (f) $\dfrac{\csc 30° + \csc 60° + \csc 90°}{\sec 0° + \sec 30° + \sec 60°}$

(d) $\cos 30° \cos 60° - \sin 30° \sin 60°$

    Ans. (a) $\frac{3}{2}$ (b) $\frac{3}{2}$ (c) 1 (d) 0 (e) $1/\sqrt{3}$ (f) 1

25.21 A man drives 500 ft along a road which is inclined 20° to the horizontal. How high above his starting point is he?

   Ans.   170 ft

25.22 A tree broken over by the wind forms a right triangle with the ground. If the broken part makes an angle of 50° with the ground and if the top of the tree is now 20 ft from its base, how tall was the tree?

   Ans.   56 ft

25.23 Two straight roads intersect to form an angle of 75°. Find the shortest distance from one road to a gas station on the other road 1000 m from the junction.

   Ans.   970 m

25.24 Two buildings with flat roofs are 60 ft apart. From the roof of the shorter building, 40 ft in height, the angle of elevation to the edge of the roof of the taller building is 40°. How high is the taller building?

   Ans.   90 ft

25.25 A ladder, with its foot in the street, makes an angle of 30° with the street when its top rests on a building on one side of the street and makes an angle of 40° with the street when its top rests on a building on the other side of the street. If the ladder is 50 ft long, how wide is the street?

   Ans.   82 ft

25.26 Find the perimeter of an isosceles triangle whose base is 40 cm and whose base angle is 70°.

   Ans.   156 cm

Chapter 26

Reduction to Functions of Positive Acute Angles

COTERMINAL ANGLES.   Let θ be any angle; then

$$\sin(\theta + n360°) = \sin\theta \qquad \cot(\theta + n360°) = \cot\theta$$
$$\cos(\theta + n360°) = \cos\theta \qquad \sec(\theta + n360°) = \sec\theta$$
$$\tan(\theta + n360°) = \tan\theta \qquad \csc(\theta + n360°) = \csc\theta$$

where n is any positive or negative integer or zero.

EXAMPLES. $\sin 400° = \sin(40° + 360°) = \sin 40°$

$$\cos 850° = \cos(130° + 2 \cdot 360°) = \cos 130°$$
$$\tan(-1000°) = \tan(80° - 3 \cdot 360°) = \tan\ 80°$$

FUNCTIONS OF A NEGATIVE ANGLE.   Let θ be an angle; then

$$\sin(-\theta) = -\sin\theta \qquad \cot(-\theta) = -\cot\ \theta$$
$$\cos(-\theta) = \cos\theta \qquad \sec(-\theta) = \sec\ \theta$$
$$\tan(-\theta) = -\tan\theta \qquad \csc(-\theta) = -\csc\ \theta$$

EXAMPLES. $\sin(-50°) = -\sin 50°, \cos(-30°) = \cos 30°, \tan(-200°) = -\tan\ 200°.$

REDUCTION FORMULAS.   Let θ be an angle; then

$$\sin(90° - \theta) = \cos\ \theta \qquad \sin(90° + \theta) = \cos\ \theta$$
$$\cos(90° - \theta) = \sin\theta \qquad \cos(90° + \theta) = -\sin\ \theta$$
$$\tan(90° - \theta) = \cot\theta \qquad \tan(90° + \theta) = -\cot\ \theta$$

$$\cot(90° - \theta) = \tan\theta \qquad \cot(90° + \theta) = -\tan\theta$$
$$\sec(90° - \theta) = \csc\theta \qquad \sec(90° + \theta) = -\csc\theta$$
$$\csc(90° - \theta) = \sec\theta \qquad \csc(90° + \theta) = \sec\theta$$

$$\sin(180° - \theta) = \sin\theta \qquad \sin(180° + \theta) = -\sin\theta$$
$$\cos(180° - \theta) = -\cos\theta \qquad \cos(180° + \theta) = -\cos\theta$$
$$\tan(180° - \theta) = -\tan\theta \qquad \tan(180° + \theta) = \tan\theta$$
$$\cot(180° - \theta) = -\cot\theta \qquad \cot(180° + \theta) = \cot\theta$$
$$\sec(180° - \theta) = -\sec\theta \qquad \sec(180° + \theta) = -\sec\theta$$
$$\csc(180° - \theta) = \csc\theta \qquad \csc(180° + \theta) = -\csc\theta$$

GENERAL REDUCTION FORMULA.   Any trigonometric function of $(n \cdot 90° \pm \theta)$, where θ is any angle, is *numerically* equal

(*a*)   To the same function of θ if n is an even integer

(*b*)   To the corresponding cofunction of θ if n is an odd integer

The algebraic sign in each case is the same as the sign of the given function for that quadrant in which $n \cdot 90° \pm \theta$ lies when θ is a positive acute angle.

EXAMPLES

(*1*) $\sin(180° - \theta) = \sin(2 \cdot 90° - \theta) = \sin\theta$ since $180°$ is an even multiple of $90°$ and, when θ is positive acute, the terminal side of $180° - \theta$ lies in quadrant II.

(*2*) $\cos(180° + \theta) = \cos(2 \cdot 90° + \theta) = -\cos\theta$ since $180°$ is an even multiple of $90°$ and, when θ is positive acute, the terminal side of $180° + \theta$ lies is quadrant III.

(*3*) $\tan(270° - \theta) = \tan(3 \cdot 90° - \theta) = \cot\theta$ since $270°$ is an odd multiple of $90°$ and, when θ is positive acute, the terminal side of $270° - \theta$ lies in quadrant III.

(*4*) $\cos(270° + \theta) = \cos(3 \cdot 90° + \theta) = \sin\theta$ since $270°$ is an odd multiple of $90°$ and, when θ is positive acute, the terminal side of $270° + \theta$ lies in quadrant IV.

Solved Problems

26.1   Express each of the following in terms of a function of θ:

(*a*) $\sin(\theta - 90°)$ (*d*) $\cos(-180° + \theta)$ (*g*) $\sin(540° + \theta)$ (*j*) $\cos(-450° - \theta)$

(*b*) $\cos(\theta - 90°)$ (*e*) $\sin(-270° - \theta)$ (*h*) $\tan(720° - \theta)$ (*k*) $\csc(-900° + \theta)$

(*c*) $\sec(-\theta - 90°)$ (*f*) $\tan(\theta - 360°)$ (*i*) $\tan(720° + \theta)$ (*l*) $\sin(-540° - \theta)$

(*a*) $\sin(\theta - 90°) = \sin(-90° + \theta) = \sin(-1 \cdot 90° + \theta) = -\cos\theta$, the sign being negative since, when θ is positive acute, the terminal side of $\theta - 90°$ lies in quadrant IV.

(*b*) $\cos(\theta - 90°) = \cos(-90° + \theta) = \cos(-1 \cdot 90° + \theta) = \sin\theta$.

(*c*) $\sec(-\theta - 90°) = \sec(-90° - \theta) = \sec(-1 \cdot 90° - \theta) = -\csc\theta$, the sign being negative since, when θ is positive acute, the terminal side of $-\theta - 90°$ lies in quadrant III.

(*d*) $\cos(-180° + \theta) = \cos(-2 \cdot 90° + \theta) = -\cos\theta$. (quadrant III)

(*e*) $\sin(-270° - \theta) = \sin(-3 \cdot 90° - \theta) = \cos\theta$. (quadrant I)

(*f*) $\tan(\theta - 360°) = \sin(-4 \cdot 90° + \theta) = \tan\theta$. (quadrant I)

(*g*) $\sin(540° + \theta) = \sin(6 \cdot 90° + \theta) = -\sin\theta$. (quadrant III)

(*h*) $\tan(720° - \theta) = \tan(8 \cdot 90° - \theta) = -\tan\theta = \tan(2 \cdot 360° - \theta) = \tan(-\theta) = -\tan\theta$.

(i) $\tan(720° + \theta) = \tan(8 \cdot 90° + \theta) = -\tan\theta = \tan(2 \cdot 360° + \theta) = \tan\theta$.

(j) $\cos(-450° - \theta) = \cos(-5 \cdot 90° - \theta) = -\sin\theta$.

(k) $\csc(-900° + \theta) = \csc(-10 \cdot 90° + \theta) = -\csc\theta$.

(l) $\sin(-540° - \theta) = \sin(-6 \cdot 90° - \theta) = \sin\theta$.

26.2 Express each of the following in terms of functions of a positive acute angle in two ways:

(a) $\sin 130°$ (c) $\sin 200°$ (e) $\tan 165°$ (g) $\sin 670°$ (i) $\csc 865°$ (k) $\cos(-680°)$

(b) $\tan 325°$ (d) $\cos 310°$ (f) $\sec 250°$ (h) $\cot 930°$ (j) $\sin(-100°)$ (l) $\tan(-290°)$

(a) $\sin 130° = \sin(2 \cdot 90° - 50°) = \sin 50° = \sin(1 \cdot 90° + 40°) = \cos 40°$.

(b) $\tan 325° = \tan(4 \cdot 90° - 35°) = -\tan 35° = \tan(3 \cdot 90° + 55°) = -\cot 55°$.

(c) $\sin 200° = \sin(2 \cdot 90° + 20°) = -\sin 20° = \sin(3 \cdot 90° - 70°) = -\cos 70°$.

(d) $\cos 310° = \cos(4 \cdot 90° - 50°) = \cos 50° = \cos(3 \cdot 90° + 40°) = \sin 40°$.

(e) $\tan 165° = \tan(2 \cdot 90° - 15°) = -\tan 15° = \tan(1 \cdot 90° + 75°) = -\cot 75°$.

(f) $\sec 250° = \sec(2 \cdot 90° + 70°) = -\sec 70° = \sec(3 \cdot 90° - 20°) = -\csc 20°$.

(g) $\sin 670° = \sin(8 \cdot 90° - 50°) = -\sin 50° = \sin(7 \cdot 90° + 40°) = -\cos 40°$
  or $\sin 670° = \sin(310° + 360°) = \sin 310° = \sin(4 \cdot 90° - 50°) = -\sin 50°$.

(h) $\cot 930° = \cot(10 \cdot 90° + 30°) = \cot 30° = \cot(11 \cdot 90° - 60°) = \tan 60°$
  or $\cot 930° = \cot(210° + 2 \cdot 360°) = \cot 210° = \cot(2 \cdot 90° + 30°) = \cot 30°$.

(i) $\csc 865° = \csc(10 \cdot 90° - 35°) = \csc 35° = \csc(9 \cdot 90° + 55°) = \sec 55°$
  or $\csc 865° = \csc(145° + 2 \cdot 360°) = \csc 145° = \csc(2 \cdot 90° - 35°) = \csc 35°$.

(j) $\sin(-100°) = \sin(-2 \cdot 90° + 80°) = -\sin 80° = \sin(-1 \cdot 90° - 10°) = -\cos 10°$
  or $\sin(-100°) = -\sin 100° = -\sin(2° + 90° - 80°) = -\sin 80°$ or $\sin(-100°) = \sin(-100° + 360°) =$
  $\sin 260° = \sin(2 \cdot 90° + 80°) = -\sin 80°$.

(k) $\cos(-680°) = \cos(-8 \cdot 90° + 40°) = \cos 40° = \cos(-7 \cdot 90° - 50°) = \sin 50°$
  or $\cos(-680°) = \cos(-680° + 2 \cdot 360°) = \cos 40°$.

(l) $\tan(-290°) = \tan(-4 \cdot 90° + 70°) = -\tan 70° = \tan(-3 \cdot 90° - 20°) = -\cot 20°$
  or $\tan(-290°) = \tan(-290° + 360°) = \tan 70°$.

26.3 Find the values of the sine, cosine, and tangent of

(a) $120°$ (b) $210°$ (c) $315°$ (d) $-135°$ (d) $-240°$ (f) $-330°$

Call θ, always positive acute, the *related angle* of ϕ when $\phi = 180° - \theta, 180° + \theta$, or $360° - \theta$. Then any function of ϕ is numerically equal to the same function of θ. The algebraic sign in each case is that of the function in the quadrant in which the terminal side of ϕ lies.

(a) $120° = 180° - 60°$. The related angle is $60°$; $120°$ is in quadrant II; $\sin 210° = \sin 60° = \sqrt{3}/2$, $\cos 120° = -\cos 60° = -\frac{1}{2}$, $\tan 120° = -\tan 60° = -\sqrt{3}$.

(b) $210° = 180° + 30°$. The related angle is $30°$; $120°$ is in quadrant III; $\sin 210° = -\sin 30° = -\frac{1}{2}$, $\cos 210° = -\cos 30° = -\sqrt{3}/2$, $\tan 210° = \tan 30° = \sqrt{3}/3$.

(c) $315° = 360° - 45°$. The related angle $45°$; $315°$ is in quadrant IV; $\sin 315° = -\sin 45° = -\sqrt{2}/2$, $\cos 315° = \cos 45° = \sqrt{2}/2$, $\tan 315° = -\tan 45° = -1$.

(d) Any function of $-135°$ is the same function of $-135° + 360° = 225° = \phi$; $225° = 180° + 45°$. The related angle is $45°$; $225°$ is in quadrant III. $\sin(-135°) = -\sin 45° = -\sqrt{2}/2$, $\cos(-135°) = -\cos 45° = -\sqrt{-2}/2$, $\tan(-135°) = 1$.

(e) Any function of $-240°$ is the same function of $-240° + 360° = 120°$; $120° = 180° - 60°$. The related angle is $60°$; $120°$ is in quadrant II; $\sin(-240°) = \sin 60° = \sqrt{3}/2$, $\cos(-240°) = -\cos 60° = -\frac{1}{2}$, $\tan(-240°) = -\tan 60° = -\sqrt{3}$.

(f) Any function of $-330°$ is the same function of $-330° + 360° = 30°$; $\sin(-330°) = \sin 30° = \frac{1}{2}$, $\cos(-330°) = \cos 30° = \sqrt{3}/2$, $\tan(-330°) = \tan(-30°) = \sqrt{3}/3$.

26.4   Using a calculator, verify that

(a) $\sin 125°14' = \sin(180° - 54°46') = \sin 54°46' = 0.8168$

(b) $\cos 169°40' = \cos(180° - 10°20') = -\cos 10°20' = -0.9838$

(c) $\tan 200°23' = \tan(180° + 20°23') = \tan 20°23' = 0.3716$

(d) $\cot 250°44' = \cot(180° + 70°44') = \cot 70°44' = 0.3495$

(e) $\cos 313°18' = \cos(360° - 46°42') = \cos 46°42' = 0.6858$

(f) $\sin 341°25' = \sin(360° - 18°8') = -\sin 18°8' = -0.3112$

26.5   If $\tan 25° = a$, find

(a) $\dfrac{\tan 155° - \tan 115°}{1 + \tan 155° \tan 115°} = \dfrac{-\tan 25° - (-\cot 25°)}{1 + (-\tan 25°)(-\cot 25°)} = \dfrac{-a + 1/a}{1 + a(1/a)} = \dfrac{-a^2 + 1}{a + a} = \dfrac{1 - a^2}{2a}.$

(b) $\dfrac{\tan 205° - \tan 115°}{\tan 245° - \tan 335°} = \dfrac{\tan 25° - (-\cot 25°)}{\cot 25° + (-\tan 25°)} = \dfrac{a + 1/a}{1/a - a} = \dfrac{a^2 + 1}{1 - a^2}.$

26.6   If $m\angle A + m\angle B + m\angle C = 180°$, then show

(a) $\sin(B + C) = \sin A$

(b) $\sin\frac{1}{2}(B + C) = \cos\frac{1}{2}A$

(a) $\sin(B + C) = \sin(180° - A) = A$

(b) $\sin\frac{1}{2}(B + C) = \sin\frac{1}{2}(180° - A) = \sin(90° - \frac{1}{2}A) = \cos\frac{1}{2}A$

26.7   Show that $\sin\theta$ and $\tan\frac{1}{2}\theta$ have the same sign.

(a)   Suppose $m\angle\theta = n \cdot 180°$ If n is even (including zero), say $2m$, then $\sin(2m \cdot 180°) = \tan(m \cdot 180°) = 0$. The case when n is odd is excluded since then $\tan\frac{1}{2}\theta$ is not defined.

(b)   Suppose $m\angle\theta = n \cdot 180° + \phi$, where $0 < m\angle\phi < 180°$. If n is even, including zero, θ is in quadrant I or quadrant II and $\sin\theta$ is positive while $\frac{1}{2}\theta$ is in quadrant I or quadrant III and $\tan\frac{1}{2}\theta$ is positive. If n is odd, θ is in quadrant III or IV and $\sin\theta$ is negative while $\frac{1}{2}\theta$ is in quadrant II or IV and $\tan\frac{1}{2}\theta$ is negative.

26.8   Find all positive values of θ less than $360°$ for which $\sin\theta = -\frac{1}{2}$.

There will be two angles (see Chapters 24 and 25), one in the third quadrant and one in the fourth quadrant. The related angle of each has its sine equal to $+\frac{1}{2}$ and is $30°$. Thus, the required angles are θ with measure $180° + 30° = 210°$ and θ with measure $360° - 30° = 330°$.

(NOTE: To obtain *all* values of θ for which $\sin\theta = -\frac{1}{2}$, and $n \cdot 360°$ to each of the above solutions; thus $\theta = 210° + n \cdot 360°$ and $\theta = 330° + n \cdot 360°$, where n is any integer.)

26.9   Find all positive values of θ less than $360°$ which satisfy $\sin 2\theta = \cos\frac{1}{2}\theta$.

Since $\cos\frac{1}{2}\theta = \sin(90° - \frac{1}{2}\theta) = \sin 2\theta$, $2\theta = 90° - \frac{1}{2}\theta$, $450° - \frac{1}{2}\theta$, $810° - \frac{1}{2}\theta, 1170° - \frac{1}{2}\theta, \ldots$. Then $\frac{5}{2}\theta = 90°, 450°, 810°, 1170°, \ldots$ and $m\angle\theta = 36°, 180°, 324°, 468°, \ldots$.

Since $\cos\frac{1}{2}\theta = \sin(90° + \frac{1}{2}\theta) = \sin 2\theta$, $2\theta = 90° + \frac{1}{2}\theta$, $450° + \frac{1}{2}\theta$, $810° + \frac{1}{2}\theta, \ldots$. Then $\frac{1}{2}\theta = 90°$, $450°, 810°, \ldots$ and $\theta = 60°, 300°, 540°, \ldots$.

The required solutions have measures $36°, 180°, 324°; 60°, 300°$.

Supplementary Problems

26.10 Express each of the following in terms of functions of a positive acute angle:

    (a) $\sin 145°$ (d) $\cot 155°$ (g) $\sin(-200°)$ (j) $\cot 610°$
    (b) $\cos 215°$ (e) $\sec 325°$ (h) $\cos(-760°)$ (k) $\sec 455°$
    (c) $\tan 440°$ (f) $\csc 190°$ (i) $\tan(-1385°)$ (l) $\csc 825°$

    Ans.   (a) $\sin 35°$ or $\cos 55°$ (g) $\sin 20°$ or $\cos 70°$
            (b) $-\cos 35°$ or $-\sin 55°$ (h) $\cos 40°$ or $\sin 50°$
            (c) $\tan 80°$ or $\cot 10°$ (i) $\tan 55°$ or $\cot 35°$
            (d) $-\cot 25°$ or $-\tan 65°$ (j) $\cot 70°$ or $\tan 20°$
            (e) $\sec 35°$ or $\csc 55°$ (k) $-\sec 85°$ or $-\csc 5°$
            (f) $-\csc 10°$ or $\sec 80°$ (l) $\csc 75°$ or $\sec 15°$

26.11 Find the exact values of the sine, cosine, and tangent of

    (a) $150°$ (b) $225°$ (c) $300°$ (d) $-120°$ (e) $-120°$ (f) $-315°$

    Ans.   (a) $\frac{1}{2}, -\sqrt{3}/2, 1/\sqrt{3}$ (d) $-\sqrt{3}/2, -\frac{1}{2}, \sqrt{3}$
           (b) $-\sqrt{2}/2, -\sqrt{2}/2, 1$ (e) $\frac{1}{2}, \sqrt{3}/2, -1\sqrt{3}$
           (c) $-\sqrt{3}/2, \frac{1}{2}, -\sqrt{3}$ (f) $\sqrt{2}/2, \sqrt{2}/2, 1$

26.12 Using a calculator, verify that

    (a) $\sin 155°13' = 0.4192$
    (b) $\cos 104°38' = -0.2526$
    (c) $\tan 305°24' = -1.4071$
    (d) $\sin 114°18' = 0.9114$
    (e) $\cos 166°51' = -0.9738$

26.13 Find all angles, $0 \le \theta < 360°$, for which

    (a) $\sin \theta = \sqrt{2}/2$ (b) $\cos \theta = -1$

    Ans.   (a) $45°, 135°$
           (b) $180°$

26.14 When θ is a second quadrant angle for which $\tan \theta = -\frac{2}{3}$, show that

    (a) $\dfrac{\sin(90° - \theta) - \cos(180° - \theta)}{\tan(270° + \theta) + \cot(360° - \theta)} = -\dfrac{2}{\sqrt{3}}$ (b) $\dfrac{\tan(90° - \theta) + \cos(180° - \theta)}{\sin(270° + \theta) - \cot(-\theta)} = \dfrac{2 + \sqrt{13}}{2 - \sqrt{13}}$

Chapter 27

Graphs of the Trigonometric Functions

LINE REPRESENTATIONS OF THE TRIGONOMETRIC FUNCTIONS. Let θ be any given angle in standard position. (See the Figs. 27-1 through 27-4 for θ in each of the quadrants.) With the vertex O as center describe a circle of radius one unit cutting the initial side $\overrightarrow{OX}$ of θ at A, the positive y axis at B, and the terminal side of θ at P. Draw $\overline{MP}$ perpendicular to $\overrightarrow{OX}$; draw also the tangents to the circle at A and B meeting the terminal side of θ or its extension through O in the points Q and R, respectively.

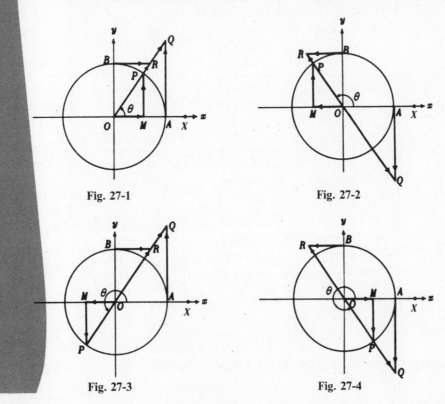

Fig. 27-1   Fig. 27-2

Fig. 27-3   Fig. 27-4

In each of the figures, the right triangles OMP, OAQ, and OBR are similar, and

$$\sin \theta = \frac{MP}{OP} = MP \qquad\qquad \cot \theta = \frac{OM}{MP} = \frac{BR}{OB} = BR$$

$$\cos \theta = \frac{OM}{OP} = OM \qquad\qquad \sec \theta = \frac{OP}{OM} = \frac{OQ}{OA} = OQ$$

$$\tan \theta = \frac{MP}{OM} = \frac{AQ}{OA} = AQ \qquad\qquad \csc \theta = \frac{OP}{MP} = \frac{OR}{OB} = OR$$

The, $\overrightarrow{MP}, \overrightarrow{OM}, \overrightarrow{AQ}$, etc., are directed line segments, the magnitude of a function being given by the length of the corresponding segment and the sign being given by the indicated direction. The directed segments $\overrightarrow{OQ}$ and $\overrightarrow{OR}$ are to be considered positive when measured on the terminal side of the angle and negative when measured on the terminal side extended.

VARIATIONS OF THE TRIGONOMETRIC FUNCTIONS. Let P move counterclockwise about the unit circle, starting at A, so that $m \angle \theta = m \angle XOP$ varies continuously from $0°$ to $360°$. Using Figs. 27-1 through 27-4, Table 27.1 is derived.

Table 27.1

As θ increases from	0° to 90°	90° to 180°	180° to 270°	270° to 360°
$\sin \theta$	I from 0 to 1	D from 1 to 0	D from 0 to -1	I from -1 to 0
$\cos \theta$	D from 1 to 0	D from 0 to -1	I from -1 to 0	I from 0 to 1
$\tan \theta$	I from 0 without limit (0 to $+\infty$)	I from large negative values to 0 ($-\infty$ to 0)	I from 0 without limit (0 to $+\infty$)	I from large negative values to 0 ($-\infty$ to 0)
$\cot \theta$	D from large positive values to 0 ($+\infty$ to 0)	D from 0 without limit (0 to $-\infty$)	D from large positive values to 0 ($+\infty$ to 0)	D from 0 without limit (0 to $-\infty$)
$\sec \theta$	I from 1 without limit (1 to $+\infty$)	I from large negative values to -1 ($-\infty$ to -1)	D from -1 without limit (-1 to $-\infty$)	D from large positive values to 1 ($+\infty$ to 1)
$\csc \theta$	D from large positive values to 1 ($+\infty$ to 1)	I from 1 without limit (1 to $+\infty$)	I from large negative values to -1 ($-\infty$ to -1)	D from -1 without limit (-1 to $-\infty$)

I = increases; D = decreases.

GRAPHS OF THE TRIGONOMETRIC FUNCTIONS. In Table 27.2, values of the angle x are given in radians.

Table 27.2

x	$y = \sin x$	$y = \cos x$	$y = \tan x$	$y = \cot x$	$y = \sec x$	$y = \csc x$
0	0	1.00	0	$\pm\infty$	1.00	$\pm\infty$
$\pi/6$	0.50	0.87	0.58	1.73	1.15	2.00
$\pi/4$	0.71	0.71	1.00	1.00	1.41	1.41
$\pi/3$	0.87	0.50	1.73	0.58	2.00	1.15
$\pi/2$	1.00	0	$\pm\infty$	0	$\pm\infty$	1.00
$2\pi/3$	0.87	−0.50	−1.73	−0.58	−2.00	1.15
$3\pi/4$	0.71	−0.71	−1.00	−1.00	−1.41	1.41
$5\pi/6$	0.50	−0.87	−0.58	−1.73	−1.15	2.00
π	0	−1.00	0	$+\infty$	−1.00	$+\infty$
$7\pi/6$	−0.50	−0.87	0.58	1.73	−1.15	−2.00
$5\pi/4$	−0.71	−0.71	1.00	1.00	−1.41	−1.41
$4\pi/3$	−0.87	−0.50	1.73	0.58	−2.00	−1.15
$3\pi/2$	−1.00	0	$\pm\infty$	0	$\pm\infty$	−1.00
$5\pi/3$	−0.87	0.50	−1.73	−0.58	2.00	−1.15
$7\pi/4$	−0.71	0.71	−1.00	−1.00	1.41	−1.41
$11\pi/6$	−0.50	0.87	−0.58	−1.73	1.15	−2.00
2π	0	1.00	0	$\pm\infty$	1.00	$\pm\infty$

Note 1.   Since $\sin(\tfrac{1}{2}\pi + x) = \cos x$, the graph of $y = \cos x$ may be obtained most easily by shifting the graph of $y = \sin x$ a distance $\tfrac{1}{2}\pi$ to the left. See Fig. 27-5.

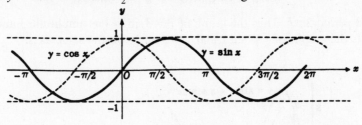

Fig. 27-5

Note 2.   Since $\csc\left(\tfrac{1}{2}\pi + x\right) = \sec x$, the graph of $y = \csc x$ may be obtained by shifting the graph of $y = \sec x$ a distant $\tfrac{1}{2}\pi$ to the right. Notice, too, the relationship between the graphs for $\tan x$ and $\cot x$. See Figs. 27-6 through 27-9.

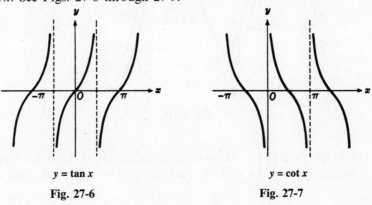

$y = \tan x$

Fig. 27-6

$y = \cot x$

Fig. 27-7

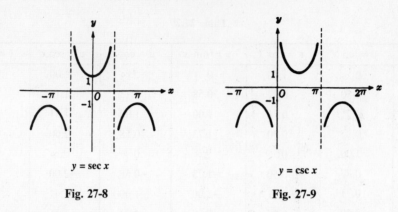

$y = \sec x$

Fig. 27-8

$y = \csc x$

Fig. 27-9

PERIODIC FUNCTIONS. Any function of a variable x, $f(x)$, which repeats its values in definite cycles, is called *periodic*. The smallest range of values of x which corresponds to a complete cycle of values of the function is called the period of the function. It is evident from the graphs of the trigonometric functions that the sine, cosine, secant, and cosecant are of period 2π while the tangent and cotangent are of period π.

THE GENERAL SINE CURVE. The *amplitude (maximum ordinate) and period (wavelength) of* $y = \sin x$ are, respectively, 1 and 2π. For a given value of x, the value of $y = a \sin x, a > 0$, is a times the value of $y = \sin x$. Thus, the amplitude of $y = a \sin x$ is a and the period is 2π. Since when $bx = 2\pi, x = 2\pi/b$, the amplitude of $y = \sin bx, b > 0$, is 1 and the period is $2\pi/b$.

The general sine curve (sinusoid) of equation

$$y = a \sin bx, \qquad a > 0, \quad b > 0,$$

has amplitude a and period $2\pi/b$. Thus the graph of $y = 3 \sin 2x$ has amplitude 3 and period $2\pi/2 = \pi$. Figure 27-10 exhibits the graphs of $y = \sin x$ and $y = 3 \sin 2x$ on the same axes.

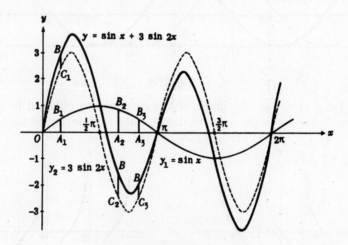

Fig. 27-10

COMPOSITION OF SINE CURVES. More complicated forms of wave motions are obtained by combining two or more sine curves. The method of adding corresponding ordinates is illustrated in the following example.

EXAMPLE.   Construct the graph of $y = \sin x + 3 \sin 2x$. See Fig. 27-10.

First the graphs of $y_1 = \sin x$ and $y_2 = 3 \sin 2x$ are constructed on the same axes. Then, corresponding to a given value $x = OA_1$, the ordinate A_1B of $y = \sin x + 3 \sin 2x$ is the *algebraic* sum of the ordinates A_1B_1 of $y_1 = \sin x$ and A_1C_1 of $y_2 = 3 \sin 2x$.

Also, $A_2B = A_2B_2 + A_2C_2, A_3B = A_3B_3 + A_3C_3$, etc.

Solved Problems

27.1   Sketch the graphs of the following for one period:

(a) $y = 4 \sin x$ (c) $y = 3 \sin \frac{1}{2}x$ (e) $y = 3 \cos \frac{1}{2}x = 3 \sin (\frac{1}{2}x + \frac{1}{2}\pi)$

(b) $y = \sin 3x$ (d) $y = 2 \cos x = 2 \sin (x + \frac{1}{2}\pi)$

In each case we use the same curve and then put in the y axis and choose the units on each axis to satisfy the requirements of amplitude and period of each curve.

(a) $y = 4 \sin x$ has amplitude $= 4$ and period $= 2\pi$. See Fig. 27-11(a).

(b) $y = \sin 3x$ has amplitude $= 1$ and period $= 2\pi/3$. See Fig. 27-11(b).

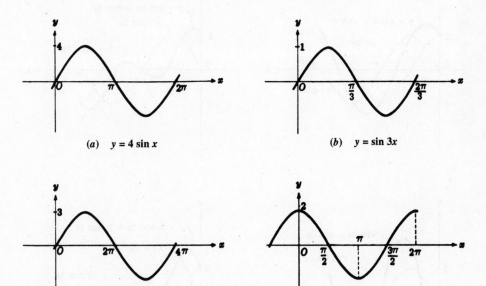

(a) $y = 4 \sin x$ (b) $y = \sin 3x$

(c) $y = 3 \sin \frac{1}{2}x$ (d) $y = 2 \cos x$

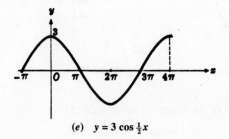

(e) $y = 3 \cos \frac{1}{2}x$

Fig. 27-11

(c) $y = 3 \sin \frac{1}{2}x$ has amplitude = 3 and period = $2\pi/\frac{1}{2} = 4\pi$. See Fig. 27-11(c).

(d) $y = 2 \cos x$ has amplitude = 2 and period = 2π. Note the position of the y axis. See Fig. 27-11(d).

(e) $y = 3 \cos \frac{1}{2}x$ has amplitude = 3 and period = 4π. See Fig. 27-11(e).

27.2 Construct the graph of each of the following:

(a) $y = \sin x + \cos x$ (c) $y = \sin 2x - \cos 3x$

(b) $y = \sin 2x - \cos 3x$ (d) $y = 3 \sin 2x + 2 \cos 3x$

See Fig. 27-12(a)–(d).

Supplementary Problems

27.3 Sketch the graph of each of the following for one period:

(a) $y = 3 \sin x$ (b) $y = \sin 2x$ (c) $y = 4 \sin (x/2)$ (d) $y = 4 \cos x$

(e) $y = 2 \cos (x/3)$

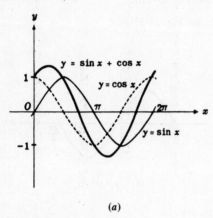

(a)

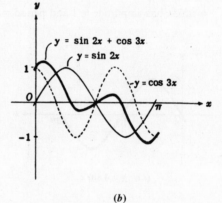

(b)

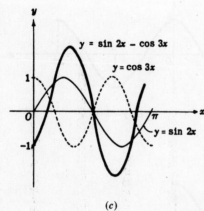

(c)

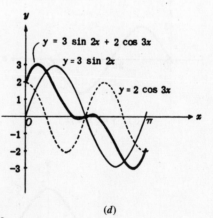

(d)

Fig. 27-12

27.4 Construct the graph of each of the following for one period:

(a) $y = \sin x + 2 \cos x$ (d) $y = \sin 2x + \sin 3x$

(b) $y = \sin 3x + \cos 2x$ (e) $y = \sin 3x - \cos 2x$

(c) $y = \sin x + \sin 2x$ (f) $y = 2 \sin 3x + 3 \cos 2x$

Chapter 28

Fundamental Relations and Identities

FUNDAMENTAL RELATIONS

Reciprocal Relations	Quotient Relations	Pythagorean Relations
$\csc\theta = \dfrac{1}{\sin\theta}$	$\tan\theta = \dfrac{\sin\theta}{\cos\theta}$	$\sin^2\theta + \cos^2\theta = 1$
$\sec\theta = \dfrac{1}{\cos\theta}$	$\cot\theta = \dfrac{\cos\theta}{\sin\theta}$	$1 + \tan^2\theta = \sec^2\theta$
$\cot\theta = \dfrac{1}{\tan\theta}$		$1 + \cot^2\theta = \csc^2\theta$

The above relations hold for every value of θ for which the functions involved are defined.

Thus, $\sin^2\theta + \cos^2\theta = 1$ holds for every value of θ while $\tan\theta = \sin\theta/\cos\theta$ holds for all values of θ for which $\tan\theta$ is defined, i.e., for all $\theta \neq n \cdot 90°$ where n is odd. Note that for the excluded values of θ, $\cos\theta = 0$ and $\sin\theta \neq 0$.

For proofs of the quotient and Pythagorean relations, see Problems 28.1–28.2. The reciprocal relations were treated in Chapter 24. (See also Problems 28.3–28.6.)

SIMPLIFICATION OF TRIGONOMETRIC EXPRESSIONS.
It is frequently desirable to transform or reduce a given expression involving trigonometric functions to a simpler form.

EXAMPLE 1

(a) Using $\csc\theta = \dfrac{1}{\sin\theta}$, $\cos\theta\csc\theta = \cos\theta\dfrac{1}{\sin\theta} = \dfrac{\cos\theta}{\sin\theta} = \cot\theta$.

(b) Using $\tan\theta = \dfrac{\sin\theta}{\cos\theta}$, $\cos\theta\tan\theta = \cos\theta\dfrac{\sin\theta}{\cos\theta} = \sin\theta$.

EXAMPLE 2. Using the relation $\sin^2\theta + \cos^2\theta = 1$,

(a) $\sin^3\theta + \sin\theta\cos^2\theta = (\sin^2\theta + \cos^2\theta)\sin\theta = (1)\sin\theta = \sin\theta$.

189

(b) $\dfrac{\cos^2 \theta}{1 - \sin \theta} = \dfrac{1 - \sin^2 \theta}{1 - \sin \theta} = \dfrac{(1 - \sin \theta)(1 + \sin \theta)}{1 - \sin \theta} = 1 + \sin \theta.$

(NOTE: The relation $\sin^2 \theta + \cos^2 \theta = 1$ may be written as $\sin^2 \theta = 1 - \cos^2 \theta$ and as $\cos^2 \theta = 1 - \sin^2 \theta$. Each form is equally useful.) (See Problems 28.8–28.9.)

TRIGONOMETRIC IDENTITIES. A relation involving the trigonometric functions which is valid for all values of the angle for which the functions are defined is called a trigonometric identity. The eight fundamental relations above are trigonometric identities; so also are

$$\cos \theta \csc \theta = \cot \theta \qquad \text{and} \qquad \cos \theta \tan \theta = \sin \theta$$

of Example 1 above.

A trigonometric identity is verified by transforming one member (your choice) into the other. In general, one begins with the more complicated side.

Success in verifying identities requires

(a) Complete familiarity with the fundamental relations
(b) Complete familiarity with the processes of factoring, adding fractions, etc.
(c) Practice

(See Problems 28.10–28.17.)

Solved Problems

28.1   Prove the quotient relations: $\tan \theta = \dfrac{\sin \theta}{\cos \theta}$, $\cot \theta = \dfrac{\cos \theta}{\sin \theta}$.

For any angle θ, $\sin \theta = y/r$, $\cos \theta = x/r$, $\tan \theta = y/x$, and $\cot \theta = x/y$, where with θ drawn in standard position, $P(x, y)$ is any point on the terminal side of θ at a distance r from the origin.

Then $\tan \theta = \dfrac{y}{x} = \dfrac{y/r}{x/r} = \dfrac{\sin \theta}{\cos \theta}$ and $\cot \theta = \dfrac{x}{y} = \dfrac{x/r}{y/r} = \dfrac{\cos \theta}{\sin \theta}.$

$\left(\text{Also,} \cot \theta = \dfrac{1}{\tan \theta} = \dfrac{\cos \theta}{\sin \theta}. \right)$

28.2   Prove the Pythagorean relations: (a) $\sin^2 \theta + \cos^2 \theta = 1$ (b) $1 + \tan^2 \theta - \sec^2 \theta$
(c) $1 + \cot^2 \theta = \csc^2 \theta.$

For $P(x, y)$ defined as in Problem 28.1, we have A) $x^2 + y^2 = r^2$.

(a) Dividing A) by r^2, $(x/r)^2 + (y/r)^2 = 1$ and $\sin^2 \theta + \cos^2 \theta = 1$.

(b) Dividing A) by x^2, $1 + (y/x)^2 = (r/x)^2$ and $1 + \tan^2 \theta = \sec^2 \theta$. Also, dividing $\sin^2 \theta + \cos^2 \theta = 1$ by $\cos^2 \theta$,

$$\left(\frac{\sin \theta}{\cos \theta} \right)^2 + 1 = \left(\frac{1}{\cos \theta} \right)^2 \qquad \text{or} \qquad \tan^2 \theta + 1 = \sec^2 \theta.$$

(c) Dividing A) by y^2, $(x/y)^2 + 1 = (r/y)^2$ and $\cot^2 \theta + 1 = \csc^2 \theta$. Also, dividing $\sin^2 \theta + \cos^2 \theta = 1$ by $\sin^2 \theta$,

$$1 + \left(\frac{\cos \theta}{\sin \theta} \right)^2 = \left(\frac{1}{\sin \theta} \right)^2 \qquad \text{or} \qquad 1 + \cot^2 \theta = \csc^2 \theta.$$

28.3 Express each of the other functions of θ in terms of $\sin \theta$.

$$\cos^2 \theta = 1 - \sin^2 \theta \quad \text{and} \quad \cos \theta = \pm\sqrt{1 - \sin^2 \theta}$$

$$\tan \theta = \frac{\sin \theta}{\cos \theta} = \frac{\sin \theta}{\pm\sqrt{1 - \sin^2 \theta}} \qquad \cot \theta = \frac{1}{\tan \theta} = \frac{\pm\sqrt{1 - \sin^2 \theta}}{\sin \theta}$$

$$\sec \theta = \frac{1}{\cos \theta} = \frac{1}{\pm\sqrt{1 - \sin^2 \theta}} \qquad \csc \theta = \frac{1}{\sin \theta}$$

Note that $\cos \theta = \pm\sqrt{1 - \sin^2 \theta}$. Writing $\cos \theta = \sqrt{1 - \sin^2 \theta}$ limits angle θ to those quadrants (first and fourth) in which the cosine is positive.

28.4 Express each of the other functions of θ in terms of $\tan \theta$.

$$\sec^2 \theta = 1 + \tan^2 \theta \quad \text{and} \quad \sec \theta = \pm\sqrt{1 + \tan^2 \theta}, \qquad \cos \theta = \frac{1}{\sec \theta} = \frac{1}{\pm\sqrt{1 + \tan^2 \theta}},$$

$$\frac{\sin \theta}{\cos \theta} = \tan \theta \quad \text{and} \quad \sin \theta = \tan \theta \cos \theta = \tan \theta \frac{1}{\pm\sqrt{1 + \tan^2 \theta}} = \frac{\tan \theta}{\pm\sqrt{1 + \tan^2 \theta}},$$

$$\csc \theta = \frac{1}{\sin \theta} = \frac{\pm\sqrt{1 + \tan^2 \theta}}{\tan \theta}, \qquad \cot \theta = \frac{1}{\tan \theta}.$$

28.5 Using the fundamental relations, find the possible values of the functions of θ, given $\sin \theta = \frac{3}{5}$.

From $\cos^2 \theta = 1 - \sin^2 \theta$, $\cos \theta = \pm\sqrt{1 - \sin^2 \theta} = \pm\sqrt{1 - \left(\frac{3}{5}\right)^2} = \pm\sqrt{\frac{16}{25}} = \pm\frac{4}{5}$.

Now $\sin \theta$ and $\cos \theta$ are both positive when θ is a first quadrant angle while $\sin \theta = +$ and $\cos \theta = -$ when θ is a second quadrant angle. Thus,

First Quadrant		Second Quadrant	
$\sin \theta = \frac{3}{5}$	$\cot \theta = \frac{4}{3}$	$\sin \theta = \frac{3}{5}$	$\cot \theta = -\frac{4}{3}$
$\cos \theta = \frac{4}{5}$	$\sec \theta = \frac{5}{4}$	$\cos \theta = -\frac{4}{5}$	$\sec \theta = -\frac{5}{4}$
$\tan \theta = \frac{\frac{3}{5}}{\frac{4}{5}} = \frac{3}{4}$	$\csc \theta = \frac{5}{3}$	$\tan \theta = -\frac{3}{4}$	$\csc \theta = \frac{5}{3}$

28.6 Using the fundamental relations, find the possible values of the functions of θ, given $\tan \theta = -\frac{5}{12}$.

Since $\tan \theta = -$, θ is either a second or fourth quadrant angle.

Second Quadrant	Fourth Quadrant
$\tan \theta = -\frac{5}{12}$	$\tan \theta = -\frac{5}{12}$
$\cot \theta = 1/\tan \theta = -\frac{12}{5}$	$\cot \theta = -\frac{12}{5}$
$\sec \theta = -\sqrt{1 + \tan^2 \theta} = -\frac{13}{12}$	$\sec \theta = \frac{13}{12}$
$\cos \theta = 1/\sec \theta = -\frac{12}{13}$	$\cos \theta = \frac{12}{13}$
$\csc \theta = \sqrt{1 + \cot^2 \theta} = \frac{13}{5}$	$\csc \theta = -\frac{13}{5}$
$\sin \theta = 1/\csc \theta = \frac{5}{13}$	$\sin \theta = -\frac{5}{13}$

28.7 Perform the indicated operations.

(a) $(\sin \theta - \cos \theta)(\sin \theta + \cos \theta) = \sin^2 \theta - \cos^2 \theta$

(b) $(\sin A + \cos A)^2 = \sin^2 A + 2 \sin A \cos A + \cos^2 A$

(c) $(\sin x + \cos y)(\sin y - \cos x) = \sin x \sin y - \sin x \cos x + \sin y \cos y - \cos x \cos y$

(d) $(\tan^2 A - \cot A)^2 = \tan^4 A - 2 \tan^2 A \cot A + \cot^2 A$

(e) $1 + \dfrac{\cos \theta}{\sin \theta} = \dfrac{\sin \theta + \cos \theta}{\sin \theta}$

(f) $1 - \dfrac{\sin \theta}{\cos \theta} + \dfrac{2}{\cos^2 \theta} = \dfrac{\cos^2 \theta - \sin \theta \cos \theta + 2}{\cos^2 \theta}$

28.8 Factor.

(a) $\sin^2 \theta - \sin \theta \cos \theta = \sin \theta (\sin \theta - \cos \theta)$

(b) $\sin^2 \theta + \sin^2 \theta \cos^2 \theta = \sin^2 \theta (1 + \cos^2 \theta)$

(c) $\sin^2 \theta + \sin \theta \sec \theta - 6 \sec^2 \theta = (\sin \theta + 3 \sec \theta)(\sin \theta - 2 \sec \theta)$

(d) $\sin^3 \theta \cos^2 \theta - \sin^2 \theta \cos^3 \theta + \sin \theta \cos^2 \theta = \sin \theta \cos^2 \theta (\sin^2 \theta - \sin \theta \cos \theta + 1)$

(e) $\sin^4 \theta - \cos^4 \theta = (\sin^2 \theta + \cos^2 \theta)(\sin^2 \theta - \cos^2 \theta) = (\sin^2 \theta + \cos^2 \theta)(\sin \theta - \cos \theta)(\sin \theta + \cos \theta)$

28.9 Simplify each of the following:

(a) $\sec \theta - \sec \theta \sin^2 \theta = \sec \theta (1 - \sin^2 \theta) = \sec \theta \cos^2 \theta = \dfrac{1}{\cos \theta} \cos^2 \theta = \cos \theta$

(b) $\sin \theta \sec \theta \cot \theta = \sin \theta \dfrac{1}{\cos \theta} \dfrac{\cos \theta}{\sin \theta} = \dfrac{\sin \theta \cos \theta}{\cos \theta \sin \theta} = 1$

(c) $\sin^2 \theta (1 + \cot^2 \theta) = \sin^2 \theta \csc^2 \theta = \sin^2 \theta \dfrac{1}{\sin^2 \theta} = 1$

(d) $\sin^2 \theta \sec^2 \theta - \sec^2 \theta = (\sin^2 \theta - 1) \sec^2 \theta = -\cos^2 \theta \sec^2 \theta = -\cos^2 \theta \dfrac{1}{\cos^2 \theta} = -1$

(e) $(\sin \theta + \cos \theta)^2 + (\sin \theta - \cos \theta)^2 = \sin^2 \theta + 2 \sin \theta \cos \theta + \cos^2 \theta + \sin^2 \theta - 2 \sin \theta \cos \theta + \cos^2 \theta$
$$= 2(\sin^2 \theta + \cos^2 \theta) = 2$$

(f) $\tan^2 \theta \cos^2 \theta + \cot^2 \theta \sin^2 \theta = \dfrac{\sin^2 \theta}{\cos^2 \theta} \cos^2 \theta + \dfrac{\cos^2 \theta}{\sin^2 \theta} \sin^2 \theta = \sin^2 \theta + \cos^2 \theta = 1$

(g) $\tan \theta + \dfrac{\cos \theta}{1 + \sin \theta} = \dfrac{\sin \theta}{\cos \theta} + \dfrac{\cos \theta}{1 + \sin \theta} = \dfrac{\sin \theta (1 + \sin \theta) + \cos^2 \theta}{\cos \theta (1 + \sin \theta)}$

$$= \dfrac{\sin \theta + \sin^2 \theta + \cos^2 \theta}{\cos \theta (1 + \sin \theta)} = \dfrac{\sin \theta + 1}{\cos \theta (1 + \sin \theta)} = \dfrac{1}{\cos \theta} = \sec \theta$$

Verify the following identities:

28.10 $\sec^2 \theta \csc^2 \theta = \sec^2 \theta + \csc^2 \theta$

$$\sec^2 \theta + \csc^2 \theta = \dfrac{1}{\cos^2 \theta} + \dfrac{1}{\sin^2 \theta} = \dfrac{\sin^2 \theta + \cos^2 \theta}{\sin^2 \theta \cos^2 \theta} = \dfrac{1}{\sin^2 \theta \cos^2 \theta} = \dfrac{1}{\sin^2 \theta} \dfrac{1}{\cos^2 \theta} = \csc^2 \theta \sec^2 \theta$$

28.11 $\sec^4 \theta - \sec^2 \theta = \tan^4 \theta + \tan^2 \theta$

$$\tan^4 \theta + \tan^2 \theta = \tan^2 \theta (\tan^2 \theta + 1) = \tan^2 \theta \sec^2 \theta = (\sec^2 \theta - 1) \sec^2 \theta = \sec^4 \theta - \sec^2 \theta$$

or $\sec^4 \theta - \sec^2 \theta = \sec^2 \theta (\sec^2 \theta - 1) = \sec^2 \theta \tan^2 \theta = (1 + \tan^2 \theta) \tan^2 \theta = \tan^2 \theta + \tan^4 \theta$

28.12 $2 \csc x = \dfrac{\sin x}{1 + \cos x} + \dfrac{1 + \cos x}{\sin x}$

$$\dfrac{\sin x}{1 + \cos x} + \dfrac{1 + \cos x}{\sin x} = \dfrac{\sin^2 x + (1 + \cos x)^2}{\sin x (1 + \cos x)} = \dfrac{\sin^2 x + 1 + 2 \cos x + \cos^2 x}{\sin x (1 + \cos x)}$$

$$= \dfrac{2 + 2 \cos x}{\sin x (1 + \cos x)} = \dfrac{2(1 + \cos x)}{\sin x (1 + \cos x)} = \dfrac{2}{\sin x} = 2 \csc x$$

28.13 $\dfrac{1 - \sin x}{\cos x} = \dfrac{\cos x}{1 + \sin x}$

$$\dfrac{\cos x}{1 + \sin x} = \dfrac{\cos^2 x}{\cos x\,(1 + \sin x)} = \dfrac{1 - \sin^2 x}{\cos x\,(1 + \sin x)} = \dfrac{(1 - \sin x)(1 + \sin x)}{\cos x\,(1 + \sin x)} = \dfrac{1 - \sin x}{\cos x}$$

28.14 $\dfrac{\sec A - \csc A}{\sec A + \csc A} = \dfrac{\tan A - 1}{\tan A + 1}$

$$\dfrac{\sec A - \csc A}{\sec A + \csc A} = \dfrac{\dfrac{1}{\cos A} - \dfrac{1}{\sin A}}{\dfrac{1}{\cos A} + \dfrac{1}{\sin A}} = \dfrac{\dfrac{\sin A}{\cos A} - 1}{\dfrac{\sin A}{\cos A} + 1} = \dfrac{\tan A - 1}{\tan A + 1}$$

28.15 $\dfrac{\tan x - \sin x}{\sin^3 x} = \dfrac{\sec x}{1 + \cos x}$

$$\dfrac{\tan x - \sin x}{\sin^3 x} = \dfrac{\dfrac{\sin x}{\cos x} - \sin x}{\sin^3 x} = \dfrac{\sin x - \sin x \cos x}{\cos x \sin^3 x} = \dfrac{\sin x\,(1 - \cos x)}{\cos x \sin^3 x}$$

$$= \dfrac{1 - \cos x}{\cos x \sin^2 x} = \dfrac{1 - \cos x}{\cos x\,(1 - \cos^2 x)} = \dfrac{1}{\cos x\,(1 + \cos x)} = \dfrac{\sec x}{1 + \cos x}$$

28.16 $\dfrac{\cos A \cot A - \sin A \tan A}{\csc A - \sec A} = 1 + \sin A \cos A$

$$\dfrac{\cos A \cot A - \sin A \tan A}{\csc A - \sec A} = \dfrac{\cos A \dfrac{\cos A}{\sin A} - \sin A \dfrac{\sin A}{\cos A}}{\dfrac{1}{\sin A} - \dfrac{1}{\cos A}} = \dfrac{\cos^3 A - \sin^3 A}{\cos A - \sin A}$$

$$= \dfrac{(\cos A - \sin A)(\cos^2 A + \cos A \sin A + \sin^2 A)}{\cos A - \sin A}$$

$$= \cos^2 A + \cos A \sin A + \sin^2 A = 1 + \cos A \sin A$$

28.17 $\dfrac{\sin \theta - \cos \theta + 1}{\sin \theta + \cos \theta - 1} = \dfrac{\sin \theta + 1}{\cos \theta}$

$$\dfrac{\sin \theta + 1}{\cos \theta} = \dfrac{(\sin \theta + 1)(\sin \theta + \cos \theta - 1)}{\cos \theta\,(\sin \theta + \cos \theta - 1)} = \dfrac{\sin^2 \theta + \sin \theta \cos \theta + \cos \theta - 1}{\cos \theta\,(\sin \theta + \cos \theta - 1)}$$

$$= \dfrac{-\cos^2 \theta + \sin \theta \cos \theta + \cos \theta}{\cos \theta\,(\sin \theta + \cos \theta - 1)} = \dfrac{\cos \theta\,(\sin \theta - \cos \theta + 1)}{\cos \theta\,(\sin \theta + \cos \theta - 1)}$$

$$= \dfrac{\sin \theta - \cos \theta + 1}{\sin \theta + \cos \theta - 1}$$

Supplementary Problems

28.18 Find the possible values of the trigonometric functions of θ, given $\sin \theta = \frac{2}{3}$.

    Ans. Quad I: $\frac{2}{3}, \sqrt{5}/3, 2/\sqrt{5}, \sqrt{5}/2, 3/\sqrt{5}, \frac{3}{2}$

             Quad II: $\frac{2}{3}, -\sqrt{5}/3, -2/\sqrt{5}, -\sqrt{5}/2, -3/\sqrt{5}, \frac{3}{2}$

28.19 Find the possible values of the trigonometric functions of θ, given $\cos \theta = -\frac{5}{6}$.

   Ans.   Quad II: $\sqrt{11}/6, -\frac{5}{6}, -\sqrt{11}/5, -5/\sqrt{11}, -\frac{6}{5}, 6/\sqrt{11}$
             Quad III: $-\sqrt{11}/6, -\frac{5}{6}, \sqrt{11}/5, 5/\sqrt{11}, -\frac{6}{5}, -6/\sqrt{11}$

28.20 Find the possible values of the trigonometric functions of θ, given $\tan \theta = \frac{5}{4}$.

   Ans.   Quad I: $5/\sqrt{41}, 4/\sqrt{41}, \frac{5}{4}, \frac{4}{5}, \sqrt{41}/4, \sqrt{41}/5$
             Quad III: $-5/\sqrt{41}, -4/\sqrt{41}, \frac{5}{4}, \frac{4}{5}, -\sqrt{41}/4, -\sqrt{41}/5$

28.21 Find the possible values of the trigonometric functions of θ, given $\cot \theta = -\sqrt{3}$.

   Ans.   Quad II: $\frac{1}{2}, -\sqrt{3}/2, -1/\sqrt{3}, -\sqrt{3}, -2/\sqrt{3}, 2$
             Quad IV: $-\frac{1}{2}, \sqrt{3}/2, -1/\sqrt{3}, -\sqrt{3}, 2/\sqrt{3}, -2$

28.22 Find the possible value of $\dfrac{\sin \theta + \cos \theta - \tan \theta}{\sec \theta + \csc \theta - \cot \theta}$ when $\tan \theta = -\frac{4}{3}$

   Ans.   Quad II: $\frac{23}{5}$
             Quad IV: $\frac{34}{35}$

Verify the following identities:

28.23 $\sin \theta \sec \theta = \tan \theta$

28.24 $(1 - \sin^2 A)(1 + \tan^2 A) = 1$

28.25 $(1 - \cos \theta)(1 + \sec \theta) \cot \theta = \sin \theta$

28.26 $\csc^2 x (1 - \cos^2 x) = 1$

28.27 $\dfrac{\sin \theta}{\csc \theta} + \dfrac{\cos \theta}{\sec \theta} = 1$

28.28 $\dfrac{1 - 2\cos^2 A}{\sin A \cos A} = \tan A - \cot A$

28.29 $\tan^2 x \csc^2 x \cot^2 x \sin^2 x = 1$

28.30 $\sin A \cos A (\tan A + \cot A) = 1$

28.31 $1 - \dfrac{\cos^2 \theta}{1 + \sin \theta} = \sin \theta$

28.32 $\dfrac{1}{\sec \theta + \tan \theta} = \sec \theta - \tan \theta$

28.33 $\dfrac{1}{1 - \sin A} + \dfrac{1}{1 + \sin A} = 2 \sec^2 A$

28.34 $\dfrac{1 - \cos x}{1 + \cos x} = \dfrac{\sec x - 1}{\sec x + 1} = (\cot x - \csc x)^2$

28.35 $\tan \theta \sin \theta + \cos \theta = \sec \theta$

28.36 $\tan \theta - \csc \theta \sec \theta (1 - 2\cos^2 \theta) = \cot \theta$

28.37 $\dfrac{\sin \theta}{\sin \theta + \cos \theta} = \dfrac{\sec \theta}{\sec \theta + \csc \theta}$

28.38 $\dfrac{\sin x + \tan x}{\cot x + \csc x} = \sin x \tan x$

28.39 $\dfrac{\sec x + \csc x}{\tan x + \cot x} = \sin x + \cos x$

28.40 $\dfrac{\sin^3 \theta + \cos^3 \theta}{\sin \theta + \cos \theta} = 1 - \sin \theta \cos \theta$

28.41 $\cot \theta + \dfrac{\sin \theta}{1 + \cos \theta} = \csc \theta$

28.42 $\dfrac{\sin \theta \cos \theta}{\cos^2 \theta - \sin^2 \theta} = \dfrac{\tan \theta}{1 - \tan^2 \theta}$

28.43 $(\tan x + \tan y)(1 - \cot x \cot y)$
           $\qquad + (\cot x + \cot y)(1 - \tan x \tan y) = 0$

28.44 $(x \sin \theta - y \cos \theta)^2 + (x \cos \theta + y \sin \theta)^2$
           $\qquad = x^2 + y^2$

28.45 $(2r \sin \theta \cos \theta)^2 + r^2 (\cos^2 \theta - \sin^2 \theta)^2 = r^2$

28.46 $(r \sin \theta \cos \phi)^2 + (r \sin \theta \sin \phi)^2 + (r \cos \theta)^2 = r^2$

Chapter 29

Trigonometric Functions of Two Angles

ADDITION FORMULAS

$$\sin(\alpha + \beta) = \sin\alpha\cos\beta + \cos\alpha\sin\beta$$
$$\cos(\alpha + \beta) = \cos\alpha\cos\beta - \sin\alpha\sin\beta$$
$$\tan(\alpha + \beta) = \frac{\tan\alpha + \tan\beta}{1 - \tan\alpha\tan\beta}$$

(For a proof of these formulas, see Problems 29.1–29.2.

SUBTRACTION FORMULAS

$$\sin(\alpha - \beta) = \sin\alpha\cos\beta - \cos\alpha\sin\beta$$
$$\cos(\alpha - \beta) = \cos\alpha\cos\beta + \sin\alpha\sin\beta$$
$$\tan(\alpha - \beta) = \frac{\tan\alpha - \tan\beta}{1 + \tan\alpha\tan\beta}$$

(For a proof of these formulas, see Problem 29.3.)

DOUBLE-ANGLE FORMULAS

$$\sin 2\alpha = 2\sin\alpha\cos\alpha$$
$$\cos 2\alpha = \cos^2\alpha - \sin^2\alpha = 1 - 2\sin^2\alpha = 2\cos^2\alpha - 1$$
$$\tan 2\alpha = \frac{2\tan\alpha}{1 - \tan^2\alpha}$$

(For a proof of these formulas, see Problem 29.9.)

195

HALF-ANGLE FORMULAS

$$\sin\tfrac{1}{2}\theta = \pm\sqrt{\frac{1-\cos\theta}{2}}$$

$$\cos\tfrac{1}{2}\theta = \pm\sqrt{\frac{1+\cos\theta}{2}}$$

$$\tan\tfrac{1}{2}\theta = \pm\sqrt{\frac{1-\cos\theta}{1+\cos\theta}} = \frac{\sin\theta}{1+\cos\theta} = \frac{1-\cos\theta}{\sin\theta}$$

(For a proof of these formulas, see Problem 29.10.)

Solved Problems

29.1   Prove (1) $\sin(\alpha+\beta) = \sin\alpha\cos\beta + \cos\alpha\sin\beta$ and (2) $\cos(\alpha+\beta) = \cos\alpha\cos\beta - \sin\alpha\sin\beta$ when α and β are positive acute angles.

Let α and β be the measures of positive acute angles such that $\alpha+\beta < 90°$ [Fig. 29-1(a)] and $\alpha+\beta < 90°$ [Fig. 29-1(b)].

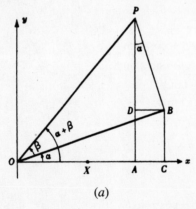
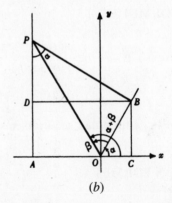

(a)   (b)

Fig. 29-1

To construct these figures, place angle α in standard position and then place angle β with its vertex at O and with its initial side along the terminal side of angle α. Let P be any point on the terminal side of angle $(\alpha+\beta)$: Draw $\overline{PA}$ perpendicular to $\overrightarrow{OX}$, $\overline{PB}$ perpendicular to the terminal side of angle α, $\overline{BC}$ perpendicular to $\overrightarrow{OX}$, and $\overline{BC}$ perpendicular to $\overline{AP}$.

Now m$\angle APB = \alpha$ since corresponding sides (OA and AP, OB and BP) are perpendicular. Then

$$\sin(\alpha+\beta) = \frac{AP}{OP} = \frac{AD+DP}{OP} = \frac{CB+DP}{OP} = \frac{CB}{OP} + \frac{DP}{OP} = \frac{CB}{OB}\cdot\frac{OB}{OP} + \frac{DP}{BP}\cdot\frac{BP}{OP}$$

$$= \sin\alpha\cos\beta + \cos\alpha\sin\beta$$

and $$\cos(\alpha+\beta) = \frac{OA}{OP} = \frac{OC-AC}{OP} = \frac{OC-DB}{OP} = \frac{OC}{OP} - \frac{DB}{OP} = \frac{OC}{OB}\cdot\frac{OB}{OP} - \frac{DB}{BP}\cdot\frac{BP}{OP}$$

29.2 Prove $\tan(\alpha + \beta) = \dfrac{\tan \alpha + \tan \beta}{1 - \tan \alpha \tan \beta}$.

$$\tan(\alpha + \beta) = \frac{\sin(\alpha + \beta)}{\cos(\alpha + \beta)} = \frac{\sin \alpha \cos \beta + \cos \alpha \sin \beta}{\cos \alpha \cos \beta - \sin \alpha \sin \beta}$$

$$= \frac{\dfrac{\sin \alpha \cos \beta}{\cos \alpha \cos \beta} + \dfrac{\cos \alpha \sin \beta}{\cos \alpha \cos \beta}}{\dfrac{\cos \alpha \cos \beta}{\cos \alpha \cos \beta} - \dfrac{\sin \alpha \sin \beta}{\cos \alpha \cos \beta}}$$

$$= \frac{\tan \alpha + \tan \beta}{1 - \tan \alpha \tan \beta}$$

29.3 Prove the subtraction formulas.

$$\sin(\alpha - \beta) = \sin[\alpha + (-\beta)] = \sin \alpha \cos(-\beta) + \cos \alpha \sin(-\beta)$$

$$= \sin \alpha (\cos \beta) + \cos \alpha (-\sin \beta) = \sin \alpha \cos \beta - \cos \alpha \sin \beta$$

$$\cos(\alpha - \beta) = \cos[\alpha + (-\beta)] = \cos \alpha \cos(-\beta) - \sin \alpha \sin(-\beta)$$

$$= \cos \alpha (\cos \beta) - \sin \alpha (-\sin \beta) = \cos \alpha \cos \beta + \sin \alpha \sin \beta$$

$$\tan(\alpha - \beta) = \tan[\alpha + (-\beta)] = \frac{\tan \alpha + \tan(-\beta)}{1 - \tan \alpha \tan(-\beta)} = \frac{\tan \alpha + (-\tan \beta)}{1 - \tan \alpha(-\tan \beta)}$$

$$= \frac{\tan \alpha - \tan \beta}{1 + \tan \alpha \tan \beta}$$

29.4 Find the values of sine, cosine, and tangent of $15°$, using (a) $15° = 45° - 30°$ and (b) $15° = 60° - 45°$.

(a) $$\sin 15° = \sin(45° - 30°) = \sin 45° \cos 30° - \cos 45° \sin 30°$$

$$= \frac{1}{\sqrt{2}} \cdot \frac{\sqrt{3}}{2} - \frac{1}{\sqrt{2}} \cdot \frac{1}{2} = \frac{\sqrt{3} - 1}{2\sqrt{2}} = \frac{\sqrt{2}}{4}(\sqrt{3} - 1)$$

$$\cos 15° = \cos(45° - 30°) = \cos 45° \cos 30° + \sin 45° \sin 30°$$

$$= \frac{1}{\sqrt{2}} \cdot \frac{\sqrt{3}}{2} + \frac{1}{\sqrt{2}} \cdot \frac{1}{2} = \frac{\sqrt{2}}{4}(\sqrt{3} + 1)$$

$$\tan 15° = \tan(45° - 30°) = \frac{\tan 45° - \tan 30°}{1 + \tan 45° \tan 30°} = \frac{1 - 1/\sqrt{3}}{1 + 1(1/\sqrt{3})} = \frac{\sqrt{3} - 1}{\sqrt{3} + 1} = 2 - \sqrt{3}$$

(b) $$\sin 15° = \sin(60° - 45°) = \sin 60° \cos 45° - \cos 60° \sin 45°$$

$$= \frac{\sqrt{3}}{2} \cdot \frac{1}{\sqrt{2}} - \frac{1}{2} \cdot \frac{1}{\sqrt{2}} = \frac{\sqrt{2}}{4}(\sqrt{3} - 1)$$

$$\cos 15° = \cos(60° - 45°) = \cos 60° \cos 45° + \sin 60° \sin 45°$$

$$= \frac{1}{2} \cdot \frac{1}{\sqrt{2}} + \frac{\sqrt{3}}{2} \cdot \frac{1}{\sqrt{2}} = \frac{\sqrt{2}}{4}(\sqrt{3} + 1)$$

$$\tan 15° = \tan(60° - 45°) = \frac{\tan 60° - \tan 45°}{1 + \tan 60° \tan 45°} = \frac{\sqrt{3} - 1}{\sqrt{3} + 1} = 2 - \sqrt{3}$$

29.5   Prove (a) $\sin(45° + \theta) - \sin(45° - \theta) = \sqrt{2}\sin\theta$,   (b) $\sin(30° + \theta) + \cos(60° + \theta) = \cos\theta$.

(a) $\sin(45° + \theta) - \sin(45° - \theta) = (\sin 45° \cos\theta + \cos 45° \sin\theta) - (\sin 45° \cos\theta - \cos 45° \sin\theta)$

$$= 2\cos 45° \sin\theta = 2\frac{1}{\sqrt{2}}\sin\theta = \sqrt{2}\sin\theta$$

(b) $\sin(30° + \theta) + \cos(60° + \theta) = (\sin 30° \cos\theta + \cos 30° \sin\theta) + (\cos 60° \cos\theta - \sin 60° \sin\theta)$

$$= \left(\frac{1}{2}\cos\theta + \frac{\sqrt{3}}{2}\sin\theta\right) + \left(\frac{1}{2}\cos\theta - \frac{\sqrt{3}}{2}\sin\theta\right) = \cos\theta$$

29.6   Simplify   (a) $\sin(\alpha + \beta) + \sin(\alpha - \beta)$,   (b) $\cos(\alpha + \beta) - \cos(\alpha - \beta)$,   (c) $\dfrac{\tan(\alpha + \beta) - \tan\alpha}{1 + \tan(\alpha + \beta)\tan\alpha}$,
(d) $(\sin\alpha\cos\beta - \cos\alpha\sin\beta)^2 + (\cos\alpha\cos\beta + \sin\alpha\sin\beta)^2$.

(a) $\sin(\alpha + \beta) + \sin(\alpha - \beta) = (\sin\alpha\cos\beta + \cos\alpha\sin\beta) + (\sin\alpha\cos\beta - \cos\alpha\sin\beta)$

$$= 2\sin\alpha\cos\beta$$

(b) $\cos(\alpha + \beta) - \cos(\alpha - \beta) = (\cos\alpha\cos\beta - \sin\alpha\sin\beta) - (\cos\alpha\cos\beta + \sin\alpha\sin\beta)$

$$= -2\sin\alpha\sin\beta$$

(c) $\dfrac{\tan(\alpha + \beta) - \tan\alpha}{1 + \tan(\alpha + \beta)\tan\alpha} = \tan[(\alpha + \beta) - \alpha] = \tan\beta$

(d) $(\sin\alpha\cos\beta - \cos\alpha\sin\beta)^2 + (\cos\alpha\cos\beta + \sin\alpha\sin\beta)^2 = \sin^2(\alpha - \beta) + \cos^2(\alpha - \beta) = 1$

29.7   Find $\sin(\alpha + \beta), \cos(\alpha + \beta), \sin(\alpha - \beta), \cos(\alpha - \beta)$ and determine the quadrants in which $(\alpha + \beta)$ and $(\alpha - \beta)$ terminate, given

(a) $\sin\alpha = \frac{4}{5}, \cos\beta = \frac{5}{13}; \alpha$ and β in quadrant I.

(b) $\sin\alpha = \frac{2}{3}, \cos\beta = \frac{3}{4}; \alpha$ in quadrant II, β in quadrant IV.

(a) $\cos\alpha = \frac{3}{5}$ and $\sin\beta = \frac{12}{13}$. See Figs. 29-2 and 29-3.

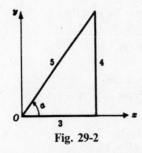

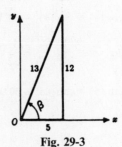

Fig. 29-2   Fig. 29-3

$$\sin(\alpha + \beta) = \sin\alpha\cos\beta + \cos\alpha\sin\beta = \tfrac{4}{5}\cdot\tfrac{5}{13} + \tfrac{3}{5}\cdot\tfrac{12}{13} = \tfrac{56}{65}$$

$$\cos(\alpha + \beta) = \cos\alpha\cos\beta - \sin\alpha\sin\beta = \tfrac{3}{5}\cdot\tfrac{5}{13} - \tfrac{4}{5}\cdot\tfrac{12}{13} = -\tfrac{33}{65}$$

$(\alpha + \beta)$ in quadrant II

$$\sin(\alpha - \beta) = \sin\alpha\cos\beta - \cos\alpha\sin\beta = \tfrac{4}{5}\cdot\tfrac{5}{13} - \tfrac{3}{5}\cdot\tfrac{12}{13} = -\tfrac{16}{65}$$

$$\cos(\alpha - \beta) = \cos\alpha\cos\beta + \sin\alpha\sin\beta = \tfrac{3}{5}\cdot\tfrac{5}{13} + \tfrac{4}{5}\cdot\tfrac{12}{13} = \tfrac{63}{65}$$

$(\alpha - \beta)$ in quadrant IV

(b) $\cos\alpha = -\sqrt{5}/3$ and $\sin\beta = -\sqrt{7}/4$. See Figs. 29-4 and 29-5.

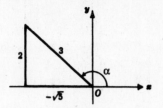

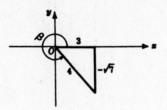

Fig. 29-4   **Fig. 29-5**

$$\sin(\alpha+\beta) = \sin\alpha\cos\beta + \cos\alpha\sin\beta = \frac{2}{3}\cdot\frac{3}{4} + \left(-\frac{\sqrt{5}}{3}\right)\left(-\frac{\sqrt{7}}{4}\right) = \frac{6+\sqrt{35}}{12}$$

$$\cos(\alpha+\beta) = \cos\alpha\cos\beta - \cos\alpha\sin\beta = \left(-\frac{\sqrt{5}}{3}\right)\frac{3}{4} - \frac{2}{3}\left(-\frac{\sqrt{7}}{4}\right) = \frac{-3\sqrt{5}+2\sqrt{7}}{12}$$

$(\alpha+\beta)$ in quadrant II

$$\sin(\alpha-\beta) = \sin\alpha\cos\beta - \cos\alpha\sin\beta = \frac{2}{3}\cdot\frac{3}{4} - \left(-\frac{\sqrt{5}}{3}\right)\left(-\frac{\sqrt{7}}{4}\right) = \frac{6-\sqrt{35}}{12}$$

$$\cos(\alpha-\beta) = \cos\alpha\cos\beta + \sin\alpha\sin\beta = \left(-\frac{\sqrt{5}}{3}\right)\frac{3}{4} + \frac{2}{3}\left(-\frac{\sqrt{7}}{4}\right) = \frac{-3\sqrt{5}-2\sqrt{7}}{12}$$

$(\alpha-\beta)$ in quadrant II

29.8 Prove (a) $\cot(\alpha+\beta) = \dfrac{\cot\alpha\cot\beta - 1}{\cot\beta + \cot\alpha}$, (b) $\cot(\alpha-\beta) = \dfrac{\cot\alpha\cot\beta + 1}{\cot\beta - \cot\alpha}$.

(a) $\cot(\alpha+\beta) = \dfrac{1}{\tan(\alpha+\beta)} = \dfrac{1-\tan\alpha\tan\beta}{\tan\alpha+\tan\beta} = \dfrac{1-\dfrac{1}{\cot\alpha\cot\beta}}{\dfrac{1}{\cot\alpha}+\dfrac{1}{\cot\beta}} = \dfrac{\cot\alpha\cot\beta-1}{\cot\beta+\cot\alpha}$

(b) $\cot(\alpha-\beta) = \cot[\alpha+(-\beta)] = \dfrac{\cot\alpha\cot(-\beta)-1}{\cot(-\beta)+\cot\alpha} = \dfrac{-\cot\alpha\cot\beta-1}{-\cot\beta+\cot\alpha} = \dfrac{\cot\alpha\cot\beta+1}{\cot\beta-\cot\alpha}$

29.9 Prove the double-angle formulas.

In $\sin(\alpha+\beta) = \sin\alpha\cos\beta + \cos\alpha\sin\beta$, $\cos(\alpha+\beta) = \cos\alpha\cos\beta - \sin\alpha\sin\beta$, and $\tan(\alpha+\beta) = \dfrac{\tan\alpha+\tan\beta}{1-\tan\alpha\tan\beta}$ put $\beta = \alpha$. Then

$$\sin 2\alpha = \sin\alpha\cos\alpha + \cos\alpha\sin\alpha = 2\sin\alpha\cos\alpha,$$

$$\cos 2\alpha = \cos\alpha\cos\alpha - \sin\alpha\sin\alpha$$

$$= \cos^2\alpha - \sin^2\alpha = (1-\sin^2\alpha) - \sin^2\alpha = 1-2\sin^2\alpha$$

$$= \cos^2\alpha - (1-\cos^2\alpha) = 2\cos^2\alpha - 1,$$

$$\tan 2\alpha = \frac{\tan\alpha+\tan\alpha}{1-\tan\alpha\tan\alpha} = \frac{2\tan\alpha}{1-\tan^2\alpha}.$$

29.10 Prove the half-angle formulas.

In $\cos 2\alpha = 1 - 2 \sin^2\alpha$, put $\alpha = \frac{1}{2}\theta$. Then

$$\cos\theta = 1 - 2\sin^2\tfrac{1}{2}\theta, \quad \sin^2\tfrac{1}{2}\theta = \frac{1-\cos\theta}{2}, \quad \text{and} \quad \sin\tfrac{1}{2}\theta = \pm\sqrt{\frac{1-\cos\theta}{2}}.$$

In $\cos 2\alpha = 2\cos^2\alpha - 1$, put $\alpha = \frac{1}{2}\theta$. Then

$$\cos\theta = 2\cos^2\tfrac{1}{2}\theta - 1, \quad \cos^2\tfrac{1}{2}\theta = \frac{1+\cos\theta}{2}, \quad \text{and} \quad \cos\tfrac{1}{2}\theta = \pm\sqrt{\frac{1+\cos\theta}{2}}.$$

Finally,

$$\tan\tfrac{1}{2}\theta = \frac{\sin\frac{1}{2}\theta}{\cos\frac{1}{2}\theta} = \pm\sqrt{\frac{1-\cos\theta}{1+\cos\theta}}$$

$$= \pm\sqrt{\frac{(1-\cos\theta)(1+\cos\theta)}{(1+\cos\theta)(1+\cos\theta)}} = \pm\sqrt{\frac{1-\cos^2\theta}{(1+\cos\theta)^2}} = \frac{\sin\theta}{1+\cos\theta}$$

$$= \pm\sqrt{\frac{(1-\cos\theta)(1-\cos\theta)}{(1+\cos\theta)(1-\cos\theta)}} = \pm\sqrt{\frac{(1-\cos\theta)^2}{1-\cos^2\theta}} = \frac{1-\cos\theta}{\sin\theta}.$$

The signs $\pm$ are not needed here since $\tan\frac{1}{2}\theta$ and $\sin\theta$ always have the same sign (Problem 29.7) and $1 - \cos\theta$ is always positive.

29.11 Using the half-angle formulas, find the exact values of (a) $\sin 15°$, (b) $\sin 292\frac{1}{2}°$.

(a)
$$\sin 15° = \sqrt{\frac{1-\cos 30°}{2}} = \sqrt{\frac{1-\sqrt{3}/2}{2}} = \frac{1}{2}\sqrt{2-\sqrt{3}}$$

(b)
$$\sin 292\tfrac{1}{2}° = -\sqrt{\frac{1-\cos 585°}{2}} = -\sqrt{\frac{1-\cos 225°}{2}}$$

$$= -\sqrt{\frac{1+1/\sqrt{2}}{2}} = -\frac{1}{2}\sqrt{2+\sqrt{2}}$$

29.12 Find the values of sine, cosine, and tangent of $\frac{1}{2}\theta$, given (a) $\sin\theta = \frac{5}{13}, \theta$ in quadrant II and (b) $\cos\theta = \frac{3}{7}$, θ in quadrant IV.

(a) $\sin\theta = \frac{5}{13}, \cos\theta = -\frac{12}{13}$, and $\frac{1}{2}\theta$ in quadrant I. See Fig. 29-6.

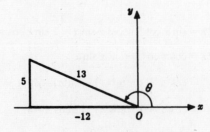

Fig. 29-6

$$\sin\tfrac{1}{2}\theta = \sqrt{\frac{1-\cos\theta}{2}} = \sqrt{\frac{1+\frac{12}{13}}{2}} = \sqrt{\frac{25}{26}} = \frac{5\sqrt{26}}{26}$$

$$\cos\tfrac{1}{2}\theta = \sqrt{\frac{1+\cos\theta}{2}} = \sqrt{\frac{1-\frac{12}{13}}{2}} = \sqrt{\frac{1}{26}} = \frac{\sqrt{26}}{26}$$

$$\tan\tfrac{1}{2}\theta = \frac{1-\cos\theta}{\sin\theta} = \frac{1+\frac{12}{13}}{\frac{5}{13}} = 5$$

(b) $\sin\theta = -2\sqrt{10}/7, \cos\theta = \frac{3}{7}$, and $\frac{1}{2}\theta$ in quadrant II. See Fig. 29-7

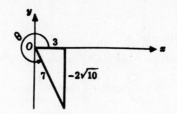

Fig. 29-7

$$\sin\tfrac{1}{2}\theta = \sqrt{\frac{1-\cos\theta}{2}} = \sqrt{\frac{1-\frac{3}{7}}{2}} = \sqrt{\frac{14}{7}}$$

$$\cos\tfrac{1}{2}\theta = -\sqrt{\frac{1+\cos\theta}{2}} = -\sqrt{\frac{1+\frac{3}{7}}{2}} = \sqrt{\frac{35}{7}}$$

$$\tan\tfrac{1}{2}\theta = \frac{1-\cos\theta}{\sin\theta} = \frac{1-\frac{3}{7}}{-2\sqrt{10}/7} = -\frac{\sqrt{10}}{5}$$

29.13 Show that (a) $\sin\theta = 2\sin\tfrac{1}{2}\theta\cos\tfrac{1}{2}\theta$, (b) $\sin A = \pm\sqrt{\dfrac{1-\cos 2A}{2}}$, (c) $\tan 4x = \dfrac{\sin 8x}{1+\cos 8x}$,

(d) $\cos 6\theta = 1 - 2\sin^2 3\theta$, (e) $\sin^2\tfrac{1}{2}\theta = \tfrac{1}{2}(1-\cos\theta), \cos^2\tfrac{1}{2}\theta = \tfrac{1}{2}(1+\cos\theta)$.

(a) This is obtained from $\sin 2\alpha = 2\sin\alpha\cos\alpha$ by putting $\alpha = \tfrac{1}{2}\theta$.

(b) This is obtained from $\sin\tfrac{1}{2}\theta = \pm\sqrt{\dfrac{(1-\cos\theta)}{2}}$ by putting $\theta = 2A$.

(c) This is obtained from $\tan\tfrac{1}{2}\theta = \dfrac{\sin\theta}{1+\cos\theta}$ by putting $\theta = 8x$.

(d) This is obtained from $\cos 2\alpha = 1 - 2\sin^2\alpha$ by putting $\alpha = 3\theta$.

(e) These formulas are obtained by squaring $\sin\tfrac{1}{2}\theta = \pm\sqrt{\dfrac{1-\cos\theta}{2}}$ and $\cos\tfrac{1}{2}\theta = \pm\sqrt{\dfrac{1+\cos\theta}{2}}$.

29.14 Express (a) $\sin 3\alpha$ in terms of $\sin\alpha$, (b) $\cos 4\alpha$ in terms of $\cos\alpha$.

(a) $\sin 3\alpha = \sin(2\alpha + \alpha) = \sin 2\alpha\cos\alpha + \cos 2\alpha\sin\alpha$

$$= (2\sin\alpha\cos\alpha)\cos\alpha + (1 - 2\sin^2\alpha)\sin\alpha = 2\sin\alpha\cos^2\alpha + (1 - 2\sin^2\alpha)\sin\alpha$$

$$= 2\sin\alpha(1 - \sin^2\alpha) + (1 - 2\sin^2\alpha)\sin\alpha = 3\sin\alpha - 4\sin^2\alpha$$

(b) $\cos 4\alpha = \cos 2(2\alpha) = 2\cos^2 2\alpha - 1 = 2(2\cos^2\alpha - 1)^2 - 1 = 8\cos^4\alpha - 8\cos^2\alpha + 1$

29.15   Prove $\cos 2x = \cos^4 x - \sin^4 x$.

$$\cos^4 x - \sin^4 x = (\cos^2 x + \sin^2 x)(\cos^2 x - \sin^2 x) = \cos^2 x - \sin^2 x = \cos 2x$$

29.16   Prove $1 - \frac{1}{2}\sin 2x = \dfrac{\sin^3 x + \cos^2 x}{\sin x + \cos x}$.

$$\frac{\sin^3 x + \cos^3 x}{\sin x + \cos x} = \frac{(\sin x + \cos x)(\sin^2 x - \sin x \cos x + \cos^2 x)}{\sin x + \cos x}$$

$$= 1 - \sin x \cos x = 1 - \tfrac{1}{2}(2 \sin x \cos x) = 1 - \tfrac{1}{2}\sin 2x$$

29.17   Prove $\cos \theta = \sin (\theta + 30°) + \cos (\theta + 60°)$.

$$\sin (\theta + 30°) + \cos (\theta + 60°) = (\sin \theta \cos 30° + \cos \theta \sin 30°) + (\cos \theta \cos 60° - \sin \theta \sin 60°)$$

$$= \frac{\sqrt{3}}{2}\sin \theta + \frac{1}{2}\cos \theta + \frac{1}{2}\cos \theta - \frac{\sqrt{3}}{2}\sin \theta = \cos \theta$$

29.18   Prove $\cos x = \dfrac{1 - \tan^2 \frac{1}{2}x}{1 + \tan^2 \frac{1}{2}x}$.

$$\frac{1 - \tan^2 \frac{1}{2}x}{1 + \tan^2 \frac{1}{2}x} = \frac{1 - \dfrac{\sin^2 \frac{1}{2}x}{\cos^2 \frac{1}{2}x}}{\sec^2 \frac{1}{2}x} = \frac{\left(1 - \dfrac{\sin^2 \frac{1}{2}x}{\cos^2 \frac{1}{2}x}\right)\cos^2 \frac{1}{2}x}{\sec^2 \frac{1}{2}x \, \cos^2 \frac{1}{2}x} = \cos^2 \tfrac{1}{2}x - \sin^2 \tfrac{1}{2}x = \cos x$$

29.19   Prove $2 \tan 2x = \dfrac{\cos x + \sin x}{\cos x - \sin x} - \dfrac{\cos x - \sin x}{\cos x + \sin x}$.

$$\frac{\cos x + \sin x}{\cos x - \sin x} - \frac{\cos x - \sin x}{\cos x + \sin x}$$

$$= \frac{(\cos x + \sin x)^2 - (\cos x - \sin x)^2}{(\cos x - \sin x)(\cos x + \sin x)}$$

$$= \frac{(\cos^2 x + 2 \sin x \cos x + \sin^2 x) - (\cos^2 x - 2 \sin x \cos x + \sin^2 x)}{\cos^2 x - \sin^2 x}$$

$$= \frac{4 \sin x \cos x}{\cos^2 x - \sin^2 x} = \frac{2 \sin 2x}{\cos 2x} = 2 \tan 2x$$

29.20   Prove $\sin^4 A = \frac{3}{8} - \frac{1}{2}\cos 2A + \frac{1}{8}\cos 4A$.

$$\sin^4 A = (\sin^2 A)^2 = \left(\frac{1 - \cos 2A}{2}\right)^2 = \frac{1 - 2\cos 2A + \cos^2 2A}{4}$$

$$= \frac{1}{4}\left(1 - 2\cos 2A + \frac{1 + \cos 4A}{2}\right) = \frac{3}{8} - \frac{1}{2}\cos 2A + \frac{1}{8}\cos 4A$$

29.21   Prove $\tan^6 x = \tan^4 x \sec^2 x - \tan^2 x \sec^2 x + \sec^2 x - 1$.

$$\tan^6 x = \tan^4 x \tan^2 x = \tan^4 x (\sec^2 x - 1) = \tan^4 x \sec^2 x - \tan^2 x \tan^2 x$$

$$= \tan^4 x \sec^2 x - \tan^2 x (\sec^2 x - 1) = \tan^4 x \sec^2 x - \tan^2 x \sec^2 x + \tan^2 x$$

$$= \tan^4 x \sec^2 x - \tan^2 x \sec^2 x + \sec^2 x - 1$$

29.22   When $A + B + C = 180°$, show that $\sin 2A + \sin 2B + \sin 2C = 4 \sin A \sin B \sin C$.

Since $C = 180° - (A + B)$,

$$\sin 2A + \sin 2B + \sin 2C = \sin 2A + \sin 2B + \sin [360° - 2(A + B)]$$

$$= \sin 2A + \sin 2B - \sin 2(A + B)$$

$$= \sin 2A + \sin 2B - \sin 2A \cos 2B - \cos 2A \sin 2B$$

$$= (\sin 2A)(1 - \cos 2B) + (\sin 2B)(1 - \cos 2A)$$

$$= 2 \sin 2A \sin^2 B + 2 \sin 2B \sin^2 A$$

$$= 4 \sin A \cos A \sin^2 B + 4 \sin B \cos B \sin^2 A$$

$$= 4 \sin A \sin B (\sin A \cos B + \cos A \sin B)$$

$$= 4 \sin A \sin B \sin (A + B)$$

$$= 4 \sin A \sin B \sin [180° - (A + B)]$$

$$= 4 \sin A \sin B \sin C$$

29.23   When $A + B + C = 180°$, show that $\tan A + \tan B + \tan C = \tan A \tan B \tan C$.

Since $C = 180° - (A + B)$,

$$\tan A + \tan B + \tan C = \tan A + \tan B + \tan [180° - (A + B)] = \tan A + \tan B - \tan (A + B)$$

$$= \tan A + \tan B - \frac{\tan A + \tan B}{1 - \tan A \tan B} = (\tan A + \tan B)\left(1 - \frac{1}{1 - \tan A \tan B}\right)$$

$$= (\tan A + \tan B)\left(-\frac{\tan A - \tan B}{1 - \tan A \tan B}\right) = -\tan A \tan B \frac{\tan A + \tan B}{1 - \tan A \tan B}$$

$$= -\tan A \tan B \tan (A + B) = \tan A \tan B \tan [180° - (A + B)]$$

$$= \tan A \tan B \tan C$$

Supplementary Problems

29.24   Find the values of sine, cosine, and tangent of (a) 75°   (b) 255°.

Ans. (a) $\dfrac{\sqrt{2}}{4}(\sqrt{3} + 1), \dfrac{\sqrt{2}}{4}(\sqrt{3} - 1), 2 + \sqrt{3}$ (b) $-\dfrac{\sqrt{2}}{4}(\sqrt{3} + 1), -\dfrac{\sqrt{2}}{4}(\sqrt{3} - 1), 2 + \sqrt{3}$

29.25   Find the values of $\sin (\alpha + \beta), \cos (\alpha + \beta)$, and $\tan (\alpha + \beta)$, given

(a) $\sin \alpha = \frac{3}{5}, \cos \beta = \frac{5}{13}, \alpha$ and β in quadrant I.

   Ans. $\frac{63}{65}, -\frac{16}{65}, -\frac{63}{16}$

(b) $\sin \alpha = \frac{8}{17}, \tan \beta = \frac{5}{12}, \alpha$ and β in quadrant I.

   Ans. $\frac{171}{221}, \frac{140}{221}, \frac{171}{140}$

(c) $\cos \alpha = -\frac{12}{13}, \cot \beta = \frac{24}{7}, \alpha$ in quadrant II, β in quadrant III.

   Ans. $-\frac{36}{325}, \frac{323}{325}, -\frac{36}{323}$

(d) $\sin \alpha = \frac{1}{3}, \sin \beta = \frac{2}{5}, \alpha$ in quadrant I, β in quadrant II.

   Ans. $\dfrac{4\sqrt{2} - \sqrt{21}}{15}, -\dfrac{2 + 2\sqrt{42}}{15}, -\dfrac{4\sqrt{2} - \sqrt{21}}{2 + 2\sqrt{42}}$

29.26 Find the values of $\sin(\alpha - \beta)$, $\cos(\alpha - \beta)$, and $\tan(\alpha - \beta)$, given

(a) $\sin\alpha = \frac{3}{5}$, $\sin\beta = \frac{5}{13}$, α and β in quadrant I.

   Ans. $\frac{16}{65}, \frac{63}{65}, \frac{16}{63}$

(b) $\sin\alpha = \frac{8}{17}$, $\tan\beta = \frac{5}{12}$, α and β in quadrant I.

   Ans. $\frac{21}{221}, \frac{220}{221}, \frac{21}{220}$

(c) $\cos\alpha = -\frac{12}{13}$, $\cot\beta = \frac{24}{7}$, α in quadrant II, β in quadrant I.

   Ans. $\frac{204}{325}, -\frac{253}{325}, -\frac{204}{253}$

(d) $\sin\alpha = \frac{1}{3}$, $\sin\beta = \frac{2}{3}$, α in quadrant II, β in quadrant I.

   Ans. $\dfrac{4\sqrt{2} + \sqrt{21}}{15}, \ -\dfrac{2\sqrt{42} - 2}{15}, \ -\dfrac{4\sqrt{2} + \sqrt{21}}{2\sqrt{42} - 2}$

29.27 Prove

(a) $\sin(\alpha + \beta) - \sin(\alpha - \beta) = 2\cos\alpha\sin\beta$

(b) $\cos(\alpha + \beta) + \cos(\alpha - \beta) = 2\cos\alpha\cos\beta$

(c) $\tan(45° - \theta) = \dfrac{1 - \tan\theta}{1 + \tan\theta}$

(d) $\dfrac{\tan(\alpha + \beta)}{\cot(\alpha - \beta)} = \dfrac{\tan^2\alpha - \tan^2\beta}{1 - \tan^2\alpha\tan^2\beta}$

(e) $\tan(\alpha + \beta + \gamma) = \tan[(\alpha + \beta) + \gamma] = \dfrac{\tan\alpha + \tan\beta + \tan\gamma - \tan\alpha\tan\beta\tan\gamma}{1 - \tan\alpha\tan\beta - \tan\beta\tan\gamma - \tan\gamma\tan\alpha}$

(f) $\dfrac{\sin(x + y)}{\cos(x - y)} = \dfrac{\tan x + \tan y}{1 + \tan x\tan y}$

(g) $\tan(45° + \theta) = \dfrac{\cos\theta + \sin\theta}{\cos\theta - \sin\theta}$

(h) $\sin(\alpha + \beta)\sin(\alpha - \beta) = \sin^2\alpha - \sin^2\beta$

29.28 If A and B are the measures of acute angles, find $A + B$ given

(a) $\tan A = \frac{1}{4}$, $\tan B = \frac{3}{5}$. **Hint:** $\tan(A + B) = 1$.

   Ans. 45°

(b) $\tan A = \frac{5}{3}$, $\tan B = 4$.

   Ans. 135°

29.29 If $\tan(x + y) = 33$ and $\tan x = 3$, show that $\tan y = 0.3$.

29.30 Find the values of $\sin 2\theta$, $\cos 2\theta$, and $\tan 2\theta$, given

(a) $\sin\theta = \frac{3}{5}$, θ in quadrant I.

   Ans. $\frac{24}{25}, \frac{7}{25}, \frac{24}{7}$

(b) $\sin\theta = \frac{3}{5}$, θ in quadrant II.

   Ans. $-\frac{24}{25}, \frac{7}{25}, -\frac{24}{7}$

(c) $\sin\theta = -\frac{1}{2}$, θ in quadrant IV.

   Ans. $-\sqrt{3}/2, \frac{1}{2}, -\sqrt{3}$

    (d) $\tan \theta = -\frac{1}{5}$, θ in quadrant II.

          Ans. $-\frac{5}{13}, \frac{12}{13}, -\frac{5}{12}$

    (e) $\tan \theta = u$, θ in quadrant I.

          Ans. $\dfrac{2u}{1+u^2}, \dfrac{1-u^2}{1+u^2}, \dfrac{2u}{1-u^2}$

29.31   Prove

    (a) $\tan \theta \sin 2\theta = 2 \sin^2 \theta$

    (b) $\cot \theta \sin 2\theta = 1 + \cos 2\theta$

    (c) $\dfrac{\sin^3 x - \cos^3 x}{\sin x - \cos x} = 1 + \dfrac{1}{2} \sin 2x$

    (d) $\dfrac{1 - \sin 2A}{\cos 2A} = \dfrac{1 - \tan A}{1 + \tan A}$

    (e) $\cos 2\theta = \dfrac{1 - \tan^2 \theta}{1 + \tan^2 \theta}$

    (f) $\dfrac{1 + \cos 2\theta}{\sin 2\theta} = \cot \theta$

    (g) $\cos 3\theta = 4\cos^3 \theta - 3 \cos \theta$

    (h) $\cos^4 x = \frac{3}{8} + \frac{1}{2} \cos 2x + \frac{1}{8} \cos 4x$

29.32   Find the values of sine, cosine, and tangent of

    (a) $30°$, given $\cos 60° = \frac{1}{2}$.

          Ans. $\frac{1}{2}, \sqrt{3}/2, 1/\sqrt{3}$

    (b) $105°$, given $\cos 210° = -\sqrt{3}/2$.

          Ans. $\frac{1}{2}\sqrt{2 + \sqrt{3}}, \ -\frac{1}{2}\sqrt{2 - \sqrt{3}}, \ -(2 + \sqrt{3})$

    (c) $\frac{1}{2}\theta$, given $\sin \theta = \frac{3}{5}$, θ in quadrant I.

          Ans. $1/\sqrt{10}, \ 3/\sqrt{10}, \ \frac{1}{3}$

    (d) θ, given $\cot 2\theta = \frac{7}{24}$, 2θ in quadrant I.

          Ans. $\frac{3}{5}, \frac{4}{5}, \frac{3}{4}$

    (e) θ, given $\cot 2\theta = -\frac{5}{12}$, 2θ in quadrant II.

          Ans. $3/\sqrt{13}, \ 2/\sqrt{13}, \ \frac{3}{2}$

29.33   Prove

    (a) $\cos x = 2 \cos^2 \frac{1}{2} x - 1 = 1 - 2 \sin^2 \frac{1}{2} x$

    (b) $\sin x = 2 \sin \frac{1}{2} x \cos \frac{1}{2} x$

    (c) $(\sin \frac{1}{2}\theta - \cos \frac{1}{2}\theta)^2 = 1 - \sin \theta$

    (d) $\tan \frac{1}{2}\theta = \csc \theta - \cot \theta$

    (e) $\dfrac{1 - \tan \frac{1}{2}\theta}{1 + \tan \frac{1}{2}\theta} = \dfrac{1 - \sin \theta}{\cos \theta} = \dfrac{\cos \theta}{1 + \sin \theta}$

    (f) $\dfrac{2 \tan \frac{1}{2} x}{1 + \tan^2 \frac{1}{2} x} = \sin x$

29.34 In the right triangle ABC, in which C is the right angle, prove

$$\sin 2A = \frac{2ab}{c^2}, \qquad \cos 2A = \frac{b^2 - a^2}{c^2}, \qquad \sin \tfrac{1}{2} A = \sqrt{\frac{c-b}{2c}}, \qquad \cos \tfrac{1}{2} A = \sqrt{\frac{c+b}{2c}}.$$

29.35 Prove (a) $\dfrac{\sin 3x}{\sin x} - \dfrac{\cos 3x}{\cos x} = 2,$ (b) $\tan 50° - \tan 40° = 2 \tan 10°.$

29.36 If $A + B + C = 180°$, prove

(a) $\sin A + \sin B + \sin C = 4 \cos \tfrac{1}{2} A \cos \tfrac{1}{2} A \cos \tfrac{1}{2} B \cos \tfrac{1}{2} C$

(b) $\cos A + \cos B + \cos C = 1 + 4 \sin \tfrac{1}{2} A \sin \tfrac{1}{2} B \sin \tfrac{1}{2} C$

(c) $\sin^2 A + \sin^2 B - \sin^2 C = 2 \sin A \sin B \cos C$

(d) $\tan \tfrac{1}{2} A \tan \tfrac{1}{2} B + \tan \tfrac{1}{2} B \tan \tfrac{1}{2} C + \tan \tfrac{1}{2} C \tan \tfrac{1}{2} A = 1$

Chapter 30

Sum, Difference, and Product Formulas

PRODUCTS OF SINES AND COSINES

$$\sin \alpha \cos \beta = \tfrac{1}{2}[\sin(\alpha + \beta) + \sin(\alpha - \beta)]$$

$$\cos \alpha \sin \beta = \tfrac{1}{2}[\sin(\alpha + \beta) - \sin(\alpha - \beta)]$$

$$\cos \alpha \cos \beta = \tfrac{1}{2}[\cos(\alpha + \beta) + \cos(\alpha - \beta)]$$

$$\sin \alpha \sin \beta = -\tfrac{1}{2}[\cos(\alpha + \beta) + \cos(\alpha - \beta)]$$

(For proofs of these formulas, see Problem 30.1.)

SUM AND DIFFERENCE OF SINES AND COSINES

$$\sin A + \sin B = 2 \sin \tfrac{1}{2}(A + B) \cos \tfrac{1}{2}(A - B)$$

$$\sin A - \sin B = 2 \cos \tfrac{1}{2}(A + B) \sin \tfrac{1}{2}(A - B)$$

$$\cos A + \cos B = 2 \cos \tfrac{1}{2}(A + B) \cos \tfrac{1}{2}(A - B)$$

$$\cos A - \cos B = -2 \sin \tfrac{1}{2}(A + B) \sin \tfrac{1}{2}(A - B)$$

(For proofs of these formulas, see Problem 30.2)

Solved Problems

30.1 Derive the product formulas.

Since $\sin(\alpha + \beta) + \sin(\alpha - \beta) = (\sin \alpha \cos \beta + \cos \alpha \cos \beta) + (\sin \alpha \cos \beta - \cos \alpha \sin \beta)$

$$= 2 \sin \alpha \cos \beta,$$

$$\sin \alpha \cos \beta = \tfrac{1}{2}[\sin(\alpha + \beta) + \sin(\alpha - \beta)]$$

Since $\sin(\alpha + \beta) - \sin(\alpha - \beta) = 2 \cos \alpha \sin \beta$,

$$\cos \alpha \sin \beta = \tfrac{1}{2}[\sin(\alpha + \beta) - \sin(\alpha - \beta)]$$

Since $\cos(\alpha+\beta)+\cos(\alpha-\beta) = (\cos\alpha\cos\beta - \sin\alpha\sin\beta) + (\cos\alpha\cos\beta + \sin\alpha\sin\beta)$

$$= 2\cos\alpha\cos\beta,$$

$$\cos\alpha\cos\beta = \tfrac{1}{2}[\cos(\alpha+\beta)+\cos(\alpha-\beta)]$$

Since $\cos(\alpha+\beta)-\cos(\alpha-\beta) = -2\sin\alpha\sin\beta,$

$$\sin\alpha\sin\beta = -\tfrac{1}{2}[\cos(\alpha+\beta)-\cos(\alpha-\beta)]$$

30.2 Derive the sum and difference formulas.

Let $\alpha+\beta = A$ and $\alpha-\beta = B$ so that $\alpha = \tfrac{1}{2}(A+B)$ and $\beta = \tfrac{1}{2}(A-B)$. Then (see Problem 30.1)

$\sin(\alpha+\beta)+\sin(\alpha-\beta) = 2\sin\alpha\cos\beta$ becomes $\sin A + \sin B = 2\sin\tfrac{1}{2}(A+B)\cos\tfrac{1}{2}(A-B)$

$\sin(\alpha+\beta)-\sin(\alpha-\beta) = 2\cos\alpha\sin\beta$ becomes $\sin A - \sin B = 2\cos\tfrac{1}{2}(A+B)\sin\tfrac{1}{2}(A-B)$

$\cos(\alpha+\beta)+\cos(\alpha-\beta) = 2\cos\alpha\cos\beta$ becomes $\cos A + \cos B = 2\cos\tfrac{1}{2}(A+B)\cos\tfrac{1}{2}(A-B)$

$\cos(\alpha+\beta)-\cos(\alpha-\beta) = -2\sin\alpha\cos\beta$ becomes $\cos A - \cos B = -2\sin\tfrac{1}{2}(A+B)\sin\tfrac{1}{2}(A-B)$

30.3 Express each of the following as a sum or difference:

(a) $\sin 40°\cos 30°$,   (b) $\cos 110°\sin 55°$,   (c) $\cos 50°\cos 35°$,   (d) $\sin 55°\sin 40°$.

(a) $\sin 40°\cos 30° = \tfrac{1}{2}[\sin(40°+30°)+\sin(40°-30°)] = \tfrac{1}{2}(\sin 70° + \sin 10°)$

(b) $\cos 110°\sin 55° = \tfrac{1}{2}[\sin(110°+55°)-\sin(110°-55°)] = \tfrac{1}{2}(\sin 165° - \sin 55°)$

(c) $\cos 50°\cos 35° = \tfrac{1}{2}[\cos(50°+35°)+\cos(50°-35°)] = \tfrac{1}{2}(\cos 85° + \cos 15°)$

(d) $\sin 55°\sin 40° = -\tfrac{1}{2}[\cos(55°+40°)-\cos(55°-40°)] = -\tfrac{1}{2}(\cos 95° - \cos 15°)$

30.4 Express each of the following as a product:

(a) $\sin 50° + \sin 40°$,   (b) $\sin 70° - \sin 20°$,   (c) $\cos 55° + \cos 25°$,   (d) $\cos 35° - \cos 75°$.

(a) $\sin 50° + \sin 40° = 2\sin\tfrac{1}{2}(50°+40°)\cos\tfrac{1}{2}(50°-40°) = 2\sin 45°\cos 5°$

(b) $\sin 70° - \sin 20° = 2\cos\tfrac{1}{2}(70°+20°)\sin\tfrac{1}{2}(70°-20°) = 2\cos 45°\sin 25°$

(c) $\cos 55° + \cos 25° = 2\cos\tfrac{1}{2}(55°+25°)\cos\tfrac{1}{2}(55°-25°) = 2\cos 40°\cos 15°$

(d) $\cos 35° - \cos 75° = -2\sin\tfrac{1}{2}(35°+75°)\sin\tfrac{1}{2}(35°-75°) = -2\sin 55°\sin(-20°)$

$$= 2\sin 55°\sin 20°$$

30.5 Prove $\dfrac{\sin 4A + \sin 2A}{\cos 4A + \cos 2A} = \tan 3A.$

$$\frac{\sin 4A + \sin 2A}{\cos 4A + \cos 2A} = \frac{2\sin\tfrac{1}{2}(4A+2A)\cos\tfrac{1}{2}(4A-2A)}{2\cos\tfrac{1}{2}(4A+2A)\cos\tfrac{1}{2}(4A-2A)} = \frac{\sin 3A}{\cos 3A} = \tan 3A$$

30.6 Prove $\dfrac{\sin A - \sin B}{\sin A + \sin B} = \dfrac{\tan\tfrac{1}{2}(A-B)}{\tan\tfrac{1}{2}(A+B)}.$

$$\frac{\sin A - \sin B}{\sin A + \sin B} = \frac{2\cos\tfrac{1}{2}(A+B)\sin\tfrac{1}{2}(A-B)}{2\sin\tfrac{1}{2}(A+B)\cos\tfrac{1}{2}(A-B)} = \cot\tfrac{1}{2}(A+B)\tan\tfrac{1}{2}(A-B); \frac{\tan\tfrac{1}{2}(A-B)}{\tan\tfrac{1}{2}(A+B)}$$

30.7 Prove $\cos^3 x \sin^2 x = \frac{1}{16}(2\cos x - \cos 3x - \cos 5x)$.

$$\cos^3 x \sin^2 x = (\sin x \cos x)^2 \cos x = \frac{1}{4}\sin^2 2x \cos x = \frac{1}{4}(\sin 2x)(\sin 2x \cos x)$$
$$= \frac{1}{4}(\sin 2x)[\frac{1}{2}(\sin 3x + \sin x)] = \frac{1}{8}(\sin 3x \sin 2x + \sin 2x \sin x)$$
$$= \frac{1}{8}\{-\frac{1}{2}(\cos 5x - \cos x) + [-\frac{1}{2}(\cos 3x - \cos x)]\}$$
$$= \frac{1}{16}(2\cos x - \cos 3x - \cos x)$$

30.8 Prove $1 + \cos 2x + \cos 4x + \cos 6x = 4\cos x \cos 2x \cos 3x$.

$$1 + (\cos 2x + \cos 4x) + \cos 6x = 1 + 2\cos 3x \cos x + \cos 6x = (1 + \cos 6x) + 2\cos 3x \cos x$$
$$= 2\cos^2 3x + 2\cos 3x \cos x = 2\cos 3x(\cos 3x + \cos x)$$
$$= 2\cos 3x(2\cos 2x \cos x) = 4\cos x \cos 2x \cos 3x$$

30.9 Transform $4\cos x + 3\sin x$ into the form $c\cos(x - \alpha)$.

Since $c\cos(x - \alpha) = c(\cos x \cos \alpha + \sin x \sin \alpha)$, set $c\cos \alpha = 4$ and $c\sin \alpha = 3$. Then $\cos \alpha = 4/c$ and $\sin \alpha = 3/c$. Since $\sin^2 \alpha + \cos^2 \alpha = 1$, $c = 5$ and -5.
Using $c = 5$, $\cos \alpha = \frac{4}{5}$, $\sin \alpha = \frac{3}{5}$, and $\alpha = 36°52'$. Thus, $4\cos x + 3\sin x = 5\cos(x - 36°52')$.
Using $c = -5$, $\alpha = 216°52'$ and $4\cos x + 3\sin x = -5\cos(x - 216°52')$.

30.10 Find the maximum and minimum values of $4\cos x + 3\sin x$ on the interval $0 \leqslant x \leqslant 2\pi$.

From Problem 30.9, $4\cos x + 3\sin x = 5\cos(x - 36°52')$.
Now on the prescribed interval, $\cos \theta$ attains its maximum value 1 when $\theta = 0$ and its minimum value -1 when $\theta = \pi$. Thus, the maximum value of $4\cos x + 3\sin x$ is 5 which occurs when $x - 36°52' = 0$ or when $x = 36°52'$, while the minimum value is -5 which occurs when $x - 36°52' = \pi$ or when $x = 216°52'$.

Supplementary Problems

30.11 Express each of the following products as a sum or difference of sines or of cosines.

(a) $\sin 35° \cos 25° = \frac{1}{2}(\sin 60° + \sin 10°)$

(b) $\sin 25° \cos 75° = \frac{1}{2}(\sin 100° - \sin 50°)$

(c) $\cos 50° \cos 70° = \frac{1}{2}(\cos 120° + \cos 20°)$

(d) $\sin 130° \sin 55° = -\frac{1}{2}(\cos 180° - \cos 75°)$

(e) $\sin 4x \cos 2x = \frac{1}{2}(\sin 6x + \sin 2x)$

(f) $\sin(x/2)\cos(3x/2) = \frac{1}{2}(\sin 2x - \sin x)$

(g) $\cos 7x \cos 4x = \frac{1}{2}(\cos 11x + \cos 3x)$

(h) $\sin 5x \sin 4x = -\frac{1}{2}(\cos 9x - \cos x)$

30.12 Show that

(a) $2\sin 45° \cos 15° = \frac{1}{2}(\sqrt{3} + 1)$ and $\cos 15° = \frac{\sqrt{2}}{4}(\sqrt{3} + 1)$

(b) $2\sin 82\frac{1}{2}° \cos 37\frac{1}{2}° = \frac{1}{2}(\sqrt{3} + \sqrt{2})$ (c) $2\sin 127\frac{1}{2}° \sin 97\frac{1}{2}° = \frac{1}{2}(\sqrt{3} + \sqrt{2})$

30.13 Express each of the following as a product:

(a) $\sin 50° + \sin 20° = 2\sin 35° \cos 15°$ (e) $\sin 4x + \sin 2x = 2\sin 3x \cos x$

(b) $\sin 75° - \sin 35° = 2\cos 55° \sin 20°$ (f) $\sin 7\theta - \sin 3\theta = 2\cos 5\theta \sin 2\theta$

(c) $\cos 65° + \cos 15° = 2\cos 40° \cos 25°$ (g) $\cos 6\theta + \cos 2\theta = 2\cos 4\theta \cos 2\theta$

(d) $\cos 80° - \cos 70° = -2\sin 75° \sin 5°$ (h) $\cos(3x/2) - \cos(9x/2) = 2\sin 3x \sin(3x/2)$

30.14 Show that

(a) $\sin 40° + \sin 20° = \cos 10°$ (c) $\cos 465° + \cos 165° = -\sqrt{6}/2$

(b) $\sin 105° + \sin 15° = \sqrt{6}/2$ (d) $\dfrac{\sin 75° - \sin 15°}{\cos 75° + \cos 15°} = \dfrac{1}{\sqrt{3}}$

30.15 Prove

(a) $\dfrac{\sin A + \sin 3A}{\cos A + \cos 3A} = \tan 2A$ (c) $\dfrac{\sin A + \sin B}{\sin A - \sin B} = \dfrac{\tan \frac{1}{2}(A+B)}{\tan \frac{1}{2}(A-B)}$

(b) $\dfrac{\sin 2A + \sin 4A}{\cos 2A + \cos 4A} = \tan 3A$ (d) $\dfrac{\cos A + \cos B}{\cos A - \cos B} = -\cot \frac{1}{2}(A-B)\cot \frac{1}{2}(A+B)$

(e) $\sin\theta + \sin 2\theta + \sin 3\theta = \sin 2\theta + (\sin\theta + \sin 3\theta) = \sin 2\theta(1 + 2\cos\theta)$

(f) $\cos\theta + \cos 2\theta + \cos 3\theta = \cos 2\theta(1 + 2\cos\theta)$

(g) $\sin 2\theta + \sin 4\theta + \sin 6\theta = (\sin 2\theta + \sin 4\theta) + 2\sin 3\theta \cos 3\theta = 4\cos\theta \cos 2\theta \sin 3\theta$

(h) $\dfrac{\sin 3x + \cos 5x + \sin 7x + \sin 9x}{\cos 3x + \cos 5x + \cos 7x + \cos 9x} = \dfrac{(\sin 3x + \sin 9x) + (\sin 5x + \sin 7x)}{(\cos 3x + \cos 9x) + (\cos 5x + \cos 7x)} = \tan 6x$

30.16 Prove

(a) $\cos 130° + \cos 110° + \cos 10° = 0$ (b) $\cos 220° + \cos 100° + \cos 20° = 0$

30.17 Prove

(a) $\cos^2\theta \sin^3\theta = \frac{1}{16}(2\sin\theta + \sin 3\theta - \sin 5\theta)$

(b) $\cos^2\theta \sin^4\theta = \frac{1}{32}(2 - \cos 2\theta - 2\cos 4\theta + \cos 6\theta)$

(c) $\cos^5 = \frac{1}{16}(10\cos\theta + 5\cos 3\theta + \cos 5\theta)$

(d) $\sin^5\theta = \frac{1}{16}(10\sin\theta - 5\sin 3\theta + \sin 5\theta)$

30.18 Transform

(a) $4\cos x + 3\sin x$ into the form $c\sin(x + \alpha)$.   *Ans.* $5\sin(x + 53°8')$

(b) $4\cos x + 3\sin x$ into the form $c\sin(x - \alpha)$.   *Ans.* $5\sin(x - 306°52')$

(c) $\sin x - \cos x$ into the form $c\sin(x - \alpha)$.   *Ans.* $\sqrt{2}\sin(x - 45°)$

(d) $5\cos 3t + 12\sin 3t$ into the form $c\cos(3t - \alpha)$. *Ans.* $13\cos(3t - 67°23')$

30.19 Find the maximum and minimum values of each sum of Problem 30.18 and a value of x or t between 0 and 2π at which each occurs.

    Ans. (a) Maximum $= 5$, when $x = 36°52'$ (i.e., when $x + 53°8' = 90°$); minimum $= -5$, when $x = 216°52'$

        (b) Same as (a)

        (c) Maximum $= \sqrt{2}$, when $x = 135°$; minimum $= -\sqrt{2}$, when $x = 315°$

        (d) Maximum $= 13$, when $t = 22°28'$; minimum $= -13$; when $t = 82°28'$

Oblique Triangles

AN OBLIQUE TRIANGLE is one which does not contain a right angle. Such a triangle contains either three acute angles or two acute angles and one obtuse angle.

The convention of denoting the measures of the angles by A, B, C and the lengths of the corresponding opposite sides by a, b, c will be used here. See Figs. 31-1 and 31-2.

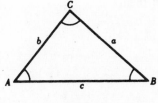

Fig. 31-1

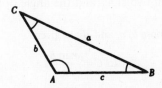

Fig. 31-2

When three parts, not all angles, are known, the triangle is uniquely determined, except in one case to be noted below. The four cases of oblique triangles are

Case I.   Given one side and two angles

Case II.   Given two sides and the angle opposite one of them

Case III.   Given two sides and the included angle

Case IV.   Given the three sides

THE LAW OF SINES.   In any triangle, the sides are proportional to the sines of the opposite angles, i.e.,

$$\frac{a}{\sin A} = \frac{b}{\sin B} = \frac{c}{\sin C}$$

The following relations follow readily:

$$\frac{a}{b} = \frac{\sin A}{\sin B}, \quad \frac{b}{c} = \frac{\sin B}{\sin C}, \quad \frac{c}{a} = \frac{\sin C}{\sin A}.$$

(For a proof of the law of sines, see Problem 31.1.)

PROJECTION FORMULAS.   In any triangle ABC,

$$a = b \cos C + c \cos B$$
$$b = c \cos A + a \cos C$$
$$c = a \cos B + b \cos A$$

(For the derivation of these formulas, see Problem 31.3.)

CASE 1.   Given one side and two angles

EXAMPLE.   Suppose a, B, and C are given.

To find A, use $A = 180° - (B + C)$.

To find b, use $\dfrac{b}{a} = \dfrac{\sin B}{\sin A}$ whence $b = \dfrac{a \sin B}{\sin A}$.

To find c, use $\dfrac{c}{a} = \dfrac{\sin C}{\sin A}$ whence $c = \dfrac{a \sin C}{\sin A}$.

(See Problems 31.4–31.6.)

CASE II.   Given two sides and the angle opposite one of them

EXAMPLE.   Suppose b, c, and B are given.

From $\dfrac{\sin C}{\sin B} = \dfrac{c}{b}$, $\sin C = \dfrac{c \sin B}{b}$.

If $\sin C > 1$, no angle C is determined.
If $\sin C = 1$, $C = 90°$ and a right triangle is determined.
If $\sin C < 1$, two angles are determined: an acute angle C and an obtuse angle $C' = 180° - C$. Thus, there may be one or two triangles determined. This is known as the "ambiguous case."

This case is discussed geometrically in Problem 31.7. The results obtained may be summarized as follows: When the given angle is *acute*, there will be

(*a*)   *One* solution if the side opposite the given angle is equal to or greater than the other given side.
(*b*)   *No* solution, *one* solution (right triangle), or *two* solutions if the side opposite the given angle is less than the other given side.

When the given angle is *obtuse*, there will be

(*c*)   *No* solution when the side opposite the given angle is less than or equal to the other given side.
(*d*)   *One* solution if the side opposite the given angle is greater than the other given side.

EXAMPLE

(*1*)   When $b = 30$, $c = 20$, and $B = 40°$, there is one solution since B is acute and $b > c$.

(*2*)   When $b = 20$, $c = 30$, and $B = 40°$, there is either no solution, one solution, or two solutions. The particular subcase is determined after computing $\sin C = \dfrac{c \sin B}{b}$.

(*3*)   When $b = 30$, $c = 20$, and $B = 140°$, there is one solution.
(*4*)   When $b = 20$, $c = 30$, and $B = 140°$, there is no solution.

This, the so-called ambiguous case, is solved by the law of sines and may be checked by the projection formulas. (See Problems 31.8–31.10.)

THE LAW OF COSINES. In any triangle ABC, the square of any side is equal to the sum of the squares of the other two sides diminished by twice the product of these sides and the cosine of their included angle; i.e.,

$$a^2 = b^2 + c^2 - 2bc \cos A$$

$$b^2 = c^2 + a^2 - 2ca \cos B$$

$$c^2 = a^2 + b^2 - 2ab \cos C$$

(For the derivation of these formulas, see Problem 31.11.)

CASE III. Given two sides and the included angle

EXAMPLE. Suppose a, b, and C are given.

To find c, use $c^2 = a^2 + b^2 - 2ab \cos C$.

To find A, use $\sin A = \dfrac{a \sin C}{c}$. To find B, use $\sin B = \dfrac{b \sin C}{c}$.

To check, use $A + B + C = 180°$.
(See Problems 31.12–31.14.)

CASE IV. Given three sides

EXAMPLE. With a, b, and c given, solve the law of cosines for each of the angles.

To find the angles, use $\cos A = \dfrac{b^2 + c^2 - a^2}{2bc}$, $\cos B = \dfrac{c^2 + a^2 - b^2}{2ca}$, $\cos C = \dfrac{a^2 + b^2 - c^2}{2ab}$.

To check, use $A + B + C = 180°$.
(See Problems 31.15–31.16.)

Solved Problems

31.1   Derive the law of sines.

Let ABC be any oblique triangle. In Fig. 31-3(a), angles A and B are acute while in Fig. 31-3(b), angle B is obtuse. Draw $\overline{CD}$ perpendicular to $\overrightarrow{AB}$ or $\overrightarrow{AB}$ extended and denote its length by h.

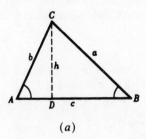

(a) (b)

Fig. 31-3

In the right triangle ACD of either figure, $h = b \sin A$ while in the right triangle BCD, $h = a \sin B$ since in Fig. 31-3(b), $h = a \sin \angle DBC = a \sin (180° - B) = a \sin B$. Thus,

$$a \sin B = b \sin A \quad \text{or} \quad \frac{a}{\sin A} = \frac{b}{\sin B}.$$

In a similar manner (by drawing a perpendicular from B to $\overline{AC}$ or a perpendicular from A to $\overline{BC}$), we obtain

$$\frac{a}{\sin A} = \frac{c}{\sin C} \quad \text{or} \quad \frac{b}{\sin B} = \frac{c}{\sin C}.$$

Thus, finally,

$$\frac{a}{\sin A} = \frac{b}{\sin B} = \frac{c}{\sin C}.$$

31.2 Derive one of the projection formulas.

Refer to Fig. 31-3. In the right triangle ACD of either figure, $AD = b \cos A$. In the right triangle BCD of Fig. 31-3(a), $DB = a \cos B$. Thus, in Fig. 31-3(a)

$$c = AB = AD + DB = b \cos A + a \cos B = a \cos B + b \cos A$$

In the right triangle BCD of Fig. 31-3(b), $BD = a \cos \angle DBC = a \cos(180° - B) = -a \cos B$. Thus, in Fig. 31-3(b),

$$c = AB = AD - BD = b \cos A - (-a \cos B) = a \cos B + b \cos A$$

CASE I

31.3 Solve the triangle ABC, given $c = 25, A = 35°$, and $B = 68°$. See Fig. 31-4.

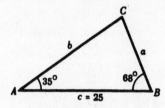

Fig. 31-4

To find C: $C = 180° - (A + B) = 180° - 103° = 77°$

To find a: $a = \dfrac{c \sin A}{\sin C} = \dfrac{25 \sin 35°}{\sin 77°} = \dfrac{25(0.5736)}{0.9744} = 15$

To find b: $b = \dfrac{c \sin B}{\sin C} = \dfrac{25 \sin 68°}{\sin 77°} = \dfrac{25(0.9272)}{0.9744} = 24$

To check by projection formula:

$$c = a \cos B + b \cos A = 15 \cos 68° + 24 \cos 35° = 15(0.3746) + 24(0.8192) = 25.3$$

The required parts are $a = 15, b = 24$, and $C = 77°$.

31.4 A and B are two points on opposite banks of a river. From A a line $AC = 275$ ft is laid off and the angles $CAB = 125°40'$ and $ACB = 48°50'$ are measured. Find the length of $\overline{AB}$. See Fig 31-5.

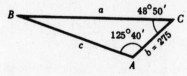

Fig. 31-5

In the triangle ABC, $B = 180° - (C + A) = 5°30'$ and

$$AB = c = \frac{b \sin C}{\sin B} = \frac{275 \sin 48°50'}{\sin 5°30'} = \frac{275(0.7528)}{0.0958} = 2160 \text{ ft}$$

31.5　A tower 125 ft high is on a cliff on the bank of a river. From the top of the tower the angle of depression of a point on the opposite shore is $28°40'$ and from the base of the tower the angle of depression of the same point is $18°20'$. Find the width of the river and the height of the cliff.

In Fig. 31-6 BC represents the tower, $\overline{DB}$ represents the cliff, and A is the point on the opposite shore.

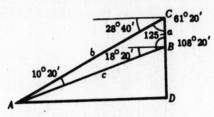

Fig. 31-6

In triangle ABC, $C = 90° - 28°40' = 61°20'$, $B = 90° + 18°20' = 108°20'$, $A = 180° - (B + C) = 10°20'$.

$$c = \frac{a \sin C}{\sin A} = \frac{125 \sin 61°20'}{\sin 10°20'} = \frac{125(0.8774)}{0.1794} = 611$$

In the right triangle ABD, $DB = c \sin 18°20' = 611(0.3145) = 192$, $AD = c \cos 18°20' = 611(0.9492) = 580$.

The river is 580 ft wide and the cliff is 192 ft high.

31.6　Discuss the several special cases when two sides and the angle opposite one of them are given.

Let b, c, and B be the given parts. Construct the given angle B and lay off the side $BA = c$. With A as center and radius equal to b (the side opposite the given angle) describe an arc. Figures 31-7(a)–(e) illustrate the special cases which may occur when the given angle B is acute, while Figs. 31-7(f)–(g) illustrate the cases when B is obtuse.

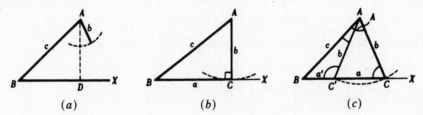

(a)　　　　　　　　　(b)　　　　　　　　　(c)

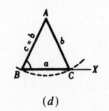　　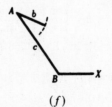　　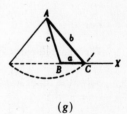

(d)　　　　　　　(e)　　　　　　　(f)　　　　　　　(g)

Fig. 31-7

The given angle B is acute.

Fig. 31-7(a).　　When $b < AD = c \sin B$, the arc does not meet BX and no triangle is determined.

Fig. 31-7(b).　　When $b = AD$, the arc is tangent to BX and one triangle—a right triangle with the right angle at C—is determined.

Fig. 31-7(c).   When $b > AD$ and $b < c$, the arc meets BX in two points C and C' on the same side of B. Two triangles ABC, in which C is acute, and ABC', in which $C' = 180° - C$ is obtuse, are determined.

Fig. 31-7(d).   When $b > AD$ and $b = c$, the arc meets $\overrightarrow{BX}$ in C and B. One triangle (isosceles) is determined.

Fig. 31-7(e).   When $b > c$, the arc meets BX in C and $\overrightarrow{BX}$ extended in C'. Since the triangle ABC' does not contain the given angle B, only one triangle ABC is determined.

The given angle is obtuse.

Fig. 31-7(f).   When $b < c$ or $b = c$, no triangle is formed.

Fig. 31-7(g).   When $b > c$, only one triangle is formed as in Fig 31.7(e).

CASE II

31.7   Solve the triangle ABC, given $c = 628$, $b = 480$, and $C = 55°10'$. Refer to Fig. 31-8.

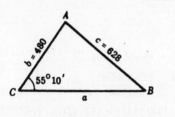
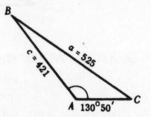

Fig. 31-8   Fig.   31-9

Since C is acute and $c > b$, there is only one solution.

For B: $\sin B = \dfrac{b \sin C}{c} = \dfrac{480 \sin 55°10'}{628} = \dfrac{480(0.8208)}{628} = 0.6274$ and $B = 38°50'$

For A: $A = 180° - (B + C) = 86°0'$

For a: $a = \dfrac{b \sin A}{\sin B} = \dfrac{480 \sin 86°0'}{\sin 38°50'} = \dfrac{480(0.9976)}{0.6271} = 764$

The required parts are $B = 38°50'$, $A = 86°0'$, and $a = 764$.

31.8   Solve the triangle ABC, given $a = 525$, $c = 421$, and $A = 130°50'$. Refer to Fig. 31-9.

Since A is obtuse and $a > c$, there is one solution.

For C: $\sin C = \dfrac{c \sin A}{a} = \dfrac{421 \sin 130°50'}{525} = \dfrac{421(0.7566)}{525} = 0.6067$ and $C = 37°20'$

For B: $B = 180° - (C + A) = 11°50'$

For b: $b = \dfrac{a \sin B}{\sin A} = \dfrac{525 \sin 11°50'}{\sin 130°50'} = \dfrac{525(0.2051)}{0.7566} = 142$

The required parts are $C = 37°20'$, $B = 11°50'$, and $b = 142$.

31.9   Solve the triangle ABC, given $a = 31.5$, $b = 51.8$, and $A = 33°40'$. Refer to Fig. 31-10.

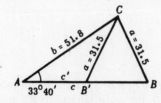

Fig. 31-10

Since A is acute and $a < b$, there is the possibility of two solutions.

For B:
$$\sin B = \frac{b \sin A}{a} = \frac{51.8 \sin 33°40'}{31.5} = \frac{51.8(0.5544)}{31.5} = 0.9117$$

There are two solutions, $B = 65°40'$ and $B' = 180° - 65°40' = 114°20'$.

For C:
$$C = 180° - (A + B) = 80°40'$$

For C':
$$C' = 180° - (A + B') = 32°0'$$

For c:
$$c = \frac{a \sin C}{\sin A} = \frac{31.5 \sin 80°40'}{\sin 33°40'} = \frac{31.5(0.9868)}{0.5544} = 56.1$$

For c':
$$c' = \frac{a \sin C'}{\sin A} = \frac{31.5 \sin 32°0'}{\sin 33°40'} = \frac{31.5(0.5299)}{0.5544} = 30.1$$

The required parts are

For triangle ABC: $B = 65°40'$, $C = 80°40'$,   and $c = 56.1$.

For triangle ABC': $B' = 114°20'$, $C' = 32°0'$,   and $c' = 30.1$.

31.10   Derive the law of cosines.

In the right triangle ABC of either figure, $b^2 = h^2 + (AD)^2$.

In the right triangle BCD of Fig. 31-11(a), $h = a \sin B$ and $DB = a \cos B$. Then $AD = AB - DB = c - a \cos$ and

$$b^2 = h^2 + (AD)^2 = a^2 \sin^2 B + c^2 - 2ca \cos B + a^2 \cos^2 B$$
$$= a^2(\sin^2 B + \cos^2 B) + c^2 - 2ca \cos B = c^2 + a^2 - 2ca \cos B$$

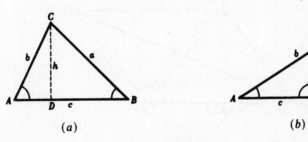

(a)   (b)

Fig. 31-11

In the right triangle BCD of Fig. 31-11 (b), $h = a \sin \angle CBD = a \sin(180° - B) = a \sin B$ and $BD = a \cos \angle CBD = a \cos(180° - B) = -a \cos B$. Then $AD = AB + BD = c - a \cos B$ and $b^2 = c^2 + a^2 - 2ca \cos B$.

The remaining equations may be obtained by cyclic changes of the letters.

CASE III

31.11   Solve the triangle ABC, given $a = 132, b = 224$, and $C = 28°40'$. See Fig. 31-12.

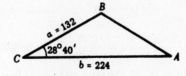

Fig. 31-12

For c:
$$c^2 = a^2 + b^2 - 2ab\cos C$$
$$= (132)^2 + (224)^2 - 2(132)(224)\cos 28°40'$$
$$= (132)^2 + (224)^2 - 2(132)(224)(0.8774)$$
$$= 15\,714 \quad \text{and} \quad c = 125$$

For A: $\sin A = \dfrac{a\sin C}{c} = \dfrac{132\sin 28°40'}{125} = \dfrac{132(0.4797)}{125} = 0.5066$ and $A = 30°30'$

For B: $\sin B = \dfrac{b\sin C}{c} = \dfrac{224\sin 28°40'}{125} = \dfrac{224(0.4797)}{125} = 0.8596$ and $B = 120°40'$

(Since $b > a$, A is acute; since $A + C < 90°$, $B > 90°$.)
Check: $A + B + C = 179°50'$. The required parts are $A = 30°30', B = 120°40', c = 125$.

31.12 Two forces of 17.5 1b and 22.5 1b act on a body. If their directions make an angle of $50°10'$ with each other, find the magnitude of their resultant and the angle which it makes with the larger force.

In the parallelogram $ABCD$ (see Fig. 31-13), $A + B = C + D = 180°$ and $B = 180° - 50°10' = 129°50'$. In the triangle ABC,
$$b^2 = c^2 + a^2 - 2ca\cos B \qquad [\cos 129°50' = -\cos(180° - 129°50') = -\cos 50°10']$$
$$= (22.5)^2 + (17.5)^2 - 2(22.5)(17.5)(-0.6406) = 1317 \quad \text{and} \quad b = 36.3$$

$$\sin A = \frac{a\sin B}{b} = \frac{17.5\sin 129°50'}{36.3} = \frac{17.5(0.7679)}{36.3} = 0.3702 \quad \text{and} \quad A = 21°40'$$

The resultant is a force of 36.3 1b; the required angle is $21°40'$.

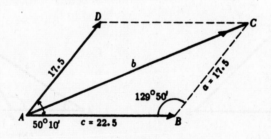

Fig. 31-13

31.13 From A a pilot flies 125 mi in the direction N 38°20' W and turns back. Through an error, he then flies 125 mi in the direction S 51°40'E. How far and in what direction must he now fly to reach his intended destination A?

Denote the turn back point as B and his final position as C. In the triangle ABC (see Fig. 31-14),
$$b^2 = c^2 + a^2 - 2ca\cos B = (125)^2 + (125)^2 - 2(125)(125)\cos 13°20'$$
$$= 2(125)^2(1 - 0.9730) = 843.7 \quad \text{and} \quad b = 29.0$$

$$\sin A = \frac{a\sin B}{b} = \frac{125\sin 13°20'}{29.0} = \frac{125(0.2306)}{29.0} = 0.9940 \quad \text{and} \quad A = 83°40'$$

Since $\angle CAN_1 = A - \angle N_1AB = 45°20'$, the pilot must fly a course S 45°20' W for 29.0 mi in going from C to A.

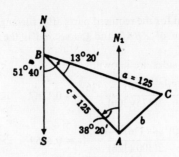

Fig. 31-14

CASE IV

31.14 Solve the triangle ABC, given $a = 30.3$, $b = 40.4$, and $c = 62.6$. Refer to Fig. 31-15.

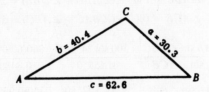

Fig. 31-15

For A: $\cos A = \dfrac{b^2 + c^2 - a^2}{2bc} = \dfrac{(40.4)^2 + (62.6)^2 - (30.3)^2}{2(40.4)(62.6)} = 0.9159$ and $A = 23°40'$

For B: $\cos B = \dfrac{c^2 + a^2 - b^2}{2ca} = \dfrac{(62.6)^2 + (30.3)^2 + (40.4)^2}{2(62.6)(30.3)} = 0.8448$ and $B = 32°20'$

For C: $\cos C = \dfrac{a^2 + b^2 - c^2}{2ab} = \dfrac{(30.3)^2 + (40.4)^2 - (62.6)^2}{2(30.3)(40.4)} = -0.5590$ and $C = 124°0'$

Check: $A + B + C = 180°$.

31.15 The distances of a point C from two points A and B, which cannot be measured directly, are required. The line CA is continued through A for a distance 175 m to D, the line $\overleftrightarrow{CB}$ is continued through B for 225 m to E, and the distances $AB = 300$ m, $DB = 326$ m, and $DE = 488$ m are measured. Find AC and BC. Refer to Fig. 31-16.

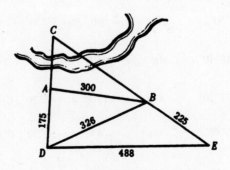

Fig. 31-16

Triangle ABC may be solved for the required parts after the angles $\angle BAC$ and $\angle ABC$ have been found. The first angle is the supplement of $\angle BAD$ and the second is the supplement of the sum of $\angle ABD$ and $\angle DBE$.

In the triangle ABD whose sides are known,

$$\cos \angle BAD = \frac{(175)^2 + (300)^2 - (326)^2}{2(175)(300)} = 0.1367 \quad \text{and} \quad \angle BAD = 82°10'$$

$$\cos \angle ABD = \frac{(300)^2 + (326)^2 - (175)^2}{2(300)(326)} = 0.8469 \quad \text{and} \quad \angle ABD = 32°10'$$

In the triangle BDE whose sides are known,

$$\cos \angle DBE = \frac{(225)^2 + (326)^2 - (488)^2}{2(225)(326)} = -0.5538 \quad \text{and} \quad \angle DBE = 123°40'$$

In the triangle ABC: $AB = 300$,

$$\angle BAC = 180° - \angle BAD = 97°50'$$
$$\angle ABC = 180° - (\angle ABD + \angle DBE) = 24°10'$$
$$\angle ACB = 180° - (\angle BAC + \angle ABC) = 58°0'$$

Then

$$AC = \frac{AB \sin \angle ABC}{\sin \angle ACB} = \frac{300 \sin 24°10'}{\sin 58°0'} = \frac{300(0.4094)}{0.8480} = 145$$

and

$$BC = \frac{AB \sin \angle BAC}{\sin \angle ACB} = \frac{300 \sin 97°50'}{\sin 58°0'} = \frac{300(0.9907)}{0.8480} = 350$$

The required distances are $AC = 145$ m and $BC = 350$ m.

Supplementary Problems

Solve each of the following oblique triangles ABC, given:

31.16 $a = 125$, $A = 54°40'$, $B = 65°10'$.   *Ans.* $b = 139$, $c = 133$, $C = 60°10'$

31.17 $b = 321$, $A = 75°20'$, $C = 38°30'$.   *Ans.* $a = 339$, $c = 218$, $B = 66°10'$

31.18 $b = 215$, $c = 150$, $B = 42°40'$.   *Ans.* $a = 300$, $A = 109°10'$, $C = 28°10'$

31.19 $a = 512$, $b = 426$, $A = 48°50'$.   *Ans.* $c = 680$, $B = 38°50'$, $C = 92°20'$

31.20 $b = 50.4$, $c = 33.3$, $B = 118°30'$.   *Ans.* $a = 25.1$, $A = 26°0'$, $C = 35°30'$

31.21 $b = 40.2$, $a = 31.5$, $B = 112°20'$.   *Ans.* $c = 15.7$, $A = 46°30'$, $C = 21°10'$

31.22 $b = 51.5$, $a = 62.5$, $B = 40°40'$.   *Ans.* $c = 78.9$, $A = 52°20'$, $C = 87°0'$, $c' = 16.0$, $A' = 127°40'$, $C' = 11°40'$

31.23 $a = 320$, $c = 475$, $B = 35°20'$.   *Ans.* $b = 552$, $B = 85°30'$, $C' = 59°10'$, $b' = 224$, $B' = 23°50'$, $C' = 120°50'$

31.24 $b = 120$, $c = 270$, $A = 118°40'$.   *Ans.* $a = 344$, $B = 17°50'$, $C = 43°30'$

31.25 $a = 24.5$, $b = 18.6$, $c = 26.4$.   *Ans.* $A = 63°10'$, $B = 42°40'$, $C = 74°10'$

31.26 $a = 6.34$, $b = 7.30$, $c = 9.98$.   *Ans.* $A = 39°20'$, $B = 46°50'$, $C = 93°50'$

31.27   Two ships have radio equipment with a range of 200 mi. One is 155 mi N 42°40' E and the other is 165 mi N 45°10' W of a shore station. Can the two ships communicate directly?

   Ans.   No; they are 222 mi apart.

31.28   A ship sails 15.0 mi on a course S 40°10' W and then 21.0 mi on a course N 28°20' W. Find the distance and direction of the last position from the first.

   Ans.   20.9 mi, N 70°40' W

31.29   A lighthouse is 10 mi northwest of a dock. A ship leaves the dock at 9 A.M. and steams west at 12 mi per hr. At what time will it be 8 mi from the lighthouse?

   Ans.   9:16 A.M. and 9:54 A.M.

31.30   Two forces of 115 lb and 215 lb acting on an object have a resultant of magnitude 275 lb. Find the angle between the directions in which the given forces act.

   Ans.   70°50'

31.31   A tower 150 m high is situated at the top of a hill. At a point 650 m down the hill the angle between the surface of the hill and the line of sight to the top of the tower is 12°30'. Find the inclination of the hill to a horizontal plane.

   Ans.   7°50'

Chapter 32

Inverse Trigonometric Functions

INVERSE FUNCTIONS.   The equation

$$y = 2x + 3$$

defines a unique value of y for each value of x. Similarly, the equation

$$y = \frac{x}{2} - 3$$

does the same; however, these two equations have an interesting, even provocative, relationship:

If
$$f(x) = 2x + 3 \quad \text{and} \quad g(x) = \frac{x}{2} - 3,$$

then
$$f(g(x)) = g(f(x)) = x$$

That is, f and g "undo" each other. We call the function g the inverse of f and we call f the inverse of g.

NOTATION.   This relationship is written as follows:

$$f = g^{-1}$$

and
$$g = f^{-1}$$

DEFINITION.   If f and g are functions and if

$$f(g(x)) = g(f(x)) = x$$

for all values of x for which these composites are defined, then we say that f and g are each other's inverses.

To determine the equation of an inverse function for $y = f(x)$, simply solve $y = f(x)$ for x and then interchange the roles of the two variables.

EXAMPLE.   If $f(x) = y = 3x - 5$,

then	$3x = y + 5$	(interchange x and y)
and	$x = \dfrac{y+5}{3}$	(solve for x)
Thus,	$f^{-1}(x) = y = \dfrac{x+5}{3}$	(interchange x and y)

Note that in Fig. 32-1 $f(x)$ and $f^{-1}(x)$ are mirror images of each other in the line $y = x$:

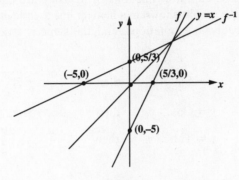

Fig. 32-1

The reader should verify that for $\qquad\qquad g(x) = x^3, \quad g^{-1}(x) = x^{1/3}.$

For

$$h(x) = x^2, \quad x > 0, \quad h^{-1}(x) = \sqrt{x} \qquad\qquad (32.1)$$

Note also that if $f(x)$ is *not* one-to-one, then $f^{-1}(x)$ is *not* a function. Thus, functions that are not one-to-one do not have inverse functions. For $f(x) = x^2$, unless we restrict the domain (as we did above in (32.1)), $f(x)$ does not possess an inverse function.

One particularly important class of inverse functions is the class of inverse trigonometric functions.

INVERSE TRIGONOMETRIC RELATIONS.   The equation

$$x = \sin y \qquad\qquad (32.2)$$

defines a unique value of x for each given angle y. But when x is given, the equation may have no solution or many solutions. For example: if $x = 2$, there is no solution, since the sine of an angle never exceeds 1; if $x = \frac{1}{2}$, there are many solutions: $y = 30°, 150°, 390°, 510°, -210°, -330°, \ldots$.

To express y in terms of x, we will write

$$y = \arcsin x \qquad\qquad (32.3)$$

In spite of the use of the word *arc*, (32.3) is to be interpreted as stating that "y is an angle whose sine is x." Similarly, we shall write $y = \arccos x$ if $x = \cos y$, $y = \arctan x$ if $x = \tan y$, etc. An alternate notation for $y = \arcsin x$ is $y = \sin^{-1} x$ (and similarly, $y = \cos^{-1} x$, etc., for the other functions). Note that $y = \arcsin x$, $\arccos x$, etc., are all relations but not functions of x.

GRAPHS OF THE INVERSE TRIGONOMETRIC RELATIONS.   The graph of $y = \arcsin x$ is the graph of $x = \sin y$ and differs from the graph $y = \sin x$ in that the roles of x and y are interchanged. Thus, the graph of $y = \arcsin x$ is a sine curve drawn on the y axis instead of the x axis.

Similarly, the graphs of the remaining inverse trigonometric relations are those of the corresponding trigonometric functions except that the roles of x and y are interchanged.

INVERSE TRIGONOMETRIC FUNCTIONS. It is at times necessary to consider the inverse trigonometric relations as single-valued (i.e., one value of y corresponding to each admissible value of x). To do this, we agree to select one out of the many angles corresponding to the given value of x. For example, when $x = \frac{1}{2}$, we shall agree to select the value $y = 30°$, and when $x = -\frac{1}{2}$, we shall agree to select the value $y = -30°$. This selected value is called the *principal value* of arcsin x. When only the principal value is called for, we shall write Arcsin x, Arccos x, etc. The portions of the graphs on which the principal values of each of the inverse trigonometric relations lie are shown in Figs. 32-2(*a*) through (*f*) by a heavier line. Note that Arcsin x, Arccos x, etc., are functions of x. They are called the *inverse trigonometric functions*. Thus, the portions of the graphs shown in a heavier line are the graphs of these functions. Note that $\text{Sin}^{-1} x$ and Arcsin x are equivalent notations, and the same is true for the other trigonometric functions.

If $\qquad\qquad\qquad y = \text{Sin}^{-1} x, \qquad$ then $\qquad -\dfrac{\pi}{2} \le y \le \dfrac{\pi}{2}.$

If $\qquad\qquad\qquad y = \text{Cos}^{-1} x, \qquad$ then $\qquad 0 \le y \le \pi;$

and if $\qquad\qquad\quad y = \text{Tan}^{-1} x, \qquad$ then $\qquad -\dfrac{\pi}{2} < y < \dfrac{\pi}{2}.$

Similarly,

if $\qquad\qquad\qquad y = \text{Sec}^{-1} x \qquad$ then $\qquad 0 \le y \le \pi, \quad y \ne \dfrac{\pi}{2};$

if $\qquad\qquad\qquad y = \text{Cot}^{-1} x \qquad$ then $\qquad 0 < y < \pi;$

and if $\qquad\qquad\quad y = \text{Csc}^{-1} x, \qquad$ then $\qquad -\dfrac{\pi}{2} \le y \le \dfrac{\pi}{2}, \quad y \ne 0.$

For example, $\text{Sin}^{-1} \dfrac{\sqrt{3}}{2} = \dfrac{\pi}{3}$, $\text{Arctan } 1 = \dfrac{\pi}{4}$, $\text{Sin}^{-1} \dfrac{-\sqrt{3}}{2} = \dfrac{-\pi}{3}$, $\text{Arccos}\left(\dfrac{-1}{2}\right) = \dfrac{2\pi}{3}$, $\text{Sec}^{-1}\left(\dfrac{-2}{\sqrt{3}}\right) = \dfrac{-5\pi}{6}$ and $\text{Arccsc}\,(-\sqrt{2}) + \dfrac{-\pi}{4}.$

Note that the inverse trigonometric functions are inverses of the trigonometric functions. For example $\sin(\text{Sin}^{-1} x) = y$.

GENERAL VALUES OF THE INVERSE TRIGONOMETRIC RELATIONS. Let y be an inverse trigonometric relation of x. Since the value of a trigonometric function of y is known, there are determined in general two positions for the terminal side of the angle y (see Fig. 24-4.). Let y_1 and y_2 respectively be angles determined by the two positions of the terminal side. Then the totality of values of y consist of the angles y_1 and y_2, together with all angles coterminal with them, that is,

$$y_1 + 2n\pi \quad \text{and} \quad y_2 + 2n\pi$$

where n is any positive or negative integer, or is zero.

One of the values y_1 or y_2 may always be taken as the principal value of the inverse trigonometric relation with the domains properly restricted.

EXAMPLE. Write expressions for the general value of (*a*) arcsin $\frac{1}{2}$, (*b*) arccos (−1), (*c*) arctan (−1).

(*a*) The principal value of arcsin $\frac{1}{2}$ is $\pi/6$, and a second value (not coterminal with the principal value) is $5\pi/6$. The general value of arcsin $\frac{1}{2}$ is given by $\pi/6 + 2n\pi$, $5\pi/6 + 2n\pi$, where n is any positive or negative integer, or is zero.

(*b*) The principal value is π and there is no other value not coterminal with it. Thus, the general value is given by $\pi + 2n\pi$, where n is a positive or negative integer, or is zero.

(*c*) The principal value is $-\pi/4$, and a second value (not coterminal with the principal value) is $3\pi/4$. Thus, the general value is given by $-\pi/4 + 2n\pi, 3\pi/4 + 2n\pi$, where n is a positive or negative integer, or is zero.

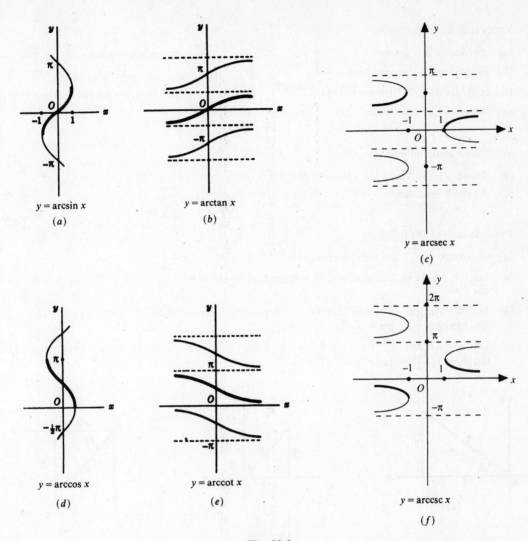

$y = \arcsin x$
(a)

$y = \arctan x$
(b)

$y = \operatorname{arcsec} x$
(c)

$y = \arccos x$
(d)

$y = \operatorname{arccot} x$
(e)

$y = \operatorname{arccsc} x$
(f)

Fig. 32-2

Solved Problems

32.1   Verify each of the following:

(a) Arcsin $0 = 0$ (e) Arcsec $2 = \pi/3$ (i) Arctan $(-1) = -\pi/4$

(b) Arccos $(-1) = \pi$ (f) Arccsc $(-\sqrt{2}) = -3\pi/4$ (j) Arccot $0 = \pi/2$

(c) Arctan $\sqrt{3} = \pi/3$ (g) Arccos $\theta = \pi/2$ (k) Arcsec $(-\sqrt{2}) = -3\pi/4$

(d) Arccot $\sqrt{3} = \pi/6$ (h) Arcsin $(-1) = -\pi/2$ (l) Arccsc $(-2) = -5\pi/6$

32.2   Verify each of the following:

(a) Arcsin $0.3333 = 19°28'$ (g) Arcsin $(-0.6439) = -40°5'$

(b) Arccos $0.4000 = 66°25'$ (h) Arccos $(-0.4519) = 116°52'$

(c) Arctan $1.5000 = 56°19'$ (i) Arctan $(-1.4400) = -55°13'$

(d) Arccot $1.1875 = 40°6'$ (j) Arccot $(-0.7340) = 126°17'$

(e) Arcsec $1.0324 = 14°24'$ (k) Arcsec $(-1.2067) = -145°58'$

(f) Arccsc $1.5082 = 41°32'$ (l) Arccsc $(-4.1923) = -166°12'$

32.3   Verify each of the following:

(a) $\sin(\text{Arcsin}\,\tfrac{1}{2}) = \sin\pi/6 = \tfrac{1}{2}$ (e) $\text{Arccos}\,[\cos(-\pi/4)] = \text{Arccos}\,\sqrt{2}/2 = \pi/4$

(b) $\cos[\text{Arccos}\,(-\tfrac{1}{2})] = \cos 2\pi/3 = -\tfrac{1}{2}$ (f) $\text{Arcsin}\,(\tan 3\pi/4) = \text{Arcsin}\,(-1) = -\pi/2$

(c) $\cos[\text{Arcsin}\,(-\sqrt{2}/2)] = \cos(-\pi/4) = \sqrt{2}/2$ (g) $\text{Arccos}\,[\tan(-5\pi/4)] = \text{Arccos}\,(-1) = \pi$

(d) $\text{Arcsin}\,(\sin\pi/3) = \text{Arcsin}\,\sqrt{3}/2 = \pi/3$

32.4   Verify each of the following:

(a) $\text{Arcsin}\,\sqrt{2}/2 - \text{Arcsin}\,\tfrac{1}{2} = \pi/4 - \pi/6 = \pi/12$

(b) $\text{Arccos}\,0 + \text{Arctan}\,(-1) = \pi/2 + (-\pi/4) = \pi/4 = \text{Arctan}\,1$

32.5   Evaluate each of the following:

(a) $\cos(\text{Arcsin}\,\tfrac{3}{5})$,   (b) $\sin[\text{Arccos}\,(-\tfrac{2}{3})]$,   (c) $\tan[\text{Arcsin}\,(-\tfrac{3}{4})]$.

(a) Let $\theta = \text{Arcsin}\,\tfrac{3}{5}$; then $\sin\theta = \tfrac{3}{5}$, θ being a first quadrant angle. From Fig. 32-3(a), $\cos(\text{Arcsin}\,\tfrac{3}{5}) = \cos\theta = \tfrac{4}{5}$.

(b) Let $\theta = \text{Arccos}\,(-\tfrac{2}{3})$; then $\cos\theta = -\tfrac{2}{3}$, θ being a second quadrant angle. From Fig. 32-3(b), $\sin[\text{Arccos}\,(-\tfrac{2}{3})] = \sin\theta = \sqrt{5}/3$.

(c) Let $\theta = \text{Arcsin}\,(-\tfrac{3}{4})$; then $\sin\theta = -\tfrac{3}{4}$, θ being a fourth quadrant angle. From Fig. 32-3(c), $\tan[\text{Arcsin}\,(-\tfrac{3}{4})] = \tan\theta = -3/\sqrt{7} = -3\sqrt{7}/7$.

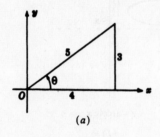   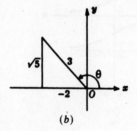   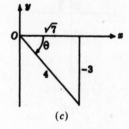

(a)   (b)   (c)

Fig. 32-3

32.6   Evaluate $\sin(\text{Arcsin}\,\tfrac{12}{13} + \text{Arcsin}\,\tfrac{4}{5})$.

Let $\theta = \text{Arcsin}\,\tfrac{12}{13}$ and $\phi = \text{Arcsin}\,\tfrac{4}{5}$. Then $\sin\theta = \tfrac{12}{13}$ and $\sin\phi = \tfrac{4}{5}$, θ and ϕ being first quadrant angles. From Fig. 32-4 and Fig. 32-5,

$$\sin(\text{Arcsin}\,\tfrac{12}{13} + \text{Arcsin}\,\tfrac{4}{5}) = \sin(\theta + \phi) = \sin\theta\cos\phi + \cos\theta\sin\phi$$
$$= \tfrac{12}{13}\cdot\tfrac{3}{5} + \tfrac{5}{13}\cdot\tfrac{4}{5} = \tfrac{56}{65}$$

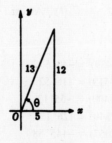   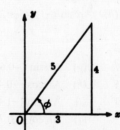

Fig. 32-4   **Fig. 32-5**

32.7   Evaluate $\cos(\text{Arctan}\ \frac{15}{8} - \text{Arcsin}\ \frac{7}{25})$.

Let $\theta = \text{Arctan}\ \frac{15}{8}$ and $\phi = \text{Arcsin}\ \frac{7}{25}$. Then $\tan\theta = \frac{15}{8}$ and $\sin\phi = \frac{7}{25}$, θ and ϕ being first quadrant angles. From Fig. 32-6 and Fig. 32-7,

$$\cos(\text{Arctan}\ \tfrac{15}{8} - \text{Arcsin}\ \tfrac{7}{25}) = \cos(\theta - \phi) = \cos\theta\cos\phi + \sin\theta\sin\phi$$
$$= \tfrac{8}{17}\cdot\tfrac{24}{25} + \tfrac{15}{17}\cdot\tfrac{7}{25} = \tfrac{297}{425}$$

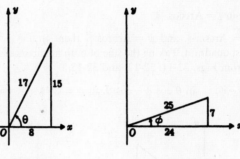

Fig. 32-6   Fig. 32-7

32.8   Evaluate $\sin(2\ \text{Arctan}\ 3)$.

Let $\theta = \text{Arctan}\ 3$; then $\tan\theta = 3$, θ being a first quadrant angle. From Fig. 32-8,

$$\sin(2\ \text{Arctan}\ 3) = \sin 2\theta = 2\sin\theta\cos\theta = 2\left(\frac{3}{\sqrt{10}}\right)\left(\frac{1}{\sqrt{10}}\right) = \frac{3}{5}$$

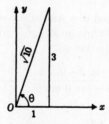

Fig. 32-8

32.9   Show that $\text{Arcsin}\ 1/\sqrt{5} + \text{Arcsin}\ 2/\sqrt{5} = \pi/2$.

Let $\theta = \text{Arcsin}\ 1/\sqrt{5}$ and $\phi = \text{Arcsin}\ 2/\sqrt{5}$; then $\sin\theta = 1/\sqrt{5}$ and $\sin\phi = 2/\sqrt{5}$, each angle terminating in the first quadrant. We are to show that $\theta + \phi = \pi/2$ or, taking the sines of both members, that $\sin(\theta + \phi) = \sin\pi/2$.

From Fig. 32-9 and Fig. 32-10,

$$\sin(\theta + \phi) = \sin\theta\cos\phi + \cos\theta\sin\phi = \frac{1}{\sqrt{5}}\cdot\frac{1}{\sqrt{5}} + \frac{2}{\sqrt{5}}\cdot\frac{2}{\sqrt{5}} = 1 = \sin\frac{\pi}{2}$$

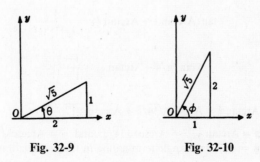

Fig. 32-9   Fig. 32-10

32.10 Show that $2 \operatorname{Arctan} \frac{1}{2} = \operatorname{Arctan} \frac{4}{3}$.

Let $\theta = \operatorname{Arctan} \frac{1}{2}$ and $\phi = \operatorname{Arctan} \frac{4}{3}$; then $\tan \theta = \frac{1}{2}$ and $\tan \phi = \frac{4}{3}$. We are to show that $2\theta = \phi$ or, taking the tangents of both members, that $\tan 2\theta = \tan \phi$. Now

$$\tan 2\theta = \frac{2 \tan \theta}{1 - \tan^2 \theta} = \frac{2(\frac{1}{2})}{1 - (\frac{1}{2})^2} = \frac{4}{3} = \tan \phi$$

32.11 Show that $\operatorname{Arcsin} \frac{77}{85} - \operatorname{Arcsin} \frac{3}{5} = \operatorname{Arccos} \frac{15}{17}$.

Let $\theta = \operatorname{Arcsin} \frac{77}{85}$, $\phi = \operatorname{Arcsin} \frac{3}{5}$, and $\psi = \operatorname{Arccos} \frac{15}{17}$ then $\sin \theta = \frac{77}{85}$, $\sin \phi = \frac{3}{5}$, and $\cos \psi = \frac{15}{17}$, each angle terminating in the first quadrant. Taking the sine of both members of the given relation, we are to show that $\sin (\theta - \phi) = \sin \psi$. From Figs. 32-11, 32-12, and 32-13,

$$\sin (\theta - \phi) = \sin \theta \cos \phi - \cos \theta \sin \phi = \frac{77}{85} \cdot \frac{4}{5} - \frac{36}{85} \cdot \frac{3}{5} = \frac{8}{17} = \sin \psi$$

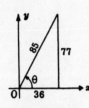   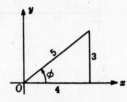   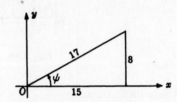

Fig. 32-11   **Fig. 32-12**   **Fig. 32-13**

32.12 Show that $\operatorname{Arccot} \frac{43}{32} - \operatorname{Arccot} \frac{1}{4} = \operatorname{Arccos} \frac{12}{13}$.

Let $\theta = \operatorname{Arccot} \frac{43}{32}$, $\phi = \operatorname{Arctan} \frac{1}{4}$, and $\psi = \operatorname{Arccos} \frac{12}{13}$ (see Fig. 32-14); then $\cot \theta = \frac{43}{32}$, $\tan \phi = \frac{1}{4}$, and $\cos \psi = \frac{12}{13}$, each angle terminating in the first quadrant. Taking the tangent of both members of the given relation, we are to show that $\tan (\theta - \phi) = \tan \psi$.

$$\tan (\theta - \phi) = \frac{\tan \theta - \tan \phi}{1 + \tan \theta \tan \phi} = \frac{\frac{32}{43} - \frac{1}{4}}{1 + (\frac{32}{43})(\frac{1}{4})} = \frac{5}{12} = \tan \psi$$

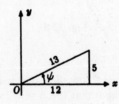

Fig. 32-14

32.13 Show that $\operatorname{Arctan} \frac{1}{2} + \operatorname{Arctan} \frac{1}{5} + \operatorname{Arctan} \frac{1}{8} = \pi/4$.

We shall show that $\operatorname{Arctan} \frac{1}{2} + \operatorname{Arctan} \frac{1}{5} = \pi/4 - \operatorname{Arctan} \frac{1}{8}$.

$$\tan(\operatorname{Arctan} \tfrac{1}{2} + \operatorname{Arctan} \tfrac{1}{5}) = \frac{\frac{1}{2} + \frac{1}{5}}{1 - (\frac{1}{2})(\frac{1}{5})} = \frac{7}{9}$$

and

$$\tan(\pi/4 - \operatorname{Arctan} \tfrac{1}{8}) = \frac{1 - \frac{1}{8}}{1 + \frac{1}{8}} = \frac{7}{9}$$

32.14 Show that $2 \operatorname{Arctan} \frac{1}{3} + \operatorname{Arctan} \frac{1}{7} = \operatorname{Arcsec} \sqrt{34}/5 + \operatorname{Arccsc} \sqrt{17}$.

Let $\theta = \operatorname{Arctan} \frac{1}{3}$, $\phi = \operatorname{Arctan} \frac{1}{7}$, $\lambda = \operatorname{Arcsec} \sqrt{34}/5$, and $\psi = \operatorname{Arccsc} \sqrt{17}$; then $\tan \theta = \frac{1}{3}$, $\tan \phi = \frac{1}{7}$, $\sec \lambda = \sqrt{34}/5$, and $\csc \psi = \sqrt{17}$, each angle terminating in the first quadrant.

Taking the tangent of both members of the given relation, we are to show that $\tan(2\theta + \phi) = \tan(\lambda + \psi)$. Now

$$\tan 2\theta = \frac{2\tan\theta}{1-\tan^2\theta} = \frac{2(\frac{1}{3})}{1-(\frac{1}{3})^2} = \frac{3}{4}$$

$$\tan(2\theta + \phi) = \frac{\tan 2\theta + \tan\phi}{1-\tan 2\theta \tan\phi} = \frac{\frac{3}{4}+\frac{1}{7}}{1-(\frac{3}{4})(\frac{1}{7})} = 1$$

and, using Fig. 32-15 and Fig. 32-16,

$$\tan(\lambda + \psi) = \frac{\frac{3}{5}+\frac{1}{4}}{1-(\frac{3}{5})(\frac{1}{4})} = 1$$

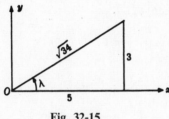

Fig. 32-15

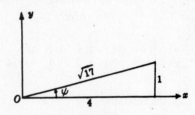

Fig. 32-16

32.15 Find the general value of each of the following:

(a) $\arcsin \sqrt{2}/2 = \pi/4 + 2n\pi, 3\pi/4 + 2n\pi$ (d) $\arcsin(-1) = -\pi/2 + 2n\pi$

(b) $\arccos \frac{1}{2} = \pi/3 + 2n\pi, 5\pi/3 + 2n\pi$ (e) $\arccos 0 = \pi/2 + 2n\pi, 3\pi/2 + 2n\pi$

(c) $\arctan 0 = 2n\pi, (2n+1)\pi$ (f) $\arctan(-\sqrt{3}) = -\pi/3 + 2n\pi, 2\pi/3 + 2n\pi$

where n is a positive or negative integer, or is zero.

32.16 Show that the general value of

(a) $\arcsin x = n\pi + (-1)^n \text{Arcsin } x$,

(b) $\arccos x = 2n\pi \pm \text{Arccos } x$,

(c) $\arctan x = n\pi + \text{Arctan } x$,

where n is any positive or negative integer, or is zero.

(a) Let $\theta = \text{Arcsin } x$. Then since $\sin(\pi - \theta) = \sin\theta$, all values of $\arcsin x$ are given by

        (1) $\theta + 2m\pi$ and   (2) $\pi - \theta + 2m\pi = (2m+1)\pi - \theta$

Now, when $n = 2m$, that is, n is an even integer, (1) may be written as $n\pi + \theta = n\pi + (-1)^n\theta$; and when $n = 2m+1$, that is, n is an odd integer, (2) may be written as $n\pi - \theta = n\pi + (-1)^n\theta$. Thus, $\arcsin x = n\pi + (-1)^n \text{Arcsin } x$, where n is any positive or negative integer, or is zero.

(b) Let $\theta = \text{Arccos } x$. Then since $\cos(-\theta) = \cos\theta$, all values of $\arccos x$ are given by $\theta + 2n\pi$ and $-\theta + 2n\pi$ or $2n\pi \pm \theta = 2n\pi \pm \text{Arccos } x$, where n is any positive or negative integer, or is zero.

(c) Let $\theta = \text{Arctan } x$. Then since $\tan(\pi + \theta) = \tan\theta$, all values of $\arctan x$ are given by $\theta + 2m\pi$ and $(\pi + \theta) + 2m\pi = \theta + (2m+1)\pi$ or, as in (a), by $n\pi + \text{Arctan } x$, where n is any positive or negative integer, or is zero.

32.17 Express the general value of each of the functions of Problem 32.15, using the form of Problem 32.16.

(a) $\arcsin \sqrt{2}/2 = n\pi + (-1)^n \pi/4$ (d) $\arcsin(-1) = n\pi + (-1)^n(-\pi/2)$

(b) $\arccos \frac{1}{2} = 2n\pi \pm \pi/3$ (e) $\arccos 0 = 2n\pi \pm \pi/2$

(c) $\arctan 0 = n\pi$ (f) $\arctan(-\sqrt{3}) = n\pi - \pi/3$

where n is any positive or negative integer, or is zero.

Supplementary Problems

32.18 Write the following in inverse function notation:

(a) $\sin \theta = \frac{3}{4}$,   (b) $\cos \alpha = -1$,   (c) $\tan x = -2$,   (d) $\cot \beta = \frac{1}{2}$.

    Ans. (a) $\theta = \arcsin \frac{3}{4}$ (b) $\alpha = \arccos(-1)$ (c) $x = \arctan(-2)$ (d) $\beta = \arccos \frac{1}{2}$

32.19 Find the value of each of the following:

(a) Arcsin $\sqrt{3}/2$ (d) Arccot 1   (g) Arctan $(-\sqrt{3})$ (j) Arccsc (-1)

(b) Arccos $(-\sqrt{2}/2)$ (e) Arcsin $(-\frac{1}{2})$ (h) Arccot 0

(c) Arctan $1/\sqrt{3}$ (f) Arccos $(-\frac{1}{2})$ (i) Arcsec $(-\sqrt{2})$

    Ans. (a) $\pi/3$ (b) $3\pi/4$ (c) $\pi/6$ (d) $\pi/4$ (e) $-\pi/6$ (f) $2\pi/3$
                  (g) $-\pi/3$ (h) $\pi/2$ (i) $-3\pi/4$ (j) $-\pi/2$

32.20 Evaluate each the following:

(a) $\sin[\text{Arcsin}(-\frac{1}{2})$ (f) $\sin(\text{Arccos} \frac{4}{5})$ (k) Arctan (cot 230°)

(b) $\cos(\text{Arccos} \sqrt{3}/2)$ (g) $\cos[\text{Arcsin}(-\frac{12}{13})]$ (l) Arccot (tan 100°)

(c) $\tan[\text{Arctan}(-1)]$ (h) $\sin(\text{Arctan} 2)$ (m) $\sin(2 \text{ Arcsin} \frac{2}{3})$

(d) $\sin[\text{Arccos}(-\sqrt{3}/2)]$ (i) Arccos (sin 220°)   (n) $\cos(2 \text{ Arcsin} \frac{3}{5})$

(e) $\tan(\text{Arcsin} 0)$ (j) Arcsin [cos $(-105°)$]   (o) $\sin(\frac{1}{2} \text{ Arccos} \frac{4}{5})$

    Ans. (a) $-\frac{1}{2}$ (f) $\frac{3}{5}$ (k) 40°

             (b) $\sqrt{3}/2$ (g) $\frac{5}{13}$ (l) 170°

             (c) -1 (h) $2/\sqrt{5}$ (m) $4\sqrt{5}/9$

             (d) $\frac{1}{2}$ (i) 130°   (n) $\frac{7}{25}$

             (e) 0   (j) $-15°$ (o) $1/\sqrt{10}$

32.21 Show that

(a) $\sin(\text{Arcsin} \frac{5}{13} + \text{Arcsin} \frac{4}{5}) = \frac{63}{65}$

(b) $\cos(\text{Arccos} \frac{15}{17} - \text{Arccos} \frac{7}{25}) = \frac{297}{425}$

(c) $\sin\left(\text{Arcsin} \frac{1}{2} - \text{Arccos} \frac{1}{3}\right) = \frac{1 - 2\sqrt{6}}{6}$

(d) $\tan(\text{Arctan} \frac{3}{4} + \text{Arccot} \frac{15}{3}) = \frac{77}{36}$

(e) $\cos\left(\text{Arctan} \frac{-4}{3} + \text{Arcsin} \frac{12}{13}\right) = \frac{63}{65}$

(f) $\tan\left(\text{Arcsin} \frac{-3}{5} - \text{Arccos} \frac{5}{13}\right) = \frac{63}{16}$

(g) $\tan\left(2 \text{ Arcsin} \frac{4}{5} + \text{Arccos} \frac{12}{13}\right) = -\frac{253}{204}$

(h) $\sin(2 \text{ Arcsin} \frac{4}{5} - \text{Arccos} \frac{12}{13}) = \frac{323}{325}$

32.22 Show that

(a) $\text{Arctan } \dfrac{1}{2} + \text{Arctan } \dfrac{1}{3} = \dfrac{\pi}{4}$ (e) $\text{Arccos } \dfrac{12}{13} + \text{Arctan } \dfrac{1}{4} = \text{Arccot } \dfrac{43}{32}$

(b) $\text{Arcsin } \dfrac{4}{5} + \text{Arctan } \dfrac{3}{4} = \dfrac{\pi}{4}$ (f) $\text{Arcsin } \dfrac{3}{5} + \text{Arcsin } \dfrac{15}{17} = \text{Arccos } \dfrac{-13}{85}$

(c) $\text{Arctan } \dfrac{4}{3} - \text{Arctan } \dfrac{1}{7} = \dfrac{\pi}{4}$ (g) $\text{Arctan } \alpha + \text{Arctan } \dfrac{1}{\alpha} = \dfrac{\pi}{2}$ $(a > 0)$

(d) $2 \text{ Arctan } \dfrac{1}{3} + \text{Arctan } \dfrac{1}{7} = \dfrac{\pi}{4}$

32.23 Prove: The area of the segment cut from a circle of radius r by a chord at a distance d from the center is given by $K = r^2 \text{ Arccos } \dfrac{d}{r} - d\sqrt{r^2 - d^2}$.

32.24 Determine whether the following functions possess an inverse function:

(a) $y = 5x - 3$ (b) $y = \sqrt{x}$ (c) $y = x^4$ (d) $y = x^5 - 6$

Ans. (a) yes   (b) yes   (c) no   (d) yes

Chapter 33

Trigonometric Equations

TRIGONOMETRIC EQUATIONS, i.e., equations involving trigonometric functions of unknown angles, are called

(a) Identical equations or *identities*, if they are satisfied by all values of the unknown angles for which the functions are defined.

(b) Conditional equations, or equations, if they are satisfied only by particular values of the unknown angles.

For example,

(a) $\sin x \csc x = 1$ is an identity, being satisfied by every value of x for which $\csc x$ is defined.

(b) $\sin x = 0$ is a conditional equation since it is not satisfied by $x = \frac{1}{4}\pi$ or $\frac{1}{2}\pi$.

Hereafter in this chapter we shall use the term "equation" instead of "conditional equation."

A SOLUTION OF A TRIGONOMETRIC EQUATION, as $\sin x = 0$, is a value of the angle x which satisfies the equation. Two solutions of $\sin x = 0$ are $x = 0$ and $x = \pi$.

If a given equation has one solution, it has in general an unlimited number of solutions. Thus, the complete solution of $\sin x = 0$ is given by

$$x = 0 + 2n\pi, \qquad x = \pi + 2\pi$$

where n is any positive or negative integer or is zero.

In this chapter we shall list only the particular solutions for which $0 \le x < 2\pi$.

PROCEDURES FOR SOLVING TRIGONOMETRIC EQUATIONS. There is no general method for solving trigonometric equations. Three standard procedures are illustrated below and other procedures are introduced in the solved problems.

(A) The equation may be factorable.

EXAMPLE 1. Solve $\sin x - 2\sin x \cos x = 0$.

Factoring, $\sin x - 2\sin x \cos x = \sin x(1 - 2\cos x) = 0$, and setting each factor equal to zero, we have

$$\sin x = 0 \quad \text{and} \quad x = 0, \pi; \quad 1 - 2\cos x = 0 \quad \text{or} \quad \cos x = \tfrac{1}{2} \quad \text{and} \quad x = \frac{\pi}{3}, \frac{5\pi}{3}.$$

Check.   For $x = 0$, $\sin x - 2 \sin x \cos x = 0 - 2(0)(1) = 0$;
         for $x = \pi/3$, $\sin x - 2 \sin x \cos x = \frac{1}{2}\sqrt{3} - 2(\frac{1}{2}\sqrt{3})(\frac{1}{2}) = 0$;
         for $x = \pi$, $\sin x - 2 \sin x \cos x = 0 - 2(0)(-1) = 0$;
         for $x = 5\pi/3$, $\sin x - 2 \sin x \cos x = -\frac{1}{2}\sqrt{3} - 2(-\frac{1}{2}\sqrt{3})(\frac{1}{2}) = 0$.

Thus, the required solutions ($0 \le x < 2\pi$) are $x = 0, \pi/3, \pi, 5\pi/3$.

(B)   The various functions occurring in the equation may be expressed in terms of a single function.

EXAMPLE 2.   Solve $2 \tan^2 x + \sec^2 x = 2$.
Replacing $\sec^2 x$ by $1 + \tan^2 x$, we have $2 \tan^2 x + (1 + \tan^2 x) = 2, 3 \tan^2 x = 1$, and $\tan x = \pm 1\sqrt{3}$. From $\tan x = 1\sqrt{3}, x = \pi/6$ and $7\pi/6$; from $\tan x = -1/\sqrt{3}, x = 5\pi/6$ and $11\pi/6$. After checking each of these values in the original equation, we find that the required solutions ($0 \le x < 2\pi$) are $x = \pi/6, 5\pi/6, 7\pi/6, 11\pi/6$.
The necessity of the check is illustrated in Example 3.

EXAMPLE 3.   Solve $\sec x + \tan x = 0$.
Multiplying the equation $\sec x + \tan x = \frac{1}{\cos x} + \frac{\sin x}{\cos x} = 0$ by $\cos x$, we have $1 + \sin x = 0$ or $\sin x = -1$; then $x = 3\pi/2$. However, neither $\sec x$ nor $\tan x$ is defined when $x = 3\pi/2$ and the equation has no solution.

(C)   Both members of the equation are squared.

EXAMPLE 4.   Solve $\sin x + \cos x = 1$.
If the procedure of (B) were used, we would replace $\sin x$ by $\pm\sqrt{1 - \cos^2 x}$ or $\cos x$ by $\pm\sqrt{1 - \sin^2 x}$ and thereby introduce radicals. To avoid this, we write the equation in the form $\sin x = 1 - \cos x$ and square both members. We have

$$\sin^2 x = 1 - 2 \cos x + \cos^2 x$$
$$1 - \cos^2 x = 1 - 2 \cos x + \cos^2 x \tag{33.1}$$
$$2 \cos^2 x - 2 \cos x = 2 \cos x(\cos x - 1) = 0$$

From $\cos x = 0, x = \pi/2, 3\pi/2$; from $\cos x = 1, x = 0$.

Check.   For $x = 0$, $\sin x + \cos x = 0 + 1 = 1$;
         for $x = \pi/2$, $\sin x + \cos x = 1 + 0 = 1$;
         for $x = 3\pi/2$, $\sin x + \cos x = -1 + 0 \ne 1$.

Thus, the required solutions are $x = 0, \pi/2$.
The value $x = 3\pi/2$, called an *extraneous solution*, was introduced by squaring the two members. Note that (33.1) is also obtained when both members of $\sin x = \cos x - 1$ are squared and that $x = 3\pi/2$ satisfies this latter relation.

Solved Problems

Solve each of the trigonometric equations 33.1–33.22 for all x such that $0 \le x < 2\pi$. (If all solutions are required, adjoin $+2n\pi$, where n is zero or any positive or negative integer, to each result given.) In a number of solutions, the details of the check have been omitted.

33.1 $2 \sin x - 1 = 0$.

Here $\sin x = \frac{1}{2}$ and $x = \pi/6, 5\pi/6$.

33.2 $\sin x \cos x = 0$.

From $\sin x = 0$, $x = 0, \pi$; from $\cos x = 0, x = \pi/2, 3\pi/2$. The required solutions are $x = 0, \pi/2, \pi, 3\pi/2$.

33.3 $(\tan x - 1)(4\sin^2 x - 3) = 0$.

From $\tan x - 1 = 0$, $\tan x = 1$ and $x = \pi/4, 5\pi/4$; from $4\sin^2 x - 3 = 0$, $\sin x = \pm\sqrt{3}/2$ and $x = \pi/3$, $2\pi/3, 4\pi/3, 5\pi/3$. The required solutions are $x = \pi/4, \pi/3, 2\pi/3, 5\pi/4, 4\pi/3, 5\pi/3$.

33.4 $\sin^2 x + \sin x - 2 = 0$.

Factoring, $(\sin x + 2)(\sin x - 1) = 0$.
From $\sin x + 2 = 0$, $\sin x = -2$ and there is no solution; from $\sin x - 1 = 0$, $\sin x = 1$ and $x = \pi/2$.
The required solution is $x = \pi/2$.

33.5 $3\cos^2 x = \sin^2 x$.

First Solution.   Replacing $\sin^2 x$ by $1 - \cos^2 x$, we have $3\cos^2 x = 1 - \cos^2 x$ or $4\cos^2 x = 1$. Then $\cos x = \pm\frac{1}{2}$ and the required solutions are $x = \pi/3, 2\pi/3, 4\pi/3, 5\pi/3$.

Second Solution.   Dividing the equation by $\cos^2 x$, we have $3 = \tan^2 x$. Then $\tan x = \pm\sqrt{3}$ and the solutions above are obtained.

33.6 $2\sin x - \csc x = 1$.

Multiplying the equation by $\sin x$, $2\sin^2 x - 1 = \sin x$, and rearranging, we have

$$2\sin^2 x - \sin x - 1 = (2\sin x + 1)(\sin x - 1) = 0$$

From $2\sin x + 1 = 0$, $\sin x = -\frac{1}{2}$ and $x = 7\pi/6, 11\pi/6$; from $\sin x = 1$, $x = \pi/2$.
Check.   For $x = \pi/2$, $2\sin x - \csc x = 2(1) - 1 = 1$;
    for $x = 7\pi/6$ and $11\pi/6$, $2\sin x - \csc x = 2(-\frac{1}{2}) - (-2) = 1$.
The solutions are $x = \pi/2, 7\pi/6, 11\pi/6$.

33.7 $2\sec x = \tan x + \cot\ x$.

Transforming to sines and cosines, and clearing fractions, we have

$$\frac{2}{\cos x} = \frac{\sin x}{\cos x} + \frac{\cos x}{\sin x} \quad \text{or} \quad 2\sin x = \sin^2 x + \cos^2 x = 1$$

Then $\sin x = \frac{1}{2}$ and $x = \pi/6, 5\pi/6$.

33.8 $\tan x + 3\cot x = 4$.

Multiplying by $\tan x$ and rearranging, $\tan^2 x - 4\tan x + 3 = (\tan x - 1)(\tan x - 3) = 0$.
From $\tan x - 1 = 0$, $\tan x = 1$ and $x = \pi/4, 5\pi/4$; from $\tan x - 3 = 0$, $\tan x = 3$ and $x = 71°34', 251°34'$.
Check.   For $x = \pi/4$ and $5\pi/4$, $\tan x + 3\cot x = 1 + 3(1) = 4$;
    for $x = 71°34'$ and $x - 251°34'$, $\tan x + 3\cot x = 3 + 3(\frac{1}{3}) = 4$.
The solutions are $45°, 71°34', 251°, 251°34'$.

33.9 $\csc\ x + \cot x = \sqrt{3}$.

First Solution.   Writing the equation in the form $\csc x = \sqrt{3} - \cot x$ and squaring, we have

$$\csc^2\ x = 3 - 2\sqrt{3}\cot x + \cot^2\ x$$

Replacing $\csc^2 x$ by $1 + \cot^2 x$ and combining, this becomes $2\sqrt{3}\cot x - 2 = 0$. Then $\cot x = 1/\sqrt{3}$ and $x = \pi/3, 4\pi/3$.

Check.   For $x = \pi/3$, $\csc x + \cot\ x = 2/\sqrt{3} + 1/\sqrt{3} = \sqrt{3}$;
    for $x = 4\pi/3$, $\csc x + \cot\ x = -2/\sqrt{3} + 1/\sqrt{3} \neq \sqrt{3}$. The required solution is $x = \pi/3$.

Second Solution.   Upon making the indicated replacement, the equation becomes

$$\frac{1}{\sin x} + \frac{\cos x}{\sin x} = \sqrt{3}$$

and, clearing of fractions, $1 + \cos x = \sqrt{3}\sin x$.

Squaring both members, we have $1 + 2\cos x + \cos^2 x = 3\sin^2 x = 3(1 - \cos^2 x)$ or

$$4\cos^2 x + 2\cos x - 2 = 2(2\cos x - 1)(\cos x + 1) = 0$$

From $2\cos x - 1 = 0$, $\cos x = \frac{1}{2}$ and $x = \pi/3, 5\pi/3$; from $\cos x + 1 = 0$, $\cos x = -1$ and $x = \pi$.

Now $x = \pi/3$ is the solution. The values $x = \pi$ and $5\pi/3$ are to be excluded since $\csc \pi$ is not defined while $\csc 5\pi/3$ and $\cot 5\pi/3$ are both negative.

33.10 $\cos x - \sqrt{3}\sin x = 1$.

First Solution. Putting the equation in the form $\cos x - 1 = \sqrt{3}\sin x$ and squaring, we have

$$\cos^2 x - 2\cos x + 1 = 3\sin^2 x = 3(1 - \cos^2 x)$$

Then, combining and factoring,

$$4\cos^2 x - 2\cos x - 2 = 2(2\cos x + 1)(\cos x - 1) = 0$$

From $2\cos x + 1 = 0, \cos x = -\frac{1}{2}$ and $x = 2\pi/3, 4\pi/3$; from $\cos x - 1, \cos x = 1$ and $x = 0$.

Check. For $x = 0$, $\cos x - \sqrt{3}\sin x = 1 - \sqrt{3}(0) = 1$;
  for $x = 2\pi/3$, $\cos x - \sqrt{3}\sin x = -\frac{1}{2} - \sqrt{3}(\sqrt{3}/2) \neq 1$;
  for $x = 4\pi/3$, $\cos x - \sqrt{3}\sin x = -\frac{1}{2} - \sqrt{3}(-\sqrt{3}/2) = 1$.

The required solutions are $x = 0, 4\pi/3$.

Second Solution. The left member of the given equation may be put in the form

$$\sin\theta\cos x + \cos\theta\sin x = \sin(\theta + x)$$

in which θ is a known angle, by dividing the given equation by $r > 0$, $\frac{1}{r}\cos x + \left(\frac{-\sqrt{3}}{r}\right)\sin x = \frac{1}{r}$, and setting $\sin\theta = \frac{1}{r}$ and $\cos\theta = \frac{-\sqrt{3}}{r}$. Since $\sin^2\theta + \cos^2\theta = 1$, $\left(\frac{1}{r}\right)^2 + \left(\frac{-\sqrt{3}}{r}\right)^2 = 1$ and $r = 2$. Now $\sin\theta = \frac{1}{2}, \cos\theta = -\sqrt{3}/2$ so that the given equation may be written as $\sin(\theta + x) = \frac{1}{2}$ with $\theta = 5\pi/6$. Then $\theta + x = 5\pi/6 + x = \arcsin\frac{1}{2} = \pi/6, 5\pi/6, 13\pi/6, 17\pi/6, \ldots$ and $x = -2\pi/3, 0, 4\pi/3, 2\pi, \ldots$. As before, the required solutions are $x = 0, 4\pi/3$.

Note that r is the positive square root of the sum of the squares of the coefficients of $\cos x$ and $\sin x$ when the equation is written in the form $a\cos x + b\sin x = c$; that is,

$$r = \sqrt{a^2 + b^2}$$

The equation will have no solution if $\dfrac{c}{\sqrt{a^2 + b^2}}$ is greater than 1 or less than -1.

33.11 $2\cos x = 1 - \sin x$.

First Solution. As in Problem 33.10, we obtain

$$4\cos^2 x = 1 - 2\sin x + \sin^2 x$$
$$4(1 - \sin^2 x) = 1 - 2\sin x + \sin^2 x$$
$$5\sin^2 x - 2\sin x - 3 = (5\sin x + 3)(\sin x - 1) = 0$$

From $5\sin x + 3 = 0$, $\sin x = -\frac{3}{5} = -0.6000$ and $x = 216°52', 323°8'$; from $\sin x - 1 = 0$, $\sin x = 1$ and $x = \pi/2$.

Check. For $x = \pi/2, 2(0) = 1 - 1$;
  for $x = 216°52', 2(-\frac{4}{5}) \neq 1 - (-\frac{3}{5})$;
  for $x = 323°8', 2(\frac{4}{5}) = 1 - (-\frac{3}{5})$.

The required solutions are $x = 90°, 323°8'$.

Second Solution. Writing the equation as $2\cos x + \sin x = 1$ and dividing by $r = \sqrt{2^2 + 1^2} = \sqrt{5}$, we have

$$\frac{2}{\sqrt{5}}\cos x + \frac{1}{\sqrt{5}}\sin x = \frac{1}{\sqrt{5}} \qquad\qquad (1)$$

Let $\sin\theta = 2/\sqrt{5}, \cos\theta = 1/\sqrt{5}$; then ($1$) becomes

$$\sin\theta\cos x + \cos\theta\sin x = \sin(\theta + x) = \frac{1}{\sqrt{5}}$$

with $\theta = 63°26'$. Now $\theta + x = 63°26' + x = \arcsin(1/\sqrt{5}) = \arcsin 0.4472 = 26°34', 153°26', 386°34', \ldots$ and $x = 90°, 323°8'$ as before.

EQUATIONS INVOLVING MULTIPLE ANGLES

33.12 $\sin 3x = -\frac{1}{2}\sqrt{2}$.

Since we require x such that $0 \le x < 2\pi, 3x$ must be such that $0 \le 3x < 6\pi$. Then $3x = 5\pi/4, 7\pi/4, 13\pi/4, 15\pi/4, 21\pi/4, 23\pi/4$ and $x = 5\pi/12, 7\pi/12, 13\pi/12, 5\pi/4, 7\pi/4, 23\pi/12$. Each of these values is a solution.

33.13 $\cos\frac{1}{2}x = \frac{1}{2}$.

Since we require x such that $0 \le x < 2\pi$, $\frac{1}{2}x$ must be such that $0 \le \frac{1}{2}x < \pi$. Then $\frac{1}{2}x = \pi/3$ and $x = 2\pi/3$.

33.14 $\sin 2x + \cos x = 0$.

Substituting for $\sin 2x$, we have $2\sin x\cos x + \cos x = \cos x(2\sin x + 1) = 0$. From $\cos x = 0, x = \pi/2, 3\pi/2$; from $\sin x = -\frac{1}{2}$, $x = 7\pi/6, 11\pi/6$. The required solutions are $x = \pi/2, 7\pi/6, 3\pi/2, 11\pi/6$.

33.15 $2\cos^2\frac{1}{2}x = \cos^2 x$.

Substituting $1 + \cos x$ for $2\cos^2\frac{1}{2}x$, the equation becomes $\cos^2 x - \cos x - 1 = 0$; then $\cos x = \dfrac{1 \pm \sqrt{5}}{2} = 1.6180, -0.6180$. Since $\cos x$ cannot exceed 1, we consider $\cos x = -0.6180$ and obtain the solutions $x = 128°10', 231°50'$.

(NOTE: To solve $\sqrt{2}\cos\frac{1}{2}x = \cos x$ and $\sqrt{2}\cos\frac{1}{2}x = -\cos x$, we square and obtain the equation of this problem. The solution of the first of these equations is $231°50'$ and the solution of the second is $128°10'$.)

33.16 $\cos 2x + \cos x + 1 = 0$.

Substituting $2\cos^2 x - 1$ for $\cos 2x$, we have $2\cos^2 x + \cos x = \cos x(2\cos x + 1) = 0$. From $\cos x = 0$, $x = \pi/2, 3\pi/2$; from $\cos x = -\frac{1}{2}, x = 2\pi/3, 4\pi/3$. The required solutions are $x = \pi/2, 2\pi/3, 3\pi/2, 4\pi/3$.

33.17 $\tan 2x + 2\sin x = 0$.

Using $\tan 2x = \dfrac{\sin 2x}{\cos 2x} = \dfrac{2\sin x\cos x}{\cos 2x}$, we have

$$\frac{2\sin x\cos x}{\cos 2x} + 2\sin x = 2\sin x\left(\frac{\cos x}{\cos 2x} + 1\right) = 2\sin x\left(\frac{\cos x + \cos 2x}{\cos 2x}\right) = 0$$

From $\sin x = 0$, $x = 0, \pi$; from $\cos x + \cos 2x = \cos x + 2\cos^2 x - 1 = (2\cos x - 1)(\cos x + 1) = 0$, $x = \pi/3, 5\pi/3$, and π. The required solutions are $x = 0, \pi/3, \pi, 5\pi/3$.

33.18 $\sin 2x = \cos 2x$.

First Solution.   Let $2x = \theta$; then we are to solve $\sin\theta = \cos\theta$ for $0 \le \theta < 4\pi$. Then $\theta = \pi/4, 5\pi/4, 9\pi/4, 13\pi/4$ and $x = \theta/2 = \pi/8, 5\pi/8, 9\pi/8, 13\pi/8$ are the solutions.

Second Solution.   Dividing by $\cos 2x$, the equation becomes $\tan 2x = 1$ for which $2x = \pi/4, 5\pi/4, 9\pi/4, 13\pi/4$ as in the first solution.

33.19 $\sin 2x = \cos 4x$.

Since $\cos 4x = \cos 2(2x) = 1 - 2\sin^2 2x$, the equation becomes

$$2\sin^2 2x + \sin 2x - 1 = (2\sin 2x - 1)(\sin 2x + 1) = 0$$

From $2\sin 2x - 1 = 0$ or $\sin 2x = \frac{1}{2}$, $2x = \pi/6, 5\pi/6, 13\pi/6, 17\pi/6$ and $x = \pi/12, 5\pi/12, 13\pi/12, 17\pi/12$; from $\sin 2x + 1 = 0$ or $\sin 2x = -1$, $2x = 3\pi/2, 7\pi/2$ and $x = 3\pi/4, 7\pi/4$. All of these values are solutions.

33.20 $\sin 3x = \cos 2x$.

To avoid the substitution for $\sin 3x$, we use one of the procedures below.
First Solution.   Since $\cos 2x = \sin(\frac{1}{2}\pi - 2x)$ and also $\cos 2x = \sin(\frac{1}{2}\pi + 2x)$, we consider

(a) $\sin 3x = \sin(\frac{1}{2}\pi - 2x)$, obtaining $3x = \pi/2 - 2x, 5\pi/2 - 2x, 9\pi/2 - 2x, \dots$.

(b) $\sin 3x = \sin(\frac{1}{2}\pi + 2x)$, obtaining $3x = \pi/2 + 2x, 5\pi/2 + 2x, 9\pi/2 + 2x, \dots$

From (a), $5x = \pi/2, 5\pi/2, 9\pi/2, 13\pi/2, 17\pi/2$ (since $5x < 10\pi$); and from (b) $x = \pi/2$. The required solutions are $x = \pi/10, \pi/2, 9\pi/10, 13\pi/10, 17\pi/10$.
Second Solution.   Since $\sin 3x = \cos(\frac{1}{2}\pi - 3x)$ and $\cos 2x = \cos(-2x)$, we consider

(c) $\cos 2x = \cos(\frac{1}{2}\pi - 3x)$, obtaining $5x = \pi/2, 5\pi/2, 9\pi/2, 13\pi/2, 17\pi/2$

(d) $\cos(-2x) = \cos(\frac{1}{2}\pi - 3x)$, obtaining $x = \pi/2$, as before

33.21 $\tan 4x = \cot 6x$.

Since $\cot 6x = \tan(\frac{1}{2}\pi - 6x)$, we consider the equation $\tan 4x = \tan(\frac{1}{2}\pi - 6x)$. Then $4x = \pi/2 - 6x$, $3\pi/2 - 6x$, $5\pi/2 - 6x, \dots$, the function $\tan\theta$ being of period π. Thus, $10x = \pi/2, 3\pi/2, 5\pi/2, 7\pi/2$, $9\pi/2, \dots, 39\pi/2$ and the required solutions are $x = \pi/20, 3\pi/20, \pi/4, 7\pi/20, \dots, 39\pi/20$.

33.22 $\sin 5x - \sin 3x - \sin x = 0$.

Replacing $\sin 5x - \sin 3x$ by $2\cos 4x \sin x$ (Chapter 28), the given equation becomes

$$2\cos 4x \sin x - \sin x = \sin x(2\cos 4x - 1) = 0$$

From $\sin x = 0$, $x = 0, \pi$; from $\cos 4x - 1 = 0$ or $\cos 4x = \frac{1}{2}$, $4x = \pi/3, 5\pi/3, 7\pi/3, 11\pi/3, 13\pi/3, 17\pi/3$, $19\pi/3, 23\pi/3$ and $x = \pi/12, 5\pi/12, 7\pi/12, 11\pi/12, 13\pi/12, 17\pi/12, 19\pi/12, 23\pi/12$. Each of the values obtained is a solution.

33.23   Solve the system

$$\begin{array}{ll} r\sin\theta = 2 & \qquad(1) \\ r\cos\theta = 3 & \quad\text{for } r > 0 \quad\text{and}\quad 0 \le \theta < 2\pi \qquad(2) \end{array}$$

Squaring the two equations and adding, $r^2\sin\theta + r^2\cos\theta = r^2 = 13$ and $r = \sqrt{13} = 3.606$.
When $r > 0$, $\sin\theta$ and $\cos\theta$ and both > 0 and θ is acute. Dividing (1) by (2), $\tan\theta = \frac{2}{3} = 0.6667$ and $\theta = 33°41'$.

33.24   Solve the system

$$\begin{array}{ll} r\sin\theta = 3 & \qquad(1) \\ r = 4(1 + \sin\theta) & \quad\text{for } r > 0 \quad\text{and}\quad 0 \le \theta < 2\pi \qquad(2) \end{array}$$

Dividing (2) by (1),

$$\frac{1}{\sin\theta} = \frac{4(1 + \sin\theta)}{3} \quad\text{or}\quad 4\sin^2\theta + 4\sin\theta - 3 = 0$$

and $(2\sin\theta + 3)(2\sin\theta - 1) = 0$

From $2\sin\theta - 1 = 0$, $\sin\theta = \frac{1}{2}$, $\theta = \pi/6$ and $5\pi/6$; using (1), $r(\frac{1}{2}) = 3$ and $r = 6$. Note that $2\sin\theta + 3 = 0$ is excluded since when $r > 0$, $\sin\theta > 0$ by (1). The required solutions are $\theta = \pi/6, r = 6$ and $\theta = 5\pi/6, r = 6$.

33.25 Solve the system

$$\sin x + \sin y = 1.2 \qquad (1)$$
$$\cos x + \cos y = 1.5 \qquad \text{for } 0 \le x, \ y < 2\pi \qquad (2)$$

Since each sum on the left is greater than 1, each of the four functions is positive and both x and y are acute.

Using the appropriate formulas of Chapter 28, we obtain

$$2 \sin\tfrac{1}{2}(x+y)\cos\tfrac{1}{2}(x-y) = 1.2 \qquad (1')$$
$$2 \cos\tfrac{1}{2}(x+y)\cos\tfrac{1}{2}(x-y) = 1.5 \qquad (2')$$

Dividing $(1')$ by $(2')$, $\dfrac{\sin\tfrac{1}{2}(x+y)}{\cos\tfrac{1}{2}(x+y)} = \tan\tfrac{1}{2}(x+y) = \dfrac{1.2}{1.5} = 0.8000$ and $\tfrac{1}{2}(x+y) = 38°40'$ since $\tfrac{1}{2}(x+y)$ is also acute. Substituting for $\sin\tfrac{1}{2}(x+y) = 0.6248$ in $(1')$, we have $\cos\tfrac{1}{2}(x-y) = \dfrac{0.6}{0.6248} = 0.9603$ and $\tfrac{1}{2}(x-y) = 16°12'$.

Then $x = \tfrac{1}{2}(x+y) + \tfrac{1}{2}(x-y) = 54°52'$ and $y = \tfrac{1}{2}(x+y) - \tfrac{1}{2}(x-y) = 22°28'$.

33.26 Solve $\operatorname{Arccos} 2x = \operatorname{Arcsin} x$.

If x is positive, $\alpha = \operatorname{Arccos} 2x$ and $\beta = \operatorname{Arcsin} x$ terminate in quadrant I; if x is negative, α terminates in quadrant II and β terminates in quadrant IV. Thus, x must be positive.

For x positive, $\sin\beta = x$ and $\cos\beta = \sqrt{1-x^2}$. Taking the cosine of both members of the given equation, we have

$$\cos(\operatorname{Arccos} 2x) = \cos(\operatorname{Arcsin} x) = \cos\beta \quad \text{or} \quad 2x = \sqrt{1-x^2}$$

Squaring, $4x^2 = 1 - x^2, 5x^2 = 1$, and $x = \sqrt{5}/5 = 0.4472$.

Check. $\operatorname{Arccos} 2x = \operatorname{Arccos} 0.8944 = 26°30' = \operatorname{Arcsin} 0.4472$, approximating the angle to the nearest $10'$.

33.27 Solve $\operatorname{Arccos}(2x^2 - 1) = 2\operatorname{Arccos}\tfrac{1}{2}$.

Let $\alpha = \operatorname{Arccos}(2x^2 - 1)$ and $\beta = \operatorname{Arccos}\tfrac{1}{2}$; then $\cos\alpha = 2x^2 - 1$ and $\cos\beta = \tfrac{1}{2}$.

Taking the cosine of both members of the given equation,

$$\cos\alpha = 2x^2 - 1 = \cos 2\beta = 2\cos^2\beta - 1 = 2(\tfrac{1}{2})^2 - 1 = -\tfrac{1}{2}$$

Then $2x^2 = \tfrac{1}{2}$ and $x = \pm\tfrac{1}{2}$.

Check. For $x = \pm\tfrac{1}{2}$, $\operatorname{Arccos}(-\tfrac{1}{2}) = 2\operatorname{Arccos}\tfrac{1}{2}$ or $120° = 2(60°)$.

33.28 Solve $\operatorname{Arccos} 2x - \operatorname{Arccos} x = \pi/3$.

If x is positive, $0 < \operatorname{Arccos} 2x < \operatorname{Arccos} x$; if x is negative, $\operatorname{Arccos} 2x > \operatorname{Arccos} x > 0$. Thus, x must be negative.

Let $\alpha = \operatorname{Arccos} 2x$ and $\beta = \operatorname{Arccos} x$; then $\cos\alpha = 2x, \sin\alpha = \sqrt{1-4x^2}, \cos\beta = x$, and $\sin\beta = \sqrt{1-x^2}$ since both α and β terminate in quadrant II.

Taking the cosine of both members of the given equation,

$$\cos(\alpha-\beta) = \cos\alpha\cos\beta + \sin\alpha\sin\beta = 2x^2 + \sqrt{1-4x^2}\,\sqrt{1-x^2} = \cos\frac{\pi}{3} = \frac{1}{2}$$

or

$$\sqrt{1-4x^2}\,\sqrt{1-x^2} = \tfrac{1}{2} - 2x^2$$

Squaring, $1 - 5x^2 + 4x^4 = \tfrac{1}{4} - 2x^2 + 4x^4, 3x^2 = \tfrac{3}{4}$, and $x = -\tfrac{1}{2}$.

Check. $\operatorname{Arccos}(-1) - \operatorname{Arccos}(-\tfrac{1}{2}) = \pi - 2\pi/3 = \pi/3$.

33.29 Solve $\text{Arcsin } 2x = \frac{1}{4}\pi - \text{Arcsin } x$.

Let $\alpha = \text{Arcsin } 2x$ and $\beta = \arcsin x$; then $\sin \alpha = 2x$ and $\sin \beta = x$. If x is negative, α and β terminate in quadrant IV; thus, x must be positive and β acute.

Taking the sine of both members of the given equation,

$$\sin \alpha = \sin(\tfrac{1}{4}\pi - \beta) = \sin\tfrac{1}{4}\pi \cos \beta - \cos\tfrac{1}{4}\pi \sin \beta$$

or $$2x = \tfrac{1}{2}\sqrt{2}\sqrt{1-x^2} - \tfrac{1}{2}\sqrt{2}x \quad \text{and} \quad (2\sqrt{2}+1)x = \sqrt{1-x^2}$$

Squaring, $(8 + 4\sqrt{2} + 1)x^2 = 1 - x^2$, $x^2 = 1/(10 + 4\sqrt{2})$, and $x = 0.2527$.

Check. $\text{Arcsin } 0.5054 = 30°22'$; $\text{Arcsin } 0.2527 = 14°38'$, and $\frac{1}{4}\pi - 14°38' = 30°22'$.

33.30 Solve $\text{Arctan } x + \text{Arctan } (1 - x) = \text{Arctan } \frac{4}{3}$.

Let $\alpha = \text{Arctan } x$ and $\beta = \text{Arctan } (1 - x)$; then $\tan \alpha = x$ and $\tan \beta = 1 - x$. Taking the tangent of both members of the given equation,

$$\tan (\alpha + \beta) = \frac{\tan \alpha + \tan \beta}{1 - \tan \alpha \tan \beta} = \frac{x + (1-x)}{1 - x(1-x)} = \frac{1}{1 - x + x^2} = \tan (\text{Arctan } \tfrac{4}{3}) = \tfrac{4}{3}$$

Then $3 = 4 - 4x + 4x^2$, $4x^2 - 4x + 1 = (2x - 1)^2 = 0$, and $x = \frac{1}{2}$.

Check. $\text{Arctan } \frac{1}{2} + \text{Arctan } (1 - \frac{1}{2}) = 2 \text{ Arctan } 0.5000 = 53°8'$ and $\text{Arctan } \frac{4}{3} = \text{Arctan } 1.3333 = 53°8'$.

Supplementary Problems

Solve each of the following equations for all x such that $0 \le x < 2\pi$:

33.31 $\sin x = \sqrt{3}/2$.   *Ans.* $\pi/3, 2\pi/3$

33.32 $\cos^2 x = \frac{1}{2}$.   *Ans.* $\pi/4, 3\pi/4, 5\pi/4, 7\pi/4$

33.33 $\sin x \cos x = 0$.   *Ans.* $0, \pi/2, \pi, 3\pi/2$

33.34 $(\tan x - 1)(2 \sin x + 1) = 0$.   *Ans.* $\pi/4, 7\pi/6, 5\pi/4, 11\pi/6$

33.35 $2 \sin^2 x - \sin x - 1 = 0$.   *Ans.* $\pi/2, 7\pi/6, 11\pi/6$

33.36 $\sin 2x + \sin x = 0$.   *Ans.* $0, 2\pi/3, \pi, 4\pi/3$

33.37 $\cos x + \cos 2x = 0$.   *Ans.* $\pi/3, \pi, 5\pi/3$

33.38 $2 \tan x \sin x - \tan x = 0$.   *Ans.* $0, \pi/6, 5\pi/6, \pi$

33.39 $2 \cos x + \sec x = 3$.   *Ans.* $0, \pi/3, 5\pi/3$

33.40 $2 \sin x + \csc x = 3$.   *Ans.* $\pi/6, \pi/2, 5\pi/6$

33.41 $\sin x + 1 = \cos x$.   *Ans.* $0, 3\pi/2$

33.42 $\sec x - 1 = \tan x$.   *Ans.* 0

33.43 $2\cos x + 3\sin x = 2$.   *Ans.* $0°, 112°37'$

33.44 $3\sin x + 5\cos x + 5 = 0$.   *Ans.* $180°, 241°56'$

33.45 $1 + \sin x = 2\cos x$.   *Ans.* $36°52', 270°$

33.46 $3\sin x + 4\cos x = 2$.   *Ans.* $103°18', 330°27'$

33.47 $\sin 2x = -\sqrt{3}/2$.   *Ans.* $2\pi/3, 5\pi/6, 5\pi/3, 11\pi/6$

33.48 $\tan 3x = 1$.   *Ans.* $\pi/12, 5\pi/12, 3\pi/4, 13\pi/12, 17\pi/12, 7\pi/4$

33.49 $\cos x/2 = \sqrt{3}/2$.   *Ans.* $\pi/3$.

33.50 $\cot x/3 = -1/\sqrt{3}$.   *Ans.*   No solution in given interval

33.51 $\sin x \cos x = \frac{1}{2}$.   *Ans.* $\pi/4, 5\pi/4$

33.52 $\sin x/2 + \cos x = 1$.   *Ans.* $0, \pi/3, 5\pi/3$

33.53 $\sin 3x + \sin x = 0$.   *Ans.* $0, \pi/2, \pi, 3\pi/2$

33.54 $\cos 2x + \cos 3x = 0$.   *Ans.* $\pi/5, 3\pi/5, \pi, 7\pi/5, 9\pi/5$

33.55 $\sin 2x + \sin 4x = 2\sin 3x$.   *Ans.* $0, \pi/3, 2\pi/3, \pi, 4\pi/3, 5\pi/3$

33.56 $\cos 5x + \cos x = 2\cos 2x$.   *Ans.* $0, \pi/4, 2\pi/3, 3\pi/4, 5\pi/4, 4\pi/3, 7\pi/4$

33.57 $\sin x + \sin 3x = \cos x + \cos 3x$.   *Ans.* $\pi/8, \pi/2, 5\pi/8, 9\pi/8, 3\pi/2, 13\pi/8$

Solve each of the following systems for $r \geq 0$ and $0 \leq \theta < 2\pi$:

33.58 $r = a\sin\theta$ *Ans.* $\theta = \pi/6, r = a/2$
            $r = a\cos 2\theta$ $\theta = 5\pi/6, r = a/2; \theta = 3\pi/2, r = -a$

33.59 $r = a\cos\theta$ *Ans.* $\theta = \pi/2, r = 0; \theta = 3\pi/2, r = 0$
            $r = a\sin 2\theta$ $\theta = \pi/6, r = \sqrt{3}a/2$
                                                                      $\theta = 5\pi/6, r = -\sqrt{3}a/2$

33.60 $r = 4(1 + \cos \theta)$ *Ans.* $\theta = \pi/3, r = 6$

   $r = 3 \sec \theta$ $\theta = 5\pi/3,\ r = 6$

Solve each of the following equations:

33.61   Arctan $2x$ + Arctan $x = \pi/4$.   *Ans.* $x = 0.281$

33.62   Arcsin x + Arctan $x = \pi/2$.   *Ans.* $x = 0.786$

33.63   Arccos x + Arctan $x = \pi/2$.   *Ans.* $x = 0$

Chapter 34

Complex Numbers

PURE IMAGINARY NUMBERS. The square root of a negative number (i.e., $\sqrt{-1}, \sqrt{-5}, \sqrt{-9}$) is called a *pure imaginary number*. Since by definition $\sqrt{-5} = \sqrt{5} \cdot \sqrt{-1}$ and $\sqrt{-9} = \sqrt{9} \cdot \sqrt{-1} = 3\sqrt{-1}$, it is convenient to introduce the symbol $i = \sqrt{-1}$ and to adopt $\sqrt{-5} = i\sqrt{5}$ and $\sqrt{-9} = 3i$ as the standard form for these numbers.

The symbol i has the property $i^2 = -1$, and for higher integral powers we have $i^3 = i^2 \cdot i = (-1)i = -i, i^4 = (i^2)^2 = (-1)^2 = 1, i^5 = i^4 \cdot i = i$, etc.

The use of the standard form simplifies the operation on pure imaginaries and eliminates the possibility of certain common errors. Thus, $\sqrt{-9} \cdot \sqrt{4} = \sqrt{-36} = 6i$ since $\sqrt{-9} \cdot \sqrt{4} = 3i(2) = 6i$ but $\sqrt{-9} \cdot \sqrt{-4} \neq \sqrt{36}$ since $\sqrt{-9} \cdot \sqrt{-4} = (3i)(2i) = 6i^2 = -6$.

Notice the cyclic nature of the powers of i. i^m equals $i, -1, -i, 1$ for every natural number, $m \cdot i^6 = -1, i^7 = -i, i^8 = 1$, etc.

COMPLEX NUMBERS. A number $a + bi$, where a and b are real numbers, is called a *complex number*. The first term a is called the *real part* of the complex number and the second term bi is called the *pure imaginary part*.

Complex numbers may be thought of as including all real numbers and all pure imaginary numbers. For example, $5 = 5 + 0i$ and $3i = 0 + 3i$.

Two complex numbers $a + bi$ and $c + di$ are said to be *equal* if and only if $a = c$ and $b = d$.

The *conjugate* of a complex number $a + bi$ is the complex number $a - bi$. Thus, $2 + 3i$ and $2 - 3i$, $-3 + 4i$ and $-3 - 4i$ are pairs of conjugate complex numbers.

ALGEBRAIC OPERATIONS

  (*1*) **ADDITION.** To add complex numbers, add the real parts and add the pure imaginary parts.

  EXAMPLE 1. $(2 + 3i) + (4 - 5i) = (2 + 4) + (3 - 5)i = 6 - 2i$.

  (*2*) **SUBTRACTION.** To subtract two complex numbers, subtract the real parts and subtract the pure imaginary parts.

  EXAMPLE 2. $(2 + 3i) - (4 - 5i) = (2 - 4) + [3 - (-5)]i = -2 + 8i$.

  (*3*) **MULTIPLICATION.** To multiply two complex numbers, carry out the multiplication as if the numbers were ordinary binomials and replace i^2 by -1.

  EXAMPLE 3. $(2 + 3i)(4 - 5i) = 8 + 2i - 15i^2 = 8 + 2i - 15(-1) = 23 + 2i$.

(4) **DIVISION.** To divide two complex numbers, multiply both numerator and denominator of the fraction by the conjugate of the denominator.

EXAMPLE 4. $\dfrac{2+3i}{4-5i} = \dfrac{(2+3i)(4+5i)}{(4-5i)(4+5i)} = \dfrac{(8-15)+(10+12)i}{16+25} = \dfrac{7}{41} + \dfrac{22}{41}i.$

$\left[\text{Note the form of the result; it is neither } \dfrac{-7+22i}{41} \text{ nor } \dfrac{1}{41}(-7+22i). \right]$

(See Problems 34.1–34.9.)

GRAPHIC REPRESENTATION OF COMPLEX NUMBERS. The complex number $x + yi$ may be represented graphically by the point P (see Fig. 34-1) whose rectangular coordinates are (x, y).

The point O, having coordinates $(0, 0)$, represents the complex number $0 + 0i = 0$. All points on the x axis have coordinates of the form $(x, 0)$ and correspond to real numbers $x + 0i = x$. For this reason, the x axis is called the *axis of reals*. All points on the y axis have coordinates of the form $(0, y)$ and correspond to pure imaginary numbers $0 + yi = yi$. The y axis is called the *axis of imaginaries*. The plane on which the complex numbers are represented is called the *complex plane*. See Fig. 34-1.

In addition to representing a complex number by a point P in the complex plane, the number may be represented by the directed line segment or vector OP. See Fig. 34-2. The vector OP is sometimes denoted by $\overrightarrow{OP}$ and is the directed line segment beginning at O and terminating at P.

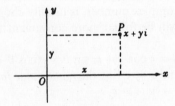

Fig. 34-1

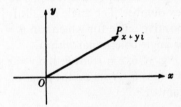

Fig. 34-2

GRAPHIC REPRESENTATION OF ADDITION AND SUBTRACTION. Let $z_1 = x_1 + iy_1$ and $z_2 = x_2 + iy_2$ be two complex numbers. The vector representation of these numbers suggests the illustrated parallelogram law for determining graphically the sum $z_1 + z_2 = (x_1 + iy_1) + (x_2 + iy_2)$, since the coordinates of the endpoint of the vector $z_1 + z_2$ must be, for each of the x coordinates and the y coordinates, the sum of the corresponding x or y values. See Fig. 34-3.

Since $z_1 - z_2 = (x_1 + iy_1) - (x_2 + iy_2) = (x_1 + iy_1) + (-x_2 - iy_2)$, the difference $z_1 - z_2$ of the two complex numbers may be obtained graphically by applying the parallelogram law to $x_1 + iy_1$ and $-x_2 - iy_2$. (See Fig. 34-4.)

In Fig. 34-5 both the sum $OR = z_1 + z_2$ and the difference $OS = z_1 - z_2$ are shown. Note that the segments $\overline{OS}$ and $\overline{P_2P_1}$ (the other diagonal of OP_2RP_1) are congruent. (See Problem 34.11.)

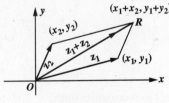

Fig. 34-3

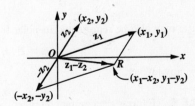

Fig. 34-4

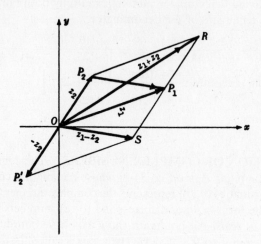

Fig. 34-5

POLAR OR TRIGONOMETRIC FORM OF COMPLEX NUMBERS.

Let the complex number $x + yi$ be represented (Fig. 34-6) by the vector OP. This vector (and hence the complex number) may be described in terms of the length r of the vector and *any* positive angle θ which the vector makes with the positive x axis (axis of positive reals). The number $r = \sqrt{x^2 + y^2}$ is called the *modulus* or *absolute value* of the complex number. The angle θ, called the *amplitude* of the complex number, is usually chosen as the smallest, positive angle for which $\tan \theta = y/x$ but at times it will be found more convenient to choose some other angle coterminal with it.

From Fig. 34-6, $x = r \cos \theta$; and $y = r \sin \theta$; then $z = x + yi = r \cos \theta + ir \sin \theta = r(\cos \theta + i \sin \theta)$. We call $z = r(\cos \theta + i \sin \theta)$ the *polar* or *trigonometric form* and $z = x + yi$ the *rectangular form* of the complex number z.

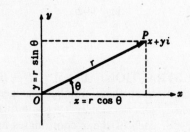

Fig. 34-6

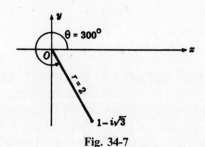

Fig. 34-7

EXAMPLE 5.   Express $z = 1 - i\sqrt{3}$ in polar form. (See Fig. 34-7.)

The modulus is $r = \sqrt{(1)^2 + (-\sqrt{3})^2} = 2$. Since $\tan \theta = y/x = -\sqrt{3}/1 = -\sqrt{3}$, the amplitude θ is either 120° or 300°. Now we know that P lies in quadrant IV; hence, $\theta = 300°$ and the required polar form is

$$z = r(\cos \theta + i \sin \theta) = 2(\cos 300° + i \sin 300°)$$

Note that z may also be represented in polar form by

$$z = 2[\cos (300° + n360°) + i \sin (300° + n360°)]$$

where n is any integer.

EXAMPLE 6.   Express the complex number $z = 8(\cos 210° + i \sin 210°)$ in rectangular form.
Since $\cos 210° = -\sqrt{3}/2$ and $\sin 210° = -\frac{1}{2}$,

$$z = 8(\cos 210° + i \sin 210°) = 8\left[-\frac{\sqrt{3}}{2} + i\left(-\frac{1}{2}\right)\right] = -4\sqrt{3} - 4i$$

is the required rectangular form.
(See Problems 34.12–34.13.)

MULTIPLICATION AND DIVISION IN POLAR FORM

MULTIPLICATION. The modulus of the product of two complex numbers is the product of their moduli, and the amplitude of the product is the sum of their amplitudes.

DIVISION. The modulus of the quotient of two complex numbers is the modulus of the dividend divided by the modulus of the divisor, and the amplitude of the quotient is the amplitude of the dividend minus the amplitude of the divisor. For a proof of these theorems, see Problem 34.14.

EXAMPLE 7. Find (a) the product $z_1 z_2$, (b) the quotient and z_1/z_2, and (c) the quotient z_2/z_1 where $z_1 = 2(\cos 300° + i \sin 300°)$ and $z_2 = 8(\cos 210° + i \sin 210°)$.

(a) The modulus of the product is $2(8) = 16$. The amplitude is $300° + 210° = 510°$, but following the convention, we shall use the smallest positive coterminal angle $510° - 360° = 150°$. Thus, $z_1 z_2 = 16(\cos 150° + i \sin 150°)$.

(b) The modulus of the quotient z_1/z_2 is $\frac{2}{8} = \frac{1}{4}$ and the amplitude is $300° - 210° = 90°$. Thus, $z_1/z_2 = \frac{1}{4}(\cos 90° + i \sin 90°)$.

(c) The modulus of the quotient z_2/z_1 is $\frac{8}{2} = 4$. The amplitude is $210° - 300° = -90°$ but we shall use the smallest positive coterminal angle $-90° + 360° = 270°$. Thus $z_2/z_1 = 4(\cos 270° + i \sin 270°)$.

[NOTE: From Examples 5 and 6 the numbers are $z_1 = 1 - i\sqrt{3}$ and $z_2 = -4\sqrt{3} - 4i$ in rectangular form. Then

$$z_1 z_2 = (1 - i\sqrt{3})(-4\sqrt{3} - 4i) = -8\sqrt{3} + 8i = 16(\cos 150° + i \sin 150°)$$

as in (a), and

$$\frac{z_2}{z_1} = \frac{-4\sqrt{3} - 4i}{1 - i\sqrt{3}} = \frac{(-4\sqrt{3} - 4i)(1 + i\sqrt{3})}{(1 - i\sqrt{3})(1 + i\sqrt{3})} = \frac{-16i}{4} = -4i$$
$$= 4(\cos 270° + i \sin 270°)$$

as in (c).
(See Problems 34.15–34.16.)

DE MOIVRE'S THEOREM.   If n is any rational number,

$$[r(\cos \theta + i \sin \theta)]^n = r^n(\cos n\theta + i \sin n\theta)$$

A proof of this theorem is beyond the scope of this book; a verification for $n = 2$ and $n = 3$ is given is Problem 34.17.

EXAMPLE 8.
$$(\sqrt{3} - i)^{10} = [2(\cos 330° + i \sin 330°)]^{10}$$
$$= 2^{10}(\cos 10 \cdot 330° + i \sin 10 \cdot 330°)$$
$$= 1024(\cos 60° + i \sin 60°) = 1024\left(\frac{1}{2} + \frac{i\sqrt{3}}{2}\right)$$
$$= 512 + 512i\sqrt{3}$$

(See Problem 34.18.)

ROOTS OF COMPLEX NUMBERS.   We state, without proof, the theorem:

A complex number $a + bi = r(\cos \theta + i \sin \theta)$ has exactly n distinct nth roots.

The procedure for determining these roots is given in Example 9.

EXAMPLE 9.   Find all fifth roots of $4 - 4i$.

The usual polar form of $4 - 4i$ is $4\sqrt{2}(\cos 315° + i \sin 315°)$, but we shall need the more general form $4\sqrt{2}[\cos (315° + k360°) + i \sin (315° + k360°)]$, where k is any integer, including zero.

Using De Moiver's theorem, a fifth root of $4 - 4i$ is given by

$$\{4\sqrt{2}[\cos (315° + k360°) + i \sin (315° + k360°)]\}^{1/5} = (4\sqrt{2})^{1/5}\left(\cos 315° + \frac{k360°}{5} + i \sin \frac{315° + k360°}{5}\right)$$

$$= \sqrt{2}[\cos (63° + k72°) + i \sin (63° + k72°)]$$

Assigning in turn the values $k = 0, 1, \ldots$, we find

$$k = 0: \quad \sqrt{2}(\cos 63° + i \sin 63°) = R_1$$
$$k = 1: \quad \sqrt{2}(\cos 135° + i \sin 135°) = R_2$$
$$k = 2: \quad \sqrt{2}(\cos 207° + i \sin 207°) = R_3$$
$$k = 3: \quad \sqrt{2}(\cos 279° + i \sin 279°) = R_4$$
$$k = 4: \quad \sqrt{2}(\cos 351° + i \sin 351°) = R_5$$
$$k = 5: \quad \sqrt{2}(\cos 423° + i \sin 423°) = \sqrt{2}(\cos 63° + i \sin 63°) = R_1, \quad \text{etc.}$$

Thus, the five fifth roots are obtained by assigning the values $0, 1, 2, 3, 4$ (i.e., $0, 1, 2, 3, \ldots, n-1$) to k. (See also Problem 34.19.)

The modulus of each of the roots is $\sqrt{2}$; hence these roots lie on a circle of radius $\sqrt{2}$ with center at the origin. The difference in amplitude of two consecutive roots is $72°$; hence the roots are equally spaced on this circle, as shown in Fig. 34-8.

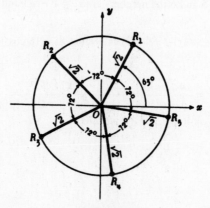

Fig. 34-8

Solved Problems

In Problems 34.1–34.6, perform the indicated operations, simplify, and write the results in the form $a + bi$.

34.1 $(3 - 4i) + (-5 + 7i) = (3 - 5) + (-4 + 7)i = -2 + 3i$

34.2 $(4 + 2i) - (-1 + 3i) = [4 - (-1)] + (2 - 3)i = 5 - i$

34.3 $(2 + i)(3 - 2i) = (6 + 2) + (-4 + 3)i = 8 - i$

34.4 $(3+4i)(3-4i) = 9+16 = 25$

34.5 $\dfrac{1+3i}{2+i} = \dfrac{(1+3i)(2-i)}{(2+i)(2-i)} = \dfrac{(2+3)+(-1+6)i}{4+1} = 1+i$

34.6 $\dfrac{3-2i}{2-3i} = \dfrac{(3-2i)(2+3i)}{(2-3i)(2+3i)} = \dfrac{(6+6)+(9-4)i}{4+9} = \dfrac{12}{13} + \dfrac{5}{13}i$

34.7   Find x and y such that $2x - yi = 4 + 3i$.

   Here $2x = 4$ and $-y = 3$; then $x = 2$ and $y = -3$.

34.8   Show that the conjugate complex numbers $2+i$ and $2-i$ are roots of the equation $x^2 - 4x + 5 = 0$.

   For $x = 2+i$: $(2+i)^2 - 4(2+i) + 5 = 4 + 4i + i^2 - 8 - 4i + 5 = 0$.

   For $x = 2-i$: $(2-i)^2 - 4(2-i) + 5 = 4 - 4i + i^2 - 8 + 4i + 5 = 0$.

   Since each number satisfies the equation, it is a root of the equation.

34.9   Show that the conjugate of the sum of two complex numbers is equal to the sum of their conjugates.

   Let the complex numbers be $a + bi$ and $c + di$. Their sum is $(a + c) + (b + d)i$ and the conjugate of the sum is $(a + c) - (b + d)i$.

   The conjugates of the two given numbers are $a - bi$ and $c - di$, and their sum is $(a + c) + (-b - d)i = (a + c) - (b + d)i$.

34.10   Represent graphically (as a vector) the following complex numbers: (a) $3 + 2i$ (b) $2 - i$ (c) $-2 + i$ (d) $-1 - 3i$.

   We locate, in turn, the points whose coordinates are $(3, 2), (2, -1), (-2, 1), (-1, -3)$ and join each to the origin O.

34.11   Perform graphically the following operations:

   (a) $(3 + 4i) + (2 + 5i)$,   (b) $(3 + 4i) + (2 - 3i)$ (c) $(4 + 3i) - (2 + i)$,   (d) $(4 + 3i) - (2 - i)$.

   For (a) and (b), draw as in Figs. 34-9(a) and 34-9(b) the two vectors and apply the parallelogram law. For (c) draw the vectors representing $4 + 3i$ and $-2 - i$ and apply the parallelogram law as in Fig. 34-9(c). For (d) draw the vectors representing $4 + 3i$ and $-2 + i$ and apply the parallelogram law as in Fig. 34-9(d).

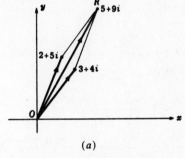

(a)

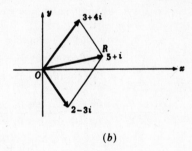

(b)

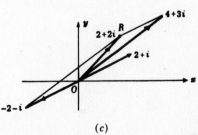

(c)

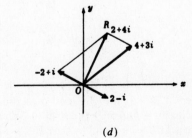

(d)

Fig. 34-9

34.12   Express each of the following complex numbers z in polar form:

(a) $-1 + i\sqrt{3}$,   (b) $6\sqrt{3} + 6i$,   (c) $2 - 2i$,   (d) $-3 = -3 + 0i$,   (e) $4i = 0 + 4i$,   (f) $-3 - 4i$.

(a) P lies in the second quadrant; $r = \sqrt{(-1)^2 + (\sqrt{3})^2} = 2$; $\tan\theta = \sqrt{3}/(-1) = -\sqrt{3}$ and $\theta = 120°$. Thus, $z = 2(\cos 120° + i \sin 120°)$.

(b) P lies in the first quadrant; $r = \sqrt{(6\sqrt{3})^2 + 6^2} = 12$; $\tan -\theta = 6/6\sqrt{3} = 1/\sqrt{3}$ and $\theta = 30°$. Thus, $z = 12(\cos 30° + i \sin 30°)$.

(c) P lies in the fourth quadrant; $r = \sqrt{2^2 + (-2)^2} = 2\sqrt{2}$; $\tan\theta = -\frac{2}{2} = -1$ and $\theta = 315°$. Thus, $z = 2\sqrt{2}(\cos 315° + i \sin 315°)$.

(d) P lies on the negative x axis and $\theta = 180°$; $r = \sqrt{(-3)^2 + 0^2} = 3$. Thus, $z = 3(\cos 180° + i \sin 180°)$.

(e) P lies on the positive y axis and $\theta = 90°$; $r = \sqrt{0^2 + 4^2} = 4$. Thus, $z = 4(\cos 90° + i \sin 90°)$.

(f) P lies in the third quadrant; $r = \sqrt{(-3)^2 + (-4)^2} = 5$; $\tan\theta = -4/(-3) = 1.3333$, $\theta = 233°8'$. Thus, $z = 5(\cos 233°8' + i \sin 233°8')$.

34.13   Express each of the following complex numbers z in rectangular form:

(a) $4(\cos 240° + i \sin 240°)$ (c) $3(\cos 90° + i \sin 90°)$

(b) $2(\cos 315° + i \sin 315°)$ (d) $5(\cos 128° + i \sin 128°)$

(a) $4(\cos 240° + i \sin 240°) = 4[-\frac{1}{2} + i(-\sqrt{3}/2)] = -2 - 2i\sqrt{3}$

(b) $2(\cos 315° + i \sin 315°) = 2[1/\sqrt{2} + i(-1/\sqrt{2})] = \sqrt{2} - i\sqrt{2}$

(c) $3(\cos 90° + i \sin 90°) = 3[0 + i(1)] = 3i$

(d) $5(\cos 128° + i \sin 128°) = 5[-0.6157 + i(0.7880)] = -3.0785 + 3.9400i$

34.14   Prove:

(a)   The modulus of the product of two complex numbers is the product of their moduli, and the amplitude of the product is the sum of their amplitudes.

(b)   The modulus of the quotient of two complex numbers is the modulus of the dividend divided by the modulus of the divisor, and the amplitude of the quotient is the amplitude of the dividend minus the amplitude of the divisor.

Let $z_1 = r_1(\cos\theta_1 + i \sin\theta_1)$ and $z_2 = r_2(\cos\theta_2 + i \sin\theta_2)$.

(a)
$$z_1 z_2 = r_1(\cos\theta_1 + i \sin\theta_1) \cdot r_2(\cos\theta_2 + i \sin\theta_2)$$
$$= r_1 r_2[(\cos\theta_1 \cos\theta_2 - \sin\theta_1 \sin\theta_2) + i(\sin\theta_1 \cos\theta_2 + \cos\theta_1 \sin\theta_2)]$$
$$= r_1 r_2[\cos(\theta_1 + \theta_2) + i \sin(\theta_1 + \theta_2)]$$

(b)
$$\frac{r_1(\cos\theta_1 + i \sin\theta_1)}{r_2(\cos\theta_2 + i \sin\theta_2)} = \frac{r_1(\cos\theta_1 + i \sin\theta_1)(\cos\theta_2 - i \sin\theta_2)}{r_2(\cos\theta_2 + i \sin\theta_2)(\cos\theta_2 - i \sin\theta_2)}$$
$$= \frac{r_1}{r_2} \cdot \frac{\cos\theta_1 \cos\theta_2 + \sin\theta_1 \sin\theta_2) + i(\sin\theta_1 \cos\theta_2 - \cos\theta_1 \sin\theta_2)}{\cos^2\theta_2 + \sin^2\theta_2)}$$
$$= \frac{r_1}{r_2}[\cos(\theta_1 - \theta_2) + i \sin(\theta_1 - \theta_2)]$$

34.15   Perform the indicated operations, giving the result in both polar and rectangular form.

(a) $5(\cos 170° + i \sin 170°) \cdot (\cos 55° + i \sin 55°)$

(b) $2(\cos 50° + i \sin 50°) \cdot 3(\cos 40° + i \sin 40°)$

(c) $6(\cos 110° + i \sin 110°) \cdot \frac{1}{2}(\cos 212° + i \sin 212°)$

(d) $10(\cos 305° + i \sin 305°) \div 2(\cos 65° + i \sin 65°)$

(e) $4(\cos 220° + i \sin 220°) \div 2(\cos 40° + i \sin 40°)$

(f) $6(\cos 230° + i \sin 230°) \div 3(\cos 75° + i \sin 75°)$

(a) The modulus of the product is $5(1) = 5$ and the amplitude is $170° + 55° = 225°$.
In polar form the product is $5(\cos 225° + i \sin 225°)$ and in rectangular form the product is $5(-\sqrt{2}/2 - i\sqrt{2}/2) = -5\sqrt{2}/2 - 5i\sqrt{2}/2$.

(b) The modulus of the product is $2(3) = 6$ and the amplitude is $50° + 40° = 90°$.
In polar form the product is $6(\cos 90° + i \sin 90°)$ and in rectangular form it is $6(0 + i) = 6i$.

(c) The modulus of the product is $6(\frac{1}{2}) = 3$ and the amplitude is $110° + 212° = 322°$.
In polar form the product is $3(\cos 322° + i \sin 322°)$ and in rectangular form it is $3(0.7880 - 0.6157i) = 2.3640 - 1.8471i$.

(d) The modulus of the quotient is $\frac{10}{2} = 5$ and the amplitude is $305° - 65° = 240°$.
In polar form the product is $5(\cos 240° + i \sin 240°)$ and in rectangular form it is $(5 - \frac{1}{2} - i\sqrt{3}/2) = -\frac{5}{2} - 5i\sqrt{3}/2$.

(e) The modulus of the quotient is $\frac{4}{2} = 2$ and the amplitude is $220° - 40° = 180°$.
In polar form the quotient is $2(\cos 180° + i \sin 180°)$ and in rectangular form it is $2(-1 + 0i) = -2$.

(f) The modulus of the quotient is $\frac{6}{3} = 2$ and the amplitude is $230° - 75° = 155°$.
In polar form the quotient is $2(\cos 155° + i \sin 155°)$ and in rectangular form it is $2(-0.9063 + 0.4226i) = -1.8126 + 0.8452i$.

34.16 Express each of the numbers in polar form, perform the indicated operation, and give the result in rectangular form

(a) $(-1 + i\sqrt{3})(\sqrt{3} + i)$ (d) $-2 \div (-\sqrt{3} + i)$ (g) $(3 + 2i)(2 + i)$

(b) $(3 - 3i\sqrt{3})(-2 - 2i\sqrt{3})$ (e) $6i \div (-3 - 3i)$ (h) $(2 + 3i) \div (2 - 3i)$

(c) $(4 - 4i\sqrt{3}) \div (-2\sqrt{3} + 2i)$ (f) $(1 + i\sqrt{3})(1 + i\sqrt{3})$

(a) $$(-1 + i\sqrt{3})(\sqrt{3} + i) = 2(\cos 120° + i \sin 120°) \cdot 2(\cos 30° + i \sin 30°)$$
$$= 4(\cos 150° + i \sin 150°) = 4(-\sqrt{3}/2 + \tfrac{1}{2}i) = -2\sqrt{3} + 2i$$

(b) $$(3 - 3i\sqrt{3})(-2 - 2i\sqrt{3}) = 6(\cos 300° + i \sin 300°) \cdot 4(\cos 240° + i \sin 240°)$$
$$= 24(\cos 540° + i \sin 540°) = 24(-1 + 0i) = -24$$

(c) $$(4 - 4i\sqrt{3}) \div (-2\sqrt{3} + 2i) = 8(\cos 300° + i \sin 300°) \div 4(\cos 150° + i \sin 150°)$$
$$= 2(\cos 150° + i \sin 150°) = 2(-\sqrt{3}/2 + \tfrac{1}{2}i) = -\sqrt{3} + i$$

(d) $$-2 \div (-\sqrt{3} + i) = 2(\cos 180° + i \sin 180°) \div 2(\cos 150° + i \sin 150°)$$
$$= \cos 30° + i \sin 30° = \tfrac{1}{2}\sqrt{3} + \tfrac{1}{2}i$$

(e) $$6i \div (-3 - 3i) = 6(\cos 90° + i \sin 90°) \div 3\sqrt{2}(\cos 225° + i \sin 225°)$$
$$= \sqrt{2}(\cos 225° + i \sin 225°) = -1 - i$$

(f) $$(1 + i\sqrt{3})(1 + i\sqrt{3}) = 2(\cos 60° + i \sin 60°) \cdot 2(\cos 60° + i \sin 60°)$$
$$= 4(\cos 120° + i \sin 120°) = 4(-\tfrac{1}{2} + \tfrac{1}{2}i\sqrt{3}) = -2 + 2i\sqrt{3}$$

(g) $$(3 + 2i)(2 + i) = \sqrt{13}(\cos 33°41' + i \sin 33°41') \cdot \sqrt{5}(\cos 26°34' + i \sin 26°34')$$
$$= \sqrt{65}(\cos 60°15' + i \sin 60°15')$$
$$= \sqrt{65}(0.4962 + 0.8682i) = 4.001 + 7.000i = 4 + 7i$$

(h) $$\frac{2 + 3i}{2 - 3i} = \frac{\sqrt{13}(\cos 56°19' + i \sin 56°19')}{\sqrt{13}(\cos 303°41' + i \sin 303°41')} = \frac{\cos 416°19' + i \sin 416°19'}{\cos 303°41'' + i \sin 303°41'}$$
$$= \cos 112°38' + i \sin 112°38' = -0.3849 + 0.9230i$$

34.17 Verify De Moivre's theorem for $n = 2$ and $n = 3$.

Let $z = r(\cos \theta + i \sin \theta)$.

For $n = 2$: $\quad z^2 = [r(\cos \theta + i \sin \theta)][r(\cos \theta + i \sin \theta)]$

$\qquad\qquad = r^2[(\cos^2 \theta - \sin^2 \theta) + i(2 \sin \theta \cos \theta)] = r^2(\cos 2\theta + i \sin 2\theta)$

For $n = 3$: $\quad z^3 = z^2 \cdot z = [r^2(\cos 2\theta + i \sin 2\theta)][r(\cos \theta + i \sin \theta)]$

$\qquad\qquad = r^3[(\cos 2\theta \cos \theta - \sin 2\theta \sin \theta) + i(\sin 2\theta \cos \theta + \cos 2\theta \sin \theta)]$

$\qquad\qquad = r^3(\cos 3\theta + i \sin 3\theta).$

The theorem may be established for n a positive integer by mathematical induction.

34.18 Evaluate each of the following using De Moivre's theorem and express each result in rectangular form:

(a) $(1 + i\sqrt{3})^4$, $\quad$ (b) $(\sqrt{3} - i)^5$, $\quad$ (c) $(-1 + i)^{10}$, $\quad$ (d) $(2 + 3i)^4$.

(a) $\qquad\qquad (1 + i\sqrt{3})^4 = [2(\cos 60° + i \sin 60°)]^4 = 2^4(\cos 4 \cdot 60° + i \sin 4 \cdot 60°)$

$\qquad\qquad\qquad = 2^4(\cos 240° + i \sin 240°) = -8 - 8i\sqrt{3}$

(b) $\qquad\qquad (\sqrt{3} - i)^5 = [2(\cos 330° + i \sin 330°)]^5 = 32(\cos 1650° + i \sin 1650°)$

$\qquad\qquad\qquad = 32(\cos 210° + i \sin 210°) = -16\sqrt{3} - 16i$

(c) $\qquad (-1 + i)^{10} = [\sqrt{2}(\cos 135° + i \sin 135°)]^{10} = 32(\cos 270° + i \sin 270°) = -32i$

(d) $\qquad\qquad (2 + 3i)^4 = [\sqrt{13}(\cos 56°19' + i \sin 56°19')]^4 = 13^2(\cos 225°16' + i \sin 225°16')$

$\qquad\qquad\qquad = 169(-0.7038 - 7104i) = -118.9 - 120.1i$

34.19 Find the indicated roots in rectangular form, except when this would necessitate the use of tables.

(a) Square roots of $2 - 2i\sqrt{3}$ $\qquad$ (e) Fourth roots of i

(b) Fourth roots of $-8 - 8i\sqrt{3}$ $\qquad$ (f) Sixth roots of -1

(c) Cube roots of $-4\sqrt{2} + 4i\sqrt{2}$ $\quad$ (g) Fourth roots of $-16i$

(d) Cube roots of 1 $\qquad\qquad\qquad$ (h) Fifth roots of $1 + 3i$

(a) $\qquad\qquad 2 - 2i\sqrt{3} = 4[\cos (300° + k360°) + i \sin (300° + k360°)]$

and $\qquad (2 - 2i\sqrt{3})^{1/2} = 2[\cos (150° + k180°) + i \sin (150° + k180°)]$

Putting $k = 0$ and 1, the required roots are

$\qquad R_1 = 2(\cos 150° + i \sin 150°) = 2(-\tfrac{1}{2}\sqrt{3} + \tfrac{1}{2}i) = -\sqrt{3} + i$

$\qquad R_2 = 2(\cos 330° + i \sin 330°) = 2(\tfrac{1}{2}\sqrt{3} - \tfrac{1}{2}i) = \sqrt{3} - i$

(b) $\qquad\qquad -8 - 8i\sqrt{3} = 16[\cos (240° + k360°) + i \sin (240° + k360°)]$

and $\qquad (-8 - 8i\sqrt{3})^{1/4} = 2[\cos (60° + k90°) + i \sin (60° + k90°)]$

Putting $k = 0, 1, 2, 3$, the required roots are

$\qquad R_1 = 2(\cos 60° + i \sin 60°) = 2(\tfrac{1}{2} + i\tfrac{1}{2}\sqrt{3}) = 1 + i\sqrt{3}$

$\qquad R_2 = 2(\cos 150° + i \sin 150°) = 2(-\tfrac{1}{2}\sqrt{3} + \tfrac{1}{2}i) = -\sqrt{3} + i$

$\qquad R_3 = 2(\cos 240° + i \sin 240°) = 2(-\tfrac{1}{2} - i\tfrac{1}{2}\sqrt{3}) = -1 - i\sqrt{3}$

$\qquad R_4 = 2(\cos 330° + i \sin 330°) = 2(\tfrac{1}{2}\sqrt{3} - \tfrac{1}{2}i) = \sqrt{3} - i$

(c) $\qquad\qquad -4\sqrt{2} + 4i\sqrt{2} = 8[\cos (135° + k360°) + i \sin (135° + k360°)]$

and $\qquad (-4\sqrt{2} + 4i\sqrt{2})^{1/3} = 2[\cos (45° + k120°) + i \sin (45° + k120°)]$

Putting $k = 0, 1, 2$, the required roots are

$$R_1 = 2(\cos 45° + i \sin 45°) = 2(1/\sqrt{2} + i/\sqrt{2}) = \sqrt{2} + i\sqrt{2}$$
$$R_2 = 2(\cos 165° + i \sin 165°)$$
$$R_3 = 2(\cos 285° + i \sin 285°)$$

(d) $1 = \cos (0° + k360°) + i \sin (0° + k360°)$ and $1^{1/3} = \cos (k120°) + i \sin (k120°)$. Putting $k = 0, 1, 2$, the required roots are

$$R_1 = \cos 0° + i \sin 0° = 1$$
$$R_2 = \cos 120° + i \sin 120° = -\tfrac{1}{2} + i\tfrac{1}{2}\sqrt{3}$$
$$R_3 = \cos 240° + i \sin 240° = -\tfrac{1}{2} - i\tfrac{1}{2}\sqrt{3}$$

Note that
$$R_2^2 = \cos 2(120°) + i \sin 2(120°) = R_3,$$
$$R_3^2 = \cos 2(240°) + i \sin 2(240°) = R_2,$$

and
$$R_2 R_3 = (\cos 120° + i \sin 120°)(\cos 240° + i \sin 240°)$$
$$= \cos 0° + i \sin 0° = R_1.$$

(e) $i = \cos (90° + k360°) + i \sin (90° + k360°)$ and $i^{1/4} = \cos (22\tfrac{1}{2}° + k90°) + i \sin (22\tfrac{1}{2}° + k90°)$. Thus, the required roots are

$$R_1 = \cos 22\tfrac{1}{2}° + i \sin 22\tfrac{1}{2}° \qquad R_3 = \cos 202\tfrac{1}{2}° + i \sin 202\tfrac{1}{2}°$$
$$R_2 = \cos 112\tfrac{1}{2}° + i \sin 112\tfrac{1}{2}° \qquad R_4 = \cos 292\tfrac{1}{2}° + i \sin 292\tfrac{1}{2}°$$

(f) $-1 = \cos (180° + k360°) + i \sin (180° + k360°)$ and $(-1)^{1/6} = \cos (30° + k60°) + i \sin (30° + k60°)$. Thus, the required roots are

$$R_1 = \cos 30° + i \sin 30° = \tfrac{1}{2}\sqrt{3} + \tfrac{1}{2}i$$
$$R_2 = \cos 90° + i \sin 90° = i$$
$$R_3 = \cos 150° + i \sin 150° = -\tfrac{1}{2}\sqrt{3} + \tfrac{1}{2}i$$
$$R_4 = \cos 210° + i \sin 210° = -\tfrac{1}{2}\sqrt{3} - \tfrac{1}{2}i$$
$$R_5 = \cos 270° + i \sin 270° = -i$$
$$R_6 = \cos 330° + i \sin 330° = \tfrac{1}{2}\sqrt{3} - \tfrac{1}{2}i$$

Note that $R_2^2 = \cos R_5^2 = \cos 180° + i \sin 180°$ and thus R_2 and R_5 are the square roots of -1; that $R_1^3 = R_3^3 = R_5^3 = \cos 90° + i \sin 90° = i$ and thus R_1, R_3, R_5 are the cube roots of i; and that $R_2^3 = R_4^3 = R_6^3 = \cos 270° + i \sin 270° = -i$ and thus R_2, R_4, R_6 are the cube roots of $-i$.

(g) $-16i = 16[\cos (270° + K360°) + i \sin (270° + k360°)]$ and
$(-16i)^{1/4} = 2[\cos (67\tfrac{1}{2}° + k90°) + i \sin (67\tfrac{1}{2}° + k90°)]$. Thus, the required roots are

$$R_1 = 2(\cos 67\tfrac{1}{2}° + i \sin 67\tfrac{1}{2}°) \qquad R_3 = 2(\cos 247\tfrac{1}{2}° + i \sin 247\tfrac{1}{2}°)$$
$$R_2 = 2(\cos 157\tfrac{1}{2}° + i \sin 157\tfrac{1}{2}°) \qquad R_4 = 2(\cos 337\tfrac{1}{2}° + i \sin 337\tfrac{1}{2}°)$$

(h) $1 + 3i = \sqrt{10}[\cos (71°34' + k360°) + i \sin (71°34' + k360°)]$ and
$(1 + 3i)^{1/5} = \sqrt[10]{10}[\cos (14°19' + k72°) + i \sin (14°19' + k72°)]$. The required roots are

$$R_1 = \sqrt[10]{10}(\cos 14°19' + i \sin 14°19')$$
$$R_2 = \sqrt[10]{10}(\cos 86°19' + i \sin 86°19')$$
$$R_3 = \sqrt[10]{10}(\cos 158°19' + i \sin 158°19')$$
$$R_4 = \sqrt[10]{10}(\cos 230°19' + i \sin 230°19')$$
$$R_5 = \sqrt[10]{10}(\cos 302°19' + i \sin 302°19').$$

Supplementary Problems

34.20   Perform the indicated operations, writing the results in the form $a + bi$.

   (a) $(6 - 2i) + (2 + 3i) = 8 + i$ (k) $(2 + \sqrt{-5})(3 - 2\sqrt{-4}) = (6 + 4\sqrt{5}) + (3\sqrt{5} - 8)i$

   (b) $(6 - 2i) - (2 + 3i) = 4 - 5i$ (l) $(1 + 2\sqrt{-3})(2 - \sqrt{-3}) = 8 + 3\sqrt{3}i$

   (c) $(3 + 2i) + (-4 - 3i) = -1 - i$ (m) $(2 - i)^2 = 3 - 4i$

   (d) $(3 - 2i) - (4 - 3i) = -1 + i$ (n) $(4 + 2i)^2 = 12 + 16i$

   (e) $3(2 - i) = 6 - 3i$ (o) $(1 + i)^2(2 + 3i) = -6 + 4i$

   (f) $2i(3 + 4i) = -8 + 6i$

   (g) $(2 + 3i)(1 + 2i) = -4 + 7i$ (p) $\dfrac{2 + 3i}{1 + i} = \dfrac{5}{2} + \dfrac{1}{2}i$

   (h) $(2 - 3i)(5 + 2i) = 16 - 11i$

   (i) $(3 - 2i)(-4 + i) = -10 + 11i$ (q) $\dfrac{3 - 2i}{3 - 4i} = \dfrac{17}{25} + \dfrac{6}{25}i$ (r) $\dfrac{3 - 2i}{2 + 3i} = -i$

   (j) $(2 + 3i)(3 + 2i) = 13i$

34.21   Show that $3 + 2i$ and $3 - 2i$ are roots of $x^2 - 6x + 13 = 0$.

34.22   Perform graphically the following operations:

   (a) $(2 + 3i) + (1 + 4i)$ (c) $(2 + 3i) - (1 + 4i)$

   (b) $(4 - 2i) + (2 + 3i)$ (d) $(4 - 2i) - (2 + 3i)$

34.23   Express each of the following complex numbers in polar form:

   (a) $3 + 3i = 32(\cos 45° + i \sin 45°)$ (e) $-8 = 8(\cos 180° + i \sin 180°)$

   (b) $1 + \sqrt{3}i = 2(\cos 60° + i \sin 60°)$ (f) $-2i = 2(\cos 270° + i \sin 270°)$

   (c) $-2\sqrt{3} - 2i = 4(\cos 210° + i \sin 210°)$ (g) $-12 + 5i = 13(\cos 157°23' + i \sin 157°23')$

   (d) $\sqrt{2} - i\sqrt{2} = 2(\cos 315° + i \sin 315°)$ (h) $-4 - 3i = 5(\cos 216°52' + i \sin 216°52')$

34.24   Perform the indicated operation and express the results in the form $a + bi$.

   (a) $3(\cos 25° + i \sin 25°)8(\cos 200° + i \sin 200°) = -12\sqrt{2} - 12\sqrt{2}i$

   (b) $4(\cos 50° + i \sin 50°)2(\cos 100° + i \sin 100°) = -4\sqrt{3} + 4i$

   (c) $\dfrac{4(\cos 190° + i \sin 190°)}{2(\cos 70° + i \sin 70°)} = -1 + i\sqrt{3}$

   (d) $\dfrac{12(\cos 200° + i \sin 200°)}{3(\cos 350° + i \sin 350°)} = -2\sqrt{3} - 2i$

34.25   Use the polar form in finding each of the following products and quotients, and express each result in the form $a + bi$:

   (a) $(1 + i)(\sqrt{2} - i\sqrt{2}) = 2\sqrt{2}$ (d) $\dfrac{4 + 4\sqrt{3}i}{\sqrt{3} + i} = 2\sqrt{3} + 2i$

   (b) $(-1 - i\sqrt{3})(-4\sqrt{3} + 4i) = 8\sqrt{3} + 8i$ (e) $\dfrac{-1 + i\sqrt{3}}{\sqrt{2} + i\sqrt{2}} = 0.2588 + 0.9659i$

   (c) $\dfrac{1 - i}{1 + i} = -i$ (f) $\dfrac{3 + i}{2 + i} = 1.4 - 0.2i$

34.26 Use De Moivre's theorem to evaluate each of the following and express each result in the form $a + bi$:

(a) $[2(\cos 6° + i \sin 6°)]^5 = 16\sqrt{3} + 16i$

(b) $[\sqrt{2}(\cos 75° + i \sin 75°)]^4 = 2 - 2\sqrt{3}i$

(c) $(1 + i)^8 = 16$

(d) $(1 - i)^6 = 8i$

(e) $(\frac{1}{2} - i\sqrt{3}/2)^{20} = -\frac{1}{2} - i\sqrt{3}/2$

(f) $(\sqrt{3}/2 + i/2)^9 = -i$

(g) $(3 + 4i)^4 = -526.9 - 336.1i$

(h) $\dfrac{(1 - i\sqrt{3})^3}{(-2 + 2i)} = \dfrac{1}{8}$

(i) $\dfrac{(1 + i)(\sqrt{3} + i)^3}{(1 - i\sqrt{3})^3} = 1 - i$

34.27 Find all the indicated roots, expressing the results in the form $a + bi$ unless tables would be needed to do so.

(a) The square roots of i.   *Ans.* $\sqrt{2}/2 + i\sqrt{2}/2, -\sqrt{2}/2 - i\sqrt{2}/2$

(b) The square roots of $1 + i\sqrt{3}$.   *Ans.* $\sqrt{6}/2 + i\sqrt{2}/2, -\sqrt{6}/2 - i\sqrt{2}/2$

(c) The cube roots of -8.   *Ans.* $1 + i\sqrt{3}, -2, 1 - i\sqrt{3}$

(d) The cube roots of $27i$.   *Ans.* $3\sqrt{3}/2 + 3i/2, -3\sqrt{3}/2 + 3i/2, -3i$

(e) The cube roots of $-4\sqrt{3} + 4i$.
     Ans. $2(\cos 50° + i \sin 50°), 2(\cos 170° + i \sin 170°), 2(\cos 290° + i \sin 290°)$

(f) The fifth roots of $1 + i$.
     Ans. $\sqrt[10]{2}(\cos 9° + i \sin 9°), \sqrt[10]{2}(\cos 81° + i \sin 81°),$ etc.

(g) The sixth roots of $-\sqrt{3} + i$.
     Ans. $\sqrt[6]{2}(\cos 25° + i \sin 25°), \sqrt[6]{2}(\cos 85° + i \sin 85°),$ *etc.*

34.28 Find the tenth roots of 1 and show that the product of any two of them is again one of the tenth roots of 1.

34.29 Show that the reciprocal of any one of the tenth roots of 1 is again a tenth root of 1.

34.30 Denote either of the complex cube roots of 1 [Problem 34.19(d)] by ω_1 and the other by ω_2. Show that $\omega_1^2\omega_2 = \omega_1$ and $\omega_1\omega_2^2 = \omega_2$.

34.31 Show that $(\cos \theta + i \sin \theta)^{-n} = \cos n\theta - i \sin n\theta$.

34.32 Use the fact that the segments OS and P_2P_1 in Fig. 34-5 are equal to devise a second procedure for constructing the differences $OS = z_1 - z_2$ of two complex numbers z_1 and z_2.

Chapter 35

The Conic Sections

I. The Parabola

THE LOCUS OF A POINT P which moves in a plane so that its distance from a fixed line of the plane and its distance from a fixed point of the plane, not on the line, are equal is called a *parabola*.

The fixed point F is called the *focus*, and the fixed line d is called the *directrix* of the parabola. The line $\overleftrightarrow{FD}$ through the focus and perpendicular to the directrix is called the *axis* of the parabola. The axis intersects the parabola in the point V, the midpoint of FD, called the *vertex*.

The line segment joining any two distinct points of the parabola is called a *chord*. A chord (as $\overline{BB'}$) which passes through the focus is called a *focal chord*, while $\overline{FB}$ and $\overline{FB'}$ are called the *focal radii* of B and B', respectively. See Fig. 35-1.

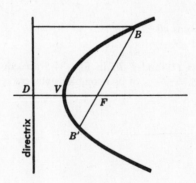

Fig. 35-1

THE EQUATION OF A PARABOLA assumes its simplest (*reduced*) form when its vertex is at the origin and its axis coincides with one of the coordinate axes.

When the vertex is at the origin and the axis coincides with the x axis, the equation of the parabola is

$$y^2 = 4px \qquad (35.1)$$

Then the focus is at $F(p, 0)$ and the equation of the directrix is $d : x = -p$. If $p > 0$, the parabola opens to the right; if $p < 0$, the parabola opens to the left. See Figs. 35-2(*a*) and (*b*).

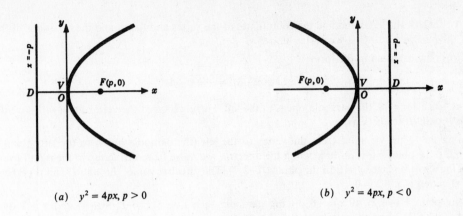

(a)　$y^2 = 4px, p > 0$　　　　　　　　　　(b)　$y^2 = 4px, p < 0$

Fig. 35-2

When the vertex is at the origin and the axis coincides with the y axis, the equation of the parabola is

$$x^2 = 4py \qquad (35.2)$$

Then the focus is at $F(0, p)$ and the equation of the directrix is $d : y = -p$. If $p > 0$, the parabola opens upward; if $p < 0$, the parabola opens downward. See Figs. 35-3(a) and (b).

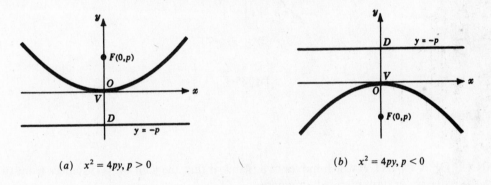

(a)　$x^2 = 4py, p > 0$　　　　　　　　　　(b)　$x^2 = 4py, p < 0$

Fig. 35-3

In either case, the distance from the directrix to the vertex and the distance from the vertex to the focus are equal to p.

[Some authors define $p > 0$ and consider the four cases $y^2 = 4px$, $y^2 = -4px$, $x^2 = 4py$, $x^2 = -4py$. Other authors label the focus $F(\frac{1}{2}p, 0)$ and directrix $d : x = -\frac{1}{2}p$ and obtain $y^2 = 2px$, etc.]

THE EQUATION OF A PARABOLA assumes the *semireduced* form

$$(y - k)^2 = 4p(x - h) \qquad (35.1')$$

or

$$(x - h)^2 = 4p(y - k) \qquad (35.2')$$

when its vertex is at the point (h, k) and its axis is parallel, respectively, to the x axis or the y axis. The distance between the directrix and vertex and the distance between the vertex and focus are the same as given in the section above.

EXAMPLE 1. Sketch the locus and find the coordinates of the vertex and focus, and the equations of the axis and directrix, of the parabola $y^2 - 6y + 8x + 41 = 0$.

We first put the equation in the form

$$(y-3)^2 = -8(x+4) \qquad\qquad (35.3)$$

and note that, since $4p = -8$, the parabola opens to the left. Having located the vertex at $V(-4,3)$, we draw in the axis through V parallel to the x axis.

In locating the focus, we move from the vertex to the left (the parabola opens to the left) along the axis a distance $p = 2$ to the point $F(-6,3)$. In locating the directrix, we move from the vertex to the right (away from the focus) along the axis a distance $p = 2$ to the point $D(-2,3)$. The directrix passes through D and is perpendicular to the axis; its equation is $x + 2 = 0$.

Using the point $F(-6,3)$, if $x = -6$ on the parabola then $(y-3)^2 = -8(-6+4)$ and $y = 7$ or -1. Then $p_1(-6,-1)$ and $p_2(-6,7)$ lie on the parabola. See Fig. 35-4.

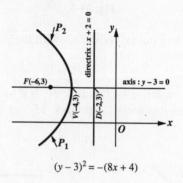

$$(y-3)^2 = -(8x+4)$$

Fig. 35-4

II. The Ellipse

THE LOCUS OF A POINT P which moves in a plane so that the sum of its distances from two fixed points in the plane is constant is called an *ellipse*.

The fixed points F and F' are called the *foci*, and their midpoint C is called the *center* of the ellipse. The line $\overleftrightarrow{FF'}$ joining the foci intersects the ellipse in the points V and V', called the *vertices*. The segment $\overline{V'V}$ intercepted on the line $\overleftrightarrow{FF'}$ by the ellipse is called its *major axis*; the segment $\overline{B'B}$ intercepted on the line through C perpendicular to $\overleftrightarrow{FF'}$ is called its *minor axis*.

A line segment whose extremities are any two points on the ellipse is called a *chord*. A chord which passes through a focus is called a focal *chord*. See Fig. 35-5.

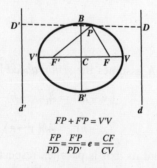

$$FP + F'P = V'V$$

$$\frac{FP}{PD} = \frac{F'P}{PD'} = e = \frac{CF}{CV}$$

Fig. 35-5

The ellipse may also be defined as the locus of a point which moves so that the ratio of its distance from a fixed point to its distance from a fixed line is equal to $e < 1$. The fixed point is a focus F or F', and the fixed line d or d' is called a *directrix*. The ratio e is called the *eccentricity* of the ellipse.

THE EQUATION OF AN ELLIPSE assumes its simplest (*reduced*) form when its center is at the origin and its major axis lies along one of the coordinate axes.

When the center is at the origin and the major axis lies along the x axis, the equation of the ellipse is

$$\frac{x^2}{a^2} + \frac{y^2}{b^2} = 1, \text{ where } a^2 \geq b^2 \qquad (35.4)$$

Then the vertices are at $V(a, 0)$ and $V'(-a, 0)$ and the length of the major axis is $V'V = 2a$. The length of the minor axis is $\overline{B'B} = 2b$. The foci are on the major axis at $F(-c, 0)$ and $F'(c, 0)$ where

$$c = \sqrt{a^2 - b^2}$$

When the center is at the origin and the major axis lies along the y axis, the equation of the ellipse is

$$\frac{x^2}{b^2} + \frac{y^2}{a^2} = 1, \text{ where } a^2 \geq b^2 \qquad (35.5)$$

See Figs. 35-6 (*a*) and (*b*).

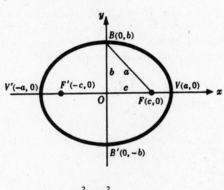

(a) $\dfrac{x^2}{a^2} + \dfrac{y^2}{b^2} = 1, a > b$

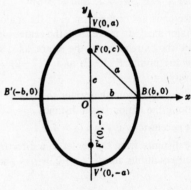

(b) $\dfrac{x^2}{b^2} + \dfrac{y^2}{a^2} = 1, a > b$

Fig. 35-6

Then the vertices are at $V(0, a)$ and $V'(0, -a)$ and the length of the major axis is $V'V = 2a$. The length of the minor axis is $B'B = 2b$. The foci are on the major axis at $F(0, c)$ and $F'(0, -c)$, where

$$c = \sqrt{a^2 - b^2}$$

In both cases, the eccentricity is

$$e = \frac{c}{a} = \frac{\sqrt{a^2 - b^2}}{a}$$

and the directrices are perpendicular to major axis at distances $\pm a^2/c = \pm a/e$ from the center.

THE EQUATION OF AN ELLIPSE with major axis vertical or horizontal assumes the *semireduced* form

$$\frac{(x-h)^2}{a^2} + \frac{(y-k)^2}{b^2} = 1 \tag{35.4'}$$

or

$$\frac{(x-h)^2}{b^2} + \frac{(y-k)^2}{a^2} = 1 \tag{35.5'}$$

Here $b \leq a$ the center is at the point (h, k), and the major axis is parallel respectively to the x axis and to the y axis. The lengths of the major and minor axes, the distance between the foci, the distance from the center to a directrix, and the eccentricity are as given in the section above.

EXAMPLE 2. Find the coordinates of the center, vertices, and foci; the lengths of the major and minor axes; the eccentricity; and the equations of the directrices of the ellipse

$$\frac{(x-4)^2}{9} + \frac{(y+2)^2}{25} = 1$$

Sketch the locus.

The center is at the point C(4, −2).

Since $a^2 \geq b^2$, $a^2 = 25$ and $b^2 = 9$.

Since a^2 is under the term in y, the major axis is parallel to the y axis. To locate the vertices (the extremities of the major axis), we move from the center parallel to the y axis a distance $a = 5$ to the points V(4, 3) and V'(4, −7). To locate the extremities of the minor axis, we move from the center perpendicular to the major axis a distance $b = 3$ to the points B'(1, −2) and B(7, −2). The lengths of the major and minor axes are $2a = 10$ and $2b = 6$, respectively.

The distance from the center to a focus is $c = \sqrt{a^2 - b^2} = \sqrt{25 - 9} = 4$. To locate the foci, we move from the center along the major axis a distance $c = 4$ to the points F(4, 2) and F'(4, −6).

The eccentricity is $e = c/a = 4/5$.

The distance from the center to a directrix is $a^2/c = 25/4$. Since the directrices are perpendicular to the major axis, their equations are $d : y = -2 + \frac{25}{4} = \frac{17}{4}$ and $d' : y = -2 - \frac{25}{4} = -\frac{33}{4}$. See Fig. 35-7.

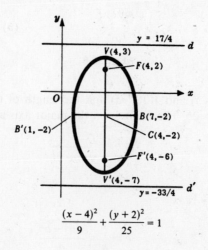

$$\frac{(x-4)^2}{9} + \frac{(y+2)^2}{25} = 1$$

Fig. 35-7

III.   The Hyperbola

THE LOCUS OF A POINT P which moves in a plane so that the absolute value of the difference of its distances from two fixed points in the plane is constant is called a *hyperbola*. (Note that the locus consists of two distinct branches, each of indefinite length.)

    The fixed points F and F' are called the *foci*, and their midpoint C is called the *center* of the hyberbola. The line $\overleftrightarrow{FF'}$ joining the foci intersects the hyperbola in the points V and V', called the *vertices*.

    The segment $\overline{V'V}$ intercepted on the line $\overleftrightarrow{FF'}$ by the hyperbola is called its *transverse axis*. The line l through C and perpendicular to $\overleftrightarrow{FF'}$ does not intersect the curve, but it will be found convenient to define a certain segment $\overline{B'B}$ on l, having C as midpoint, as the *conjugate axis*. See Fig. 35-8.

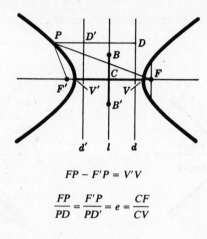

$$FP - F'P = V'V$$

$$\frac{FP}{PD} = \frac{F'P}{PD'} = e = \frac{CF}{CV}$$

Fig. 35-8

    A line segment whose extremities are any two points (both on the same branch or one on each branch) on the hyperbola is called a *chord*. A chord which passes through a focus is called a *focal chord*.

    The hyperbola may also be defined as the locus of a point which moves so that the ratio of its distance from a fixed point and its distance from a fixed line is equal to $e > 1$. The fixed point is a focus F or F' and the fixed line d or d' is called a *directrix*. The ratio e is called the *eccentricity* of the hyperbola.

THE EQUATION OF A HYPERBOLA assumes its simplest (*reduced*) form when its center is at the origin and its transverse axis lies along one of the coordinate axes.

    When the center is at the origin and the transverse axis lies along the x axis, the equation of the hyperbola is

$$\frac{x^2}{a^2} - \frac{y^2}{b^2} = 1 \tag{35.6}$$

Then the vertices are at $V(a, 0)$ and $V'(-a, 0)$ and the length of the transverse axis is $V'V = 2a$. The extremities of the conjugate axis are $B'(0, -b)$ and $B(0, b)$, and its length is $B'B = 2b$. The foci are on the transverse axis at $F'(-c, 0)$ and $F(c, 0)$, where

$$c = \sqrt{a^2 + b^2}$$

    When the center is at the origin and the transverse axis lies along the y axis, the equation of the hyperbola is

$$\frac{y^2}{a^2} - \frac{x^2}{b^2} = 1 \tag{35.7}$$

Then the vertices are at $V(0, a)$ and $V'(0, -a)$, and the length of the transverse axis is $V'V = 2a$. The extremities of the conjugate axis are $B'(-b, 0)$ and $B(b, 0)$, and its length is $B'B = 2b$. The foci are on the transverse axis at $F(0, c)$ and $F'(0, -c)$, where

$$c = \sqrt{a^2 + b^2}$$

In both cases, the eccentricity is

$$e = \frac{c}{a} = \frac{\sqrt{a^2 + b^2}}{a}$$

and the directrices are perpendicular to the transverse axis at distances $\pm a^2/c = \pm a/e$ from the center. See Figs. 35-9(a) and (b).

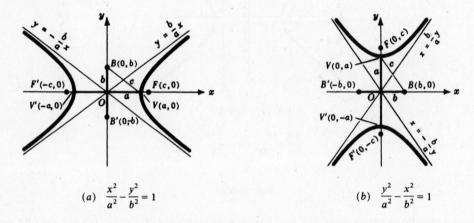

$$(a) \quad \frac{x^2}{a^2} - \frac{y^2}{b^2} = 1 \qquad\qquad\qquad (b) \quad \frac{y^2}{a^2} - \frac{x^2}{b^2} = 1$$

Fig. 35-9

THE STRAIGHT LINES with equations $y = \pm(b/a)x$ are called the asymptotes of the hyperbola (35.6), and the lines with equations $x = \pm(b/a)y$ are called the asymptotes of the hyperbola (35.7).

These lines have the property that the perpendicular distance from a point on a hyperbola to one of them approaches zero as the point moves indefinitely far from the center.

THE EQUATION OF A HYPERBOLA with horizontal or vertical transverse axes assumes the *semireduced* form

$$\frac{(x-h)^2}{a^2} - \frac{(y-k)^2}{b^2} = 1 \tag{35.6'}$$

or

$$\frac{(y-k)^2}{a^2} - \frac{(x-h)^2}{b^2} = 1 \tag{35.7'}$$

Here the center is at the point $C(h, k)$, and the transverse axis is parallel, respectively, to the x axis or to the y axis. The lengths of the transverse and conjugate axes, the distance between the foci, the distance from the center to a directrix, the slope of the asymptotes, and the eccentricity are as given in the sections above.

EXAMPLE 3. Find the coordinates of the center, vertices, and foci, the lengths of the transverse and conjugate axes; the eccentricity; and the equations of the directrices and asymptotes of the hyperbola

$$\frac{(x+3)^2}{4} - \frac{(y-1)^2}{25} = 1$$

Sketch the locus.

The center is at the point $C(-3, 1)$. Since, when the equation is put in the reduced or semireduced form, a^2 is always in the positive term on the left, $a^2 = 4$ and $b^2 = 25$; then $a = 2$ and $b = 5$.

The transverse axis is parallel to the x axis (the positive term contains x). To locate the vertices, we move from the center along the transverse axis a distance $a = 2$ to the points $V(-1, 1)$ and $V'(-5, 1)$. To locate the extremities of the conjugate axis, we move from the center perpendicular to the transverse axis a distance $b = 5$ to the points $B'(-3, -4)$ and $B(-3, 6)$. The lengths of the transverse and conjugate axes are $2a = 4$ and $2b = 10$, respectively.

The distance from the center to a focus is $c = \sqrt{a^2 + b^2} = \sqrt{4 + 25} = \sqrt{29}$. To locate the foci, we move from the center along the transverse axis a distance $c = \sqrt{29}$ to the points $F(-3 + \sqrt{29}, 1)$ and $F'(-3 - \sqrt{29}, 1)$.

The eccentricity is $e = c/a = \sqrt{29}/2$.

The distance from the center to a directrix is $a^2/c = 4\sqrt{29} = 4\sqrt{29}/29$. Since the directrices are perpendicular to the transverse axis, their equations are $d : x = -3 + 4\sqrt{29}/29$ and $d' : x = -3 - 4\sqrt{29}/29$.

The asymptotes pass through C with slopes $\pm b/a = \pm 5/2$; thus, their equations are $y - 1 = \pm 5/2(x + 3)$. Combining these two equations, $(x + 3)^2/4 - (y - 1)^2/25 = 0$. Hence, they may be obtained most readily by the simple trick of changing the right member of the equation of the hyperbola from 1 to 0. See Fig. 35-10. (See Problems 35.6–35.8.)

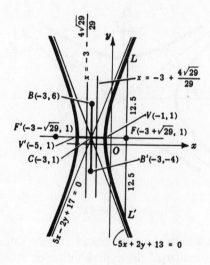

Fig. 35-10

THE HYPERBOLAS OF EQUATION $x^2 - y^2 = a^2$ and $y^2 - x^2 = a^2$, whose transverse and conjugate axes are of equal length $2a$, are called *equilateral hyperbolas*. Since their asymptotes $x \pm y = 0$ are mutually perpendicular, the equilateral hyperbola is also called the *rectangular hyperbola*. See Figs. 35-11(*a*) and (*b*).

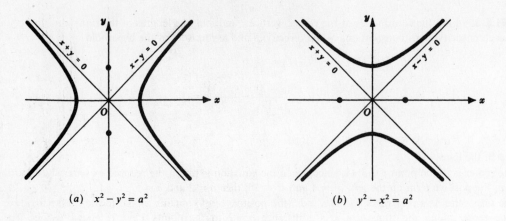

(a) $x^2 - y^2 = a^2$ (b) $y^2 - x^2 = a^2$

Fig. 35-11

TWO HYPERBOLAS such that the transverse axis of each is the conjugate axis of the other, as

$$\frac{x^2}{16} - \frac{y^2}{9} = 1$$

and

$$\frac{y^2}{9} - \frac{x^2}{16} = 1$$

are called *conjugate hyperbolas,* each being the conjugate of the other. See Fig. 35-12.

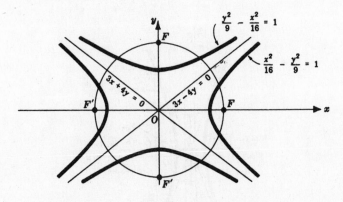

Fig. 35-12

A pair of conjugate hyperbolas have the same center and the same asymptotes. Their foci lie on a circle whose center is the common center of the hyperbolas. (See Problem 35.9.)

Solved Problems

35.1   For each of the following parabolas, sketch the curve, find the coordinates of the vertex and focus, and find the equations of the axis and directrix.

(a) $y^2 = 16x$ (b) $x^2 = -9y$ (c) $x^2 - 2x - 12y + 25 = 0$
(d) $y^2 + 4y + 20x + 4 = 0$

Ans.

(a) The parabola opens to the right ($p > 0$) with vertex at $V(0,0)$. The equation of its axis is $y = 0$. Moving from V to the right along the axis a distance $p = 4$, we locate the focus at $F(4,0)$. Moving from V to the left along the axis a distance $p = 4$, we locate the point $D(-4,0)$. Since the directrix passes through D perpendicular to the axis, its equation is $x + 4 = 0$. If $x = 4$, then $y = \pm 8$, so $P_2(4,8)$ and $P_1(4,-8)$ are points on the parabola. See Fig. 35-13(a).

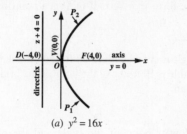

(a) $y^2 = 16x$ (b) $x^2 = -9y$

Fig. 35-13

(b) The parabola opens downward ($p < 0$) with vertex at $V(0,0)$. The equation of its axis is $x = 0$. Moving from V downward along the axis a distance $p = 9/4$, we locate the focus at $F(0,-9/4)$. Moving from V upward along the axis a distance $p = 9/4$, we locate the point $D(0,-9/4)$; the equation of the directrix is $4y - 9 = 0$. P_2 is the point $(-9/2,-9/4)$ and P_1 is $(9/2,-9/4)$. See Fig. 35-13 (b).

(c) Here $(x-1)^2 = 12(y-2)$. The parabola opens upward ($p > 0$) with vertex at $V(1,2)$. The equation of its axis is $x - 1 = 0$. See Fig. 35-14 (a).

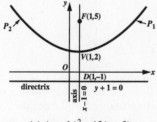
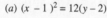
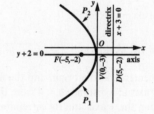

(a) $(x-1)^2 = 12(y-2)$ (b) $(y+2)^2 = -20x$

Fig. 35-14

(d) Here $(y+2)^2 = -20x$. The parabola opens to the left ($p < 0$) with vertex at $V(0,-2)$. The equation of its axis is $y + 2 = 0$. Since $p = 5$, the focus is at $F(-5,-2)$ and the equation of the directrix is $x - 5 = 0$. See Fig. 35-14 (b)

35.2   Find the equation of the parabola, given

(a) $V(0,0)$; $F(0,-4)$

(b) $V(0,0)$; directrix: $x = -5$

(c) $V(1,4)$; $F(-2,4)$

(d) $F(2,3)$; directrix: $y = -1$

(e) $V(0,0)$; axis: $y = 0$; passing through (4,5).
    Ans.
    (a) Since the directed distance $p = VF = -4$, the parabola opens downward. Its equation is $x^2 = -16y$.

    (b) The parabola opens to the right (away from the directrix). Since $p = DV = 5$, the equation is $y^2 = 20x$.

(c) Here the focus lies to the left of the vertex, and the parabola opens to the left. The directed distance $p = VF = -3$ and the equation is $(y - 4)^2 = -12(x - 1)$.

(d) Here the focus lies above the directrix and the parabola opens upward. The axis of the parabola meets the directrix in $D(2, -1)$ and the vertex is at the midpoint $V(2, 1)$ of $\overline{FD}$. Then $p = VF = 2$ and the equation is $(x - 2)^2 = 8(y - 1)$.

(e) The equation of this parabola is of the form $y^2 = 4px$. If $(4, 5)$ is a point on it, then $(5)^2 = 4p(4)$, $4p = 25/4$, and the equation is $y^2 = 25/4x$.

35.3 The cable of a suspension bridge has supporting towers which are 50 ft high and 400 ft apart and is in the shape of a parabola. If the lowest point of the cable is 10 ft above the floor of the bridge, find the length of a supporting rod 100 ft from the center of the span.

Ans.
Take the origin of coordinates at the lowest point of the cable and the positive y axis directed upward along the axis of symmetry of the parabola. Then the equation of the parabola has the form $x^2 = 4py$. Since $(200, 40)$ is a point on the parabola, $(200)^2 = 4p \cdot 40$ or $4p = 1000$, and the equation is $x^2 = 1000y$.
When $x = 100$, $(100)^2 = 1000y$, and $y = 10$ ft. The length of the supporting rod is $10 + 10 = 20$ ft.

35.4 For each of the following ellipses find the coordinates of the center, vertices, and foci; the lengths of the major and minor axes; the eccentricity; and the equations of the directrices. Sketch the curve.

(a) $x^2/16 + y^2/4 = 1$
(b) $25x^2 + 9y^2 = 25$
(c) $x^2 + 9y^2 + 4x - 18y - 23 = 0$

Ans.
(a) Here $a^2 = 16$, $b^2 = 4$, and $c = \sqrt{a^2 - b^2} = \sqrt{16 - 4} = 2\sqrt{3}$.

The center is at the origin and the major axis is along the x axis (a^2 under x^2). The vertices are on the major axis at a distance $a = 4$ from the center; their coordinates are $V(4, 0)$ and $V'(-4, 0)$. The minor axis is along the y axis and its extremities, being at a distance $b = 2$ from the center, are at $B(0, 2)$ and $B'(0, -2)$. The foci are on the major axis at a distance $2\sqrt{3}$ from the center; their coordinates are $F(2\sqrt{3}, 0)$ and $F'(-2\sqrt{3}, 0)$.
The lengths of the major and minor axes are $2a = 8$ and $2b = 4$, respectively.
The eccentricity is $e = c/a = 2\sqrt{3}/4 = 1/2\sqrt{3}$.
The directrices are perpendicular to the major axis and at a distance $a^2/c = (16/2)\sqrt{3} = 8\sqrt{3}/3$ from the center; their equations are $x = \pm 8\sqrt{3}/3$.
See Fig. 35-15(a).

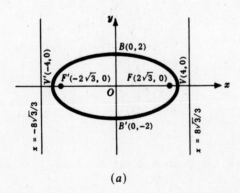

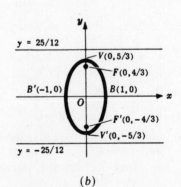

(a)   (b)

Fig. 35-15

(b) When the equation is put in the form

$$\frac{x^2}{1} + \frac{y^2}{\frac{25}{9}} = 1$$

we find $a^2 = 25/9$, $b^2 = 1$, and $c = \sqrt{a^2 - b^2} = \sqrt{25/9 - 1} = 4/3$.

The center is at the origin and the major axis is along the y axis (a^2 under y^2). The vertices are on the major axis at a distance $a = 5/3$ from the center; their coordinates are $V(0,5/3)$ and $V'(0,-5/3)$. The extremities of the minor axis are on the x axis at a distance $b = 1$ from the center; their coordinates are $B(1,0)$ and $B'(-1,0)$. The foci are on the major axis at a distance $c = 4/3$ from the center; their coordinates are $F(0,4/3)$ and $F'(0,-4/3)$.

The lengths of the major and minor axes are $2a = 10/3$ and $2b = 2$, respectively.

The eccentricity is

$$e = \frac{c}{a} = \frac{4/3}{5/3} = \frac{4}{5}$$

The directrices are perpendicular to the major axis and at a distance $a^2/c = 25/12$ from the center; their equations are $y = \pm 25/12$. See Fig. 35-15(b).

(c) When the equation is put in the form

$$\frac{(x+2)^2}{36} + \frac{(y-1)^2}{4} = 1$$

we have $a^2 = 36$, $b^2 = 4$, and $c = \sqrt{a^2 - b^2} = 4\sqrt{2}$.

The center is at the point $C(-2, 1)$ and the major axis is along the line $y = 1$. The vertices are on the major axis at a distance $a = 6$ from the center; their coordinates are $V(4,1)$ and $V'(-8,1)$. The extremities of the minor axis are on the line $x = -2$ at a distance $b = 2$ from the center; their coordinates are $B(-2,3)$ and $B'(-2,-1)$. The foci are on the major axis at a distance $c = 4\sqrt{2}$ from the center; their coordinates are $F(-2+4\sqrt{2},1)$ and $F'(-2-4\sqrt{2},1)$.

The lengths of the major and minor axes are $2a = 12$ and $2b = 4$, respectively. The eccentricity is $c/a = 4\sqrt{2}/6 = 2\sqrt{2}/3$.

The directrices are perpendicular to the major axis and at a distance $a^2/c = 9\sqrt{2}/2$ from the center; their equations are $x = -2 \pm 9\sqrt{2}/2$. See Fig. 35-16.

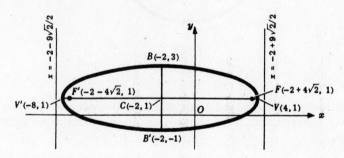

Fig. 35-16

35.5   Find the equation of the ellipse, given

(a) Vertices $(\pm 8, 0)$, minor axis $= 6$.

(b) One vertex at $(0,13)$, one focus at $(0,-12)$, center at $(0,0)$.

(c) Foci $(\pm 10, 0)$, eccentricity $= 5/6$.

(d) Vertices $(8,3)$ and $(-4,3)$, one focus at $(6,3)$.

(e) Directrices $4y - 33 = 0$, $4y + 17 = 0$; major axis on $x + 1 = 0$; eccentricity $= 4/5$.

Ans.

(a) Here $2a = V'V = 16$, $2b = 6$, and the major axis is along the x axis. The equation of the ellipse is $x^2/a^2 + y^2/b^2 = x^2/64 + y^2/9 = 1$.

(b) The major axis is along the y axis, $a = 13$, $c = 12$, and $b^2 = a^2 - c^2 = 25$. The equation of the ellipse is $x^2/b^2 + y^2/a^2 = x^2/25 + y^2/169 = 1$.

(c) Here the major axis is along the x axis and $c = 10$. Since $e = c/a = 10/a = 5/6$, $a = 12$ and $b^2 = a^2 - c^2 = 44$. The equation of the ellipse is $x^2/144 + y^2/44 = 1$.

(d) The center is at the midpoint of $\overline{V'V}$, that is, at $C(2,3)$. Then $a = CV = 6$, $c = CF = 4$, and $b^2 = a^2 - c^2 = 20$. Since the major axis is parallel to the x axis, the equation of the ellipse is

$$\frac{(x-2)^2}{36} + \frac{(y-3)^2}{20} = 1$$

(e) The major axis intersects the directrices in $D(-1, 33/4)$ and $D'(-1, -17/4)$. The center of the ellipse bisects $\overline{D'D}$ and, hence, is at $C(-1, 2)$. Since $CD = a/e = 25/4$ and $e = 4/5$, $a = 5$ and $c = ae = 4$. Then $b^2 = a^2 - c^2 = 9$. Since the major axis is parallel to the y axis, the equation of the ellipse is

$$\frac{(x+1)^2}{9} + \frac{(y-2)^2}{25} = 1$$

The locus passes through the center $C(0,0)$ and is called a diameter of the ellipse. See Fig. 35-17.

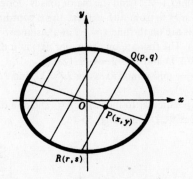

Fig. 35-17

35.6   For each of the following hyperbolas, find the coordinates of the center, vertices, and the foci; the lengths of the transverse and conjugate axes; the eccentricity; and the equations of the directrices and asymptotes. Sketch each locus.

(a) $x^2/16 + y^2/4 = 1$

(b) $25y^2 - 9y^2 = 225$

(c) $9x^2 - 4y^2 - 36x + 32y + 8 = 0$

Ans.

(a) Here $a^2 = 16$, $b^2 = 4$, and $c = \sqrt{a^2 + b^2} = 2\sqrt{5}$.

The center is at the origin and the transverse axis is along the x axis (a^2 under x^2). The vertices are on the transverse axis at a distance $a = 4$ from the center; their coordinates are $V(4,0)$ and $V'(-4,0)$. The extremities of the conjugate axis are on the y axis at a distance $b = 2$ from the center; their coordinates are $B(0,2)$ and $B'(0,-2)$.

The foci are on the transverse axis at a distance $c = 2\sqrt{5}$ from the center; their coordinates are $F(2\sqrt{5}, 0)$ and $F'(-2\sqrt{3}, 0)$.

The lengths of the transverse and conjugate axes are $2a = 8$ and $2b = 4$, are respectively.

The eccentricity is $e = c/a = 2\sqrt{5}/4 = 1/2\sqrt{5}$.

The directrices are perpendicular to the transverse axis and at a distance $a^2/c = 16/2\sqrt{5} = 8\sqrt{5}/5$ from the center; their equations are $x = \pm 8\sqrt{5}/5$.

The equations for the asymptotes are $x^2/16 - y^2/4 = 0$ or $x = \pm 2y$. See Fig. 35-18.

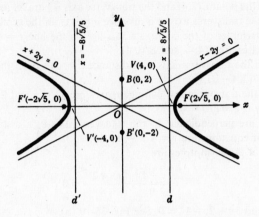

Fig. 35-18

(b) When the equation is put in the form

$$\frac{y^2}{9} - \frac{x^2}{25} = 1$$

we find $a^2 = 9$, $b^2 = 25$, and $c = \sqrt{a^2 + b^2} = 34$.

The center is at the origin and the transverse axis is along the y axis. The vertices are on the transverse axis at a distance $a = 3$ from the center; their coordinates are $V(0,3)$ and $V'(0,-3)$. The extremities of the conjugate axis are on the x axis and at a distance $b = 5$ from the center; their coordinates are $B(5,0)$ and $B'(-5,0)$.

The foci are on the transverse axis at a distance $c = \sqrt{34}$ from the center; their coordinates are $F(0,\sqrt{34})$ and $F'(0,-\sqrt{34})$.

The lengths of the transverse and conjugate axes are $2a = 6$ and $2b = 10$, respectively.

The eccentricity is $e = c/a = \sqrt{34}/3$.

The directrices are perpendicular to the transverse axis and at a distance $a^2/c = 9/\sqrt{34} = 9\sqrt{34}/34$ from the center; their equations are $y = \pm 9\sqrt{34}/34$.

The equations of the asymptotes are $y^2/9 - x^2/25 = 0$ or $5y = \pm 3x$. See Fig. 35-19(a).

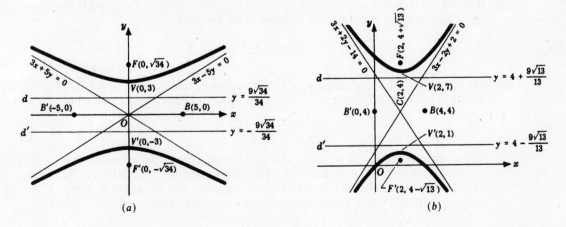

Fig. 35-19

(c) Putting the equation in the form

$$\frac{(y-4)^2}{9} - \frac{(x-2)^2}{4} = 1$$

we have $a^2 = 9$, $b^2 = 4$, and $c = \sqrt{a^2 + b^2} = \sqrt{13}$.

The center is at the point $C(2,4)$ and the transverse axis is parallel to the y axis along the line $x = 2$. The vertices are on the transverse axis at a distance $a = 3$ from the center; their coordinates are $V(2,7)$ and $V'(2,1)$. The extremities of the conjugate axis are on the line $y = 4$ at a distance $b = 2$ from the center; their coordinates are $B(4,4)$ and $B'(0,4)$.

The foci are on the transverse axis at a distance $c = \sqrt{13}$ from the center; their coordinates are $F(2, 4 + \sqrt{13})$ and $F'(2, 4 - \sqrt{13})$.

The lengths of the transverse and conjugate axes are $2a = 6$ and $2b = 4$, respectively.

The eccentricity is $e = c/a = \sqrt{13}/3$.

The directrices are perpendicular to the transverse axis at a distance $a^2/c = 9/\sqrt{13} = 9\sqrt{13}/13$ from the center; their equations are $y = 4 \pm 9\sqrt{13}/13$.

The equations of the asymptotes are

$$\frac{(y-4)^2}{9} - \frac{(x-2)^2}{4} = 0$$

or $3x - 2y + 2 = 0$ and $3x + 2y - 14 = 0$. See Fig. 35-19 (b).

35.7 Find the equation of the hyperbola, given

(a) Center $(0,0)$, vertex $(4,0)$, focus $(5,0)$.

(b) Center $(0,0)$, focus $(0,-4)$, eccentricity $= 2$.

(c) Center $(0,0)$, vertex $(5,0)$, one asymptote $5y + 3x = 0$.

(d) Center $(-5,4)$, vertex $(-11,4)$, eccentricity $= 5/3$.

(e) Vertices $(-11, 1)$ and $(5,1)$, one asymptote $x - 4y + 7 = 0$.

(f) Transverse axis parallel to the x axis, asymptotes $3x + y - 7 = 0$ and $3x - y - 5 = 0$, passes through $(4,4)$.

Ans.

(a) Here $a = CV = 4$, $c = CF = 5$, and $b^2 = c^2 - a^2 = 25 - 16 = 9$. The transverse axis is along the x axis and the equation of the hyperbola is $x^2/16 - y^2/9 = 1$.

(b) Since $c = F'C = 4$ and $e = c/a = 2$, $a = 2$ and $b^2 = c^2 - a^2 = 12$. The transverse axis is along the y axis and the equation of the hyperbola is $y^2/4 - x^2/12 = 1$.

(c) The slope of the asymptote is $-b/a = -3/5$ and, since $a = CV = 5$, $b = 3$. The transverse axis is along the x axis and the equation of the hyperbola is $x^2/25 - y^2/9 = 1$.

(d) Here $a = V'C = 6$ and $e = c/a = 5/3 = 10/6$; then $c = 10$ and $b^2 = c^2 - a^2 = 64$. The transverse axis is parallel to the x axis and the equation of the hyperbola is

$$\frac{(x+5)^2}{36} - \frac{(y-4)^2}{64} = 1$$

(e) The center is at $(-3, 1)$, the midpoint of VV'. The slope of the asymptote is $b/a = 1/4$ and, since $a = CV = 8$, $b = 2$. The transverse axis is parallel to the x axis and the equation of the hyperbola is

$$\frac{(x+3)^2}{64} + \frac{(y-1)^2}{4} = 1$$

(f) The asymptotes intersect in the center $C(2, 1)$. Since the slope of the asymptote $3x - y - 5 = 0$ is $b/a = 3/1$, we take $a = m$ and $b = 3m$. The equation of the hyperbola may be written as $(x-2)^2/m^2 - (y-1)^2/9m^2 = 1$. In order that the hyperbola pass through $(4,4)$, $4/m^2 - 9/9m^2 = 1$ and $m = 3$. Then $a = m = 3$, $b = 3m = 3\sqrt{3}$, and the required equation is

$$\frac{(x-2)^2}{3} + \frac{(y-1)^2}{27} = 1$$

35.8   Find the distance from the right-hand focus of $9x^2 - 4y^2 + 54x + 16y - 79 = 0$ to one of its asymptotes.

Ans. Putting the equation in the form

$$\frac{(x+3)^2}{16} - \frac{(y-2)^2}{36} = 0$$

we find $a^2 = 16$, $b^2 = 36$, and $c = \sqrt{a^2 + b^2} = 2\sqrt{13}$. The right-hand focus is at $F(-3 + 2\sqrt{13}, 2)$ and the equations of the asymptotes are

$$\frac{(x+3)^2}{16} - \frac{(y-2)^2}{36} = 0 \qquad \text{or} \qquad 3x + 2y + 5 = 0 \qquad \text{and} \qquad 3x - 2y + 13 = 0$$

The distance from F to the first asymptote is

$$\left| \frac{3(-3 + 2\sqrt{13}) + 2.2 + 5}{-\sqrt{13}} \right| = 6$$

35.9   Write the equation of the conjugate of the hyperbola $25x^2 - 16y^2 = 400$ and sketch both curves.

Ans.

The equation of the conjugate hyperbola is $16y^2 - 25x^2 = 400$. The common asymptotes have equations $y = \pm 5x/4$. The vertices of $25x^2 - 16y^2 = 400$ are at $(\pm 4, 0)$. The vertices of $16y^2 - 25x^2 = 400$ are at $(0, \pm 5)$. The curves are shown in Fig. 35-20.

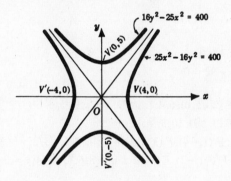

Fig. 35-20

Supplementary Problems

35.10   For each of the following parabolas, sketch the curve, find the coordinates of the vertex and focus, and find the equations of the axis and directrix.

(a) $x^2 = 12y$ *Ans.* $V(0,0), F(0,3); x = 0, y + 3 = 0$

(b) $y^2 = -10x$ *Ans.* $V(0,0), F(-5/2,0); y = 0, 2x - 5 = 0$

(c) $x^2 - 6 + 8y + 25 = 0$ *Ans.* $V(3,-2), F(3,-4); x - 3 = 0, y = 0.$

(d) $y^2 - 16x + 2y + 49 = 0$ *Ans.* $V(3,-1), F(7,-1); y + 1 = 0, x + 1 = 0$

(e) $x^2 - 2x - 6y - 53 = 0$ *Ans.* $V(1,-9), F(1,-15/2); x - 1 = 0, 2y + 21 = 0$

(f) $y^2 + 20x + 4y - 60 = 0$ *Ans.* $V(16/5,-2), F(-9/5,-2); y + 2 = 0, 5x - 41 = 0$

35.11 Find the equation of the parabola, given

    (a) $V(0,0), F(-2,0)$

    (b) $V(0,0), F(0,5)$

    (c) $V(0,0), d: y+3=0$

    (d) $V(0,0), F$ on x axis, passes through $(-2,6)$

    (e) $V(1,3), F(-1,3)$

    (f) $F(3,2), d: y+4=0$

    Ans.

    (a) $y^2 = -8x$

    (b) $x^2 = 20y$

    (c) $x^2 = 12y$

    (d) $y^2 = -18x$

    (e) $y^2 + 8x - 6y + 1 = 0$

    (f) $x^2 - 6x - 12y - 3 = 0$

35.12 For each of the following ellipses, find the coordinates of the center, vertices, and foci; the lengths of the major and minor axes; the eccentricity; and the equations of the directrices. Sketch each curve.

    (a) $4x^2 + 9y^2 = 36$

    (b) $25x^2 + 16y^2 = 400$

    (c) $x^2 + 4y^2 - 6x + 32y + 69 = 0$

    (d) $16x^2 + 9y^2 + 32x - 36y - 92 = 0$

    Ans.

    (a) $C(0,0), V(\pm 3,0); F(\pm\sqrt{5},0); 6,4; \frac{8}{3}; \sqrt{5}/3; x = \pm 9\sqrt{5}/5$

    (b) $C(0,0), V(0,\pm 5); F(0,\pm 3); 10,8; \frac{32}{5}; \frac{3}{5}; y = \pm\frac{25}{3}$

    (c) $C(3,-4), V(5,-4), V'(1,-4), F(3\pm\sqrt{3},-4); 4, 2; 1; \sqrt{3}/2; x = 3 \pm 4\sqrt{3}/3$

    (d) $C(-1,2), V(-1,6); V'(-1,-2), F(-1,2\pm\sqrt{7}); 8,6; \frac{9}{2}; \sqrt{7}/4; y = 2 \pm 16\sqrt{7}/7$

35.13 Find the equation of the ellipse, given

    (a) $V(\pm 13,0), F(12,0)$

    (b) $C(0,0), a = 5, F(0,4)$

    (c) $C(0,0), b = 2, d: x = 16\sqrt{7}/7$

    (d) $V(7,3), V'(-3,3), F(6,3)$

    (e) $F(5,4), F'(5,-2), e = \sqrt{3}/3$

    (f) Ends of minor axis $(-2,4), (-2,2); d: x = 0$

    (g) Directrices: $y = 11/5, y = -61/5$; major axis on $x = 3, e = 5/6$

    Ans.

    (a) $25x^2 + 169y^2 = 4225$

    (b) $25x^2 + 9y^2 = 225$

    (c) $x^2 + 8y^2 = 32, 7x^2 + 8y^2 = 32$

    (d) $9x^2 + 25y^2 - 36x - 150y + 36 = 0$

    (e) $3x^2 + 2y^2 - 30x - 4y + 23 = 0$

    (f) $x^2 + 2y^2 + 4x - 12y + 20 = 0$

    (g) $36x^2 + 11y^2 - 216x + 110y + 203 = 0$

35.14 For each of the following hyperbolas, find the coordinates of the center, the vertices, and the foci; the lengths of the transverse and conjugate axes; the eccentricity; the equations of the directrices; and the equations of the asymptotes. Sketch each curve.

   (a) $4x^2 - 9y^2 = 36$

   (b) $16y^2 - 9x^2 = 144$

   (c) $x^2 - 4y^2 + 6x + 16y - 11 = 0$

   (d) $144x^2 - 25y^2 - 576x + 200y + 3776 = 0$

   Ans.

   (a) $C(0,0), V(\pm 3, 0); \ F(\pm\sqrt{13}, 0); \ 6, 4, \frac{8}{3}; \ \sqrt{13}/3; \ x = \pm 9\sqrt{13}/13; \ 2x \pm 3y = 0.$

   (b) $C(0,0), V(0, \pm 3); \ F(0, \pm 5); \ 6, 8; \frac{32}{3}; \frac{5}{3}; \ y = \pm\frac{9}{5}; \ 3x \pm 4y = 0$

   (c) $C(-3, 2), V(-1, 2); \ V'(-5, 2), F(-3 \pm \sqrt{5}, 2); \ 4, 2; 1; \sqrt{5}/2; \ x = -3 \pm 4\sqrt{5}/5; \ x + 2y - 1 = 0,$
        $x - 2y + 7 = 0.$

   (d) $C(2, 4), V(2, 16); \ V'(2, -8), F(2, 17); \ F'(2, -9); \ 24, 10; \frac{25}{6}; \frac{13}{12}; \ y = 4 \pm \frac{144}{13};$
        $12x - 5y - 4 = 0, \ 12x + 5y - 44 = 0.$

35.15 Find the equation of the hyperbola, given

   (a) $V(\pm 5, 0), F(13, 0)$

   (b) $C(0,0), a = 5, F(0,6)$

   (c) $C(0,0), b = 5, d : y = \pm 16\sqrt{41}/41$

   (d) $C(2, -3), V(7, -3),$ asymptote $3x - 5y - 21 = 0$

   (e) $C(-3, 1), F(-3, 5),$ eccentricity equal to 2

   (f) $C(2, 4);$ asymptotes $x + 2y - 10 = 0, x - 2y + 6 = 0;$ passes through $(2, 0)$

   (g) $C(-3, 2), F(2, 2),$ asymptote $4x + 3y + 6 = 0$

   Ans.

   (a) $144x^2 - 25y^2 = 3600$

   (b) $25x^2 - 11y^2 + 275 = 0$

   (c) $16x^2 - 25y^2 + 400 = 0$

   (d) $9x^2 - 25y^2 - 36x - 150y - 414 = 0$

   (e) $x^2 - 3y^2 + 6x + 6y + 18 = 0$

   (f) $x^2 - 4y^2 - 4x + 32y + 4 = 0$

   (g) $16x^2 - 9y^2 + 96x + 36y - 36 = 0$

Chapter 36

Transformation of Coordinates

THE MOST GENERAL EQUATION of the second degree in x and y has the form

$$Ax^2 + 2Bxy + Cy^2 + 2Dx + 2Ey + F = 0 \qquad (36.1)$$

If (36.1) can be factored so that we have $(ax + by + c)(dx + ey + f) = 0$, the locus consists of two straight lines; if $B = 0, A = C$, the locus of (36.1) is a circle; otherwise, the locus is one of the conics of Chapter 35.

The locus of the equation

$$20x^2 - 24xy + 27y^2 + 24x - 54y - 369 = 0 \qquad (1)$$

and the locus of the equation

$$11x^2 + 36y^2 - 369 = 0 \qquad (2)$$

are identical ellipses. The difference in equations is due to their positions with respect to the coordinate axes.

In order to make a detailed study of the loci represented by (36.1), say (1), it will be necessary to introduce some device to change (1) into (2). The operations by which (1) is eventually replaced by (2) are two *transformations*. The general effect of these transformations may be interpreted as follows: Each point (x, y) of the plane remains fixed but changes its name, i.e., its coordinates, in accordance with a stated law, called the *equations of the tranformation*.

TRANSLATION OF THE COORDINATE AXES. Recall that the transformation which moves the coordinate axes to a new position while keeping them always parallel to their original position is called a *translation*. In Fig. 36-1, $\overrightarrow{Ox}$ and $\overrightarrow{Oy}$ are the axes and O is the origin of the original system of coordinates, while $\overrightarrow{O'x'}$ and $\overrightarrow{O'y'}$ are the axes and O' is the origin of the new (translated) system.

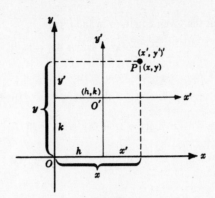

Fig. 36-1

Each point in the plane will now have two sets of coordinates, the original set being the directed distances in proper order of the point from the original axes and the new set being the directed distances from the new axes. In order to avoid errors, we propose to write the coordinates of a point when referred to the original system as, for example, $A(a, b)$ and the coordinates when referred to the new system as $A(c, d)'$. Also, we shall find it convenient at times to speak of the unprimed and primed systems.

If the axes with the origin O are translated to a new position with origin O' having coordinates (h, k) when referred to the original system and if the coordinates of any point are (x, y) before and $(x', y')'$ after the translation, then the equations of transformation are

$$x = x' + h \qquad y = y' + k \qquad\qquad (36.2)$$

EXAMPLE 1.　By means of a translation, transform $3x^2 + 4y^2 - 12x + 16y - 8 = 0$ into another equation which lacks terms of the first degree.

　　　　　First Solution.　When the values of x and y from (36.2) are substituted in the given equation, we obtain

$$3(x' + h)^2 + 4(y' + k)^2 - 12(x' + h) + 16(y' + k) - 8 = 0$$

or　　　　$3x'^2 + 4y'^2 + (6h - 12)x' + (8k + 16)y' + 3h^2 + 4k^2 - 12h + 16k - 8 = 0 \qquad (3)$

The equation will lack terms of the first degree provided $6h - 12 = 0$ and $8k + 16 = 0$, that is, provided $h = 2$ and $k = -2$. Thus, the translation $x = x' + 2, y = y' - 2$ reduces the given equation to $3x'^2 + 4y'^2 36 = 0$. The locus, an ellipse, together with the original and new system of coordinates, is shown in Fig. 36-2.

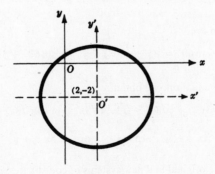

Fig. 36-2

Second Solution.   We put the given equation in the form

$$3(x^2 - 4x) + 4(y^2 + 4y) = 8$$

and complete the squares to obtain

$$3(x^2 - 4x + 4) + 4(y^2 + 4y + 4) = 8 + 3(4) + 4(4) = 36$$

or

$$3(x - 2)^2 + 4(y + 2)^2 = 36 \tag{4}$$

The transformation $x - 2 = x', y + 2 = y'$ or $x = x' + 2, y = y' - 2$ reduces (4) to $3x'^2 + 4y'^2 = 36$ as before. (See Problem 36.2–36.3.)

ROTATION OF THE COORDINATE AXES.   Recall that the transformation which holds the origin fixed while rotating the coordinate axes through a given angle is called a *rotation*.

If, while the origin remains fixed, the coordinate axes are rotated counterclockwise through an angle of measure θ, and if the coordinates of any point P are (x, y) before and (x', y') after the rotation, the equations of transformation are

$$x = x'\cos\theta - y'\sin\theta \qquad y = x'\sin\theta + y'\cos\theta \tag{36.3}$$

since, from Fig. 36-3,

$$x = OM = ON - MN = ON - RQ = OQ\cos\theta - QP\sin\theta = x'\cos\theta - y'\sin\theta$$

and

$$y = MP = MR + RP = NQ + RP = OQ\sin\theta + QP\cos\theta = x'\sin\theta + y'\cos\theta$$

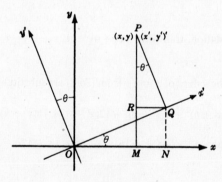

Fig. 36-3

EXAMPLE 2.   Transform the equation $x^2 + \sqrt{3}xy + 2y^2 - 5 = 0$ by rotating the coordinate axes through the angle $60°$.

The equations of transformation are

$$x = x'\cos 60° - y'\sin 60° = \tfrac{1}{2}(x' - \sqrt{3}y'), \quad y = x'\sin 60° + y'\cos 60° = \tfrac{1}{2}(\sqrt{3}x' + y').$$

Substituting for x and y in the given equation, we obtain

$$\tfrac{1}{4}(x' - \sqrt{3}y')^2 + \tfrac{1}{4}\sqrt{3}(x' - \sqrt{3}y')(\sqrt{3}x' + y) + \tfrac{1}{4}(\sqrt{3}x' + y')^2 - 5 = 0$$

$$\tfrac{5}{2}x'^2 + \tfrac{1}{2}y'^2 - 5 = 0 \qquad \text{or} \qquad 5x'^2 + y'^2 = 10$$

The locus, an ellipse, together with the original and new systems of coordinates, is shown in Fig. 36-4. (See Problems 36.4–36.6)

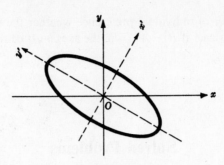

Fig. 36-4

THE SEMIREDUCED FORM OF THE SECOND-DEGREE EQUATION.   Under a rotation of the coordinate axes with equations of transformation (*36.3*), the general equation of the second degree

$$Ax^2 + 2Bxy + Cy^2 + 2Dx + 2Ey + F = 0 \qquad (36.1)$$

becomes

$$A'x'^2 + 2B'x'y' + C'y'^2 + 2D'x' + 2E'y' + F' = 0 \qquad (36.1')$$

where

$$A' = A\cos^2\theta + 2B\sin\theta\cos\theta + C\sin^2\theta$$
$$B' = (C-A)\sin\theta\cos\theta + B(\cos^2\theta - \sin^2\theta)$$
$$= \tfrac{1}{2}(C-A)\sin 2\theta + B\cos 2\theta \qquad (36.4)$$
$$C' = A\sin^2\theta - 2B\sin\theta\cos\theta + C\cos^2\theta$$
$$D' = D\cos\theta + E\sin\theta \qquad E' = E\cos\theta - D\sin\theta \qquad F' = F$$

If $B \neq 0$, (*36.1'*) will lack the term in $x'y'$ if θ is such that $0 < m\measuredangle\theta < 90°$ and $\tan 2\theta = \dfrac{2B}{A-C}$ when $A \neq C$, and $m\measuredangle\theta = 45°$ when $A = C$. Under this transformation, the general equation (*36.1*), takes the form

$$A'x'^2 + C'y'^2 + 2D'x' + 2E'y' + F' = 0 \qquad (36.1'')$$

which will be called the *semireduced form* of the second-degree equation.

Under *any* rotation of the coordinate axes, the quantities $A + C$ and $B^2 - AC$ are unchanged or *invariant*; that is, $A + C = A' + C'$ and $B^2 - AC = B'^2 - A'C'$. When (*36.1*) is transformed into (*36.1''*) $B' = 0$ and $B^2 - AC = -A'C'$. Then

If $B^2 - AC < 0$, A' and C' agree in sign and (*36.1*) represents a real ellipse, a point ellipse, or an imaginary ellipse.

If $B^2 - AC = 0$, either $A' = 0$ or $C' = 0$; now (*36.1''*) contains either a term in x'^2 or y'^2 (but not both), and (*36.1*) represent a parabola or a pair of parallel lines.

If $B^2 - AC > 0$, A' and C' differ in sign and (*36.1*) represents a hyperbola or a pair of intersecting lines.

THE REDUCED FORM OF THE SECOND-DEGREE EQUATION.   Under a suitable translation the semireduced form (*36.1''*) of the second-degree equation takes the *reduced form*

$$A'x''^2 + C'y''^2 = F'' \quad \text{when} \quad A'C' \neq 0$$

and

$$x''^2 = Gy'' \quad \text{or} \quad y''^2 = Hx'' \quad \text{when} \quad A'C' = 0$$

If $B^2 - AC \neq 0$, it is a matter of individual preference whether the rotation is performed before or after the translation of axes; however, if $B^2 - AC = 0$, the axes *must* be rotated first. (See Problems 36.7–36.9.)

Solved Problems

36.1 If the equation, of translation are $x = x' - 3, y = y' + 4$, find

(a) The coordinates of $O(0,0)$ when referred to the primed system of coordinates

(b) The coordinates of $O'(0,0)'$ when referred to the unprimed system

(c) The coordinates of $P(5,-3)$ when referred to the primed system

(d) The coordinates of $P(5,-3)'$ when referred to the unprimed system

(e) The equation of $l : 2x - 3y + 18 = 0$ when referred to the primed system

   (a) For $x = 0, y = 0$ the equations of transformation yield $x' = 3, y' = -4$; thus, in the primed system we have $O(3,-4)'$.

   (b) For $x' = 0, y' = 0$ the equations of transformation yield $x = -3, y = 4$; thus, in the unprimed system we have $O'(-3,4)$.

   (c) Here $x = 5, y = -3$; then $x' = 8, y' = -7$. Thus, we have $P(8,-7)'$.

   (d) Here $x' = 5, y' = -3$; then $x = 2, y = 1$. Thus, we have $P(2,1)$.

   (e) When the values for x and y are substituted in the given equation, we have $2(x' - 3) - 3(y' + 4) + 18 = 2x' - 3y = 0$ as the equation of l in the primed system. Note that the new origin was chosen on the line l.

36.2 Transform each of the following equations into another lacking terms of the first degree:

(a) $x^2 + 4y^2 - 2x - 12y + 1 = 0$,   (b) $9x^2 - 16y^2 - 36x - 96y - 252 = 0$,   (c) $xy + 4x - y - 8 = 0$.

(a) Since the given equation lacks a term in xy, we use the second method of Example 1. We have

$$(x^2 - 2x) + 4(y^2 - 3y) = -1 \qquad (x^2 - 2x + 1) + 4\left(y^2 - 3y + \tfrac{9}{4}\right) = -1 + 1 + 4\left(\tfrac{9}{4}\right) = 9$$

and

$$(x - 1)^2 + 4\left(y - \tfrac{3}{2}\right)^2 = 9$$

This equation taken the form $x'^2 + 4y'^2 = 9$ under the transformation $x - 1 = x', y - \tfrac{3}{2} = y'$ or $x = x' + 1, y = y' + \tfrac{3}{2}$.

(b) We have

$$9(x^2 - 4x) - 16(y^2 + 6y) = 252 \qquad 9(x^2 - 4x + 4) - 16(y^2 + 6y + 9) = 252 + 36 - 144 = 144$$

and

$$9(x - 2)^2 - 16(y + 3)^2 = 144$$

This equation taken the form $9x'^2 - 16y'^2 = 144$ under the transformation $x - 2 = x', y + 3 = y'$ or $x = x' + 2, y = y' - 3$.

(c) Since the given equation contains a term in xy, we must use the first method of Example 1. We have, using Equations (*36.2*),

$$(x' + h)(y' + k) + 4(x' + h) - (y' + k) - 8 = x'y' + (k + 4)x' + (h - 1)y' + hk + 4h - k - 8 = 0$$

The first-degree terms will disappear provided $k + 4 = 0$ and $h - 1 = 0$, that is, provided we take $h = 1$ and $k = -4$. For this choice, the equation becomes $x'y' - 4 = 0$.

36.3 By a translation of the axes, simplify each of the following:

(a) $x^2 + 6x - 4y + 1 = 0$,   (b) $y^2 + 4y + 8x - 2 = 0$.

(a) Since the given equation lacks a term in y^2, it is not possible to use the second method of Example 1. Using the transformation (36.2), we find

$$(x' + h)^2 + 6(x' + h) - 4(y' + k) + 1 = x'^2 + 2(h + 3)x' - 4y' + h^2 + 6h - 4k + 1 = 0$$

If we take $h = -3$, the term in x' disappears but it is clear that we cannot make the term in y' disappear. However, in this case, we make the constant term

$$h^2 + 6h - 4k + 1 = (-3)^2 + 6(-3) - 4k + 1 = -8 - 4k$$

disappear by taking $k = -2$. Thus, the transformed equation becomes

$$x'^2 - 4y' = 0 \quad\text{or}\quad x'^2 = 4y'$$

It is now clear that this simplification may be effected by the following variation of the second method of Example 1:

$$x^2 + 6x = 4y - 1 \qquad x^2 + 6x + 9 = 4y - 1 + 9 = 4y + 8$$

or $$(x + 3)^2 = 4(y + 2)$$

Then the transformation $x + 3 = x', y + 2 = y'$ or $x = x' - 3, y = y' - 2$ reduces the equation to $x'^2 = 4y'$.

(b) We have $y^2 + 4y = -8x + 2$, $y^2 + 4y + 4 = -8x + 2 + 4 = -8x + 6, (y + 2)^2 = -8(x - \frac{3}{4})$, and finally $y'^2 = -8x'$ under the transformation $x = x' + \frac{3}{4}, y = y' - 2$.

36.4 Write the equations of transformation for a rotation of the coordinate axes through an angle of 45° and use them to find

(a) The coordinates of $P(\sqrt{2}, 3\sqrt{2})'$ when referred to the original (unprimed) system

(b) The coordinates of $O(0, 0)$ when referred to the (primed) system

(c) The coordinates of $P(\sqrt{2}, 3\sqrt{2})$ when referred to the primed system

(d) The equation of the line $l: x + y + 3\sqrt{2} = 0$ when referred to the primed system

(e) The equation of the line $l: 3x - 3y + 4 = 0$ when referred to the prime system

The equations of transformation are

$$x = x'\cos 45° - y'\sin 45° = \frac{1}{\sqrt{2}}(x' - y')$$

$$y = x'\sin 45° + y'\cos 45° = \frac{1}{\sqrt{2}}(x' + y')$$

(a) For $x' = \sqrt{2}, y' = 3\sqrt{2}$, the equations of transformation yield

$$x = \frac{1}{\sqrt{2}}(\sqrt{2} - 3\sqrt{2}) = -2 \qquad y = \frac{1}{\sqrt{2}}(\sqrt{2} + 3\sqrt{2}) = 4$$

Thus, in the unprimed system the coordinates are $P(-2, 4)$.

(b) When the equations of transformation are solved for x' and y', we have

$$x' = \frac{1}{\sqrt{2}}(x + y) \qquad y' = -\frac{1}{\sqrt{2}}(x - y)$$

For $x = 0, y = 0$ these equations yield $x' = 0, y' = 0$; thus, in the primed system, we have $O(0, 0)'$. Since the coordinates are unchanged, the origin is called an *invariant point* of the transformation.

(c)   For $x = \sqrt{2}, y = 3\sqrt{2}$ the equations of (b) yield $x' = 4, y' = 2$. In the primed system, we have $P(4, 2)'$.

(d)   When the values for x and y from the equations of transformation are substituted in the given equation of the line, we have

$$\frac{1}{\sqrt{2}}(x' - y') + \frac{1}{\sqrt{2}}(x' + y') + 3\sqrt{2} = \frac{2}{\sqrt{2}}x' + 3\sqrt{2} = 0 \quad \text{or} \quad x' + 3 = 0$$

Note that the x' axis is perpendicular to the given line.

(e)   Here

$$\frac{3}{\sqrt{2}}(x' - y') - \frac{3}{\sqrt{2}}(x' + y') + 4 = -\frac{6}{\sqrt{2}}y' + 4 = 0 \qquad \text{or} \qquad 3y' - 2\sqrt{2} = 0$$

Note that the x' axis is parallel to the given line.

36.5   Transform the equation $2x^2 - 4xy + 5y^2 - 18x + 12y - 24 = 0$ by rotating the coordinate axes through the angle θ where $\sin \theta = 1/\sqrt{5}$ and $\cos \theta = 2/\sqrt{5}$.

The equations of transformation are $x = \frac{1}{\sqrt{5}}(2x' - y')$, $y = \frac{1}{\sqrt{5}}(x' + 2y)'$ and when the values of x and y are substituted in the given equation, we find

$$x'^2 + 6y'^2 - \frac{24}{\sqrt{5}}x' + \frac{42}{\sqrt{5}}y' - 24 = 0$$

Thus, the effect of the transformation is to produce an equation in which the cross-product term $x'y'$ is missing.

36.6   After a rotation of axes with equations of transformation

$$x = \tfrac{1}{13}(12x' - 5y') \qquad y = \tfrac{1}{13}(5x' + 12y')$$

followed by a translation with equations of transformation

$$x' = x'' + \tfrac{16}{13} \qquad y' = y'' - \tfrac{63}{13}$$

a certain equation of the second degree is reduced to $y''^2 = -8x''$. Sketch the locus, showing each set of coordinate axes.

In order to distinguish between the three coordinate systems, we shall use the term *unprimed* for the original system, *primed* for the system after the rotation, and *double-primed* for the system after the translation. We begin with the original (unprimed) axes in the usual position.

Now any pair of number, as (12,5), which are proportional respectively to $\cos \theta = \tfrac{12}{13}$ and $\sin \theta = \tfrac{5}{13}$ are the coordinates of a point on the x' axis. With the x' and y' axes in position, we next seek the origin O'' of the double-primed system. Using first the equations of translation and then the equations of rotation, we find $O''(0,0)'' = O''(\tfrac{16}{13}, -\tfrac{63}{13})' = O''(3, -4)$. Locating $O''(3, -4)$ with reference to the original system, we draw the x'' and y'' axes through O'' parallel to the x' and y' axes. Finally, on this latter set of axes, we sketch the parabola $y''^2 = -8x''$. See Fig. 36-5.

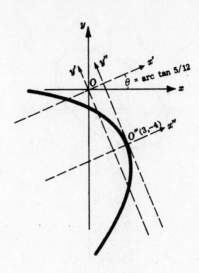

Fig. 36-5

In Problems 36.7–36.9 determine the nature of the locus, obtain the reduced form of the equation, and sketch the locus showing all sets of coordinate axes.

36.7 $20x^2 - 24xy + 27y^2 + 24x - 54y - 369 = 0$

Since $B^2 - AC = (-12)^2 - 20 \cdot 27 < 0$, the locus is an ellipse.

First Solution. The angle θ through which the axes must be rotated to eliminate the term in xy is given by

$$\tan 2\theta = \frac{2B}{A - C} = \frac{-24}{20 - 27} = \frac{24}{7}$$

Then $\cos 2\theta = \dfrac{7}{\sqrt{(24)^2 + (7)^2}} = \dfrac{7}{25}$, $\sin \theta = \sqrt{\dfrac{1 - \cos 2\theta}{2}} = \dfrac{3}{5}$, $\cos \theta = \sqrt{\dfrac{1 + \cos 2\theta}{2}} = \dfrac{4}{5}$,

and the equations of rotation are $x = \frac{1}{5}(4x' - 3y')$, $y = \frac{1}{5}(3x' + 4y')$. When this transformation is applied to the given equation, we find

$$11x'^2 + 36y'^2 - \tfrac{66}{5}x' - \tfrac{288}{2}y' - 369 = 0 \qquad \text{as the semireduced form}$$

Completing squares, we have

$$11\big(x'^2 - \tfrac{6}{5}x' + \tfrac{9}{25}\big) + 36\big(y'^2 - \tfrac{8}{5}y' + \tfrac{16}{25}\big) = 369 + 11\big(\tfrac{9}{25}\big) + 36\big(\tfrac{16}{25}\big) = 396$$

or $$11\big(x' - \tfrac{3}{5}\big)^2 + 36\big(y' - \tfrac{4}{5}\big)^2 = 396$$

The translation $x' = x'' + \frac{3}{5}, y' = y'' + \frac{4}{5}$ gives the reduced form $11x''^2 + 36y''^2 = 396$.

The x' axis passes through the point $(4, 3)$ and the coordinates of the new origin are $O''(0, 0) = O''(\frac{3}{5}, \frac{4}{5})' = O''(0, 1)$. See Fig. 36-6

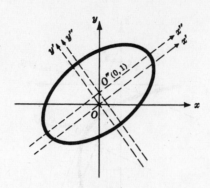

Fig. 36-6

Second Solution. Since some may prefer to eliminate the first-degree terms before rotating the axes, we give the details for this locus.

Applying the transformation (*36.2*) to the given equation, we obtain

$$20(x' + h)^2 - 24(x' + h)(y' + k) + 27(y + k)^2 + 24(x' + h) - 54(y' + k) - 369$$
$$= 20x'^2 - 24x'y' + 27y'^2 + (40h - 24k + 24)x' - (24h - 54k + 54)y'$$
$$+ 20h^2 - 24hk + 27k^2 + 24h - 54k - 369 = 0.$$

If the terms of first degree are to disappear, h and k must be chosen so that $40h - 24k + 24 = 0$, $24h - 54k + 54 = 0$. Then $h = 0, k = 1$, and the equation of the locus becomes $20x'^2 - 24x'y' + 27y'^2 - 396 = 0$.

As in the first solution, the equations of rotation to eliminate the term in $x'y'$ are $x' = \frac{1}{5}(4x'' - 3y'')$, $y' = \frac{1}{5}(3x'' + 4y'')$. When this transformation is made, we have

$$\tfrac{20}{25}(4x'' - 3y'')^2 - \tfrac{24}{25}(4x'' - 3y'')(3x'' + 4y'') + \tfrac{27}{25}(3x'' + 4y'') - 396 = 0$$

or
$$11x''^2 + 36y''^2 = 396, \quad \text{as before.}$$

36.8 $x^2 + 2xy + y^2 + 10\sqrt{2}x - 2\sqrt{2}y + 8 = 0$

Here $B^2 - AC = 1 - 1 \cdot 1 = 0$; the locus is either a parabola or a pair of parallel lines. Since $A = C = 1$, we rotate the axes through the angle $\theta = 45°$ to obtain the semireduced form. When the transformation

$$x = \frac{1}{\sqrt{2}}(x' - y'); \qquad y = \frac{1}{\sqrt{2}}(x' + y')$$

is applied to the given equation, we find $x'^2 + 4x' - 6y' + 4 = (x' + 2)^2 - 6y' = 0$. Then the translation $x' = x'' - 2, y' = y''$ produces the reduced form $x''^2 - 6y'' = 0$.

The locus is a parabola. See Fig. 36-7. The x' axis passes through the point $(1,1)$ and $O''(0,0)'' = O''(-2,0)' = O''(\sqrt{-2}, -\sqrt{-2})$.

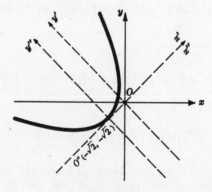

Fig. 36-7

36.9 $27x^2 + 120xy + 77y^2 + 234x + 858y + 117 = 0$.

Here $B^2 - AC = (60)^2 - 27 \cdot 77 > 0$; the locus is a hyperbola or a pair of intersecting lines. From $\tan 2\theta = 120/(27 - 77) = -\frac{12}{5}$, we have $\cos 2\theta = -\frac{5}{13}$; then $\sin \theta = \sqrt{\frac{1}{2}(1 + \frac{5}{13})} = 3/\sqrt{13}$, $\cos \theta = \sqrt{\frac{1}{2}(1 - \frac{5}{13})} = 2/\sqrt{13}$, and the equation of rotations are

$$x = \frac{1}{\sqrt{13}}(2x' - 3y'), \qquad y = \frac{1}{\sqrt{13}}(3x' + 2y')$$

Applying the transformation, we obtain $9x'^2 - y'^2 + 18\sqrt{13}x' + 6\sqrt{13}y' + 9 = 0$.

Completing the squares, we obtain $9(x' + \sqrt{13})^2 - (y' - 3\sqrt{13})^2 + 9 = 0$ which after the translation $x' = x'' - \sqrt{13}, y' = y'' + 3\sqrt{13}$ becomes $9x''^2 - y''^2 + 9 = 0$ or $y''^2 - 9x''^2 = 9$.

The locus is a hyperbola. See Fig. 36-8. The x' axis passes through the point $(2, 3)$ and $O''(0, 0)'' = O''(-\sqrt{13}, 3\sqrt{13})' = O''(-11, 3)$.

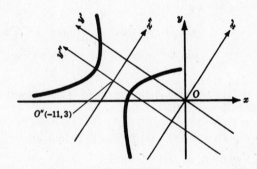

Fig. 36-8

Supplementary Problems

36.10   If the equations of translation are $x = x' + 2, y = y' - 5$, find

    (a) The coordinates of $O(0, 0)$ when referred to the primed system

    (b) The coordinates of $P(-2, 4)$ when referred to the primed system

    (c) The coordinates of $P(-2, 4)'$ when referred to the unprimed system

    (d) The equation of $l: 5x + 2y = 0$ when referred to the primed system

    (e) The equation of $l: x - 2y + 4 = 0$ when referred to the primed system

        Ans.   (a) $(-2, 5)'$ (b) $(-4, 9)'$ (c) $(0, -1)$ (d) $5x' + 2y' = 0$ (e) $x' - 2y' + 16 = 0$

36.11   Simplify each of the following equations by a suitable translation. Draw the figure showing both sets of axes.

    (a) $4x^2 + y^2 - 16x + 6y - 11 = 0$ (d) $x^2 - 12x - 8y - 4 = 0$

    (b) $9x^2 - 4y^2 - 36x + 48y - 144 = 0$ (e) $16y^2 + 5x + 32y + 6 = 0$

    (c) $9x^2 - 4y^2 - 36x + 48y - 72 = 0$ (f) $xy - 4x + 3y + 24 = 0$

        Ans.   (a) $4x'^2 + y'^2 = 36$ (c) $4y'^2 - 9x'^2 = 36$ (e) $16y'^2 = -5x'$

                (b) $9x'^2 - 4y'^2 = 36$ (d) $x'^2 - 8y' = 0$ (f) $x'y' + 36 = 0$

36.12 Simplify each of the following equations by rotating the axes through the indicated angle. Draw the figure showing both sets of axes.

    (a) $x^2 - y^2 = 16$; $45°$ (c) $16x^2 + 24xy + 9y^2 + 60x - 80y = 0$; Arccos $\frac{4}{5}$

    (b) $9x^2 + 24xy + 16y^2 = 25$; Arccos $\frac{3}{5}$ (d) $31x^2 - 24xy + 21y^2 = 39$; Arccos $2/\sqrt{13}$

         Ans. (a) $x'y' + 8 = 0$ (b) $x'^2 = 1$ (c) $x'^2 = 4y'$ (d) $x'^2 + 3y'^2 = 3$

36.13 Simplify each equation by suitable transformations and draw a figure showing all sets of axes.

    (a) $3x^2 + 2xy + 3y^2 - 8x + 16y + 30 = 0$

    (b) $13x^2 + 12xy - 3y^2 - 15x - 15y = 0$

    (c) $25x^2 - 120xy + 144y^2 + 1300x + 1274y - 2704 = 0$

    (d) $108x^2 - 312xy + 17y^2 + 750y + 225 = 0$

    (e) $16x^2 + 24xy + 9y^2 - 60x - 170y - 175 = 0$

    (f) $37x^2 + 32xy + 13y^2 - 42\sqrt{5}x - 6\sqrt{5}y = 0$

        Ans. (a) $2x''^2 + y''^2 = 4$ (c) $y''^2 = -10x''$ (e) $x''^2 = 4y''$

             (b) $30x''^2 - 10y''^2 = 3$ (d) $4x''^2 - 9y''^2 = 36$ (f) $9x''^2 + y'' = 18$

36.14 Apply the equations of transformation (*36.2*) directly to (*36.1*) and show that the first-degree terms may be made to disappear provided $B^2 - AC \neq 0$.

36.15 Use (*36.3*) to show: (a) $A' + C' = A + C$, (b) $B'^2 - A'C' = B^2 - AC$.

36.16 Solve (*36.1*) to obtain $x = \dfrac{-(By + D) \pm \sqrt{H}}{A}$, where $H = (B^2 - AC)y^2 + 2(BD - AE)y + D^2 - AF$. Show that

H is a perfect square when $\Delta = \begin{vmatrix} A & B & D \\ B & C & E \\ D & E & F \end{vmatrix} = 0$. Thus, prove (*36.1*) represents a degenerate locus if and only

if $\Delta = 0$.

36.17 Prove that Δ of Problem 36.16 is invariant under translation and rotation of the axes.

Chapter 37

Points in Space

RECTANGULAR COORDINATES IN SPACE. Consider the three mutually perpendicular planes of Fig. 37-1. These three planes (the xy plane, the xz plane, the yz plane) are called the *coordinate planes*; their three lines of intersection are called the *coordinate axes* (the x axis, the y axis, the z axis); and their common point O is called the *origin*. Positive direction is indicated on each axis by an arrow-tip.

(NOTE: The coordinate system of Fig. 37-1 is called a left-handed system. When the x and y axes are interchanged, the system becomes right-handed.)

The coordinate planes divide the space into eight regions, called *octants*. The octant whose edges are $\overrightarrow{Ox}, \overrightarrow{Oy}, \overrightarrow{Oz}$ is called the *first octant*; the other octants are not numbered.

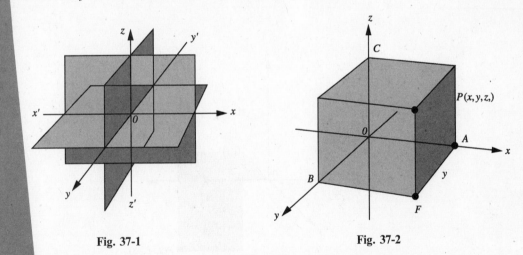

| Fig. 37-1 | Fig. 37-2 |

Let P be any point in space, not in a coordinate plane, and through P pass planes parallel to the coordinate planes meeting the coordinate axes in the points A, B, C and forming the rectangular parallelepiped of Fig. 37-2. The directed distances $x = OA$, $y = OB$, $z = OC$ are called, respectively, the x coordinate, the y coordinate, the z coordinate of P and we write $P(x, y, z)$.

Since $\overline{AF} \cong \overline{OB}$ and $\overline{FP} \cong \overline{OC}$, it is preferable to use the three edges $\overline{OA}, \overline{AF}, \overline{FP}$ instead of the complete parallelepiped in locating a given point.

EXAMPLE 1. Locate the points:
(*a*) (2, 3, 4)   (*b*) (−2, −2, 3)   (*c*) (2, −2, −3)

As standard procedure in representing on paper the left-handed system, we shall draw $\angle xOz$ measuring $90°$ and $\angle xOy$ measuring $135°$. Then distances on parallels to the x and z axes will be drawn to full scale while distances parallel to the y axis will be drawn about $\frac{7}{10}$ of full scale.

(a)   From the origin move 2 units to the right along the x axis to $A(2, 0, 0)$, from A move 3 units forward parallel to the y axis to $F(2, 3, 0)$, and from F move 4 units upward parallel to the z axis to $P(2, 3, 4)$. Se Fig. 37-3(a).

(b)   From the origin move 2 units to the left along the x axis to $A(-2, 0, 0)$, from A move 2 units backward parallel to the y axis to $F(-2, -2, 0)$, and from F move 3 units upward parallel to the z axis to $P(-2, -2, 3)$. See Fig. 37-3(b).

(c)   From the origin move 2 units to the right along the x axis to $A(2, 0, 0)$, from A move 2 units backward parallel to the y axis to $F(2, -2, 0)$, and from F move 3 units downward parallel to the z axis to $P(2, -2, -3)$. See Fig. 37-3(c).

(See Problem 37.1.)

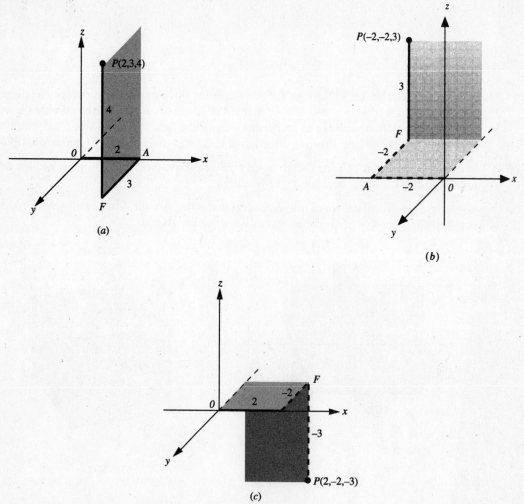

Fig. 37-3

THE DISTANCE BETWEEN TWO POINTS $P_1(x_1, y_1, z_1)$ and $P_2(x_2, y_2, z_2)$ is from Fig. 37-4

$$d = P_1 P_2 = \sqrt{(P_1 R)^2 + (R P_2)^2} = \sqrt{(P_1 S)^2 + (S R)^2 + (R P_2)^2} = \sqrt{(x_2 - x_1)^2 + (y_2 - y_1)^2 + (z_2 - z_1)^2}$$

$$(37.1)$$

(See Problem 37.2.)

IF $P_1(x_1, y_1, z_1)$ **AND** $P_2(x_1, y_2, z_2)$ are the end points of a line segment and if $P(x, y, z)$ divides the segment in the ratio $P_1P/PP_2 = r_1/r_2$, then

$$x = \frac{r_2 x_1 + r_1 x_2}{r_1 + r_2}, \quad y = \frac{r_2 y_1 + r_1 y_2}{r_1 + r_2}, \quad z = \frac{r_2 z_1 + r_1 z_2}{r_1 + r_2} \tag{37.2}$$

The coordinates of the midpoint of $\overline{P_1 P_2}$ are $\left(\frac{1}{2}(x_1 + x_2), \frac{1}{2}(y_1 + y_2), \frac{1}{2}(z_1 + z_2)\right)$. (See Problems 37.3–37.4.)

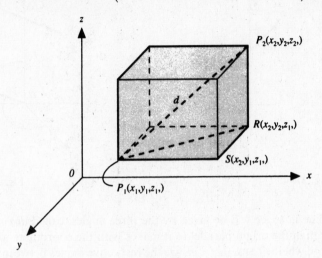

Fig. 37-4

TWO STRAIGHT LINES IN SPACE which intersect or are parallel lie in the same plane; two lines which are not coplanar are called *skew*. By definition, the angle between two directed skew lines as b and c in Fig. 37-5 is the angle between any two intersecting lines as b' and c' which are respectively parallel to the skew lines and similarly directed.

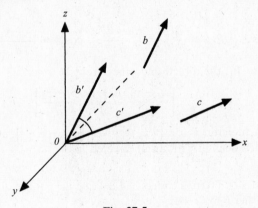

Fig. 37-5

DIRECTION COSINES OF A LINE. In the plane a directed line l [positive direction upward in Figs. 37-6(a) and (b)] forms the angles α and β with the positive directions on the x and y axes. However, in our study of the line in the plane we have favored the angle α over the angle β, calling it the angle of inclination of the line and its tangent the slope of the line.

In Fig. 37-6(a), $\alpha + \beta = \frac{1}{2}\pi$ and $m = \tan \alpha = \dfrac{\sin \alpha}{\cos \alpha} = \dfrac{\sin\left(\frac{1}{2}\pi - \beta\right)}{\cos \alpha} = \dfrac{\cos \beta}{\cos \alpha}$, and in Fig. 37-6(b),

$\alpha = \frac{1}{2}\pi + \beta$ and $\tan \alpha = \dfrac{\sin\left(\frac{1}{2}\pi + \beta\right)}{\cos \alpha} = \dfrac{\cos \beta}{\cos \alpha}$. Now the angles α and β, called *direction angles* of the line, or their cosines, $\cos \alpha$ and $\cos \beta$, called *direction cosines* of the line, might have been used instead of the slope to give the direction of the line l. Indeed, it will be the direction cosines which will be generalized in our study of the straight line in space.

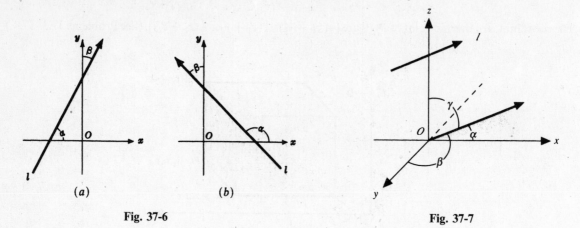

(a)   (b)

Fig. 37-6   Fig. 37-7

The direction of a line in space will be given by the three angles, called *direction angles* of the line, which it or that line through the origin parallel to it makes with the coordinate axes. If, as in Fig. 37-7, the direction angles α, β, γ, where $O \leq \alpha, \beta, \gamma < \pi$, are the respective angles between the positive directions on the x, y, z axis and the directed line l (positive direction upward), the direction angles of this line when oppositely directed are $\alpha' = \pi - \alpha, \beta' = \pi - \beta, \gamma' = \pi - \gamma$. Thus, an undirected line in space has two sets of direction angles α, β, γ and $\pi - \alpha, \pi - \beta, \pi - \gamma$, and two sets of direction cosines $[\cos \alpha, \cos \beta, \cos \gamma]$ and $[-\cos \alpha, -\cos \beta, -\cos \gamma]$ since $\cos(\pi - \phi) = -\cos \phi$. To avoid confusion with the coordinates of a point, the triples of direction cosines of a line will be enclosed in a bracket. Thus we shall write $l: [A, B, C]$ to indicate the line whose direction cosines are the triple A, B, C.

The direction cosines of the l determined by points $P_1(x_1, y_1, z_1)$ and $P_2(x_2, y_2, z_2)$ and directed from P_1 to P_2 are (see Fig. 37-4)

$$\cos \alpha = \cos \angle P_2 P_1 S = \frac{P_1 S}{P_1 P_2} = \frac{x_2 - x_1}{d}, \quad \cos \beta = \frac{y_2 - y_1}{d}, \quad \cos \gamma = \frac{z_2 - z_1}{d}.$$

when l is directed from P_2 to P_1, the direction cosines are

$$\left[\frac{x_1 - x_2}{d}, \frac{y_1 - y_2}{d}, \frac{z_1 - z_2}{d} \right]$$

Except for the natural preference for $[\frac{1}{3}, \frac{2}{3}, \frac{2}{3}]$ over $[-\frac{1}{3}, -\frac{2}{3}, -\frac{2}{3}]$, it is immaterial which set of direction cosines is used when dealing with an undirected line.

EXAMPLE 2. Find the two sets of direction cosines and indicate the positive direction along the line passing through the points $P_1(3, -1, 2)$ and $P_2(5, 2, -4)$.

We have $d = \sqrt{(2)^2 + (3)^2 + (-6)^2} = 7$. One set of direction cosines is

$$\left[\frac{x_2 - x_1}{d}, \frac{y_2 - y_1}{d}, \frac{z_2 - z_1}{d} \right] = \left[\frac{2}{7}, \frac{3}{7}, -\frac{6}{7} \right]$$

the positive direction being from P_1 to P_2. When the line is directed from P_2 to P_1, the direction cosines are $[-\frac{2}{7}, -\frac{3}{7}, \frac{6}{7}]$.

The sum of the squares of the direction cosines of any line is equal to 1; i.e.,

$$\cos^2 \alpha + \cos^2 \beta + \cos^2 \gamma = 1$$

It follows immediately that at least one of the direction cosines of any line is different from 0.

DIRECTION NUMBERS OF A LINE.   Instead of the direction cosines of a line, it is frequently more convenient to use any triple of numbers, preferably small integers when possible, which are proportional to the direction cosines. Any such triple is called a set of *direction numbers* of the line. For example, if the direction cosines are $[\frac{2}{3}, -\frac{2}{3}, -\frac{1}{3}]$, sets of direction numbers are $[2, -2, -1], [-2, 2, 1], [4, -4, -2]$, etc.; if the direction cosines are $[\frac{1}{2}, 1/\sqrt{2}, -\frac{1}{2}]$, a set of direction numbers is $[1, \sqrt{2}, -1]$.

Sets of direction numbers for the line through points $P_1(x_1, y_1, z_1)$ and $P_2(x_2, y_2, z_2)$ are $[x_2 - x_1, y_2 - y_1, z_2 - z_1]$ and $[x_1 - x_2, y_1 - y_2, z_1 - z_2]$.

If $[a, b, c]$ is a set of direction numbers of a line, then the direction cosines of the line are given by

$$\cos\alpha = \pm\frac{a}{\sqrt{a^2 + b^2 + c^2}}, \quad \cos\beta = \pm\frac{b}{\sqrt{a^2 + b^2 + c^2}}, \quad \cos\gamma = \pm\frac{c}{\sqrt{a^2 + b^2 + c^2}} \tag{37.3}$$

where the usual convention of first reading the upper signs and then the lower signs holds. (See Problems 37.5–37.8.)

THE ANGLE θ BETWEEN TWO DIRECTED LINES

$$l_1: \quad [\cos\alpha_1, \cos\beta_1, \cos\gamma_1] \quad and \quad l_2: \quad [\cos\alpha_2, \cos\beta_2, \cos\gamma_2]$$

is given by

$$\cos\theta = \cos\alpha_1\cos\alpha_2 + \cos\beta_1\cos\beta_2 + \cos\gamma_1\cos\gamma_2 \tag{37.4}$$

(For a proof see Problem 37.9.)

If the two lines are parallel then $\theta = 0$ or π, according as the lines are similarly or oppositely directed, and $\cos\alpha_1\cos\alpha_2 + \cos\beta_1\cos\beta_2 + \cos\gamma_1\cos\gamma_2 = \pm 1$. If the sign is $+$, then $\cos\alpha_1 = \cos\alpha_2, \cos\beta_1 = \cos\beta_2, \cos\gamma_1 = \cos\gamma_2$; if the sign is $-$, then $\cos\alpha_1 = -\cos\alpha_2, \cos\beta_1 = -\cos\beta_2, \cos\gamma_1 = -\cos\gamma_2$. Thus, two undirected lines are parallel if and only if their direction cosines are the same or differ only in sign. In terms of direction numbers, *two lines are parallel if and only if corresponding direction numbers are proportional.*

If the two lines are perpendicular, then $\theta = \frac{1}{2}\pi$ or $3\pi/2$, according as the lines are similarly or oppositely directed, and

$$\cos\alpha_1\cos\alpha_2 + \cos\beta_1\cos\beta_2 + \cos\gamma_1\cos\gamma_2 = 0 \tag{37.5}$$

In terms of direction numbers, *two lines with direction number $[a_1, b_1, c_1]$ and $[a_2, b_2, c_2]$, respectively, are perpendicular if and only if*

$$a_1 \cdot a_2 + b_1 \cdot b_2 + c_1 \cdot c_2 = 0. \tag{37.5'}$$

(See Problems 37.10–37.12.)

THE DIRECTION NUMBER DEVICE.   If $l_1: [a_1, b_1, c_1]$ and $l_2: [a_2, b_2, c_2]$ are two nonparallel lines, then a set of direction number $[a, b, c]$ of any line perpendicular to both l_1 and l_2 is given by

$$a = \begin{vmatrix} b_1 & c_1 \\ b_2 & c_2 \end{vmatrix}, \quad b = \begin{vmatrix} c_1 & a_1 \\ c_2 & a_2 \end{vmatrix}, \quad c = \begin{vmatrix} a_1 & b_1 \\ a_2 & b_2 \end{vmatrix}.$$

These three determinants can be obtained readily as follows:

(1)   Write the two sets of direction numbers in three columns $\begin{matrix} a_1 & b_1 & c_1 \\ a_2 & b_2 & c_2 \end{matrix}$.

(2)   Repeat the first two columns to obtain $\begin{matrix} a_1 & b_1 & c_1 & a_1 & b_1 \\ a_2 & b_2 & c_2 & a_2 & b_2 \end{matrix}$ and strike out the first column to have $\begin{matrix} a_1 & b_1 & c_1 & a_1 & b_1 \\ a_2 & b_2 & c_2 & a_2 & b_2 \end{matrix}$.

Then a is the determinant of the first and second columns remaining, b is the determinant of the second and third columns, and c is the determinant of the third and fourth columns. This procedure will be called the *direction number device*. Note, however, that it is a mechanical procedure for obtaining one solution of two homogeneous equations in three unknowns and thus has other applications.

EXAMPLE 3.   Find a set of direction numbers $[a, b, c]$ of any line perpendicular to $l_1 : [2, 3, 4]$ and $l_2 : [1, -2, -3]$.

Using the direction number device, we write $\begin{matrix} 2 & 3 & 4 & 2 & 3 \\ 1 & -2 & -3 & 1 & -2 \end{matrix}$. Then

$$a = \begin{vmatrix} 3 & 4 \\ -2 & -3 \end{vmatrix} = -1, \quad b = \begin{vmatrix} 4 & 2 \\ -3 & 1 \end{vmatrix} = 10, \quad c = \begin{vmatrix} 2 & 3 \\ 1 & -2 \end{vmatrix} = -7.$$

A set of direction numbers is $[-1, 10, -7]$ or, if preferred, $[1, -10, 7]$. (See Problem 37.13.)

Solved Problems

37.1   What is the locus of a point:

    (a)   Whose z coordinate is always 0?

    (b)   Whose z coordinate is always 3?

    (c)   Whose x coordinate is always -5?

    (d)   Whose x and y coordinates are always 0?

    (e)   Whose x coordinate is always 2 and whose y coordinate is always 3?

        (a)   All points $(a, b, 0)$ lie in the xy plane; the locus is that plane.

        (b)   Every point is 3 units above the xy plane; the locus is the plane parallel to the xy plane and 3 units above it.

        (c)   A plane parallel to the yz plane and 5 units to the left of it.

        (d)   All points $(0, 0, c)$ lie on the z axis; the locus is that line.

        (e)   In locating the point $P(2, 3, c)$, the x and y coordinates are used to locate the point $F(2, 3, 0)$ in the xy plane and then a distance $|c|$ is measured from F parallel to the z axis. The locus is the line parallel to the z axis passing through the point $(2, 3, 0)$ in the xy plane.

37.2   (a)   Find the distance between the points $P_1(-1, -3, 3)$ and $P_2(2, -4, 1)$.

    (b)   Find the perimeter of the triangle whose vertices are $A(-2, -4, -3), B(1, 0, 9), C(2, 0, 9)$.

    (c)   Show that the points $A(1, 2, 4), B(4, 1, 6)$, and $C(-5, 4, 0)$ are collinear.

        (a)   Here $d = \sqrt{(x_2 - x_1)^2 + (y_2 - y_1)^2 + (z_2 - z_1)^2} = \sqrt{[2 - (-1)]^2 + [-4 - (-3)]^2 + (1 - 3)^2} = \sqrt{14}$.

        (b)   We find $AB = \sqrt{[1 - (-2)]^2 + [0 - (-4)]^2 + [9 - (-3)]^2} = 13$,

        $BC = \sqrt{(2 - 1)^2 + (0 - 0)^2 + (9 - 9)^2} = 1$, and $CA = \sqrt{(-2 - 2)^2 + (-4 - 0)^2 + (-3 - 9)^2} = 4\sqrt{11}$. The perimeter is $13 + 1 + 4\sqrt{11} = 14 + 4\sqrt{11}$.

        (c)   Here $AB = \sqrt{(3)^2 + (-1)^2 + (2)^2} = \sqrt{14}$, $BC = \sqrt{(-9)^2 + (3)^2 + (-6)^2} = 3\sqrt{14}$,

        and $CA = \sqrt{(6)^2 + (-2)^2 + (4)^2} = 2\sqrt{14}$. Since $BC = CA + AB$, the points are collinear.

37.3   Find the coordinates of the point P of division for each pair of points and given ratio. Find also the midpoint of the segment. (a) $P_1(3, 2, -4)$, $P_2(6, -1, 2)$; $1:2$ (b) $P_1(2, 5, 4)$, $P_2(-6, 3, 8)$; $-3:5$.

(a) Here $r_1 = 1$ and $r_2 = 2$. Then

$$x = \frac{r_2 x_1 + r_1 x_2}{r_1 + r_2} = \frac{2 \cdot 3 + 1 \cdot 6}{1 + 2} = 4, \qquad y = \frac{r_2 y_1 + r_1 y_2}{r_1 + r_2} = \frac{2 \cdot 2 + 1(-1)}{1 + 2} = 1, \qquad z = \frac{r_2 z_1 + r_1 z_2}{r_1 + r_2} = -2$$

and the required point is $P(4, 1, -2)$. The midpoint has coordinates $(\frac{1}{2}(x_1 + x_2), \frac{1}{2}(y_1 + y_2), \frac{1}{2}(z_1 + z_2)) = (\frac{9}{2}, \frac{1}{2}, -1)$.

(b) Here $r_1 = -3$ and $r_2 = 5$. Then

$$x = \frac{5 \cdot 2 + (-3)(-6)}{-3 + 5} = 14, \qquad y = \frac{5 \cdot 5 + (-3)3}{-3 + 5} = 8, \qquad z = \frac{5 \cdot 4 + (-3)8}{-3 + 5} = -2$$

and the required point is $P(14, 8, -2)$. The midpoint has coordinates $(-2, 4, 6)$.

37.4   Prove: The three lines joining the midpoints of the opposite edges of a tetrahedron pass through a point P which bisects each of them.

Let the tetrahedron, shown in Fig. 37-8, have vertices $O(0, 0, 0)$, $A(a, 0, 0)$, $B(b, c, 0)$, and $C(d, e, f)$. The midpoints of $\overline{OB}$ and $\overline{AC}$ are, respectively, $D(\frac{1}{2}b, \frac{1}{2}c, 0)$ and $E(\frac{1}{2}(a + d), \frac{1}{2}e, \frac{1}{2}f)$, and the midpoint of DE is $P(\frac{1}{4}(a + b + d), \frac{1}{4}(c + e), \frac{1}{4}f)$. The midpoints of OA and BC are, respectively, $F(\frac{1}{2}a, 0, 0)$ and $G(\frac{1}{2}(b + d), \frac{1}{2}(c + e), \frac{1}{2}f)$, and the midpoint of FG is P. It is left for the reader to find the midpoints H and I of $\overline{OC}$ and $\overline{AB}$, and show that P is the midpoint of $\overline{HI}$.

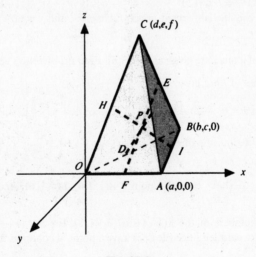

Fig. 37-8

37.5   Find the direction cosines of the line:

(a) Passing through $P_1(3, 4, 5)$ and $P_2(-1, 2, 3)$ and directed from P_1 to P_2

(b) Passing through $P_1(2, -1, -3)$ and $P_2(-4, 2, 1)$ and directed from P_2 to P_1

(c) Passing through $O(0, 0, 0)$ and $P(a, b, c)$ and directed from O to P

(d) Passing through $P_1(4, -1, 2)$ and $P_2(2, 1, 3)$ and directed so that γ is acute

(a) We have

$$\cos\alpha = \frac{x_2 - x_1}{d} = \frac{-4}{2\sqrt{6}}, \qquad \cos\beta = \frac{y_2 - y_1}{d} = \frac{-2}{2\sqrt{6}}, \qquad \cos\gamma = \frac{z_2 - z_1}{d} = \frac{-2}{2\sqrt{6}}.$$

The direction cosines are $\left[-\dfrac{2}{\sqrt{6}}, -\dfrac{1}{\sqrt{6}}, -\dfrac{1}{\sqrt{6}}\right]$.

(b) $\cos\alpha = \dfrac{x_1 - x_2}{d} = \dfrac{6}{\sqrt{61}}, \qquad \cos\beta = \dfrac{y_1 - y_2}{d} = \dfrac{-3}{\sqrt{61}}, \qquad \cos\gamma = \dfrac{z_1 - z_2}{d} = \dfrac{-4}{\sqrt{61}}.$

The direction cosines are $\left[\dfrac{6}{\sqrt{61}}, -\dfrac{3}{\sqrt{61}}, -\dfrac{4}{\sqrt{61}}\right]$.

(c) $\cos\alpha = \dfrac{a-0}{d} = \dfrac{a}{\sqrt{a^2 + b^2 + c^2}}, \quad \cos\beta = \dfrac{b-0}{d} = \dfrac{b}{\sqrt{a^2 + b^2 + c^2}}, \quad \cos\gamma = \dfrac{c-0}{d} = \dfrac{c}{\sqrt{a^2 + b^2 + c^2}}$

The direction cosines are $\left[\dfrac{a}{\sqrt{a^2 + b^2 + c^2}}, \dfrac{b}{\sqrt{a^2 + b^2 + c^2}}, \dfrac{c}{\sqrt{a^2 + b^2 + c^2}}\right]$.

(d) The two sets of direction cosines of the undirected line are

$$\cos\alpha = \pm\tfrac{2}{3}, \quad \cos\beta = \mp\tfrac{2}{3}, \quad \cos\gamma = \mp\tfrac{1}{3}$$

one set being given by the upper signs and the other by the lower signs. When γ is acute, $\cos\gamma > 0$; hence the required set is $[-\tfrac{2}{3}, \tfrac{2}{3}, \tfrac{1}{3}]$.

37.6   Given the direction angles α measuring $120°$ and β measuring $45°$, find γ if the line is directed upward.

$\cos^2\alpha + \cos^2\beta + \cos^2\gamma = \cos^2 120° + \cos^2 45° + \cos^2\gamma = (-\tfrac{1}{2})^2 + (1\sqrt{2})^2 + \cos^2\gamma = 1$. Then $\cos^2\gamma = \tfrac{1}{4}$ and $\cos\gamma = \pm\tfrac{1}{2}$. When the line is directed upward, $\cos\gamma = \tfrac{1}{2}$ and $\gamma = 60°$.

37.7   The direction numbers of a line l are given as $[2, -3, 6]$. Find the direction cosines of l when directed upward.

The direction cosines of l are given by

$$\cos\alpha = \pm\frac{a}{\sqrt{a^2 + b^2 + c^2}} = \pm\frac{2}{7}, \quad \cos\beta = \pm\left(\frac{-3}{7}\right), \quad \cos\gamma = \pm\frac{6}{7}.$$

When γ is acute, $\cos\gamma > 0$, and the direction cosines are $[\tfrac{2}{7}, -\tfrac{3}{7}, \tfrac{6}{7}]$.

37.8   Use direction numbers to show that the points $A(1,2,4)$, $B(4,1,6)$, and $C(-5,4,0)$ are collinear. [See Problem 37.2(c).]

A set if direction numbers of the line $\overleftrightarrow{AB}$ is $[3, -1, 2]$, for BC is $[-9, 3, -6]$. Since the two sets are proportional, the lines are parallel; since the lines have a point in common they are coincident and the points are collinear.

37.9   Prove: the angle θ between two directed lines l_t: $[\cos\alpha_1, \cos\beta_1, \cos\gamma_1]$ and l_2: $[\cos\alpha_2, \cos\beta_2, \cos\gamma_2]$ is given by $\cos\theta = \cos\alpha_1\cos\alpha_2 + \cos\beta_1\cos\beta_2 + \cos\gamma_1\cos\gamma_2$.

The angle θ is by definition the angle between two lines issuing from the origin parallel, respectively, to the given lines l_t and l_2 and similarly directed.

Consider the triangle OP_1P_2, in Fig. 37-9, whose vertices are the origin and the points $P_1(\cos\alpha_1, \cos\beta_1, \cos\gamma_t)$ and $P_2(\cos\alpha_2, \cos\beta_2, \cos\gamma_2)$. The line segment OP_1 is of length 1 (why?) and is parallel to l_1; similarly, OP_2 is of length 1 and is parallel to l_2. Thus, $\angle P_1OP_2 = \theta$. By the Law of Cosines, $(P_1P_2)^2 = (OP_1)^2 + (OP_2)^2 - 2(OP_1)(OP_2)\cos\theta$ and $\cos\theta = \cos\alpha_1\cos\alpha_2 + \cos\beta_1\cos\beta_2 + \cos\gamma_1\cos\gamma_2$.

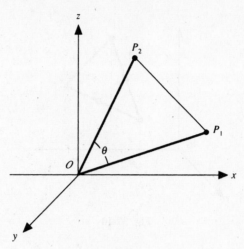

Fig. 37-9

37.10 (a) Find the angle between the directed lines l_1: $[\frac{2}{7}, \frac{3}{7}, \frac{6}{7}]$ and l_2: $[\frac{2}{3}, -\frac{1}{3}, \frac{2}{3}]$.

(b) Find the acute angle between the lines l_1: $[-2, 1, 2]$ and l_2: $[2, -6, -3]$.

(c) The line l_1 passes through $A(5, -2, 3)$ and $B(2, 1, -4)$, and the line l_2 passes through $C(-4, 1, -2)$ and $D(-3, 2, 3)$. Find the acute angle between them.

    (a) We have $\cos\theta = \cos\alpha_1\cos\alpha_2 + \cos\beta_1\cos\beta_2 + \cos\gamma_1\cos\gamma_2 = \frac{2}{7}\cdot\frac{2}{3} + \frac{3}{7}(-\frac{1}{3}) + \frac{6}{7}\cdot\frac{2}{3} = \frac{13}{21} = 0.619$ and $\theta = 51°50'$.

    (b) Since $\sqrt{(-2)^2 + (1)^2 + (2)^2} = 3$, we take $[-\frac{2}{3}, \frac{1}{3}, \frac{2}{3}]$ as direction cosines of l_1. Since $\sqrt{(2)^2 + (-6)^2 + (-3)^2} = 7$, we take $[\frac{2}{7}, -\frac{6}{7}, -\frac{3}{7}]$ as direction cosines of l_2. Then $\cos\theta = -\frac{2}{3}\cdot\frac{2}{7} + \frac{1}{3}(-\frac{6}{7}) + \frac{2}{3}(-\frac{3}{7}) = -\frac{16}{21} = -0.762$, and $\theta = 139°40'$. The required angle is $40°20'$.

    (c) Take $[3\sqrt{67}, -3\sqrt{67}, 7/\sqrt{67}]$ as direction cosines of l_1 and $[1/3\sqrt{3}, 1/3\sqrt{3}, 5/3\sqrt{3}]$ as direction cosines of l_2. Then

$$\cos\theta = \frac{1}{\sqrt{3\cdot 67}} - \frac{1}{\sqrt{3\cdot 67}} + \frac{35}{3\sqrt{3\cdot 67}} = \frac{35}{3\sqrt{201}} = 0.823 \quad \text{and} \quad \theta = 34°40'$$

37.11 (a) Show that the line joining $A(9, 2, 6)$ and $B(5, -3, 2)$ and the line joining $C(-1, -5, -2)$ and $D(7, 5, 6)$ are parallel.

(b) Show that the line joining $A(7, 2, 3)$ and $B(-2, 5, 2)$ and the line joining $C(4, 10, 1)$ and $D(1, 2, 4)$ are mutually perpendicular.

    (a) Here $[9 - 5, 2 - (-3), 6 - 2] = [4, 5, 4]$ is a set of direction numbers of AB and $[-1 - 7, -5 - 5, -2 - 6] = [-8, -10, -8]$ is a set of direction number of CD. Since the two sets are proportional, the two lines are parallel.

    (b) Here $[9, -3, 1]$ is a set of direction numbers of AB and $[3, 8, -3]$ is a set of direction numbers of CD. Since [see Equation (37.5)] $9\cdot 3 + (-3)8 + 1(-3) = 0$, the lines are perpendicular.

37.12 Find the area of the triangle whose vertices are $A(4, 2, 3)$, $B(7, -2, 4)$ and $C(3, -4, 6)$.

    The area of triangle ABC is given by $\frac{1}{2}(AB)(AC)\sin A$. We have $AB = \sqrt{26}$ and $AC = \sqrt{46}$.

    To find $\sin A$, we direct the sides $\overline{AB}$ and $\overline{AC}$ away from the origin as in Fig. 37-10. Then $\overline{AB}$ has direction cosines $[3/\sqrt{26}, -4/\sqrt{26}, 1/\sqrt{26}]$, $\overline{AC}$ has direction cosines $[-1/\sqrt{46}, -6/\sqrt{46}, 3/\sqrt{46}]$,

$$\cos A = \frac{3}{\sqrt{26}}\cdot\frac{-1}{\sqrt{46}} + \frac{-4}{\sqrt{26}}\cdot\frac{-6}{\sqrt{46}} + \frac{1}{\sqrt{26}}\cdot\frac{3}{\sqrt{46}} = \frac{24}{\sqrt{26}\sqrt{46}}$$

and

$$\sin A = \sqrt{1 - \cos^2 A} = \frac{2\sqrt{155}}{\sqrt{26}\sqrt{46}}$$

    The required area is $\frac{1}{2}\sqrt{26}\cdot\sqrt{46}\cdot\frac{2\sqrt{155}}{\sqrt{26}\sqrt{46}} = \sqrt{155}$.

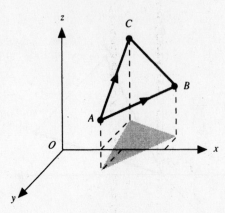

Fig. 37-10

37.13 Find a set of direction numbers for any line which is perpendicular to

   (a) $l_1: [1-2,-3]$ and $l_2: [4,-1,-5]$

   (b) The triangle whose vertices are $A(4,2,3)$, $B(7,-2,4)$, and $C(3,-4,6)$

      (a) Using the direction number device

$$\begin{matrix} 1 & -2 & -3 & 1 & -2 \\ 4 & -1 & -5 & 4 & -1 \end{matrix}, \quad \text{we obtain} \quad a=\begin{vmatrix} -2 & -3 \\ -1 & -5 \end{vmatrix}=7, \quad b=\begin{vmatrix} -3 & 1 \\ -5 & 4 \end{vmatrix}=-7, \quad c=\begin{vmatrix} 1 & -2 \\ 4 & -1 \end{vmatrix}=7.$$

      Thus, a set of direction numbers is $[7,-7,7]$; a simpler set is $[1,-1,1]$.

   (b) Since the triangle lies in a plane determined by the lines AB and AC, we seek direction numbers for any line perpendicular to these lines. For AB and AC, respective sets of direction numbers are $[3,-4,1]$ and $[-1,-6,3]$. Using the direction number device

$$\begin{matrix} 3 & -4 & 1 & 3 & -4 \\ -1 & -6 & 3 & -1 & -6 \end{matrix}, \quad a=\begin{vmatrix} -4 & 1 \\ -6 & 3 \end{vmatrix}=-6, \quad b=\begin{vmatrix} 1 & 3 \\ 3 & -1 \end{vmatrix}=-10, \quad c=\begin{vmatrix} 3 & -4 \\ -1 & -6 \end{vmatrix}=-22.$$

      Then $[-6,-10,-22]$ is a set of direction numbers and $[3,5,11]$ is a simpler set.

Supplementary Problems

37.14 Find the undirected distance between each pair of points:

   (a) $(4, 1, 5)$ and $(2, -1, 4)$ (b) $(9, 7, -2)$ and $(6, 5, 4)$ (c) $(9, -2, -3)$ and $(-3, 4, 0)$.

   Ans. (a) 3   (b) 7   (c) $3\sqrt{21}$

37.15 Find the undirected distance of each of the following points from (i) the origin, (ii) the x axis, (iii) the y axis, and (iv) the z axis: (a) $(2,6,-3)$ (b) $(2,-\sqrt{3},3)$.

   Ans. (a) $7,2,6,3$ (b) $4,2,\sqrt{3},3$

37.16 For each pair of points, find the coordinates of the point dividing $\overline{P_1 P_2}$ in the given ratio; find also the coordinates of the midpoint.

     (a) $P_1(4,1,5), P_2(2,-1,4), 3:2$ *Ans.* $(\frac{14}{5},-\frac{1}{5},\frac{22}{5}),(3,0,\frac{9}{2})$

     (b) $P_1(9,7,-2), P_2(6,5,4)1:4$ *Ans.* $(\frac{42}{5},\frac{33}{5},-\frac{4}{5}),(\frac{15}{2},6,1)$

     (c) $P_1(9,-2,-3), P_2(-3,4,0),-1:3$ *Ans.* $(15,-5,-\frac{9}{2}),(3,1,-\frac{3}{2})$

     (d) $P_1(0,0,0), P_2(2,3,4), 2:-3$ *Ans.* $(-4-6,-8),(1,\frac{3}{2},2)$

37.17 Find the equation of the locus of a point which is (a) always equidistant from the points (4,1,5) and (2,−1,4) (b) always at a distance 6 units from (4,1,5) (c) always two-thirds as far from the y axis as from the origin.

Ans.   (a) $4x + 4y + 2z − 21 = 0$ (b) $x^2 + y^2 + z^2 − 8x − 2y − 10z + 6 = 0$

(c) $5x^2 − 4y^2 + 5z^2 = 0$

37.18 Find a set of direction cosines and a set of direction numbers for the line joining P_1 and P_2, given

(a) $P_1(0,0,0), P_2(4,8,−8)$ Ans. $[\frac{1}{3}, \frac{2}{3}, −\frac{2}{3}], [1,2,−2]$

(b) $P_1(1,3,5), P_2(−1,0,−1)$ Ans. $[−\frac{2}{7}, −\frac{3}{7}, −\frac{6}{7}], [2,3,6]$

(c) $P_1(5,6,−3), P_2(1,−6,3)$ Ans. $[−\frac{2}{7}, −\frac{6}{7}, \frac{3}{7}], [2,6,−3]$

(d) $P_1(4,2,−6), P_2(−2,1,3)$ Ans. $\left[\dfrac{6}{\sqrt{118}}, \dfrac{1}{\sqrt{118}}, \dfrac{−9}{\sqrt{118}}\right], [6,1,−9]$

37.19 Find $\cos \gamma$, given (a) $\cos \alpha = \frac{14}{15}, \cos \beta = −\frac{1}{3}$ (b) $\alpha = 60°, \beta = 135°$

Ans.   (a) $\pm \frac{2}{15}$ (b) $\pm \frac{1}{2}$

37.20 Find

(a)   The acute angle between the line having direction numbers $[−4,−1,−8]$ and the line joining the points (6,4,−1) and (4,0,3).

(b)   The interior angles of the triangle whose vertices are $A(2,−1,0), B(4,1,−1), C(5,−1,−4)$.

Ans.   (a) $68°20'$ (b) $A = 48°10', B = 95°10', C = 36°40'$

37.21 Find the coordinates of the point P in which the line joining $A(5,−1,4)$ and $B(−5,7,0)$ pierces the yz plane.

Hint: Let P have coordinates $(0,b,c)$ and express the condition (see Problem 37.8) that A, B, P be collinear.

Ans. $P(0,3,2)$

37.22 Find relations which the coordinates of $P(x,y,z)$ must satisfy if P is to be collinear with (2,3,1) and (1,−2,−5).

Ans. $x − 2 : y − 3 : z − 1 = 1 : 5 : 6$ or $\dfrac{x − 2}{1} = \dfrac{y − 3}{5} = \dfrac{z − 1}{6}$

37.23 Find a set of direction numbers for any line perpendicular to

(a)   Each of the lines l_1: $[1,2,−4]$ and l_2: $[2, −1,3]$

(b)   Each of the lines joining $A(2,−1,5)$ to $B(−1,3,4)$ and $C(0,−5,4)$

Ans.   (a) $[2,−11,−5]$ (b) $[8,1,−20]$

37.24 Find the coordinates of the point P in which the line joining the points $A(4,11,18)$ and $B(−1,−4,−7)$ intersects the line joining the points $C(3,1,5)$ and $D(5,0,7)$.

Hint: Let P divide $\overline{AB}$ in the ratio $1 : r$ and $\overline{CD}$ in the ratio $1 : s$, and obtain relations $rs − r − 4s − 6 = 0$, etc.

Ans.   (1,2,3)

37.25 Prove that the four line segments joining each vertex of a tetrahedron to the point of intersection of the medians of the opposite face have a point G in common. Prove that each of the four line segments is divided in the ratio 1:3 by G.

(NOTE: G, the point P of Problem 37.4, is called the centroid of the tetrahedron.)

Chapter 38

Simultaneous Equations Involving Quadratics

ONE LINEAR AND ONE QUADRATIC EQUATION

Procedure: Solve the linear equation for one of the two unknowns (your choice) and substitute in the quadratic equation. Since this results in a quadratic equation in one unknown, the system can always be solved.

EXAMPLE 1. Solve the system $\begin{cases} 4x^2 + 3y^3 = 16 \\ 5x + y = 7 \end{cases}$

Solve the linear equation for y: $y = 7 - 5x$. Substitute in the quadratic equation:

$$4x^2 + 3(7 - 5x)^2 = 16$$

$$4x^2 + 3(49 - 70 + 25x^2) = 16$$

$$79x^2 - 210x + 131 = (x - 1)(79x - 131) = 0$$

and $x = 1, \frac{131}{79}$.

When $x = 1, y = 7 - 5x = 2$; when $x = \frac{131}{79}, y = -\frac{102}{79}$. The solutions are $x = 1, y = 2$ and $x = \frac{131}{79}, y = -\frac{102}{79}$.

The locus of the linear equation is the straight line and the locus of the quadratic equation is the ellipse in Fig. 38-1. (See Problems 38.1–38.2.)

TWO QUADRATIC EQUATIONS. In general, solving a system of two quadratic equations in two unknowns involves solving an equation of the fourth degree in one of the unknowns. Since the solution of the general equation of the fourth degree in one unknown is beyond the scope of this book, only those systems which require the solution of a quadratic equation in one unknown will be treated here.

TWO QUADRATIC EQUATIONS OF THE FORM $ax^2 + by^2 = c$

Procedure: Eliminate one of the unknowns by the method of addition for simultaneous equations in Chapter 5.

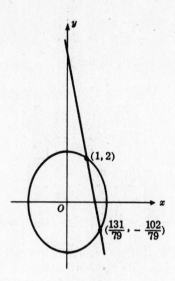

Fig. 38-1

EXAMPLE 2. Solve the system $\begin{cases} 4x^2 + 9y^2 = 72 & (38.1) \\ 3x^2 - 2y^2 = 19 & (38.2) \end{cases}$

Multiply (38.1) by 2: $8x^2 + 18y^2 = 144$

Multiply (38.2) by 9: $27x^2 - 18y^2 = 171$

Add: $\overline{35x^2 \qquad\quad = 315}$

Then $x^2 = 9$ and $x = \pm 3$.

When $x = 3$, (38.1) gives $9y^2 = 72 - 4x^2 = 72 - 36 = 36, y^2 = 4$, and $y = \pm 2$.

When $x = -3$, (38.1) gives $9y^2 = 72 - 36 = 36$, $y^2 = 4$, and $y = \pm 2$.

The four solutions $x = 3, \ y = 2; x = 3, \ y = -2; \ x = -3, \ y = 2; \ x = -3, \ y = -2$ may also be written as $x = \pm 3, \ y = \pm 2; \ x = \pm 3, \ y = \pm 2$. By convention, we read the two upper signs and the two lower signs in the latter form.

The ellipse and the hyperbola intersect in the points $(3, 2), (3, -2), (-3, 2), (-3, -2)$. See Fig. 38-2. (See Problems 38.3–38.4.)

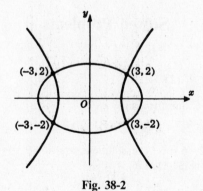

Fig. 38-2

TWO QUADRATIC EQUATIONS, ONE HOMOGENEOUS. An expression, as $2x^2 - 3xy + y^2$, whose terms are all of the same degree in the variables, is called *homogeneous*. A homogeneous expression equated to zero is called a *homogeneous equation*. A homogeneous quadratic equation in two unknowns can always be solved for one of the unknowns in terms of the other.

EXAMPLE 3. Solve the system $\begin{cases} x^2 - 3xy + 2y^2 = 0 & (38.3) \\ 2x^2 + 3xy - y^2 = 13 & (38.4) \end{cases}$

Solve (38.3) for x in terms of y: $(x-y)(x-2y) = 0$ and $x = y$, $x = 2y$.
Solve the systems (see Example 1):

$$\begin{cases} 2x^2 + 3xy - y^2 = 13 \\ \quad\quad\quad x = y \end{cases} \quad\quad \begin{cases} 2x^2 + 3xy - y^2 = 13 \\ \quad\quad\quad x = 2y \end{cases}$$

$$2y^2 + 3y^2 - y^2 = 4y^2 = 13 \quad\quad 8y^2 + 6y^2 - y^2 = 13y^2 = 13$$

$$y^2 = \tfrac{13}{4}, \; y = \pm\frac{\sqrt{13}}{2} \quad\quad\quad y^2 = 1, \; y = \pm 1$$

Then $x = y = \pm\sqrt{13}/2.$ Then $x = 2y = \pm 2.$

The solutions are $x = \sqrt{13}/2,\; y = \sqrt{13}/2;\; x = -\sqrt{13}/2,\; y = -\sqrt{13}/2;\; x = 2,\; y = 1;\; x = -2,\; y = -1$ or $x = \pm\sqrt{13}/2,\; y = \pm\sqrt{13}/2;\; x = \pm 2,\; y = \pm 1.$ (See Problem 38.5.)

TWO QUADRATIC EQUATIONS OF THE FORM $ax^2 + bxy + cy^2 = d$

Procedure: Combine the two given equations to obtain a homogeneous equation. Solve, as in Example 3, the system consisting of this homogeneous equation and either of the given equations. (See Problems 38.6–38.7.)

TWO QUADRATIC EQUATIONS, EACH SYMMETRICAL IN x AND y.
An equation, as $2x^2 - 3xy + 2y^2 + 5x + 5y = 1$, which is unchanged when the two unknowns are interchanged is called a *symmetrical equation*.

Procedure: Substitute $x = u + v$ and $y = u - v$ and then eliminate v^2 from the resulting equations. (See Problem 38.8.)

Frequently, a careful study of a given system will reveal some special device for solving it. (See Problems 38.9–38.13.)

Solved Problems

38.1   Solve the system $\begin{cases} 2y^2 - 3x = 0 & (1) \\ 4y - x = 6 & (2) \end{cases}$

Solve (2) for x: $x = 4y - 6$. Substitute in (1):

$$2y^2 - 3(4y - 6) = 2(y - 3)^2 = 0 \quad\quad y = 3, 3$$

When $y = 3$: $x = 4y - 6 = 12 - 6 = 6$. The solutions are $x = 6, y = 3; x = 6, y = 3$. The straight line is tangent to the parabola (see Fig. 38-3) at $(6, 3)$.

38.2   Solve the system $\begin{cases} y^2 - 4y - 3x + 1 = 0 & (1) \\ 3y - 4x = 7 & (2) \end{cases}$

Solve (2) for x: $x = \tfrac{1}{4}(3y - 7)$. Substitute in (1):

$$y^2 - 4y - \tfrac{3}{4}(3y - 7) + 1 = 0$$

$$4y^2 - 16y - 9y + 21 + 4 = 4y^2 - 25y + 25 = (y - 5)(4y - 5) = 0 \quad \text{or} \quad y = 5 \quad \text{and} \quad y = \tfrac{5}{4}$$

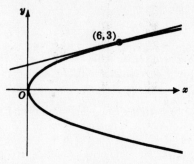

Fig. 38-3

When $y = 5, x = \frac{1}{4}(3y - 7) = 2$; when $y = \frac{5}{4}, x = \frac{1}{4}(3y - 7) = -\frac{13}{16}$. The solutions are $x = 2, y = 5$; $x = -\frac{13}{16}$, $y = \frac{5}{4}$. The straight line intersects the parabola in the points $(2, 5)$ and $(-\frac{13}{16}, \frac{5}{4})$.

38.3   Solve the system $\begin{cases} 3x^2 - y^2 = 27 & (1) \\ x^2 - y^2 = -45 & (2) \end{cases}$

Subtract: $2x^2 = 72, x^2 = 36$, and $x = \pm 6$.
When $x = 6, y^2 = x^2 + 45 = 36 + 45 = 81$, and $y = \pm 9$.
When $x = -6$, $y^2 = x^2 + 45 = 36 + 45 = 81$, and $y = \pm 9$. The solutions are $x = \pm 6$, $y = \pm 9$; $x = \pm 6$, $y = \pm 9$. The two hyperbolas intersect in the points $(6, 9), (-6, 9), (-6, -9)$ and $(6, -9)$.

38.4   Solve the system $\begin{cases} 5x^2 + 3y^2 = 92 & (1) \\ 2x^2 + 5y^2 = 52 & (2) \end{cases}$

Multiply (1) by 5: $25x^2 + 15y^2 = 460$
Multiply (2) by -3: $-6x^2 - 15y^2 = -156$
Add: $\overline{19x^2 = 304} \qquad x^2 = 16 \qquad$ and $\qquad x = \pm 4$

When $x = \pm 4$: $3y^2 = 92 - 5x^2 = 92 - 80 = 12$; $y^2 = 4$ and $y = \pm 2$. The solutions are $x = \pm 4$, $y = \pm 2$; $x = \pm 4$, $y = \pm 2$. See Fig. 38-4.

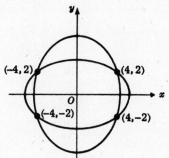

Fig. 38-4

38.5   Solve the system $\begin{cases} x^2 + 4xy = 0 & (1) \\ x^2 - xy + y^2 = 21 & (2) \end{cases}$

Solve (1) for x: $x(x + 4y) = 0$ and $x = 0, x = -4y$.
Solve the systems

$$\begin{cases} x^2 - xy + y^2 = 21 \\ x = 0 \end{cases} \qquad\qquad \begin{cases} x^2 - xy + y^2 = 21 \\ x = -4y \end{cases}$$

$ y^2 = 21, \qquad y = \pm\sqrt{21} \qquad\qquad y^2 = 1, \quad y = \pm 1; \quad x = -4y \mp 4$

The solutions are $x = 0$, $y = \pm\sqrt{21}$; $x = \pm 4$, $y = \mp 1$.

38.6   Solve the system $\begin{cases} 3x^2 + 8y^2 = 140 & (1) \\ 5x^2 + 8xy = 84 & (2) \end{cases}$

Multiply (1) by -3: $-9x^2 - 24y^2 = -420$
Multiply (2) by 5: $\underline{25x^2 + 40xy = 420}$
Add: $16x^2 + 40xy - 24y^2 = 0$

Then

$$8(2x^2 + 5xy - 3y^2) = 8(2x - y)(x + 3y) = 0 \quad \text{and} \quad x = \tfrac{1}{2}y, \ x = -3y.$$

Solve the systems

$$\begin{cases} 3x^2 + 8y^2 = 140 \\ x = \tfrac{1}{2}y \end{cases} \qquad\qquad \begin{cases} 3x^2 + 8y^2 = 140 \\ x = -3y \end{cases}$$

$$\tfrac{3}{4}y^2 + 8y^2 = \tfrac{35}{4}y^2 = 140 \qquad\qquad 27y^2 + 8y^2 = 35y^2 = 140$$

$$y^2 = 16, \quad y = \pm 4, \quad x = \tfrac{1}{2}y = \pm 2 \qquad y^2 = 4, \quad y = \pm 2; \quad x = -3y = \mp 6$$

The solutions are $x = \pm 2, \ y = \pm 4; \ x = \mp 6, \ y = \pm 2$.

38.7   Solve the system $\begin{cases} x^2 - 3xy + 2y^2 = 15 & (1) \\ 2x^2 + y^2 = 6 & (2) \end{cases}$

Multiply (1) by -2: $-2x^2 + 6xy - 4y^2 = -30$
Multiply (2) by 5: $\underline{10x^2 + 5y^2 = 30}$
Add: $8x^2 + 6xy + y^2 = (4x + y)(2x + y) = 0.$ Then $y = -4x$ and $y = -2x$.

Solve the systems

$$\begin{cases} 2x^2 + y^2 = 6 \\ y = -4x \end{cases} \qquad\qquad \begin{cases} 2x^2 + y^2 = 6 \\ y = -2x \end{cases}$$

$$2x^2 + 16x^2 = 18x^2 = 6, x^2 = 1/3; \qquad 2x^2 + 4x^2 = 6x^2 = 6, x^2 = 1;$$

$$x = \pm\sqrt{3}/3 \text{ and } y = -4x = \mp 4\sqrt{3}/3 \qquad x = \pm 1 \text{ and } y = -2x = \mp 2$$

The solutions are $x = \pm\sqrt{3}/3, y = \mp 4\sqrt{3}/3; x \pm 1, y = \mp 2$.

38.8   Solve the system $\begin{cases} x^2 + y^2 + 3x + 3y = 8 \\ xy + 4x + 4y = 2 \end{cases}$

Substitute $x = u + v, y = u - v$ in the given system:

$$(u + v)^2 + (u - v)^2 + 3(u + v) + 3(u - v) = 2u^2 + 2v^2 + 6u = 8 \quad (1)$$

$$(u + v)(u - v) + 4(u + v) + 4(u - v) = u^2 - v^2 + 8u = 2 \quad (2)$$

Add (1) and $2(2)$:

$$4u^2 + 22u - 12 = 2(2u - 1)(u + 6) = 0; \quad u = \tfrac{1}{2}, -6.$$

For $u = \tfrac{1}{2}$, (2) yields $v^2 = u^2 + 8u - 2 = \tfrac{1}{4} + 4 - 2 = \tfrac{9}{4}; v = \pm\tfrac{3}{2}$.

When $u = \tfrac{1}{2}, v = \tfrac{3}{2}: x = u + v = 2, y = u - v = -1$.
When $u = \tfrac{1}{2}, v = -\tfrac{3}{2}: x = u + v = -1, y = u - v = 2$.

For $u = -6$, (2) yields $v^2 = u^2 + 8u - 2 = 36 - 48 - 2 = -14; v = \pm i\sqrt{14}$.

When $u = -6, v = i\sqrt{14}: x = u + v = -6 + i\sqrt{14}, y = u - v = -6 - i\sqrt{14}$.
When $u = -6, v = -i\sqrt{14}: x = u + v = -6 - i\sqrt{14}, y = u - v = -6 + i\sqrt{14}$.

The solutions are $x = 2, y = -1; x = -1, y = 2; x = -6 \pm i\sqrt{14}, y = -6 \mp i\sqrt{14}$.

38.9   Solve the system $\begin{cases} x^2 + y^2 = 25 & (1) \\ \quad xy = 12 & (2) \end{cases}$

Multiply (2) by 2 and add to (1): $x^2 + 2xy + y^2 = 49$ or $x + y = \pm 7$.
Multiply (2) by -2 and add to (1): $x^2 - 2xy + y^2 = 1$ or $x - y = \pm 1$.
Solve the systems

$$\begin{cases} x + y = 7 \\ \underline{x - y = 1} \\ \quad 2x = 8; \quad x = 4 \end{cases} \qquad \begin{cases} x + y = 7 \\ \underline{x - y = -1} \\ \quad 2x = 6; \quad x = 3 \end{cases}$$

$$y = 7 - x = 3 \qquad\qquad y = 7 - x = 4$$

$$\begin{cases} x + y = -7 \\ \underline{x - y = \ \ 1} \\ \quad 2x = -6; \quad x = -3 \end{cases} \qquad \begin{cases} x + y = -7 \\ \underline{x - y = -1} \\ \quad 2x = -8; \quad x = -4 \end{cases}$$

$$y = -7 - x = -4 \qquad\qquad y = -7 - x = -3$$

The solutions are $x = \pm 4, y = \pm 3; x = \pm 3, y = \pm 4$. See Fig. 38-5.

Alternate Solutions.   Solve (2) for $y = 12/x$ and substitute (1). The resulting quartic $x^4 - 25x^2 + 144 = 0$ can be factored readily.

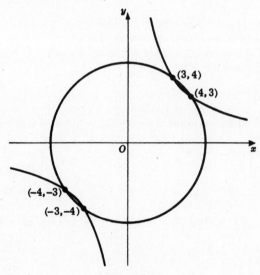

Fig. 38-5

38.10   Solve the system $\begin{cases} x^2 - xy - 12y^2 = 8 & (1) \\ x^2 + xy - 10y^2 = 20 & (2) \end{cases}$

This system may be solved by the procedure used in Problems 38.6 and 38.7. Here we give an alternate solution.

Procedure: Substitute $y = mx$ in the given equations to obtain a system in the unknowns m and x: then eliminate x to obtain a quadratic in m.

Put $y = mx$ in (1) and (2):

$$x^2 - mx^2 - 12m^2x^2 = x^2(1 - m - 12m^2) = 8$$
$$x^2 + mx^2 - 10m^2x^2 = x^2(1 + m - 10m^2) = 20$$

Now

$$x^2 = \frac{8}{1 - m - 12m^2} \quad \text{and} \quad x^2 = \frac{20}{1 + m - 10m^2}$$

so that

$$\frac{8}{1 - m - 12m^2} = \frac{20}{1 + m - 10m^2}$$

$$8 + 8m - 80m^2 = 20 - 20m - 240m^2$$

$$160m^2 + 28m - 12 = 4(5m - 1)(8m + 3) = 0 \quad \text{and} \quad m = \tfrac{1}{5}, m = -\tfrac{3}{8}$$

then $m = \tfrac{1}{5} : x^2 = \dfrac{8}{1 - m - 12m^2} = 25; x = \pm 5, y = mx - \dfrac{1}{5}(\pm 5) = \pm 1.$

then $m = -\tfrac{3}{8}$:

$$x^2 = \frac{8}{1 - m - 12m^2} = -\frac{128}{5}; \quad x = \pm \frac{8i\sqrt{10}}{5}, \quad y = mx = \mp \frac{3i\sqrt{10}}{5}.$$

The solutions are $x = \pm 5, y = \pm 1; x = \pm \dfrac{8i\sqrt{10}}{5}, y = \mp \dfrac{3i\sqrt{10}}{5}.$

38.11 Solve the system $\begin{cases} x^3 - y^3 = 19 & (1) \\ x^2 + xy + y^2 = 19 & (2) \end{cases}$

Divide (1) by (2): $x - y = 1$.

Solve the system $\begin{cases} x^2 + xy + y^2 = 19 & (2) \\ x - y = 1 & (3) \end{cases}$

Solve (3) for x: $x = y + 1$.
Substitute in (2): $(y + 1)^2 + (y + 1)y + y^2 = 3y^2 + 3y + 1 = 19$.
Then $3y^2 + 3y - 18 = 3(y + 3)(y - 2) = 0$ and $y = -3, 2$.
When $y = -3, x = y + 1 = -2$; when $y = 2, x = y + 1 = 3$.
The solutions are $x = -2, y = -3; x = 3, y = 2$.

38.12 Solve the system $\begin{cases} (2x - y)^2 - 4(2x - y) = 5 & (1) \\ x^2 - y^2 = 3 & (2) \end{cases}$

Factor (1): $(2x - y)^2 - 4(2x - y) - 5 = (2x - y - 5)(2x - y + 1) = 0$. Then $2x - y = 5$ and $2x - y = -1$.
Solve the systems

$\begin{cases} (x^2 - y^2 = 3 \\ 2x - y = 5 \end{cases}$ $\qquad\qquad$ $\begin{cases} x^2 - y^2 = 3 \\ 2x - y = -1 \end{cases}$

$y = 2x - 5$ $\qquad\qquad\qquad\qquad$ $y = 2x + 1$
$x^2 - (2x - 5)^2 = 3$ $\qquad\qquad\quad$ $x^2 - (2x + 1)^2 = 3$
$3x^2 - 20x + 28 = (x - 2)(3x - 14) = 0$ $\qquad$ $3x^2 + 4x + 4 = 0$

$x = 2, \tfrac{14}{3}$ $\qquad\qquad\qquad\qquad$ $x = \dfrac{-4 \pm \sqrt{16 - 48}}{6} = \dfrac{-2 \pm 2i\sqrt{2}}{3}$

When $x = 2, y = 2x - 5 = -1$. $\qquad\qquad$ $y = 2x + 1 = \dfrac{-1 \pm 4i\sqrt{2}}{3}$

When $x = \tfrac{14}{3}, y = 2x - 5 = \tfrac{13}{3}$.

The solutions are $x = 2, y = -1; x = \tfrac{14}{3}, y = \tfrac{13}{3}; x = \dfrac{-2 \pm 2i\sqrt{2}}{3}, y = \dfrac{-1 \pm 4i\sqrt{2}}{3}.$

38.13 Solve the system $\begin{cases} 5/x^2 + 3/y^2 = 32 & (1) \\ 4xy = 1 & (2) \end{cases}$

Write (1) as $3x^2 + 5y^2 = 32x^2y^2 = 2(4xy)^2$. Substitute (2): $3^2 + 5y^2 = 2(1)^2 = 2$ (3)
Subtract $2(2)$ from (3): $3x^2 - 8xy + 5y^2 = 0$. Then $(x - y)(3x - 5y) = 0$ and $x = y, x = 5y/3$.
Solve the systems

$\begin{cases} 4xy = 1 \\ x = y \end{cases}$ $\qquad\qquad$ $\begin{cases} 4xy = 1 \\ x = 5y/3 \end{cases}$

$4y^2 = 1; \quad y = \pm \tfrac{1}{2}$ $\qquad$ $\dfrac{20}{3}y^2 = 1, \qquad y^2 = \dfrac{3}{20} = \dfrac{15}{100};$

and $x = y = \pm \tfrac{1}{2}$.

$$y = \pm \frac{\sqrt{15}}{10} \quad \text{and} \quad x = \frac{5}{3}y = \pm \frac{\sqrt{15}}{6}.$$

The solutions are $x = \pm \frac{1}{2}, y = \pm \frac{1}{2}; x = \pm \frac{\sqrt{15}}{6}, y = \pm \frac{\sqrt{15}}{10}$.

Supplementary Problems

Solve.

38.14 $\begin{cases} xy + y^2 = 5 \\ 2x + 3y = 7 \end{cases}$ *Ans.* $\begin{array}{l} x = -4, y = 5 \\ x = \frac{1}{2}, y = 2 \end{array}$

38.15 $\begin{cases} y = x^2 - x - 1 \\ y = 2x + 3 \end{cases}$ *Ans.* $\begin{array}{l} x = -1, y = 1 \\ x = 4, y = 11 \end{array}$

38.16 $\begin{cases} 3x^3 - 7y^2 = 12 \\ x - 3y = -2 \end{cases}$ *Ans.* $\begin{array}{l} x = -2, y = 0 \\ x = \frac{17}{5}, y = \frac{9}{5} \end{array}$

38.17 $\begin{cases} x^2 + 3y^2 = 43 \\ 3x^2 + y^2 = 57 \end{cases}$ *Ans.* $\begin{array}{l} x = \pm 4, y = \pm 3 \\ x = \pm 4, y = \mp 3 \end{array}$

38.18 $\begin{cases} 9x^2 + y^2 = 90 \\ x^2 + 9y^2 = 90 \end{cases}$ *Ans.* $\begin{array}{l} x = \pm 3, y = \pm 3 \\ x = \pm 3, y = \mp 3 \end{array}$

38.19 $\begin{cases} 2/x^2 - 3/y^2 = 5 \\ 1/x^2 + 2/y^2 = 6 \end{cases}$ *Ans.* $\begin{array}{l} x = \pm \frac{1}{2}, y = \pm 1 \\ x = \pm \frac{1}{2}, y = \mp 1 \end{array}$

38.20 $\begin{cases} x^2 - xy^2 + y^2 = 28 \\ 2x^2 + 3xy - 2y^2 = 0 \end{cases}$ *Ans.* $\begin{array}{l} x = \pm 4, y = \mp 2 \\ x = \pm \sqrt{21}/3, y = \pm 4\sqrt{21}/3 \end{array}$

38.21 $\begin{cases} x^2 - xy^2 - 12y^2 = 0 \\ x^2 + xy - 10y^2 = 20 \end{cases}$ *Ans.* $\begin{array}{l} x = \pm 4\sqrt{2}, y = \pm \sqrt{2} \\ x = \mp 3i\sqrt{5}, y = \pm i\sqrt{5} \end{array}$

38.22 $\begin{cases} 6x^2 + 3xy^2 + 2y^2 = 24 \\ 3x^2 + 2xy + 2y^2 = 18 \end{cases}$ *Ans.* $\begin{array}{l} x = \pm 2, y = \mp 3 \\ x = \pm \sqrt{30}/5, y = \pm 2\sqrt{30}/5 \end{array}$

38.23 $\begin{cases} y^2 = 4x - 8 \\ y^2 = -6x + 32 \end{cases}$ *Ans.* $x = 4, y = \pm 2\sqrt{2}$

38.24 $\begin{cases} x^2 - y^2 = 16 \\ y^2 = 2x - 1 \end{cases}$ *Ans.* $\begin{array}{l} x = 5, y = \pm 3 \\ x = -3, y = \pm i\sqrt{7} \end{array}$

38.25 $\begin{cases} 2x^2 + y^2 = 6 \\ x^2 + y^2 + 2x = 3 \end{cases}$ *Ans.* $\begin{array}{l} x = -1, y = \pm 2 \\ x = 3, y = \pm 2i\sqrt{3} \end{array}$

38.26 $\begin{cases} x^2 + y^2 - 2x - 2y = 12 \\ xy = 6 \end{cases}$ *Ans.* $\begin{array}{l} x = 3 \pm \sqrt{3}, y = 3 \mp \sqrt{3} \\ x = -2 \pm i\sqrt{2}, y = -2 \mp i\sqrt{2} \end{array}$

38.27 $\begin{cases} x^3 - y^3 = 28 \\ x - y = 4 \end{cases}$ *Ans.* $\begin{array}{l} x = 1, y = -3 \\ x = 3, y = -1 \end{array}$

38.28 $\begin{cases} x + y + 3\sqrt{x + y} = 18 \\ x - y - 2\sqrt{x - y} = 15 \end{cases}$ **Hint:** Let $\sqrt{x + y} = u, \sqrt{x - y} = v$.   *Ans.* $x = 17, \quad y = -18$

38.29 Two numbers differ by 2 and their squares differ by 48. Find the numbers.   *Ans.*   11, 13

38.30 The sum of the circumference of two circles is 88 cm and the sum of their areas is $\frac{2200}{7}$ cm^2, when $\frac{22}{7}$ is used for π. Find the radius of each circle.   *Ans.*   6 cm, 8 cm

38.31 A party costing $30 is planned. It is found that by adding three more to the group, the cost per person would be reduced by 50 cents. For how many was the party originally planned?   *Ans.*   12

Chapter 39

Logarithms

THE LOGARITHM OF A POSITIVE NUMBER N to a given base b (written $\log_b N$) is the exponent of the power to which b must be raised to produce N. It will be understood throughout this chapter that b is positive and different from 1.

EXAMPLE 1.

(a) Since $9 = 3^2$, $\log_3 9 = 2$.
(b) Since $64 = 4^3$, $\log_4 64 = 3$.
(c) Since $64 = 2^6$, $\log_2 64 = 6$.
(d) Since $1000 = 10^3$, $\log_{10} 1000 = 3$.
(e) Since $0.01 = 10^{-2}$, $\log_{10} 0.01 = -2$.

(See Problems 39.1–39.3.) Note that if $f(x) = b^x$ and $g(x) = \log_b x$ (where $b > 0, b \neq 1$), Then $f(g(x)) = b^{\log_b x} = x$, and $g(f(x)) = \log_b(b^x) = x$. Thus in, f and g are inverse functions.

FUNDAMENTAL LAWS OF LOGARITHMS

(1) The logarithm of the product of two or more positive numbers is equal to the sum of the logarithms of the several numbers. For example,

$$\log_b(P \cdot Q \cdot R\cdot) = \log_b P + \log_b Q + \log_b R$$

(2) The logarithm of the quotient of two positive numbers is equal to the logarithm of the dividend minus the logarithm of the divisor. For example,

$$\log_b \frac{P}{Q} = \log_b P - \log_b Q$$

(3) The logarithm of a power of a positive number is equal to the logarithm of the number, multiplied by the exponent of the power. For example,

$$\log_b P^n = n \log_b P$$

(4) The logarithm of a root of a positive number is equal to the logarithm of the number, divided by the index of the root. For example,

$$\log_b \sqrt[n]{P} = \frac{1}{n} \log_b P$$

(See Problems 39.4–39.7.)

IN NUMERICAL COMPUTATIONS a widely used base for a system of logarithms is 10. Such logarithms are called *common logarithms*. The common logarithm of a positive number $N/1$ is written $\log N$. These interact well with the decimal number system; for example, $\log 1000 = \log 10^3 = 3$.

AN EXPONENTIAL EQUATION is an equation involving one or more unknowns in an exponent. For example, $2^x = 7$ and $(1.03)^{-x} = 2.5$ are exponential equations. Such equations are solved by means of logarithms.

EXAMPLE 2.   Solve the exponential equation $2^x = 7$.
Take logarithms of both sides: $x \log 2 = \log 7$
Solve for x : $x = \dfrac{\log 7}{\log 2} = \dfrac{0.8451}{0.3010} = 2.808$, approximately.
(See Problem 39.16.)

IN THE CALCULUS the most useful logarithmic function is the *natural logarithm* in which the base is a certain irrational number $e = 2.71828$, approximately.
   The natural logarithm of N, $\ln N$, and the common logarithm of N, $\log N$, are related by the formula

$$\ln N \cong 2.3026 \log N$$

THE CALCULATOR can be used to do logarithmic calculation with extreme ease.

EXAMPLE 3.   Evaluate log 82,734 rounded to six decimal places.

   Press: $\boxed{8}\ \boxed{2}\ \boxed{7}\ \boxed{3}\ \boxed{4}\ \boxed{\log}$. On screen: 4.917684.

EXAMPLE 4.   Solve for x to four significant digits: $\ln x = -0.3916$. Press: .3916 $\boxed{\pm}\ \boxed{e^x}$. On screen: 0.6759745. Then $x \approx 0.6760$.
   Note that we use the inverse function of $\ln x$, e^x, to find the number corresponding to a given $\ln x$. For $\log x$, we use its inverse, 10^x.

Solved Problems

39.1   Change the following from exponential to logarithmic form:
   (a) $7^2 = 49$,   (b) $6^{-1} = \frac{1}{6}$,   (c) $10^0 = 1$,   (d) $4^0 = 1$,   (e) $\sqrt[3]{8} = 2$.

   Ans.   (a) $\log_7 49 = 2$,   (b) $\log_6 \frac{1}{6} = -1$,   (c) $\log_{10} 1 = 0$,   (d) $\log_4 1 = 0$,
   (e) $\log_8 2 = \frac{1}{3}$.

39.2   Change the following from logarithmic to exponential form:
   (a) $\log_3 81 = 4$,   (b) $\log_5 \frac{1}{625} = -4$,   (c) $\log_{10} 10 = 1$,   (d) $\log_9 27 = \frac{3}{2}$.
   Ans.   (a) $3^4 = 81$,   (b) $5^{-4} = \frac{1}{625}$,   (c) $10^1 = 10$,   (d) $9^{3/2} = 27$.

39.3 Evaluate x, given:

(a) $x = \log_5 125$ (d) $x = \log_2 \frac{1}{16}$ (g) $\log_x \frac{1}{16} = -2$

(b) $x = \log_{10} 0.001$ (e) $x = \log_{1/2} 32$ (h) $\log_6 x = 2$

(c) $x = \log_8 2$ (f) $\log_x 243 = 5$ (i) $\log_a x = 0$

Ans. (a) 3, since $5^3 = 125$ (d) -4, since $2^{-4} = \frac{1}{16}$ (g) 4, since $4^{-2} = \frac{1}{16}$

(b) -3, since $10^{-3} = 0.001$ (e) -5, since $(\frac{1}{2})^{-5} = 32$ (h) 36, since $6^2 = 36$

(c) $\frac{1}{3}$, since $8^{1/3} = 2$ (f) 3, since $3^5 = 243$ (i) 1, since $a^0 = 1$

39.4 Prove the four laws of logarithms.

Let $P = b^p$ and $Q = b^q$; then $\log_b P = p$ and $\log_b Q = q$.

(1) Since $P \cdot Q = b^p \cdot b^q = b^{p+q}, \log_b PQ = p + q = \log_b P + \log_b Q$; that is, the logarithm of the product of two positive numbers is equal to the sum of the logarithms of the numbers.

(2) Since $P/Q = b^p/b^q = b^{p-q}, \log_b(P/Q) = p - q = \log_b P - \log_b Q$; that is, the logarithm of the quotient of two positive numbers is the logarithm of the numerator minus the logarithm of the denominator.

(3) Since $P^n = (b^p)^n = b^{np}, \log_b P^n = np = n\log_b P$; that is, the logarithm of a power of a positive number is equal to the product of the exponent and the logarithm of the number.

(4) Since $\sqrt[n]{P} = P^{1/n} = b^{p/n}, \log_b \sqrt[n]{P} = \frac{p}{n} = \frac{1}{n}\log_b P$; that is, the logarithm of a root of a positive number is equal to the logarithm of the number divided by the index of the root.

39.5 Express the logarithms of the given expressions in terms of the logarithms of the individual letters involved.

(a) $\log_b \dfrac{P \cdot Q}{R} = \log_b(P \cdot Q) - \log_b R = \log_b P + \log_b Q - \log_b R$

(b) $\log_b \dfrac{P}{Q \cdot R} = \log_b P - \log_b(Q \cdot R) = \log_b P - (\log_b Q + \log_b R) = \log_b P - \log_b Q - \log_b R$

(c) $\log_b P^2 \cdot \sqrt[3]{Q} = \log_b P^2 + \log_b \sqrt[3]{Q} = 2\log_b P + \frac{1}{3}\log_b Q$

(d) $\log_b \sqrt{\dfrac{P \cdot Q^3}{R^{1/2} \cdot S}} = \frac{1}{2}\log_b \dfrac{P \cdot Q^3}{R^{1/2} \cdot S} = \frac{1}{2}[\log_b(P \cdot Q^3) - \log_b(R^{1/2} \cdot S)]$

$= \frac{1}{2}(\log_b P + 3\log_b Q - \frac{1}{2}\log_b R - \log_b S)$

39.6 Express each of the following as a single logarithm:

(a) $\log_b x - 2\log_b y + \log_b z = (\log_b x + \log_b z) - 2\log_b y = \log_b xz - \log_b y^2 = \log_b \dfrac{xz}{y^2}$

(b) $\log_b 2 + \log_b \pi + \frac{1}{2}\log_b l - \frac{1}{2}\log_b g = (\log_b 2 + \log_b \pi) + \frac{1}{2}(\log_b l - \log_b g)$

$= \log_b(2\pi) + \frac{1}{2}\log_b \dfrac{l}{g} = \log_b\left(2\pi\sqrt{\dfrac{l}{g}}\right)$

39.7 Show that $b^{3\log_b x} = x^3$.

Let $3\log_b x = t$. Then $\log_b x^3 = t$ and $x^3 = b^t = b^{3\log_b x}$. Alternatively, $b^{3\log_b x} = b^{\log_b x^3} = x^3$ (using the inverse function property).

39.8 Verify that

(a) $\log 3860 = 3.5866$ (e) $\log 5.463 = 0.7374$

(b) $\log 52.6 = 1.7210$ (f) $\log 77.62 = 1.8900$

(c) $\log 7.84 = 0.8943$ (g) $\log 2.866 = 0.4573$

(d) $\log 728\,000 = 5.8621$

39.9   Find (*a*) $\log 2.864^3$, (*b*) $\log \sqrt{2.864}$.

Since $\log 2.864 = 0.4570$,

(*a*) $\log 2.864^3 = 3 \log 2.864 = 3(0.4570) = 1.3710$
(*b*) $\log \sqrt{2.864} = \frac{1}{2} \log 2.864 = \frac{1}{2}(0.4570) = 0.2285$

39.10 Find the number whose log is 1.4232. Using a calculator, press 1.4232 $\boxed{10^x}$. The result on the screen is 26.50 (rounded to two places).

39.11 Solve.

(*a*) $(1.06)^x = 3$.
Taking logarithms, $x \log 1.06 = \log 3$.

$$x = \frac{\log 3}{\log 1.06} = \frac{0.4771}{0.0253} \qquad x = 18.86$$

(*b*) $12^{2x+5} = 55(7^{3x})$.
Taking logarithms, $(2x + 5) \log 12 = \log 55 + 3x \log 7$

$$2x \log 12 - 3x \log 7 = \log 55 - 5 \log 12$$
$$x = \frac{\log 55 - 5 \log 12}{2 \log 12 - 3 \log 7} = \frac{1.7404 - 5(1.0792)}{2(1.0792) - 3(0.8451)} = \frac{3.6556}{0.3769} = 9.700$$

(*c*) $41.2^x = 12.6^{x-1}$.
Taking logarithms, $x \log 41.2 = (x - 1) \log 12.6$.

$$x \log 41.2 - x \log 12.6 = -\log 12.6 \quad \text{or} \quad x = \frac{-\log 12.6}{\log 41.2 - \log 12.6}$$
$$y = -x = \frac{\log 12.6}{\log 41.2 - \log 12.6} = \frac{1.1004}{0.5145} = 2.138$$
$$x = -2.138$$

Supplementary Problems

39.12 Solve for *x* using a calculator.

(*a*) $3^x = 30$ (*b*) $1.07^x = 3$ (*c*) $5.72^x = 8.469$ (*d*) $38.5^x = 6.5^{x-2}$

Ans.   (*a*) 3.096   (*b*)   16.23   (*c*)   1.225   (*d*) −2.104

39.13 Show that $10^{n \log(a+b)} = (a + b)^n$.

39.14 Find   (*a*) $\log 3.64$,   (*b*) $\log 36.4$,   (*c*) $\log 364$.

39.15 Find the number whose natural logarithm is   (*a*) 10,   (*b*) 130,   (*c*) 407.1.

Chapter 40

Power, Exponential, and Logarithmic Curves

POWER FUNCTIONS in x are of the form x^n. If $n > 0$, the graph of $y = x^n$ is said to be of the *parabolic* type (the curve is a parabola for $n = 2$). If $n < 0$, the graph of $y = x^n$ is said to be of the *hyperbolic* type (the curve is a hyperbola for $n = -1$).

EXAMPLE 1. Sketch the graphs of (*a*) $y = x^{3/2}$, (*b*) $y = -x^{-3/2}$.

Table 40.1 has been computed for selected values of x. We shall assume that the points corresponding to intermediate values of x lie on a smooth curve joining the points given in the table. See Figs. 40-1 and 40-2. (See Problems 40.1–40.3.)

Table 40.1

x	$y = x^{3/2}$	$y = -x^{-3/2}$
9	27	$-\frac{1}{27}$
4	8	$-\frac{1}{8}$
1	1	-1
$\frac{1}{4}$	$\frac{1}{8}$	-8
$\frac{1}{9}$	$\frac{1}{27}$	-27
0	0	—

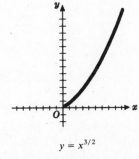

$y = x^{3/2}$

Fig. 40-1

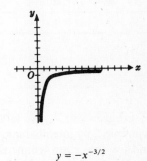

$y = -x^{-3/2}$

Fig. 40-2

EXPONENTIAL FUNCTIONS in x are of the form b^x where b is a constant. The discussion will be limited here to the case $b > 1$.

The curve whose equation is $y = b^x$ is called an *exponential curve*. The general properties of such curves are

(a) The curve passes through the point (0, 1).

(b) The curve lies above the x axis and has that axis as an asymptote.

EXAMPLE 2. Sketch the graphs of (a) $y = 2^x$, (b) $y = 3^x$. (See Problem 12.4.)

Table 40.2

x	$y = 2^x$	$y = 3^x$
3	8	27
2	4	9
1	2	3
0	1	1
−1	$\frac{1}{2}$	$\frac{1}{3}$
−2	$\frac{1}{4}$	$\frac{1}{9}$
−3	$\frac{1}{8}$	$\frac{1}{27}$

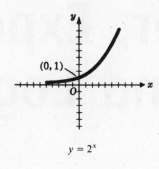

$y = 2^x$

Fig. 40-3

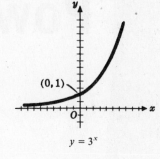

$y = 3^x$

Fig. 40-4

The exponential equation appears frequently in the form $y = c\, e^{kx}$ where c and k are nonzero constants and $e = 2.71828\ldots$ is the natural logarithmic base. See Table 40.2 and Figs. 40-3 and 40-4. (See Problems 40.5–40.6.)

THE CURVE WHOSE EQUATION IS $y = \log_b x, b > 1$, is called a *logarithmic curve*. The general properties are

(a) The curve passes through the point (1, 0).

(b) The curve lies to the right of the y axis and has that axis as an asymptote.

EXAMPLE 3. Sketch the graph of $y = \log_2 x$.

Table 40.3

x	8	4	2	1	$\frac{1}{2}$	$\frac{1}{4}$	$\frac{1}{8}$
y	3	2	1	0	−1	−2	−3

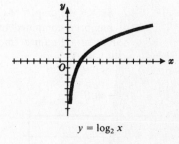

$y = \log_2 x$

Fig. 40-5

Since $x = 2^y$, the table of values in Table 40.3 may be obtained from the table for $y = 2^x$ of Example 2 by interchanging x and y. See Fig. 40-5. (See Problem 40.7.) Note that the graphs of $y = 2^x$ and $y = \log_2 x$ are symmetric about the line $y = x$ since they are graphs of inverse functions. The same will be true for $y = a^x$ and $y = \log_a x$ for all $a > 0, a \neq 1$.

Solved Problems

40.1 Sketch the graph of the *semicubic parabola* $y^2 = x^3$.

Since the given equation is equivalent to $y = \pm x^{3/2}$, the graph consists of the curve of Example 1(a) together with its reflection in the x axis. See Fig. 40-6.

40.2   Sketch the graph of $y^3 = x^2$.

Refer to Fig. 40-7 and Table 40.4.

Table 40.4

x	± 3	± 2	± 1	0
y	2.1	1.6	1	0

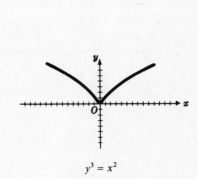

$y^2 = x^3$

Fig. 40-6

$y^3 = x^2$

Fig. 40-7

40.3   Sketch the graph of $y = x^{-2}$.

See Fig 40-8 and Table 40.5.

Table 40.5

x	± 3	± 2	± 1	$\pm \frac{1}{2}$	$\pm \frac{1}{4}$	0
y	$\frac{1}{9}$	$\frac{1}{4}$	1	4	16	—

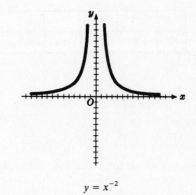

$y = x^{-2}$

Fig. 40-8

40.4   Sketch the graph of $y = 3^{-x}$.

See Fig. 40-9 and Table 40.6.
Note that the graph of $y = b^{-x}$ is a reflection in the y axis of the graph of $y = b^x$.

Table 40.6

x	3	2	1	0	-1	-2	-3
y	$\frac{1}{27}$	$\frac{1}{9}$	$\frac{1}{3}$	1	3	9	27

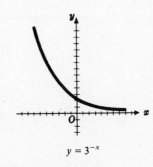

$y = 3^{-x}$

Fig. 40-9

40.5   Sketch the graph of $y = e^{2x}$.

See Fig. 40-10 and Table 40.7.

Table 40.7

x	2	1	$\frac{1}{2}$	0	$-\frac{1}{2}$	-1	-2
$y = e^{2x}$	54.6	7.4	2.7	1	0.4	0.14	0.02

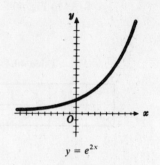

$y = e^{2x}$

Fig. 40-10

40.6   Sketch the graph of $y = e^{-x^2}$.

Refer to Fig. 40-11 and Table 40.8.

Table 40.8

x	± 2	$\pm \frac{3}{2}$	± 1	$\pm \frac{1}{2}$	0
y	0.02	0.1	0.4	0.8	1

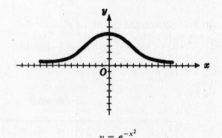

$y = e^{-x^2}$

Fig. 40-11

This is a simple form of the *normal probability curve* used in statistics.

40.7   Sketch the graphs of (a) $y = \log x$, (b) $y = \log x^2 = 2 \log x$.

See Table 40.9 and Figs. 40-12 and 40-13.

Table 40.9

x	10	5	4	3	2	1	0.5	0.25	0.1	0.01
$y = \log x$	1	0.7	0.6	0.5	0.3	0	-0.3	-0.6	-1	-2
$y = \log x^2$	2	1.4	1.2	1	0.6	0	-0.6	-1.2	-2	-4

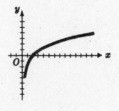

$y = \log x$

Fig. 40-12

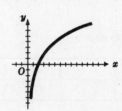

$y = \log x^2$

Fig. 40-13

Supplementary Problems

40.8 Sketch the graphs of (a) $y^2 = x^{-3}$, (b) $y^3 = x^{-2}$, (c) $y^2 = \frac{1}{x}$, (d) the cubical parabola $y = x^3$.

40.9 Sketch the graphs of

(a) $y = (2.5)^x$ (c) $y = 2^{-1/x}$ (e) $y = e^{x/2}$ (g) $y = e^{x+2}$

(b) $y = 2^{x+1}$ (d) $y = \frac{1}{2}e^x$ (f) $y = e^{-x/2}$ (h) $y = xe^{-x}$

40.10 Sketch the graphs of   (a) $y = \frac{1}{2}\log x$,   (b) $y = \log(3x + 2)$,   (c) $y = \log(x^2 + 1)$.

40.11 Show that the curve $y^q = x^p$, where p and q are positive integers, lies entirely in

(a) Quadrants I and III if p and q are both odd

(b) Quadrants I and IV if p is odd and q is even

(c) Quadrants I and II if p is even and q is odd

40.12 Show that the curve $y^q = x^{-p}$, where p and q are positive integers, lies entirely in

(a) Quadrants I and III if p and q are both odd

(b) Quadrants I and II if p is even and q is odd

(c) Quadrants I and IV if p is odd and q is even

Chapter 41

Polynomial Equations, Rational Roots

A POLYNOMIAL EQUATION (or rational integral equation) is obtained when any polynomial in one variable is set equal to zero. We shall work with polynomials having integral coefficients although many of the theorems will be stated for polynomial equations with weaker restrictions on the coefficients.

A polynomial equation is said to be in *standard form* when written as

$$a_0 x^n + a_1 x^{n-1} + a_2 x^{n-2} + \cdots + a_{n-2} x^2 + a_{n-1} x + a_n = 0 \qquad (41.1)$$

where the terms are arranged in descending powers of x, a zero has been inserted as coefficient on each missing term, the coefficients have no common factor except ± 1, and $a_0 \neq 0$. (See Problem 41.1.)

A NUMBER r IS CALLED A ROOT of $f(x) = 0$ if and only if $f(r) = 0$. It follows that the abscissas of the points of intersection of the graph of $y = f(x)$ and the x axis are roots of $f(x) = 0$.

THE FUNDAMENTAL THEOREM OF ALGEBRA. Every polynomial equation $f(x) = 0$ has at least one root, real or complex.

A polynomial equation of degree n has exactly n roots. These n roots may not all be distinct. If r is one of the roots and occurs just once, it is called a *simple root*; if r occurs exactly $m > 1$ times among the roots, it is called a *root of multiplicity m* or an *m-fold root*. If $m = 2, r$ is called a *double root*; if $m = 3$, a *triple root*; and so on. (See Problems 41.2–41.3.)

COMPLEX ROOTS. If the polynomial equation $f(x) = 0$ has real coefficients and if the complex $a + bi$ is a root of $f(x) = 0$, then the *complex conjugate $a - bi$* is also a root. (For a proof, see Problem 41.11.)

IRRATIONAL ROOTS. Given the polynomial equation $f(x) = 0$, if the irrational number $a + \sqrt{b}$, where a and b are rational, is a root of $f(x) = 0$, then the conjugate irrational $a - \sqrt{b}$ is also a root. (See Problem 41.4.)

LIMITS TO THE REAL ROOTS. A real number L is called an *upper limit* of the real roots of $f(x) = 0$ if no (real) root is greater than L; a real number l is called a *lower limit* if no (real) root is smaller than l.

If $L > 0$ and if, when $f(x)$ is divided by $x - L$ by synthetic division, every number in the third line is nonnegative, then L is an upper limit of the real roots of $f(x) = 0$.

If $l < 0$ and if, when $f(x)$ is divided by $x - l$ by synthetic division, the numbers in the third line alternate in sign, then l is a lower limit of the real roots of $f(x) = 0$.

RATIONAL ROOTS.   A polynomial equation has 0 as a root if and only if the constant term of the equation is zero.

EXAMPLE 1.   The roots of $x^5 - 2x^4 + 6x^3 - 5x^2 = x^2(x^3 - 2x^2 + 6x - 5) = 0$ are 0, 0, and the three roots of $x^3 - 2x^2 + 6x - 5 = 0$.

If a rational fraction p/q, expressed in lowest terms, is a root of (41.1) in which $a_n \neq 0$, then p is a divisor of the constant term a_n and q is a divisor of the leading coefficient of a_0 of (41.1). (For a proof, see Problem 41.12.)

EXAMPLE 2.   The theorem permits us to say that $\frac{2}{3}$ is a *possible* root of the equation $9x^4 - 5x^2 + 8x + 4 = 0$ since the numerator 2 divides the constant term 4 and the denominator 3 divides the leading coefficient 9. It does *not* assure that $\frac{2}{3}$ is a root. However, the theorem does assure that neither $\frac{1}{2}$ nor $-\frac{4}{3}$ is a root. In each case the denominator does not divide the leading coefficient.

If p, an integer, is a root of (41.1) then p is a divisor of its constant term.

EXAMPLE 3.   The *possible* rational roots of the equation

$$12x^4 - 40x^3 - 5x^2 + 45x + 18 = 0$$

are all numbers $\pm p/q$ in which the values of p are the positive divisors 1, 2, 3, 6, 9, 18 of the constant term 18 and the values of q are the positive divisors 1, 2, 3, 4, 6, 12 of the leading coefficient 12. Thus the rational roots, if any, of the equation are among the numbers

$$\pm 1, \ \pm 2, \ \pm 3, \ \pm 6, \ \pm 9, \ \pm 18, \ \pm \tfrac{1}{2}, \ \pm \tfrac{3}{2}, \ \pm \tfrac{9}{2}, \ \pm \tfrac{1}{3}, \ \pm \tfrac{2}{3}, \ \pm \tfrac{1}{4}, \ \pm \tfrac{3}{4}, \ \pm \tfrac{1}{6}, \ \pm \tfrac{1}{12}$$

THE PRINCIPAL PROBLEM OF THIS CHAPTER is to find the rational roots of a given polynomial equation. The general procedure is this: Test the possible rational roots by synthetic division, accepting as roots all those for which the last number in the third line *is* zero and rejecting all those for which it is not. Certain refinements, which help to shorten the work, are pointed out in the examples and solved problems below. (See Chapter 43 for a full description of synthetic division.)

EXAMPLE 4.   Find the rational roots of $x^5 + 2x^4 - 18x^3 - 8x^2 + 41x + 30 = 0$.
   Since the leading coefficient is 1, all rational roots p/q are integers. The possible integral roots, the divisors (both positive and negative) of the constant term 30, are

$$\pm 1, \ \pm 2, \ \pm 3, \ \pm 5, \ \pm 6, \ \pm 10, \ \pm 15, \ \pm 30$$

$$
\begin{array}{r}
1 + 2 - 18 - \ 8 + 41 + 30 \ \ \underline{\lfloor 1} \\
\text{Try 1:} \quad \underline{\ \ \ \ 1 + \ 3 - 15 - 23 + 18} \\
1 + 3 - 15 - 23 + 18 + 48
\end{array}
$$

Then 1 is not a root. This number ($+1$) should be removed from the list of possible roots lest we forget and try it again later on.

$$
\begin{array}{r}
1 + 2 - 18 - \ 8 + 41 + 30 \ \ \underline{\lfloor 2} \\
\text{Try 2:} \quad \underline{\ \ \ \ \ 2 + \ 8 - 20 - 56 - 30} \\
1 + 4 - 10 - 28 + 15 + \ \ 0
\end{array}
$$

Then 2 is a root and the remaining rational roots of the given equation are the rational roots of the *depressed equation*

$$x^4 + 4x^3 - 10x^2 - 28x - 15 = 0$$

Now ± 2, ± 6, ± 10, and ± 30 cannot be roots of this equation (they are not divisors of 15) and should be removed from the list of possibilities. We return to the depressed equation.

$$
\begin{array}{r}
1+4-10-28-15 \quad \underline{|3} \\
\text{Try 3:}\quad 3+21+33+15 \\
\hline
1+7+11+\ 5+\ 0
\end{array}
$$

Then 3 is a root and the new depressed equation is

$$x^3 + 7x^2 + 11x + 5 = 0$$

Since the coefficients of this equation are nonnegative, it has no positive roots. We now remove $+3, +5, +15$ from the original list of possible roots and return to the new depressed equation.

$$
\begin{array}{r}
1+7+11+5 \quad \underline{|-1} \\
\text{Try } -1:\quad -1-\ 6-5 \\
\hline
1+6+\ 5+0
\end{array}
$$

Then -1 is a root and the depressed equation

$$x^2 + 6x + 5 = (x+1)(x+5) = 0$$

has -1 and -5 as roots.

The necessary computations may be neatly displayed as follows:

$$
\begin{array}{l}
1+2-18-\ 8+41+30 \quad \underline{|2} \\
\quad\ 2+\ 8-20-56-30 \\
\hline
1+4-10-28-15 \qquad\qquad \underline{|3} \\
\qquad 3+21+33+15 \\
\hline
1+7+11+\ 5 \qquad\qquad\qquad\quad \underline{|-1} \\
\qquad\ -1-\ 6-\ 5 \\
\hline
1+6+\ 5
\end{array}
$$

$$x^2 + 6x + 5 = (x+1)(x+5) = 0 \qquad x = -1, -5$$

The roots are 2, 3, -1, -1, -5.

Note that the roots here are numerically small numbers; that is, 3 is a root but 30 is not, -1 is a root but -15 is not. Hereafter we shall not list integers which are large numerically or fractions with large numerator or denominator among the possible roots. (See Problems 41.5–41.9.)

Solved Problems

41.1 Write each of the following in standard form.

(a) $4x^2 + 2x^3 - 6 + 5x = 0$ *Ans.* $2x^3 + 4x^2 + 5x - 6 = 0$

(b) $-3x^2 + 6x - 4x^2 + 2 = 0$ *Ans.* $3x^3 + 4x^2 - 6x - 2 = 0$

(c) $2x^5 + x^3 + 4 = 0$ *Ans.* $2x^5 + 0 \cdot x^4 + x^3 + 0 \cdot x^2 + 0 \cdot x + 4 = 0$

(d) $x^2 + \frac{1}{2}x^2 - x + 2 = 0$ *Ans.* $2x^3 + x^2 - 2x + 4 = 0$

(e) $4x^4 + 6x^3 - 8x^2 + 12x - 10 = 0$ *Ans.* $2x^4 + 3x^3 - 4x^2 + 6x - 5 = 0$

41.2   (*a*)   Show that -1 and 2 are roots of $x^4 - 9x^2 + 4x + 12 = 0$.

Using synthetic division,

$$
\begin{array}{r}
1+0-9+\ 4+12 \ \underline{|-1} \\
-\ 1+1+\ 8-12 \\
\hline
1-1-8+12+\ 0
\end{array}
\qquad
\begin{array}{r}
1+0-9+\ 4+12 \ \underline{|2} \\
-\ 2+4-10-12 \\
\hline
1+2-5-\ 6+\ 0
\end{array}
$$

Since $f(-1) = 0$ and $f(2) = 0$, both -1 and 2 are roots.

(*b*)   Show that the equation in (*a*) has at least two other roots by finding them.
From the synthetic division in (*a*) and the Factor Theorem (Chapter 43),

$$x^4 - 9x^2 + 4x + 12 = (x+1)(x^3 - x^2 - 8x + 12)$$

Since 2 is also root of the given equation, 2 is a root of $x^3 - x^2 - 8x + 12 = 0$.
Using synthetic division

$$
\begin{array}{r}
1-1-8+12 \ \underline{|2} \\
2+2-12 \\
\hline
1+1-6+\ 0
\end{array}
$$

we obtain $x^3 - x^2 - 8x + 12 = (x-2)(x^2 + x - 6)$. Then

$$x^4 - 9x^2 + 4x + 12 = (x+1)(x-2)(x^2 + x - 6) = (x+1)(x-2)(x+3)(x-2)$$

Thus, the roots of the given equation are -1, 2, -3, 2.

(NOTE: Since $x-2$ appears twice among the factors of $f(x)$, 2 appears twice among the roots of $f(x) = 0$ and is a double root of the equation.)

41.3   (*a*)   Find all of the roots of $(x+1)(x-2)^3(x+4)^2 = 0$.
The roots are -1, 2, 2, 2, -4, -4; thus, -1 is a simplest root, 2 is a root of multiplicity three or a triple root, and -4 is a root of multiplicity two or a double root.

(*b*)   Find all the roots of $x^2(x-2)(x-5) = 0$.
The roots are 0, 0, 2, 5; 2 and 5 are simple roots and 0 is a double root.

41.4   Form the equation of lowest degree with integral coefficients having the roots $\sqrt{5}$ and $2 - 3i$.

To ensure integral coefficients, the conjugates $-\sqrt{5}$ and $2 + 3i$ must also be roots. Thus the required equation is

$$(x - \sqrt{5})(x + \sqrt{5})[x - (2-3i)][x - (2+3i)] = x^4 - 4x^3 + 8x^2 + 20x - 65 = 0$$

In Problems 41.5–41.9 find the rational roots; when possible, find all the roots.

41.5 $2x^4 - x^3 - 11x^2 + 4x + 12 = 0$

The possible rational roots are ± 1, ± 2, ± 3, ± 4, $\pm \frac{1}{2}$, $\pm \frac{3}{2}$,.... Discarding all false trials, we find

$$
\begin{array}{r}
2-1-11+\ 4+12 \ \underline{|2} \\
4+\ 6-10-12 \\
\hline
2+3-\ 5-\ 6 \qquad \underline{|-1} \\
-2-\ 1+6 \\
\hline
2+1-\ 6
\end{array}
\qquad 2x^2 + x - 6 = (2x-3)(x+2) = 0; \qquad x = \tfrac{3}{2}, -2
$$

The roots are $2, -1, \frac{3}{2}, -2$.

41.6 $4x^4 - 3x^3 - 4x + 3 = 0$

The possible rational roots are: ± 1, ± 3, $\pm \frac{1}{2}$, $\pm \frac{3}{2}$, $\pm \frac{1}{4}$, $\pm \frac{3}{4}$. By inspection the sum of the coefficients is 0; then $+1$ is a root. Discarding all false trials, we find

$$
\begin{array}{l}
4 - 3 + 0 - 4 + 3 \quad \underline{|1} \qquad x^2 + x + 1 = 0; \qquad x = \dfrac{-1 \pm i\sqrt{3}}{2} \\[2pt]
\; 4 + 1 + 1 - 3 \\[2pt]
\hline
4 + 1 + 1 - 3 \qquad \underline{|\tfrac{3}{4}} \\[2pt]
\; 3 + 3 + 3 \\[2pt]
\hline
4 + 4 + 4 \qquad\qquad \text{(Factor out 4.)} \\[2pt]
1 + 1 + 1
\end{array}
$$

The roots are 1, $\dfrac{3}{4}$, $-\dfrac{1}{2} \pm \dfrac{i\sqrt{3}}{2}$.

41.7 $24x^6 - 20x^5 - 6x^4 + 9x^3 - 2x^2 = 0$

Since $24x^6 - 20x^5 - 6x^4 + 9x^3 - 2x^2 = x^2(24x^4 - 20x^3 - 6x^2 + 9x - 2)$, the roots of the given equation are 0, 0 and the roots of $24x^4 - 20x^3 - 6x^2 + 9x - 2 = 0$. Possible rational roots are: ± 1, ± 2, $\pm \frac{1}{2}$, $\pm \frac{1}{3}$, $\pm \frac{2}{3}$, $\pm \frac{1}{4}$, ... Discarding all false trials, we find

$$
\begin{array}{l}
24 - 20 - 6 + 9 - 2 \quad \underline{|\tfrac{1}{2}} \qquad 6x^2 + x - 2 = (2x-1)(3x+2) = 0; \quad x = \tfrac{1}{2}, -\tfrac{2}{3} \\[2pt]
\; 12 - 4 - 5 + 2 \\[2pt]
\hline
24 - 8 - 10 + 4 \qquad \text{(Factor out 2.)} \\[2pt]
12 - 4 - 5 + 2 \qquad \underline{|\tfrac{1}{2}} \\[2pt]
\; 6 + 1 - 2 \\[2pt]
\hline
12 + 2 - 4 \qquad\qquad \text{(Factor out 2.)} \\[2pt]
6 + 1 - 2
\end{array}
$$

The roots are 0, 0, $\frac{1}{2}$, $\frac{1}{2}$, $\frac{1}{2}$, $-\frac{2}{3}$.

41.8 $4x^5 - 32x^4 + 93x^3 - 119x^2 + 70x - 25 = 0$

Since the signs of the coefficients alternate, the rational roots (if any) are positive. Possible rational roots are $1, 5, \frac{1}{2}, \frac{5}{2}, \frac{1}{4}, \ldots$. Discarding all false trials, we find

$$
\begin{array}{l}
4 - 32 + 93 - 119 + 70 - 25 \quad \underline{|\tfrac{5}{2}} \\[2pt]
\; 10 - 55 + 95 - 60 + 25 \\[2pt]
\hline
4 - 22 + 38 - 24 + 10 \qquad \text{(Factor out 2.)} \\[2pt]
2 - 11 + 19 - 12 + 5 \qquad \underline{|\tfrac{5}{2}} \\[2pt]
\; 5 - 15 + 10 - 5 \\[2pt]
\hline
2 - 6 + 4 - 2 \qquad \text{(Factor out 2.)} \\[2pt]
1 - 3 + 2 - 1
\end{array}
$$

The rational roots are $\frac{5}{2}, \frac{5}{2}$.

The equation $f(x) = x^3 - 3x^2 + 2x - 1 = 0$ has at least one real (irrational) root since $f(0) < 0$ while for sufficiently large $x(x > 3)$, $f(x) > 0$.

The only possible rational root of $x^3 - 3x^2 + 2x - 1 = 0$ is 1; it is not a root.

41.9 $6x^4 + 13x^3 - 11x^2 + 5x + 1 = 0$

The possible rational roots are ± 1, $\pm \frac{1}{2}$, $\pm \frac{1}{3}$, $\pm \frac{1}{6}$. After testing each possibility, we conclude that the equation has no rational roots.

41.10 If $ax^3 + bx^2 + cx + d = 0$, with integral coefficients, has a rational root r and if $c + ar^2 = 0$, then all of the roots are rational.

Using synthetic division to remove the known root, the depressed equation is

$$ax^2 + (b + ar)x + c + br + ar^2 = 0$$

which reduces to $ax^2 + (b + ar)x + br = 0$ when $c + ar^2 = 0$. Since its discriminant is $(b - ar)^2$, a perfect square, its roots are rational. Thus, all the roots of the given equation are rational.

41.11 Prove: If the polynomial $f(x)$ has real coefficients and if the imaginary $a + bi, b \neq 0$, is a root of $f(x) = 0$, then the conjugate imaginary $a - bi$ is also a root.

Since $a + bi$ is a root of $f(x) = 0$, $x - (a + bi)$ is a factor of $f(x)$. Similarly, if $a - bi$ is to be a root of $f(x) = 0$, $x - (a - bi)$ must be a factor of $f(x)$. We need to show then that when $a + bi$ is a root of $f(x) = 0$, it follows that

$$[x - (a + bi)][x - (a - bi)] = x^2 - 2ax + a^2 + b^2$$

is a factor of $f(x)$. By division we find

$$\begin{aligned} f(x) &= [x^2 - 2ax + a^2 + b^2] \cdot Q(x) + Mx + N \\ &= [x - (a + bi)][x - (a - bi)] \cdot Q(x) + Mx + N \end{aligned} \tag{1}$$

where $Q(x)$ is a polynomial of degree 2 less than that of $f(x)$ and the remainder $Mx + N$ is of degree at most 1 in x; that is, M and N are constants.

Since $a + bi$ is a root of $f(x) = 0$, we have from (1),

$$f(a + bi) = 0 \cdot Q(a + bi) + M(a + bi) + N = (aM + N) + bMi = 0$$

Then $aM + N = 0$ and $bM = 0$. Now $b \neq 0$; hence, $bM = 0$ requires $M = 0$ and then $aM + N = 0$ requires $N = 0$. Since $M = N = 0$, (1) becomes

$$f(x) = (x^2 - 2ax + a^2 + b^2) \cdot Q(x)$$

Then $x^2 - 2ax + a^2 + b^2$ is a factor of $f(x)$ as was to be proved.

41.12 Prove: If a rational fraction p/q, expressed in lowest terms, is a root of the polynomial equation (41.1) whose constant terms $a_n \neq 0$, then p is a divisor of the constant term a_n, and q is a divisor of the leading coefficient a_0.

Let the given equation be

$$a_0 x^n + a_1 x^{n-1} + a_2 x^{n-2} + \cdots + a_{n-2} x^2 + a_{n-1} x + a_n = 0, \quad a_0 a_n \neq 0$$

If p/q is a root, then

$$a_0 (p/q)^n + a_1 (p/q)^{n-1} + a_2 (p/q)^{n-2} + \cdots + a_{n-2}(p/q)^2 + a_{n-2}(p/q) + a_n = 0$$

Multiplying both members by q^n, this becomes

$$a_0 p^n + a_1 p^{n-1} q + a_2 p^{n-2} q^2 + \cdots + a_{n-2} p^2 q^{n-2} + a_{n-1} pq^{n-1} + a_n q^n = 0 \tag{2}$$

When (2) is written as

$$a_0 p^n + a_1 p^{n-1} q + a_2 p^{n-2} q^2 + \cdots + a_{n-2} p^2 q^{n-2} + a_{n-1} pq^{n-1} = -a_n q^n$$

it is clear that p, being a factor of every term of the left member of the equality, must divide $a_n q^n$. Since p/q is expressed in lowest terms, no factor of p will divide q. Hence, p must divide a_n as was to be shown.

Similarly, when (2) is written as

$$a_1 p^{n-1} q + a_2 p^{n-2} q^2 + \cdots + a_{n-2} p^2 q^{n-2} + a_{n-1} pq^{n-1} + a_n q^n = -a_0 p^n$$

it follows that q must divide a_0.

Supplementary Problems

41.13 Find all the roots.

     (a) $x^5 - 2x^4 - 9x^3 + 22x^2 + 4x - 24 = 0$ (d) $6x^4 + 5x^3 - 16x^2 - 9x - 10 = 0$

     (b) $18x^4 - 27x^3 + x^2 + 12x - 4 = 0$ (e) $9x^4 - 19x^2 - 6x + 4 = 0$

     (c) $12x^4 - 40x^3 - 5x^2 + 45x + 18 = 0$ (f) $2x^5 + 3x^4 - 3x^3 - 2x^2 = 0$

        Ans.   (a) $2, 2, 2 - 1, -3$ (c) $-\frac{2}{3}, -\frac{1}{2}, \frac{3}{2}, 2$ (e) $-1, \frac{1}{3}, (1 \pm \sqrt{13})/3$

                  (b) $1, \frac{1}{2}, \frac{2}{3}, -\frac{2}{3}$ (d) $-2, \frac{5}{3}, (-1 \pm l\sqrt{7})/4$ (f) $0, 0, 1, -2, -\frac{1}{2}$

4.14 Solve the inequalities.

     (a) $x^3 - 5x^2 + 2x + 8 > 0$ (c) $x^4 - 3x^3 + x^2 + 4 > 0$

     (b) $6x^2 - 17x^2 - 5x + 6 < 0$ (d) $x^5 - x^4 - 2x^3 + 2x^2 + x - 1 > 0$

        Ans.   (a) $x > 4, -1 < x < 2$ (b) $x < -\frac{2}{3}, \frac{1}{2} < x < 3$ (c)   all $x \neq 2$ (d) $x > 1$

41.15 Prove: if $r \neq 1$ is a root of $f(x) = 0$, then $r - 1$ divides $f(1)$.

Chapter 42

Irrational Roots of Polynomial Equations

IF $f(x) = 0$ IS A POLYNOMIAL EQUATION, the equation $f(-x) = 0$ has as roots the negatives of the roots of $f(x) = 0$. When $f(x) = 0$ is written in standard form, the equation whose roots are the negatives of the roots of $f(x) = 0$ may be obtained by changing the signs of alternate terms, beginning with the second.

EXAMPLE 1

(a) The roots $x^3 + 3x^2 - 4x - 12 = 0$ are $2, -2, -3$; the roots of $x^3 - 3x^2 - 4x + 12 = 0$ are $-2, 2, 3$.

(b) The equation $6x^4 + 13x^3 - 13x - 6 = 0$ has roots $1, -1, -\frac{2}{3}, -\frac{3}{2}$; the equation $6x^4 - 13x^3 + 13x - 6 = 0$ has roots $-1, 1, \frac{2}{3}, \frac{3}{2}$.

(See Problem 42.1.)

VARIATION OF SIGN.   If, when a polynomial is arranged in descending powers of the variable, two successive terms differ in sign, the polynomial is said to have a *variation of sign*.

EXAMPLE 2

(a) The polynomial $x^3 - 3x^2 - 4x + 12$ has two variations of sign, one from $+x^3$ to $-3x^2$ and one from $-4x$ to $+12$; the polynomial $x^3 + 3x^2 - 4x - 12$ has one variation of sign.

(b) The polynomial $6x^4 + 13x^3 - 13x - 6$ has one variation of sign; the polynomial $6x^4 - 13x^3 + 13x - 6$ has three. Note that here the term with zero coefficient has not been considered.

DESCARTES' RULE OF SIGNS.   The number of positive roots of a polynomial equation $f(x) = 0$, with real coefficients, is equal either to the number of variations of sign in $f(x)$ or to the number diminished by an even number.

The number of negative roots of $f(x) = 0$ is equal to the number of positive roots of $f(-x) = 0$.

EXAMPLE 3.   Since $f(x) = x^3 - 3x^2 - 4x + 12$ of Example 2(a) has two variations of sign, $f(x) = 0$ has either two or no positive roots.

Since $x^3 + 3x^2 - 4x - 12$ has one variation of sign, $f(-x) = 0$ has one positive root and $f(x) = 0$ has one negative root. (See Problem 42.2.)

DIMINISHING THE ROOTS OF AN EQUATION.   Let

$$f(x) = a_0 x^n + a_1 x^{n-1} + \cdots + a_{n-1} x + a_n = 0 \qquad (42.1)$$

be a polynomial equation of degree n. Let

$$f(x) = (x - h) \cdot q_1(x) + R_1,$$

$$q_1(x) = (x - h) \cdot q_2(x) + R_2,$$

$$\vdots \qquad\qquad (42.2)$$

$$q_{n-2}(x) = (x - h) \cdot q_{n-1}(x) + R_{n-1},$$

$$q_{n-1}(x) = (x - h) \cdot q_n(x) + R_n,$$

where each R is a constant and $q_n(x) = a_0$. Then the roots of

$$g(y) = a_0 y^n + R_n y^{n-1} + R_{n-1} y^{n-2} + \cdots + R_{2y} + R_1 = 0 \qquad (42.3)$$

are the roots of $f(x) = 0$ diminished by h.

We shall show that if r is any root of $f(x) = 0$ then $r - h$ is a root of $g(y) = 0$. Since $f(r) = 0$,

$$R_1 = -(r - h) \cdot q_1(r),$$

$$R_2 = q_1(r) - (r - h) \cdot q_2(r),$$

$$\vdots$$

$$R_{n-1} = q_{n-2}(r) - (r - h) \cdot q_{n-1}(r),$$

$$R_n = q_{n-1}(r) - (r - h)a_0.$$

When these replacements are made in (42.3), we have

$$a_0 y^n + [q_{n-1}(r) - (r - h)a_0]y^{n-1} + [q_{n-2}(r) - (r - h) \cdot q_{n-1}(r)]y^{n-2} + \cdots$$

$$+ [q_1(r) - (r - h) \cdot q_2(r)]y - (r - h) \cdot q_1(r)$$

$$= a_0[y - (r - h)]y^{n-1} + q_{n-1}(r)[y - (r - h)]y^{n-2} + \cdots + q_1(r)[y - (r - h)] = 0.$$

It is clear that $r - h$ is a root of (42.3) as was to be proved.

EXAMPLE 4.   Find the equation each of whose roots is 4 less than the roots of $x^3 + 3x^2 - 4x - 12 = 0$.

$$
\begin{array}{ll}
\begin{array}{r}
1 + 3 - 4 - 12 \quad \underline{|4} \\
\underline{4 + 28 + 96} \\
1 + 7 + 24 + \textcircled{84} \\
\underline{4 + 44} \\
1 + 11 + \textcircled{68} \\
\underline{4} \\
1 + \textcircled{15}
\end{array}
&
\begin{array}{r}
1 + 15 + 68 + 84 \quad \underline{|-2} \\
\underline{-2 - 26 - 84} \\
1 + 13 + 42 \qquad\quad \underline{|-6} \\
\underline{-6 - 42} \\
1 + 7 \qquad\qquad\quad \underline{|-7} \\
\underline{-7} \\
1
\end{array}
\end{array}
$$

On the left the successive remainders have been found and circled. The resulting equation is $y^3 + 15y^2 + 68y + 84 = 0$. The given equation has roots $x = 2, -2 - 3$; on the right, it is shown that $2 - 4 = -2, -2 - 4 = -6, -3 - 4 = -7$ are roots of the newly formed equation. (See Problem 42.3.)

APPROXIMATION OF IRRATIONAL ROOTS

LET $f(x) = 0$ BE A POLYNOMIAL EQUATION having no rational roots. If the given equation had rational roots, we suppose that they have been found using synthetic division and that $f(x)$ is then the last third line in process. (See Chapter 41.)

THE METHOD OF SUCCESSIVE LINEAR APPROXIMATIONS will be explained by means of examples.

EXAMPLE 5. The equation $f(x) = x^3 + x - 4 = 0$ has no rational roots. By Descartes' rule of signs, it has one positive (real) root and two imaginary roots. To approximate the real root, we shall first isolate it as lying between two consecutive intgers. Since $f(1) = -2$ and $f(2) = 6$, the root lies between $x = 1$ and $x = 2$. Figure 42-1 exhibits the portion of the graph of $f(x)$ between $(1, -2)$ and $(2, 6)$.

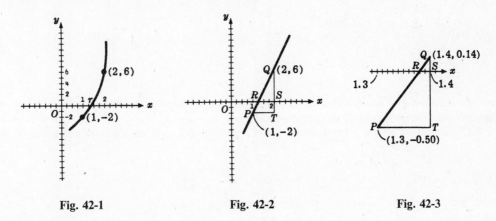

Fig. 42-1   Fig. 42-2   Fig. 42-3

In Fig. 42-2, the curve joining the two points has been replaced by a straight line which meets the x axis at R. We shall take OR, measured to the nearest tenth of a unit, as the first approximation of required root and use it to isolate the root between successive tenths. From the similar triangles RSQ and PTQ,

$$\frac{RS}{PT} = \frac{SQ}{TQ} \qquad \text{or} \qquad RS = \frac{SQ}{TQ}(PT) = \frac{6}{8}(1) = \frac{3}{4} = 0.7$$

The first approximation of the root is given by $OR = OS - RS = 2 - 0.7 = 1.3$. Since $f(1.3) = -0.50$ and $f(1.4) = 0.14$, the required root lies between $x = 1.3$ and $x = 1.4$.

We now repeat the above process using the points $(1.3, -0.50)$ and $(1.4, 0.14)$ and isolate the root between successive hundredths. From Fig. 42-3,

$$RS = \frac{SQ}{TQ}(PT) = \frac{0.14}{0.64}(0.1) = 0.02 \qquad \text{and} \qquad OR = OS - RS = 1.4 - 0.02 = 1.38$$

is the next approximation. Since $f(1.38) = 0.008$ (hence, too large) and $f(1.37) = -0.059$, the root lies between $x = 1.37$ and $x = 1.38$.

Using the points $(1.37, -0.059)$ and $(1.38, 0.008)$, we isolate the root between successive thousandths. We find (no diagram needed)

$$RS = \frac{0.008}{0.067}(0.01) = 0.001 \qquad \text{and} \qquad OR = 1.38 - 0.001 = 1.379$$

Since $f(1.379) = 0.0012$ and $f(1.378) = -0.0054$, the root lies between $x = 1.378$ and $x = 1.379$.

For the next approximation

$$RS = \frac{0.0012}{0.0066}(0.001) = 0.0001 \quad \text{and} \quad OR = 1.379 - 0.0001 = 1.3789$$

The root correct to three decimal places is 1.379. (See Problem 42.4.)

HORNER'S METHOD OF APPROXIMATION. This method will be explained by means of examples.

EXAMPLE 6. The equation $x^3 + x^2 + x - 4 = 0$ has no rational roots. By Descartes' rule of signs, it has one positive root. Since $f(1) = -1$ and $f(2) = 10$, this root is between $x = 1$ and $x = 2$. We first diminish the roots of the given equation by 1.

$$
\begin{array}{rrrr|l}
1 & 1 & 1 & -4 & \underline{1} \\
  & 1 & 2 & 3 & \\
\hline
1 & 2 & 3 & -1 & \\
  & 1 & 3 & & \\
\hline
1 & 3 & 6 & & \\
  & 1 & & & \\
\hline
1 & 4 & & &
\end{array}
$$

and obtain the equation $g(y) = y^3 + 4y^2 + 6y - 1 = 0$ having a root between $y = 0$ and $y = 1$. To approximate it, we disregard the first two terms of the equation and solve $6y - 1 = 0$ for $y = 0.1$. Since $g(0.1) = -0.359$ and $g(0.2) = 0.368$, the root of $g(y) = 0$ lies between $y = 0.1$ and $y = 0.2$, and we diminish the roots of $g(y) = 0$ by 0.1.

$$
\begin{array}{rrrr|l}
1 & 4 & 6 & -1 & \underline{0.1} \\
  & 0.1 & 0.41 & 0.641 & \\
\hline
1 & 4.1 & 6.41 & -0.359 & \\
  & 0.1 & 0.42 & & \\
\hline
1 & 4.2 & 6.83 & & \\
  & 0.1 & & & \\
\hline
1 & 4.3 & & &
\end{array}
$$

We obtain the equation $h(z) = z^3 + 4.3z^2 + 6.83z - 0.359 = 0$ having a root between 0 and 0.01. Disregarding the first two terms of this equation and solving $6.83z - 0.359 = 0$, we obtain $z = 0.05$ as an approximation of the root. Since $h(0.05) = -0.007$ and $h(0.06) = 0.07$, the root of $h(z) = 0$ lies between $z = 0.05$ and $z = 0.06$ and we diminish the roots by 0.05.

$$
\begin{array}{rrrr|l}
1 & 4.3 & 6.83 & -0.359 & \underline{0.05} \\
  & 0.05 & 0.2175 & 0.352375 & \\
\hline
1 & 4.35 & 7.0475 & -0.006625 & \\
  & 0.05 & 0.2200 & & \\
\hline
1 & 4.40 & 7.2675 & & \\
  & 0.05 & & & \\
\hline
1 & 4.45 & & &
\end{array}
$$

and obtain the equation $k(w) = w^3 + 4.45w^2 + 7.2675w - 0.006625 = 0$ having a root between $w = 0$ and $w = 0.001$. An approximation of this root, obtained by solving $7.2675w - 0.006625 = 0$ is $w = 0.0009$.

Without further computation, we are safe in stating the root of the given equation to be

$$x = 1 + 0.1 + 0.05 + 0.0009 = 1.1509$$

The complete solution may be exhibited more compactly, as follows:

1	1	1	−4	|1
	1	.2	3	
1	2	3	−1	
	1	3		
1	3	6		
	1			
1	4	6	−1	|0.1
	0.1	0.41	0.641	
1	4.1	6.41	−0.359	
	0.1	0.42		
1	4.2	6.83		
	0.1			
1	4.3	6.83	−0.359	|0.05
	0.05	0.2175	0.352375	
1	4.35	7.0475	−0.006625	
	0.05	0.2200		
1	4.40	7.2675		
	0.05			
1	4.45	7.2675	−0.006625	

$y = \frac{1}{6} = 0.1$

$z = \dfrac{0.359}{6.83} = 0.05$

$w = \dfrac{0.006625}{7.2675} = 0.0009$

(See Problems 42.5–42.6.)

Solved Problems

42.1   For each of the equations $f(x) = 0$, write the equation where roots are the negatives of those of $f(x) = 0$.

    (a) $x^3 - 8x^2 + x - 1 = 0$ Ans. $x^3 + 8x^2 + x + 1 = 0$

    (b) $x^4 + 3x^2 + 2x + 1 = 0$ Ans. $x^4 + 3x^2 - 2x + 1 = 0$

    (c) $2x^4 - 5x^2 + 8x - 3 = 0$ Ans. $2x^4 - 5x^2 - 8x - 3 = 0$

    (d) $x^5 + x + 2 = 0$ Ans. $x^5 + x - 2 = 0$

42.2   Give all the information obtainable from Descartes' rule of signs about the roots of the following equations:

    (a) $f(x) = x^3 - 8x^2 + x - 1 = 0$ [Problem 42.1(a)].

    Since there are three variations of sign in $f(x) = 0$ and no variation of sign in $f(-x) = 0$, the given equation has either three positive roots or one positive root or one positive root and two imaginary roots.

    (b) $f(x) = 2x^4 - 5x^2 + 8x - 3 = 0$ [Problem 42.1(c)].

    Since there are three variations of sign in $f(x) = 0$ and one variation of sign in $f(-x) = 0$, the given equation has either three positive and one negative root or one positive, one negative, and two imaginary roots.

    (c) $f(x) = x^5 + x + 2 = 0$ [Problem 42.1(d)].

    Since there is no variation of sign in $f(x) = 0$ and one in $f(-x) = 0$, the given equation has one negative and four imaginary roots.

42.3   Form the equation whose roots are equal to the roots of the given equation diminished by the indicated number.

(a) $x^3 - 4x^2 + 8x - 5 = 0; 2.$ (b) $2x^3 + 9x^2 - 5x - 8 = 0; -3.$ (c) $x^4 - 8x^3 + 5x^2 + x + 8 = 0; 2.$

```
 1 - 4 + 8 - 5  |2            2 + 9 -  5 -  8   |-3            1 - 8 +  5 +  1 +  8   |2
     2 - 4 + 8                   - 6 -  9 + 42                    2 - 12 - 14 - 26
 1 - 2 + 4 + 3                2 + 3 - 14 + 34                 1 - 6 -  7 - 13 - 18
     2 + 0                       - 6 +  9                         2 -  8 - 30
 1 + 0 + 4                    2 - 3 -  5                      1 - 4 - 15 - 43
     2                           - 6                             2 -  4
 1 + 2                        2 - 9                          1 - 2 - 19
                                                                 2
                                                            1 + 0
```

The required equation is   The required equation is
$y^3 + 2y^2 + 4y + 3 = 0.$ $2y^3 - 9y^2 - 5y + 34 = 0.$

The required equation is
$y^4 - 19y^2 - 43y - 18 = 0.$

42.4   Use the method of successive linear approximation to approximate the irrational roots of

$$f(x) = x^3 + 3x^2 - 2x - 5 = 0$$

By Descartes' rule of signs the equation has either one positive and two negative roots or one positive and two imaginary roots. By the location principle (see Table 42.1), there are roots between $x = 1$ and $x = 2$, $x = -1$ and $x = -2$, and $x = -3$ and $x = -4$.

Table 42.1

x	2	1	0	−1	−2	−3	−4
$f(x)$	11	−3	−5	−1	3	1	−13

(a)   To approximate the positive root, use Fig. 42-4. Then

$$RS = \frac{SQ}{TQ}(PT) = \frac{11}{14}(1) = 0.7 \quad \text{and} \quad OR = 2 - 0.7 = 1.3$$

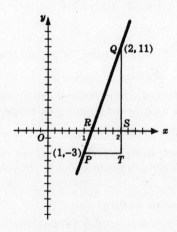

Fig. 42-4

Since $f(1.3) = -0.33$ and $f(1.4) = 0.82$, the root lies between $x = 1.3$ and $x = 1.4$.
For the next approximation,

$$RS = \frac{0.82}{1.15}(0.1) = 0.07 \qquad \text{and} \qquad OR = 1.4 - 0.07 = 1.33$$

Since $f(1.33) = -0.0006$ and $f(1.34) = 0.113$, the root lies between $x = 1.33$ and $x = 1.34$.
For the next approximation,

$$RS = \frac{0.113}{0.1136}(0.01) = 0.009 \qquad \text{and} \qquad OR = 1.34 - 0.009 = 1.331$$

Now $f(1.331) > 0$ so that this approximation is too large; in fact, $f(1.3301) > 0$. Thus, the root to three decimal places is $x = 1.330$.

In approximating a negative root of $f(x) = 0$, it is more convenient to approximate the equally positive root of $f(-x) = 0$.

(b) To approximate the root of $f(x) = 0$ between $x = -1$ and $x = -2$, we shall approximate the positive root between $x = 1$ and $x = 2$ of $g(x) = x^3 - 3x^2 - 2x + 5 = 0$. Since $g(1) = 1$ and $g(2) = -3$, we obtain from Fig. 42-5

$$SR = \frac{SP}{TP}(TQ) = \frac{1}{4}(1) = 0.2 \qquad \text{and} \qquad OR = OS + SR = 1.2$$

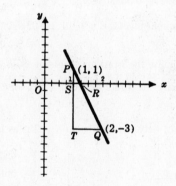

Fig. 42-5

Since $g(1.2) = 0.01$ and $g(1.3) = -0.47$, the root is between $x = 1.2$ and $x = 1.3$.
For the next approximation,

$$SR = \frac{0.01}{0.48}(0.1) = 0.002 \qquad \text{and} \qquad OR = 1.2 + 0.002 = 1.202$$

Since $g(1.202) = -0.0018$ (hence, too large) and $g(1.201) = 0.0031$, the root is between $x = 1.201$ and $x = 1.202$.
For the next approximation,

$$SR = \frac{0.0031}{0.0049}(0.001) = 0.0006 \qquad \text{and} \qquad OR = 1.201 + 0.0006 = 1.2016$$

Since $g(1.2016) = -0.00281$ and $g(1.2015) = 0.00007$, the root of $g(x) = 0$ to three decimal places is $x = 1.202$. The corresponding root of the given equation is $x = -1.202$.

(c) The approximation of the root -3.128 between $x = -3$ and $x = -4$ is left as an exercise.

42.5   Use Horner's method to approximate the irrational roots of $x^3 + 2x^2 - 4 = 0$.

By Descartes' rule of signs the equation has either one positive and two negative roots or one positive and two imaginary roots. By the location principle there is one root, between $x = 1$ and $x = 2$.

Arranged in compact form, the computation is as follows:

```
1 + 2     + 0       - 4           |1
    1     + 3       + 3
1 + 3     + 3       - 1
    1     + 4
1 + 4     + 7                            y = 1/7 = 0.1
    1
1 + 5     + 7       - 1           |0.1
    0.1   + 0.51    + 0.751
1 + 5.1   + 7.51    - 0.249
    0.1   + 0.52                          z = 0.249/8.03 = 0.03
1 + 5.2   + 8.03
    0.1
1 + 5.3   + 8.03    - 0.249       |0.03
    0.03  + 0.1599  + 0.245697
1 + 5.33  + 8.1899  - 0.003303           w = 0.003303/8.3507 = 0.00039
    0.03  + 0.1608
1 + 5.36  + 8.3507
    0.03
1 + 5.39  + 8.3507  - 0.003303
```

The root, correct to four decimal places, is $1 + 0.1 + 0.03 + 0.00039 = 1.1304$.

42.6   Use Horner's method to approximate the irrational roots of $x^3 + 3x^2 - 2x - 5 = 0$.

By the location principle there are roots between $x = 1$ and $x = 2$, $x = -1$ and $x = -2$, $x = -3$ and $x = -4$.

(*a*)   The computation for the root between $x = 1$ and $x = 2$ is as follows:

```
1 + 3     - 2       - 5           |1      y = 3/7 = 0.4
    1     + 4       + 2                   but is too large since, when used,
1 + 4     + 2       - 3                   the last number in the third line
    1     + 5                             is positive.
1 + 5     + 7
    1
1 + 6     + 7       - 3           |0.3
    0.3   + 1.89    + 2.667
1 + 6.3   + 8.89    - 0.333
    0.3   + 1.98                          z = 0.333/10.87 = 0.03
1 + 6.6   + 10.87
    0.3
1 + 6.9   + 10.87   - 0.333       |0.03
    0.03  + 0.2079  + 0.332337
1 + 6.93  + 11.0779 - 0.000663
    0.03  + 0.2088                        w = 0.000663/11.2867 = 0.000058
1 + 6.96  + 11.2867
    0.03
1 + 6.99  + 11.2867 - 0.000663
```

The root is 1.330.

When approximating a negative root of $f(x) = 0$ using Horner's method, it is more convenient to approximate the equally positive root of $f(-x) = 0$.

(b) To approximate the root between $x = -1$ and $x = -2$ of the given equation, we approximate the root between $x = 1$ and $x = 2$ of the equation $x^3 - 3x^2 - 2x + 5 = 0$. The computation is as follows:

$$
\begin{array}{rrrr}
1 - 3 & -2 & +5 & \underline{|1} \\
\underline{1} & \underline{-2} & \underline{-4} & \\
1 - 2 & -4 & +1 & \\
\underline{1} & \underline{-1} & & \\
1 - 1 & -5 & & \\
\underline{1} & & & \\
\end{array}
$$

$y = \frac{1}{5} = 0.2$

$$
\begin{array}{rrrr}
1 + 0 & -5 & +1 & \underline{|0.2} \\
\underline{0.2} & \underline{+0.04} & \underline{-0.992} & \\
1 + 0.2 & -4.96 & +0.008 & \\
\underline{0.2} & \underline{+0.08} & & \\
1 + 0.4 & -4.88 & & \\
\underline{0.2} & & & \\
1 + 0.6 & -4.88 & +0.008 & \underline{|0.001} \\
\end{array}
$$

$z = \dfrac{0.008}{4.88} = 0.001$

$$
\begin{array}{rrr}
\underline{0.001 + 0.000601} & \underline{-0.004879399} & \\
1 + 0.601 - 4.879399 & +0.003120601 & \\
\underline{0.001 + 0.000602} & & \\
1 + 0.602 - 4.878797 & & \\
\underline{0.001} & & \\
1 + 0.603 - 4.878797 & +0.003120601 & \\
\end{array}
$$

$w = \dfrac{0.003120601}{4.878797} = 0.00063$

To four decimal places the root is $x = 1.2016$; thus, the root of the given equation is $x = -1.2016$

(c) The approximation of the root $x = -3.1284$ between $x = -3$ and $x = -4$ is left as an exercise.

Supplementary Problems

42.7   Use Descartes' rule of signs to show

(a) $x^4 + 5x^2 + 24 = 0$ has only complex roots.

(b) $x^n - 1 = 0$ has exactly two real roots if n is even and only one real root if n is odd.

(c) $x^3 + 3x + 2 = 0$ has exactly one real root.

(d) $x^7 - x^5 + 2x^4 + 3x^2 + 5 = 0$ has at least four complex roots.

(e) $x^7 - 2x^4 + 3x^3 - 5 = 0$ has at most three real roots.

42.8   Find all the irrational roots of the following equations:

(a) $x^3 + x - 3 = 0$ (c) $x^2 - 9x + 3 = 0$ (e) $x^4 + 4x^3 + 6x^2 - 15x - 40 = 0$

(b) $x^3 - 3x + 1 = 0$ (d) $x^3 + 6x^2 + 7x - 3 = 0$

   Ans. (a) 1.2134   (c) $0.3376, 2.8169, -3.1546$ (e) $2.7325, -0.7325$

          (b) $0.3473, 1.5321, -1.8794$ (d) $0.3301, -2.2016, -4.1284$

42.9   Show that

(a)   The equation $f(x) = x^n p_1 x^{n-1} + \cdots + p_{n-1}x + p_n = 0$, with integral coefficients, has no rational root if $f(0)$ and $f(1)$ are odd integers.

Hint: Suppose r is an integral root; then r is odd and $r - 1$ does not divide $f(1)$.

(b)   The equation $x^4 - 301x - 1275 = 0$ has no rational roots.

42.10   In a polynomial equation of the form

$$x^n + p_1 x^{n-1} + \cdots + p_{n-1}x + p_n = 0$$

the following relations exist between the coefficients and roots:

(1)   The sum of the roots is $-p_1$.
(2)   The sum of the products of the roots taken two at a time is p_2.
(3)   The sum of the products of the roots taken three at a time is $-p_3$.
$\vdots$
(n)   The product of the roots is $(-1)^n p_n$.

If a, b, c are the roots of $x^3 - 3x^2 + 4x + 2 = 0$, find

(a) $a + b + c$

(b) $ab + bc + ca$

(c) abc

(d) $a^2 + b^2 + c^2 = (a + b + c)^2 - 2(ab + bc + ca)$

(e) $a^3 + b^3 + c^3$

(f) $\dfrac{1}{a} + \dfrac{1}{b} + \dfrac{1}{c} = \dfrac{ab + bc + ca}{abc}$

(g) $\dfrac{1}{ab} + \dfrac{1}{bc} + \dfrac{1}{ca}$

Ans.   (a) 3,   (b) 4,   (c) -2,   (d) 1,   (e) -15,   (f) -2,   (g) $-\frac{3}{2}$

42.11   For a, b, c defined as in Problem 42.10, find an equation whose roots are $1/a, 1/b, 1/c$.

Hint: Write $x = 1/y$.   *Ans.* $2y^3 + 4y^2 - 3y + 1 = 0$

42.12   Two of the roots of $2x^2 - 11x^2 + \cdots = 0$ are 2, 3. Find the third root and complete the equation.

Ans. $2x^3 - 11x^2 + 17x - 6 = 0$

Chapter 43

Graphs of Polynomials

THE GENERAL POLYNOMIAL (or rational integral function) of the nth degree in x has the form

$$f(x) = a_0 x^n + a_1 x^{n-1} + a_2 x^{n-2} + \cdots + a_{n-2} x^2 + a_{n-1} x + a_n \qquad (43.1)$$

in which n is a positive integer and the a's are constants, real or complex, with $a_0 \neq 0$. Then term $a_0 x^n$ is called the *leading term*, a_n the *constant term*, and a_0 the *leading coefficient*.

Although most of the theorems and statements below apply to the general polynomial, attention in this chapter will be restricted to polynomials whose coefficients (the a's) are integers.

REMAINDER THEOREM. If a polynomial $f(x)$ is divided by $x - h$ until a remainder free of x is obtained, this remainder is $f(h)$. (For a proof, see Problem 43.1.)

EXAMPLE 1. Let $f(x) = x^3 + 2x^2 - 3x - 4$ and $x - h = x - 2$; then $h = 2$. By actual division

$$\frac{x^3 + 2x^2 - 3x - 4}{x - 2} = x^2 + 4x + 5 + \frac{6}{x - 2}$$

or $x^3 + 2x^2 - 3x - 4 = (x^2 + 4x + 5)(x - 2) + 6$, and the remainder is 6.

By the remainder theorem, the remainder is

$$f(2) = 2^3 + 2 \cdot 2^2 - 3 \cdot 2 - 4 = 6$$

FACTOR THEOREM. If $x - h$ is a factor of $f(x)$ then $f(h) = 0$, and conversely. (For a proof, see Problem 43.2.)

SYNTHETIC DIVISION. By a process known as synthetic division, the necessary work in dividing a polynomial $f(x)$ by $x - h$ may be displayed in three lines, as follows:

(1) Arrange the dividend $f(x)$ in descending powers of x (as usual in division) and set down in the first line the coefficients, supplying zero as coefficient whenever a term is missing.

(2) Place h, the synthetic divisor, in the first line to the right of the coefficients.

(3) Recopy the leading coefficient a_0 directly below it in the third line.

(4) Multiply a_0 by h; place the product $a_0 h$ in the second line under a_1 (in the first line), add to a_1, and place the sum $a_0 h + a_1$ in the third line under a_1.

(5) Multiply the sum in Step 4 by h; place the product in the second line under a_2, add to a_2, and place the sum in the third line under a_2.

(6) Repeat the process of Step 5 until a product has been added to the constant term a_n.

329

The first n numbers in the third line are the coefficients of the quotient, a polynomial of degree $n-1$, and the last number of the third line is the remainder $f(h)$.

EXAMPLE 2.   Divide $5x^4 - 8x^2 - 15x - 6$ by $x - 2$, using synthetic division.
     Following the procedure outlined above, we have

$$5 + 0 - 8 - 15 - 6 \quad \underline{|2}$$
$$\underline{10 + 20 + 24 + 18}$$
$$5 + 10 + 12 + 9 + 12$$

The quotient is $Q(x) = 5x^3 + 10x^2 + 12x + 9$ and the remainder is $f(2) = 12$.

(See Problem 43.4.)

DESCARTES' RULE OF SIGNS can be used to discuss the number of positive and negative zeros of a given polynomial (see Chapter 42).

EXAMPLE 3.   Consider the polynomial $P(x) = 2x^2 + x - 4$. $P(x)$ has *one* variation of sign (from + to −); $P(-x) = 2x^2 - x - 4$ has *one* variation of sign (from + to −). Descrates' rule of signs tells us that there will be one positive zero [one variation in sign for $P(x)$] and one negative zero [one variation in sign for $P(-x)$].

EXAMPLE 4.   Suppose $S(x) = x^5 + x^4 + x^3 - x^2 + 1$. $S(x)$ has two variations; $S(-x)$ has three. Thus, $S(x)$ has 2 or 0 positive and 3 or 1 negative zeros.

THE GRAPH OF A POLYNOMIAL $y = f(x)$ may be obtained by computing a table of values, locating the several points (x, y), and joining them by a smooth curve. In order to avoid unnecessary labor in constructing the table, the following systematic procedure is suggested:

(1) When $x = 0, y = f(0)$ is the constant term of the polynomial.
(2) Use synthetic division to find $f(1), f(2), f(3), \ldots$ stopping as soon as the numbers in the third line of the synthetic division have the same sign.
(3) Use synthetic division to find $f(-1), f(-2), f(-3), \ldots$ stopping as soon as the numbers in the third line of the synthetic division have alternating signs.

In advanced mathematics it is proved:

(a)   The graph of a polynomial in x with integral coefficients is always a smooth curve without breaks or sharp corners.
(b)   The number of *real* intersections of the graph of a polynomial of degree n with the x axis is *never* greater than n.
(c)   If a and b are real numbers such that $f(a)$ and $f(b)$ have opposite signs, the graph has an *odd* number of real intersections with the x axis between $x = a$ and $x = b$.
(d)   If a and b are real numbers such that $f(a)$ and $f(b)$ have the same signs, the graph either does not intersect the x axis or intersects it an *even* number of times between $x = a$ and $x = b$. See Fig. 43-1.

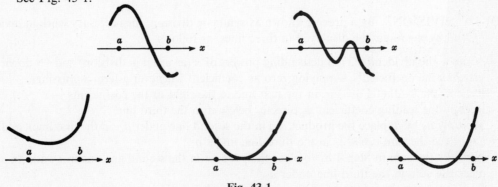

Fig. 43-1

EXAMPLE 5.   Construct the graph of $y = 2x^3 - 7x^2 - 7x + 5$. Form the table in Table 43.1.

Table 43.1

x	-2	-1	0	1	2	3	4	5
y	-25	3	5	-7	-21	-25	-7	45

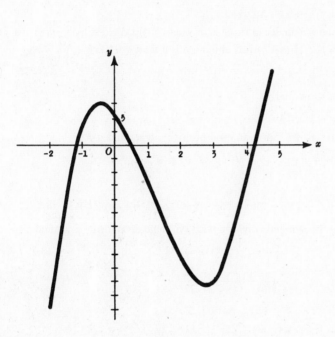

Fig. 43-2

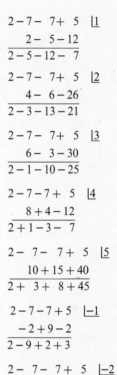

It is to be noted that

(a)   The numbers in the third line are all nonnegative for the first time in finding $f(5)$; they alternate for the first time in finding $f(-2)$.

(b)   The graph intersects the x axis between $x = -2$ and $x = -1$ since $f(-2)$ and $f(-1)$ have opposite signs, between $x = 0$ and $x = 1$, and between $x = 4$ and $x = 5$. Since the polynomial is of degree three, there are no other intersections.

(c)   Reading from the graph, the x intercepts are *approximately* $x = -1.2$, $x = 0.5$, and $x = 4.2$.

(d)   Moving from left to right the graph rises for a time, then falls for a time, and then rises thereafter. The problem of locating the point where a graph ceases to rise or ceases to fall will be considered in a later chapter. See Fig. 43-2.

(See Problems 43.5–43.9.)

Solved Problems

43.1   Prove the Remainder Theorem: If a polynomial $f(x)$ is divided by $x - h$ until a constant remainder is obtained, that remainder is $f(h)$.

In the division let the quotient be denoted by $Q(x)$ and the constant remainder by R. Then, since

$$\text{Dividend} = \text{divisor times quotient} + \text{remainder},$$
$$f(x) = (x - h)Q(x) + R$$

is true for *all* values of x. When $x = h$, we have

$$f(h) = (h - h)Q(h) + R = R$$

43.2 Prove the Factor Theorem: If $x - h$ is a factor of $f(x)$ then $f(h) = 0$, and conversely.

By the Remainder Theorem, $f(x) = (x - h)Q(x) + f(h)$.

If $x - h$ is a factor of $f(x)$, the remainder must be zero when $f(x)$ is divided by $x - h$. Thus, $f(h) = 0$.

Conversely, if $f(h) = 0$, then $f(x) = (x - h)Q(x)$ and $x - h$ is a factor of $f(x)$.

43.3 Without performing the division, show that

(a) $x - 2$ is a factor of $f(x) = x^3 - x^2 - 14x + 24$.

(b) $x + a$ is not a factor of $f(x) = x^n + a^n$ for n an even positive integer and $a \neq 0$.

(a) $f(2) = 2^3 - 2^2 - 14 \cdot 2 + 24 = 0$ (b) $f(-a) = (-a)^n + a^n = 2a^n \neq 0$.

43.4 Use synthetic division to divide $4x^4 + 12x^3 - 21x^2 - 65x + 9$ by (a) $2x - 1$, (b) $2x + 3$.

(a) Write the divisor as $2(x - \frac{1}{2})$. By synthetic division with synthetic divisor $h = \frac{1}{2}$, we find

$$
\begin{array}{r}
4 + 12 - 21 - 65 + 9 \quad \underline{|\tfrac{1}{2}} \\
2 + 7 - 7 - 36 \\
\hline
4 + 14 - 14 - 72 - 27
\end{array}
$$

Now

$$4x^4 + 12x^3 - 21x^2 - 65x + 9 = (4x^3 + 14x^2 - 14x - 72)(x - \tfrac{1}{2}) - 27$$
$$= (2x^3 + 7x^2 - 7x - 36)(2x - 1) - 27$$

Thus, when dividing $f(x)$ by $h = m/n$, the coefficients of the quotient have n as a common factor.

(b) Here $h = -\frac{3}{2}$. Then from

$$
\begin{array}{r}
4 + 12 - 21 - 65 + 9 \quad \underline{|-\tfrac{3}{2}} \\
- 6 - 9 + 45 + 30 \\
\hline
4 + 6 - 30 - 20 + 39
\end{array}
$$

we have

$$4x^4 + 12x^3 - 21x^2 - 65x + 9 = (2x^3 + 3x^2 - 15x - 10)(2x + 3) + 39$$

43.5 For the polynomial $y = f(x) = 4x^4 + 12x^3 - 31x^2 - 72x + 42$ from a table of values for integral values of x from $x = -5$ to $x = 3$.

Table 43.2 is the required table.

Table 43.2

x	-5	-4	-3	-2	-1	0	1	2	3
y	627	90	-21	30	75	42	-45	-66	195

$$\begin{array}{r|r} 4+12-31-72+42 & \underline{-5} \\ -20+40-45+585 \\ \hline 4-8+9-117+627 \end{array} \qquad \begin{array}{r|r} 4+12-31-72+42 & \underline{-4} \\ -16+16+60+48 \\ \hline 4-4-15-12+90 \end{array} \qquad \begin{array}{r|r} 4+12-31-72+42 & \underline{-3} \\ -12+0+93-63 \\ \hline 4+0-31+21-21 \end{array}$$

$$\begin{array}{r|r} 4+12-31-72+42 & \underline{-2} \\ -8-8+78-12 \\ \hline 4+4-39+6+30 \end{array} \qquad \begin{array}{r|r} 4+12-31-72+42 & \underline{-1} \\ -4-8+39+33 \\ \hline 4+8-39-33+75 \end{array} \qquad \begin{array}{l} f(0) = 42, \text{ the constant} \\ \text{term of the polynomial.} \end{array}$$

$$\begin{array}{r|r} 4+12-31-72+42 & \underline{1} \\ 4+16-15-87 \\ \hline 4+16-15-87-45 \end{array} \qquad \begin{array}{r|r} 4+12-31-72+42 & \underline{2} \\ 8+40+18-108 \\ \hline 4+20+9-54-66 \end{array} \qquad \begin{array}{r|r} 4+12-31-72+42 & \underline{3} \\ 12+72+123+153 \\ \hline 4+24+41+51+195 \end{array}$$

43.6 Sketch the graph of $y = f(x) = x^4 - 9x^2 + 7x + 4$.

From Table 43.3 it is seen that the graph crosses the x axis between $x = -4$ and $x = -3$, $x = -1$ and $x = 0$, $x = 1$ and $x = 2$, and $x = 2$ and $x = 3$. From the graph in Fig. 43-3, the points of crossing are approximately $x = -3.1, -0.4, 1.4$, and 2.1.

Table 43.3

x	-4	-3	-2	-1	0	1	2	3
y	88	-17	-30	-11	4	3	-2	25

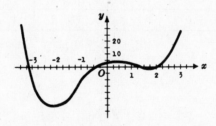

Fig. 43-3

43.7 Sketch the graph of $y = f(x) = x^3 - 5x^2 + 8x - 3$.

From Table 43.4 (formed for integral values of x from $x = -1$ to $x = 5$) it is seen that the graph crosses the x axis between $x = 0$ and $x = 1$. If there are two other real intersections, they are both between $x = 1$ and $x = 3$ since on this interval the graph rises to the right of $x = 1$, then falls, and then begins to rise to the left of $x = 3$.

By computing additional points on the interval (some are shown in the table), we are led to suspect that there are no further intersections (see Fig. 43-4). Note that were we able to locate the exact points at which the graph ceases to rise or fall these additional points would not have been necessary.

Table 43.4

x	-1	0	1	$\frac{5}{4}$	$\frac{3}{2}$	$\frac{7}{4}$	2	$\frac{9}{4}$	$\frac{5}{2}$	$\frac{11}{4}$	3	4	5
y	-17	-3	1	$\frac{73}{64}$	$\frac{9}{8}$	$\frac{67}{64}$	1	$\frac{69}{64}$	$\frac{11}{8}$	$\frac{127}{64}$	3	13	37

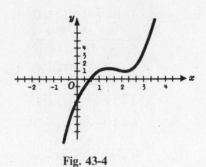

Fig. 43-4

43.8   Sketch the graph $y = f(x) = x^3 - 4x^2 - 3x + 18$.

From Table 43.5 it is evident that the graph crosses the x axis at $x = -2$ and meets it again at $x = 3$. If there is a third distinct intersection, it must be between $x = 2$ and $x = 3$ or between $x = 3$ and $x = 4$. By computing additional points for x on these intervals, we are led to suspect that no such third intersection exists.

This function has been selected so that the question of intersections can be definitely settled. When $f(x)$ is divided by $x + 2$, the quotient is $x^2 - 6x + 9 = (x - 3)^2$ and the remainder is zero. Thus, $f(x) = (x + 2)(x - 3)^2$ in factored form. It is now clear that the function is positive for $x > -2$; that is, the graph never falls below the x axis on this interval. Thus, the graph is tangent to the x axis at $x = 3$, the point of tangency accounting for two of its intersections with the x axis. (See Fig. 43-5.)

Table 43.5

x	-3	-2	-1	0	1	2	3	4	5
y	-36	0	16	18	12	4	0	6	28

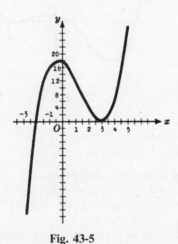

Fig. 43-5

43.9   Sketch the graph of $y = f(x) = x^3 - 6x^2 + 12x - 8$.

From Table 43.6 it is evident that the graph crosses the x axis at $x = 2$ and is symmetrical with respect to that point [i.e., $f(2 + h) = -f(2 - h)$]. Suppose the graph intersects the x axis to the right of $x = 2$. Then, since $f(x)$ is positive for $x = 3, 4, 5$ and $x \geq 6$, the graph is either tangent at the point or crosses the axis twice between some two consecutive values of x shown in the table. But, by symmetry, the graph would then have $2 + 2 + 1 = 5$ intersections with the x axis and this is impossible.

When $f(x)$ is divided by $x - 2$, the quotient is $x^2 - 4x + 4 = (x - 2)^2$ and the remainder is zero. Thus, $f(x) = (x - 2)^3$. Now it is clear that the graph lies above the x axis when $x > 2$ and below the axis when

$x < 2$. The graph crosses the x axis at $x = 2$ and is also tangent there. (See Fig. 43-6.) In determining the intersections of the graph and the x axis, the point $x = 2$ is to be counted three times.

Table 43.6

x	-2	-1	0	1	2	3	4	5	6
y	-64	-27	-8	-1	0	1	8	27	64

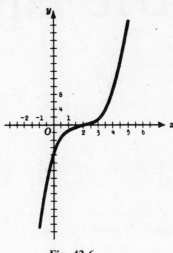

Fig. 43-6

Supplementary Problems

43.10 Given $f(x) = x^4 + 5x^3 - 7x^2 - 13x - 2$, use synthetic division to find

  (a) $f(2) = 0$ (b) $f(4) = 410$ (c) $f(-2) = -28$ (d) $f(-3) = -80$

43.11 Using synthetic division, find the quotient and remainder when

  (a) $2x^4 - 3x^3 - 6x^2 + 11x - 10$ is divided by $x - 2$;   (b) $3x^4 + 11x^3 + 7x^2 - 8$ is divided by $x + 2$.
  Ans. (a) $2x^3 + x^2 - 4x + 3; -4$ (b) $3x^3 + 5x^2 - 3x + 6; -20$

43.12 Use synthetic division to show

  (a) $x + 2$ and $3x - 2$ are factors of $3x^4 - 20x^3 + 80x - 48$.
  (b) $x - 7$ and $3x + 5$ are not factors of $6x^4 - x^3 - 94x^2 + 74x + 35$.

43.13 Use synthetic division to form a table of values and sketch the graph of

  (a) $y = x^3 - 13x + 12$ (c) $y = 3x^3 + 5x^2 - 4x - 3$ (e) $y = -x^3 + 13x + 12$
  (b) $y = 2x^3 + x^2 - 12x - 5$ (d) $y = x^4 - x^3 - 7x^2 + 13x - 6$

43.14 Sketch the graph of

  (a) $y = x(x^2 - 4)$ (b) $y = x(4 - x^2)$ (c) $y = x(x - 2)^2$ (d) $y = x(2 - x)^2$.

43.15 Use Descartes' rule of signs to discuss positive and negative zeros for

  (a) $P(x) = x^4 + x^2 + 1$ (none)
  (b) $M(x) = 7x^2 + 2x + 4$ (0 positive zeros, 0 or 2 negative)

Chapter 44

Parametric Equations

IN THIS CHAPTER we consider the analytic representation of a plane curve by means of a pair of equations, as $x = t, y = 2t + 3$, in which the coordinates of a variable point (x, y) on the curve are expressed as a function of a third variable or *parameter*. Such equations are called *parametric equations* of the curve.

A table of values of x and y is readily obtained from the given parametric equations by assigning values to the parameter. After plotting the several points (x, y), the locus may be sketched in the usual manner.

EXAMPLE. Sketch the locus of $x = t$, $y = 2t + 3$.

We form the table of values (Table 44.1), plot the points (x, y), and join these points to obtain the straight line shown in Fig. 44-1. (See Problem 44-1.)

Table 44.1

t	2	0	-3
x	2	0	-3
y	7	3	-3

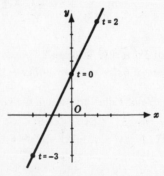

Fig. 44-1

At times it will be possible to eliminate the parameter between the two equations and thus obtain the rectangular equation of the curve. In the example above, the elimination of t is easy and results in $y = 2x + 3$. At the other times, however, it will be impractical or impossible to eliminate the parameter. (See Problem 44.2.)

Parametric representation of a curve is *not* unique. For example, $x = t, y = 2t + 3; x = \frac{1}{2}u, y = u + 3;$ $x = v - 1, y = 2v + 1$ are parametric representations with parameters t, u, v, respectively, of the straight line whose rectangular equation is $y = 2x + 3$. (See Problems 44.3–44.4.)

PATH OF A PROJECTILE. If a body is projected from the origin with initial velocity v_0 ft/s at an angle α with the positive x axis and if all forces acting on the body after projection, excepting the force of gravity, are ignored, the coordinates of the body t seconds thereafter are given by

$$x = v_0 t \cos \alpha, \qquad y = v_0 t \sin \alpha - \tfrac{1}{2}gt^2$$

where g is the acceleration due to gravity. For convenience, we take $g = 32 \text{ ft}/s^2$. (See Problems 44.5–44.7.)

Solved Problems

44.1   Sketch the locus of each of the following:

(a) $x = t, y = t^2$ (b) $x = 4t, y = 1/t$ (c) $x = 5\cos\theta, \ y = 5\sin\theta$ (d) $x = 2 + \cos\theta, y = \cos 2\theta$

The table of values and sketch of each are given in Figs. 44-2 to 44-5. In Fig. 44-3, $t = 0$; hence, we must examine the locus for values of t near 0. In Fig. 44-4 the complete locus is described on the interval $0 \le \theta \le 2\pi$. In Fig. 44-5 the complete locus is described on the interval $0 \le \theta \le \pi$. Note that only that part of the parabola below $y = 1$ is obtained; thus the complete parabola is not defined by the parametric equations.

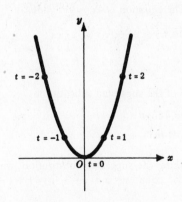

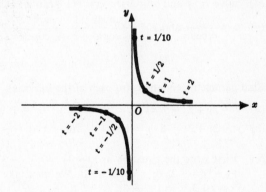

t	2	1	0	−1	−2
x	2	1	0	−1	−2
y	4	1	0	1	4

Fig. 44-2

t	2	1	1/2	1/10	−1/10	−1/2	−1	−2
x	8	4	2	2/5	−2/5	−2	−4	−8
y	1/2	1	2	10	−10	−2	−1	−1/2

Fig. 44-3

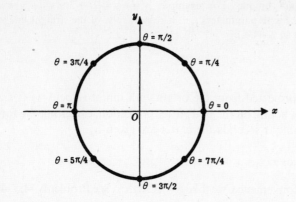 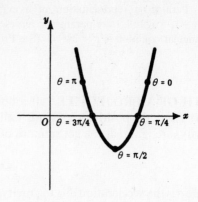

θ	0	$\pi/4$	$\pi/2$	$3\pi/4$	π	$5\pi/4$	$3\pi/2$	$7\pi/4$	2π
x	5.00	3.55	0	−3.55	−5.00	−3.55	0	3.55	5.00
y	0	3.55	5.00	3.55	0	−3.55	−5.00	−3.55	0

θ	0	$\pi/4$	$\pi/2$	$3\pi/4$	π
x	3.00	2.71	2.00	1.29	1.00
y	1.00	0	−1.00	0	1.00

Fig. 44-4   Fig. 44-5

44.2 In each of the following, eliminate the parameter to obtain the equation of the locus in rectangular coordinates: (a)–(d), Problem 44.1 (e) $x = t^2 + t$, $y = t^2 - t$ (f) $x = 3\sec\phi$, $y = 2\tan\phi$ (g) $x = v_0(\cos\alpha)t$, $y = v_0(\sin\alpha)t - \frac{1}{2}gt^2$, t being the parameter.

(a) Here $t = x$ and $y = t^2 = x^2$. The required equation is $y = x^2$.

(b) Since $t = 1/y$, $x = 4t = 4/y$ and the equation is $xy = 4$.

(c) $x^2 + y^2 = (5\cos\theta)^2 + (5\sin\theta)^2 = 25(\cos^2\theta + \sin^2\theta)$. The required equation is $x^2 + y^2 = 25$.

(d) $\cos\theta = x - 2$ and $y = \cos 2\theta = 2\cos^2\theta - 1 = 2(x-2)^2 - 1$. The equation is $(x-2)^2 = \frac{1}{2}(y+1)$.

(e) Subtracting one of the equations from the other, $t = \frac{1}{2}(x - y)$. Then $x = \frac{1}{4}(x-y)^2 + \frac{1}{2}(x-y)$ and the required equation is $x^2 - 2xy + y^2 - 2x - 2y = 0$.

(f) $\tan\phi = \frac{1}{2}y$ and $x^2 = 9\sec^2\phi = 9(1 + \tan^2\phi) = 9(1 + \frac{1}{4}y^2)$. The equation is $4x^2 - 9y^2 = 36$.

(g) $t = \dfrac{x}{v_0\cos\alpha}$ and $y = \dfrac{v_0(\sin\alpha)x}{v_0\cos\alpha} - \dfrac{1}{2}\cdot\dfrac{gx^2}{(v_0\cos\alpha)^2}$. Then $y = x\tan\alpha - \dfrac{g}{2\,v_0^2\cos^2\alpha}x^2$.

44.3 Find parametric equations for each of the following, making use of the suggested substitution:

(a) $y^2 - 2y + 2x - 5 = 0$, $x = t + 3$ (c) $9x^2 + 16y^2 = 144$, $x = 4\cos\theta$

(b) $y^2 - 2y + 2x = 5$, $y = 1 + t$ (d) $x^3 + y^3 - 3axy = 0$, $y = mx$

(a) First write the equation as $(y - 1)^2 = -2(x - 3)$. Upon making the suggested substitution, we have $(y - 1)^2 = -2t$ or $y = 1 \pm \sqrt{-2t}$. We may take as parametric equations $x = t + 3$, $y = 1 + \sqrt{-2t}$ or $x = t + 3$, $y = 1 - \sqrt{-2t}$.

(b) From $(y - 1)^2 = -2(x - 3)$, we obtain $t^2 = -2(x - 3)$ or $x = 3 - \frac{1}{2}t^2$. The parametric equations are $x = 3 - \frac{1}{2}t^2$, $y = 1 + t$.

(c) We have $9(16\cos^2\theta) + 16y^2 = 144$ or $y^2 = 9(1 - \cos^2\theta) = 9\sin^2\theta$. The parametric equations are $x = 4\cos\theta$, $y = 3\sin\theta$ or $x = 4\cos\theta$, $y = -3\sin\theta$.

(d) Substituting, we have $x^3 + m^3x^3 - 3amx^2 = 0$. Dividing by x^2, we obtain $x = \dfrac{3am}{1 + m^3}$. Then $y = mx = \dfrac{3am^2}{1 + m^3}$ and the parametric equations are $x = \dfrac{3am}{1 + m^3}$, $y = \dfrac{3am^2}{1 + m^3}$.

44.4 (a) Show that $x = a\cos\theta$, $y = b\sin\theta$, $a > b$, are parametric equations of an ellipse.

(b) Show that these equations indicate the following method for constructing an ellipse whose major and minor axes $2a$ and $2b$ are given: With the origin as common center draw two circles having radii a and b. Through the origin draw a half line l meeting the smaller circle in B and the larger circle in A. From A drop a perpendicular to the x axis meeting it in R; from B drop a perpendicular to the x axis meeting it in S and a perpendicular to RA meeting it in P. Then as l revolves about O, P describes the ellipse.

(a) We have $\cos\theta = x/a$ and $\sin\theta = y/b$; then $x^2/a^2 + y^2/b^2 = \cos^2\theta + \sin^2\theta = 1$ is the equation of an ellipse.

(b) In Fig. 44-6, let P have coordinates (x, y) and denote by θ the angle which l makes with the positive x axis. Then $x = OR = OA\cos\theta = a\cos\theta$ and $y = RP = SB = OB\sin\theta = b\sin\theta$ are parametric equations of the ellipse whose major and minor axes are a and b.

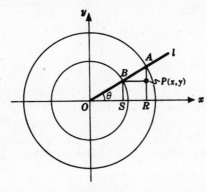

Fig. 44-6

44.5 A body is projected from the origin with initial velocity v_0 ft/s at an angle α with the positive x axis. Assuming that the only force acting upon the body after projection is the attraction of the earth, obtain parametric equations of the path of the body with t (the number of seconds after projection) as parameter.

If a body is released near the surface of the earth and all forces acting upon it other than gravity are neglected, the distance s ft through which it will fall in t seconds is given by $s = \frac{1}{2}gt^2$. where $g = 32$ ft/s^2 approximately.

If a body is given motion as stated in the problem and if after projection *no other force* acts on it, the motion is in a straight line (Newtons's first law of motion) and after t seconds the body has coordinates $(v_0 t\cos\alpha, v_0 t\sin\alpha)$.

Since, when small distances are involved, the force of gravity may be assumed to act vertically, the coordinates of the projected body after t seconds of motion are given by

$$x = v_0 t\cos\alpha, \quad y = v_0 t\sin\alpha - \tfrac{1}{2}gt^2. \tag{1}$$

In rectangular coordinates, the equation of the path is [see Problem 44.2(g)]

$$y = x\tan\alpha - \frac{g}{2 v_0^2\cos^2\alpha}x^2 \tag{2}$$

44.6 A bird is shot when flying horizontally 120 ft directly above the hunter. If its speed is 30 mi/hr, find the time during which it falls and the distance it will be from the hunter when it strikes the earth.

As in Fig. 44-7, take the bird to be at the origin when shot. Since $30\,\text{mi/hr} = 30 \cdot 5280/(60 \cdot 60) = 44\,\text{ft/s} = v_0$ and $\alpha = 0°$, the equations of motion are

(a) $x = v_0 t \cos \alpha = 44t \cos 0° = 44t$
(b) $y = v_0 t \sin \alpha - \frac{1}{2}gt^2 = -16t^2$

When the bird reaches the ground, its coordinates are $(x, -120)$.

From (b), $-120 = -16t^2$ and $t = \frac{1}{2}\sqrt{30}$; from (a), $x = 44(\frac{1}{2}\sqrt{30}) = 22\sqrt{30}$. Thus the bird will fall for $\frac{1}{2}\sqrt{30}$ s and will reach the ground $22\sqrt{30}$ ft from the hunter.

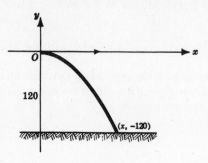

Fig. 44-7

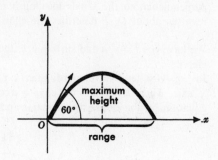

Fig. 44-8

44.7   A ball is projected upward from the ground at an angle 60° from the horizontal with initial velocity 60 ft/s. Find (a) the time it will be in the air (b) its range, that is, the horizontal distance it will travel, and (c) its maximum height attained. Refer to Fig. 44-8. (Also, see Part III for a calculus solution to maxima/minima problems.)

Let the ball be projected from the origin; then the equations of motion are

$$x = v_0 t \cos \alpha = 60t \cos 60° = 30t, \quad y = v_0 t \sin \alpha - \frac{1}{2}gt^2 = 30t\sqrt{3} - 16t^2.$$

(a)   When $y = 0$, $16t^2 - 30t\sqrt{3} = 0$ and $t = 0, 15\sqrt{3}/8$. Now $t = 0$ is the time when the ball was projected and $t = 15\sqrt{3}/8$ is the time when it reaches the ground again. Thus the ball was in the air for $15\sqrt{3}/8$ s.

(b)   When $t = 15\sqrt{3}/8$, $x = 30 \cdot 15\sqrt{3}/8 = 225\sqrt{3}/4$. The range is $225\sqrt{3}/4$ ft.

(c)   **First Solution.**   The ball will attain its maximum height when $t = \frac{1}{2}(15\sqrt{3}/8) = 15\sqrt{3}/16$. Then
$$y = 30t\sqrt{3} - 16t^2 = 30(15\sqrt{3}/16)\sqrt{3} - 16(15\sqrt{3}/16)^2 = \frac{675}{16} \text{ ft, the maximum height.}$$

Second Solution.   The maximum height is attained where the horizontal distance of the ball is one-half the range, i.e., when $x = 225\sqrt{3}/8$. Using the rectangular equation $y = -\frac{4}{225}x^2 + x\sqrt{3}$, we obtain $\frac{675}{16}$ ft as before.

44.8   The locus of a fixed point P on the circumference of a circle of radius a as the circle rolls without slipping along a straight line is called a *cycloid*. Obtain parametric equations of this locus.

Take the x axis to be the line along which the circle is to roll and place the circle initially with its center C on the y axis and P at the origin. Figure 44-9 shows the position of P after the circle has rolled through an angle θ. Drop perpendiculars $\overline{PR}$ and $\overline{CS}$ to the x axis and PA to SC. Let P have coordinates (x, y). Then

$$x = OR = OS - RS = arc\ PS - PA = a\theta - a\sin\theta$$

and $$y = RP = SA = SC - AC = a - a\cos\theta$$

Thus, the equations of the locus are $x = a(\theta - \sin\theta)$, $y = a(1 - \cos\theta)$.

The maximum height of an arch is $2a$, the diameter of the circle, and the span of an arch or the distance between two consecutive cusps is $2\pi a$, the circumference of the circle.

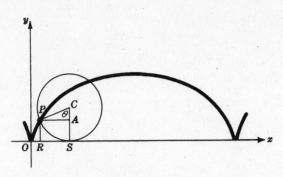

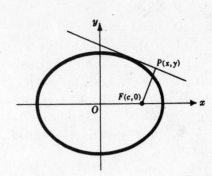

Fig. 44-9   **Fig. 44-10**

44.9 Find the equation of the locus of the feet of perpendiculars drawn from a focus to the tangents of the ellipse $b^2x^2 + a^2y^2 = a^2b^2$. This curve is called the *pedal curve* of the ellipse with respect to the focus. See Fig. 44-10.

Let $P(x,y)$ be any point on the locus. The equations of the tangents of slope m to the ellipse are

$$y = mx \pm \sqrt{a^2m^2 + b^2} \qquad \text{or} \qquad (1') \quad y - mx = \pm\sqrt{a^2m^2 + b^2}$$

The perpendicular to these tangents through $F(c,0)$ has equation

$$y = -\frac{1}{m}x + \frac{c}{m} \qquad \text{or} \qquad (2') \quad my + x = \sqrt{a^2 - b^2}$$

Squaring $(1')$ and $(2')$, and adding, we obtain

$$(1 + m^2)y^2 + (1 + m^2)x^2 = a^2(1 + m^2) \qquad \text{or, since } 1 + m^2 \neq 0, \quad x^2 + y^2 = a^2$$

as the desired equation.

Supplementary Problems

44.10 Sketch the locus and find the rectangular equation of the curve whose parametric equations are

(a) $x = t + 2, \ y = 3t + 5$ *Ans.* $\ y = 3x - 1$

(b) $x = \tan\theta, \ y = 4\cot\theta$ *Ans.* $\ xy = 4$

(c) $x = 2t^2 + 3, \ y = 3t + 2$ *Ans.* $\ 2y^2 - 8y - 9x + 35 = 0$

(d) $x = 3 + 2\tan\theta, \ y = -1 + 5\sec\theta$ *Ans.* $\ 4y^2 - 25x^2 + 8y + 150x - 321 = 0$

(e) $x = 2\sin^4\theta, \ y = 2\cos^4\theta$ *Ans.* $\ (x - y)^2 - 4(x + y) + 4 = 0$

(f) $x = \tan\theta, \ y = \tan 2\theta$ *Ans.* $\ x^2y - y + 2x = 0$

(g) $x = \sqrt{\cos t}, \ y = \tan\frac{1}{2}t$ *Ans.* $\ y^2(1 + x^2) + x^2 - 1 = 0$

(h) $x = a\cos^3\theta, \ y = a\sin^3\theta$ *Ans.* $\ x^{2/3} + y^{2/3} = a^{2/3}$

(i) $x = \dfrac{2t}{1 + t^2}, \ y = \dfrac{1 - t^2}{1 + t^2}$ *Ans.* $\ x^2 + y^2 = 1$

(j) $x = \dfrac{2at^2}{1 + t^2}, \ y = \dfrac{2at^3}{1 + t^2}$ *Ans.* $\ y^2(2a - x) = x^3$

44.11 Find parametric equations for each of the following, using the suggested value for x or y:

(a) $x^3 = 4y^2, \ x = t^2$ (c) $4x^2 - 9y^2 = 36, \ x = 3\sec\theta$ (e) $x(x^2 + y^2) = x^2 - y^2, \ y = tx$

(b) $y = x^2 + x - 6, \ x = t + 2$ (d) $y = 2x^2 - 1, \ x = \cos t$ (f) $(x^2 + 16)y = 64, \ x = 4\tan\theta$

Ans. (a) $x = t^2, y = \frac{1}{2}t^3$

(b) $x = t + 2, y = t^2 + 5t$

(c) $x = 3 \sec \theta, y = 2 \tan \theta$

(d) $x = \cos t, y = \cos 2t$

(e) $x = \dfrac{1 - t^2}{1 + t^2}, y = \dfrac{(1 - t^2)}{1 + t^2}$

(f) $x = 4 \tan \theta, y = 4 \cos^2 \theta$

44.12 A 30-ft ladder with base on a smooth horizontal surface leans against a house. A man is standing $\frac{2}{3}$ the way up the ladder when its foot begins to slide away from the house. Find the path of the man.

Ans. $x = 10 \cos \theta, y = 20 \sin \theta$ where θ is the angle at the foot of the ladder.

44.13 A stone is thrown upward with initial speed 48 ft/s at an angle measuring 60° with the horizontal from the top of a cliff 100 ft above the surface of a lake. Find (a) its greatest distance above the lake, (b) when it will strike the surface of the lake, and (c) its horizontal distance from the point where thrown when it strikes the surface.

Hint: Take the origin at the top of the cliff.

Ans. (a) 127 ft (b) $\dfrac{3\sqrt{3} + \sqrt{127}}{4}$ s later (c) $6(3\sqrt{3} + \sqrt{127})$ ft

44.14 Find the locus of the vertices of all right triangles having hypotenuse of length $2a$.

Hint: Take the hypotenuse along the x axis with its midpoint at the origin and let θ be an acute angle of the triangle.

Ans. $x = a \cos 2\theta, \ y = a \sin 2\theta$ or $x = a \sin 2\theta, \ y = a \cos 2\theta$

44.15 From the two-point form of the equation of a straight line derive the parametric equations $x = x_1 + t(x_2 - x_1), y = y_1 + t(y_1 - y_1)$. What values of the parameter t give the points on the segment $P_1 P_2$? Identify the points corresponding to $t = \frac{1}{2}, \frac{1}{3}, \frac{2}{3}$.

44.16 Verify the following method for constructing a hyperbola with transverse axis $2a$ and conjugate axis $2b$, where $a \neq b$. With the origin as common center draw two circles having radii a and b. Through the origin pass a half line l making an angle θ with the positive x axis and intersecting the larger circle in A. Let the tangent at A to the circle meet the x axis in B. Through C, the intersection of the smaller circle and the positive x axis, erect a perpendicular meeting l in D. Through D pass a line parallel to the x axis and through B a line perpendicular to the x axis, and denote their intersection by P. Then P is a point on the hyperbola.

44.17 Obtain the rectangular equation $x = a \arccos \dfrac{a - y}{a} \mp \sqrt{2ay - y^2}$ of the cycloid of Problem 44.8.

PART IV

INTRODUCTION TO CALCULUS

Chapter 45

The Derivative

IN THIS AND SUBSEQUENT CHAPTERS, it will be understood that number refers always to a real number, that the range of any variable (such as x) is a set of real numbers, and that a function of one variable [such as $f(x)$] is a single-valued function.

In this chapter a procedure is given by which from a given function $y = f(x)$ another function, denoted by y' or $f'(x)$ and called the derivative of y or of $f(x)$ with respect to x, is obtained. Depending upon the quantities denoted by x and $y = f(x)$, the derivative may be interpreted as the slope of a tangent line to a curve, as velocity, as acceleration, etc.

LIMIT OF A FUNCTION. A given function $f(x)$ is said to have a *limit M* as x approaches c [in symbols, $\lim_{x \to c} f(x) = M$] if $f(x)$ can be made as close to M as we please for all values of $x \neq c$ but sufficiently near to c, by having x get sufficiently close to c (approaching both from the left and right).

EXAMPLE 1. Consider $f(x) = x^2 - 2$ for x near 3.

If x is near to 3, say $2.99 < x < 3.01$, then $(2.99)^2 - 2 < f(x) < (3.01)^2 - 2$ or $6.9401 < f(x) < 7.0601$.

If x is nearer to 3, say $2.999 < x < 3.001$, then $(2.999)^2 - 2 < f(x) < (3.001)^2 - 2$ or $6.994001 < f(x) < 7.006001$.

If x is still nearer to 3, say $2.9999 < x < 3.0001$, then $(2.9999)^2 - 2 < f(x) < (3.0001)^2 - 2$ or $6.99940001 < f(x) < 7.00060001$.

It appears reasonable to conclude that as x is taken in a smaller and smaller interval about 3, the corresponding $f(x)$ will lie a smaller and smaller interval about 7. Conversely, it seems reasonable to conclude that if we demand that $f(x)$ have values lying in smaller and smaller intervals about 7, we need only to choose x in sufficiently smaller and smaller intervals about 3. Thus we conclude

$$\lim_{x \to 3} (x^2 - 2) = 7$$

EXAMPLE 2. Consider $f(x) = \dfrac{x^2 - x - 6}{x - 3}$ for x near 3.

When $x \neq 3$, $f(x) = \dfrac{x^2 - x - 6}{x - 3} = x + 2$. Thus, for x near 3, $x + 2$ is near to 5 and

$$\lim_{x \to 3} \frac{x^2 - x - 6}{x - 3} = 5$$

ONE-SIDED LIMITS. We say that the limit of $f(x)$ as x approaches a from the left is $L[\lim_{x \to a^-} f(x) = L]$ when $f(x)$ gets arbitrarily close to L as x approaches a from the left-hand side of a. [Similarly, $\lim_{x \to b^+} f(x) = M$ is the right-hand limit.] Clearly, if $\lim_{x \to a^-} f(x) = \lim_{x \to a^+} f(x) = L$, then $\lim_{x \to a} f(x) = L$.

345

THEOREMS ON LIMITS. If $\lim\limits_{x \to c} f(x) = M$ and $\lim\limits_{x \to c} g(x) = N$, then

I. $\lim\limits_{x \to c} [f(x) \pm g(x)] = \lim\limits_{x \to c} f(x) \pm \lim\limits_{x \to c} g(x) = M \pm N.$

II. $\lim\limits_{x \to c} [kf(x)] = k \lim\limits_{x \to c} f(x) = kM$, where k is a constant.

III. $\lim\limits_{x \to c} [f(x) \cdot g(x)] = \lim\limits_{x \to c} f(x) \cdot \lim\limits_{x \to c} g(x) = MN.$

IV. $\lim\limits_{x \to c} \dfrac{f(x)}{g(x)} = \dfrac{\lim\limits_{x \to c} f(x)}{\lim\limits_{x \to c} g(x)} = \dfrac{M}{N}$, provided $N \neq 0.$

Note that in these four statements, the assumption that M and N exist is essential. (See Problem 45.2.)

CONTINUOUS FUNCTIONS.   A function $f(x)$ is called continuous at $x = c$, provided

(1) $f(c)$ is defined,

(2) $\lim\limits_{x \to c} f(x)$ exists,

(3) $\lim\limits_{x \to c} f(x) = f(c).$

EXAMPLE 3.

(a) The function $f(x) = x^2 - 2$ of Example 1 is continuous at $x = 3$ since (1) $f(3) = 7$, (2) $\lim\limits_{x \to 3}(x^2 - 2) = 7$, (3) $\lim\limits_{x \to 3}(x^2 - 2) = f(3).$

(b) The function $f(x) = \dfrac{x^2 - x - 6}{x - 3}$ of Example 2 is not continuous at $x = 3$, since $f(3)$ is not defined. (See Problem 45.2.)

A function $f(x)$ is said to be *continuous* on the interval (a, b) if it is continuous for every value of x of the interval. A polynomial in x is continuous since it continuous for all values of x. A rational function in x, $f(x) = P(x)/Q(x)$, where $P(x)$ and $Q(x)$ are polynomials, is continuous for all values of x except those for which $Q(x) = 0$. Thus, $f(x) = \dfrac{x^2 + x + 1}{(x - 1)(x^2 + 2)}$ is continuous for all values of x, except $x = 1$.

CONTINUITY ON A CLOSED INTERVAL.   If a function $y = g(x)$ is continuous for all values of x in $[a, b]$, then it is continuous on (a, b) and also at a and b.

However, in this case, $g(x)$ is continuous at a means $\lim\limits_{x \to a^+} g(x) = g(a)$; similarly, $\lim\limits_{x \to b^-} g(x) = g(b)$.

(See Problem 45.4.)

INCREMENTS.   Let x_0 and x_1 be two distinct values of x. It is customary to denote their difference $x_1 - x_0$ by Δx (read, delta x) and to write $x_0 + \Delta x$ for x_1.

Now if $y = f(x)$ and if x changes in value from $x = x_0$ to $x = x_0 + \Delta x$, y will change in value from $y_0 = f(x_0)$ to $y_0 + \Delta y = f(x_0 + \Delta x)$. The change in y due to a change in x from $x = x_0$ to $x = x_0 + \Delta x$ is $\Delta y = f(x_0 + \Delta x) - f(x_0).$

EXAMPLE 4.   Compute the change in $y = f(x) = x^2 - 2x + 5$ when x changes in value from (a) $x = 3$ to $x = 3.2$, (b) $x = 3$ to $x = 2.9$.

(a) Take $x_0 = 3$ and $\Delta x = 0.2$. Then $y_0 = f(x_0) = f(3) = 8$, $y_0 + \Delta y = f(x_0 + \Delta x) = f(3.2) = 8.84$, and $\Delta y = 8.84 - 8 = 0.84.$

(b) Take $x_0 = 3$ and $\Delta x = -0.1$. Then $y_0 = f(3) = 8$, $y_0 + \Delta y = f(2.9) = 7.61$, and $\Delta y = 7.61 - 8 = -0.39.$

(See Problems 45.5–45.6.)

THE DERIVATIVE.   The *derivative* of $y = f(x)$ at $x = x_0$ is

$$\lim_{\Delta x \to 0} \frac{\Delta y}{\Delta x} = \lim_{\Delta x \to 0} \frac{f(x_0 + \Delta x) - f(x_0)}{\Delta x}$$

provided the limit exists. $\dfrac{\Delta y}{\Delta x}$ is called the *difference quotient*.

To find derivatives, we shall use the following five-step rule:

(*1*)   Write $y_0 = f(x_0)$.
(*2*)   Write $y_0 + \Delta y = f(x_0 + \Delta x)$.
(*3*)   Obtain $\Delta y = f(x_0 + \Delta x) - f(x_0)$.
(*4*)   Obtain $\Delta y / \Delta x$.
(*5*)   Evaluate $\lim\limits_{\Delta x \to 0} \dfrac{\Delta x}{\Delta y}$. The result is the derivative of $y = f(x)$ at $x = x_0$.

EXAMPLE 5.   Find the derivative of $y = f(x) = 2x^2 - 3x + 5$ at $x = x_0$.

(*1*)
$$y_0 = f(x_0) = 2x_0^2 - 3x_0 + 5$$

(*2*)
$$y_0 + \Delta y = f(x_0 + \Delta x) = 2(x_0 + \Delta x)^2 - 3(x_0 + \Delta x) + 5$$
$$= 2x_0^2 + 4x_0 \cdot \Delta x + 2(\Delta x)^2 - 3x_0 - 3 \cdot \Delta x + 5$$

(*3*)
$$\Delta y = f(x_0 + \Delta x) - f(x_0) = 4x_0 \cdot \Delta x - 3 \cdot \Delta x + 2(\Delta x)^2$$

(*4*)
$$\frac{\Delta y}{\Delta x} = 4x_0 - 3 + 2 \cdot \Delta x$$

(*5*)
$$\lim_{\Delta x \to 0} \frac{\Delta y}{\Delta x} = \lim_{\Delta x \to 0} (4x_0 - 3 + 2 \cdot \Delta x) = 4x_0 - 3.$$

If in the example above the subscript 0 is deleted, the five-step rule yields a function of x (here, $4x - 3$) called the derivative with respect to x of the given function. The derivative with respect to x of the function $y = f(x)$ is denoted by one of the symbols y', $\dfrac{dy}{dx}$, $f'(x)$, or $D_x y$.

Provided it exists, the value of the derivative for any given value of x, say x_0, will be denoted by $y'\Big|_{x=x_0}$, $\dfrac{dy}{dx}\Big|_{x=x_0}$, or $f'(x_0)$. (See Problems 45.8–45.15.)

HIGHER-ORDER DERIVATIVES.   The process of finding the derivative of a given function is called *differentiation*.

By differentiation, we obtain from a given function $y = f(x)$ another function $y' = f'(x)$ which will now be called the *first derivative* of y or of $f(x)$ with respect to x. If, in turn, $y' = f'(x)$ is differentiated with respect to x, another function $y'' = f''(x)$, called the *second derivative* of y or of $f(x)$ is obtained. Similarly, a third derivative may be found, and so on.

EXAMPLE 6.   Let $y = f(x) = x^4 - 3x^2 + 8x + 6$. Then $y' = f'(x) = 4x^3 - 6x + 8$, $y'' = f''(x) = 12x^2 - 6$, and $y''' = f'''(x) = 24x$. (See Problem 45.16.)

Solved Problems

45.1   Investigate $f(x) = 1/x$ for values of x near $x = 0$.

If x is near 0, say $-.01 < x < .01$, then $\dfrac{1}{-.01} < \dfrac{1}{x} < \dfrac{1}{.01}$ or $-100 < \frac{1}{x} < 100$.

If x is nearer to 0, say $-.0001 < x < .0001$, then $\dfrac{1}{-.0001} < \dfrac{1}{x} < \dfrac{1}{.0001}$ or $-10\,000 < \frac{1}{x} < 10\,000$.

It is clear that as x is taken in smaller and smaller intervals about 0, the corresponding $f(x)$ does *not* lie in smaller and smaller intervals about any number M. Hence, $\lim\limits_{x \to 0}(1/x)$ does not exist.

45.2   Evaluate when possible:

(a) $\lim\limits_{x \to 2}(4x^2 - 5x)$,   (b) $\lim\limits_{x \to 1}(x^2 - 4x + 10)$,   (c) $\lim\limits_{x \to 2}\dfrac{x^2 + 6x + 5}{x^2 - 2x - 3}$,

(d) $\lim\limits_{x \to 3}\dfrac{x^2 + 6x + 5}{x^2 - 2x - 3}$,   (e) $\lim\limits_{x \to -1}\dfrac{x^2 + 6x + 5}{x^2 - 2x - 3}$.

(a) $\lim\limits_{x \to 2}(4x^2 - 5x) = \lim\limits_{x \to 2}4x^2 - \lim\limits_{x \to 2}5x = 4\lim\limits_{x \to 2}x^2 - 5\lim\limits_{x \to 2}x = 4 \cdot 4 - 5 \cdot 2 = 6$.

(b) $\lim\limits_{x \to 1}(x^2 - 4x + 10) = (1)^2 - 4 \cdot 1 + 10 = 7$.

(c) $\lim\limits_{x \to 2}(x^2 + 6x + 5) = 21$ and $\lim\limits_{x \to 2}(x^2 - 2x - 3) = -3$; hence

$$\lim\limits_{x \to 2}\frac{x^2 + 6x + 5}{x^2 - 2x - 3} = \frac{\lim\limits_{x \to 2}(x^2 + 6x + 5)}{\lim\limits_{x \to 2}(x^2 - 2x - 3)} = \frac{21}{-3} = -7$$

(d) $\lim\limits_{x \to 3}(x^2 + 6x + 5) = 32$ and $\lim\limits_{x \to 3}(x^2 - 2x - 3) = 0$; hence $\lim\limits_{x \to 3}\dfrac{x^2 + 6x + 5}{x^2 - 2x - 3}$ does not exist.

(e) $\lim\limits_{x \to -1}(x^2 + 6x + 5) = 0$ and $\lim\limits_{x \to -1}(x^2 - 2x - 3) = 0$. Then, when $x \neq -1$,

$$\frac{x^2 + 6x + 5}{x^2 - 2x - 3} = \frac{(x+5)(x+1)}{(x-3)(x+1)} = \frac{x+5}{x-3} \quad \text{and} \quad \lim\limits_{x \to -1}\frac{x^2 + 6x + 5}{x^2 - 2x - 3} = \lim\limits_{x \to -1}\frac{x+5}{x-3} = \frac{4}{-4} = -1$$

45.3   Tell why each graph in Fig. 45-1 is not continuous at a.

(a) $y = g(x)$ exists at a and $\lim\limits_{x \to a}g(x)$ exists, but $\lim\limits_{x \to a}g(x) \neq g(a)$.

(b) $\lim\limits_{x \to a}h(x)$ does not exist since $\lim\limits_{x \to a^+}h(x) \neq \lim\limits_{x \to a^-}h(x)$.

(c) $f(a)$ is not defused.

45.4   Discuss the continuity of $y = \sqrt{x - 1}$. See Fig. 45-2.

Here, $f(1) = \sqrt{1 - 1} = 0$; $\lim\limits_{x \to 1^+}\sqrt{x - 1} = 0$, thus $f(x)$ is continuous on $[1, \infty]$.

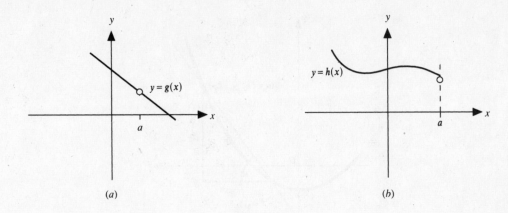

(a)   *(b)*

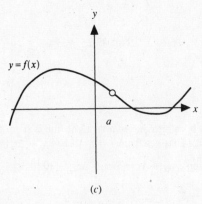

(c)

Fig. 45-1

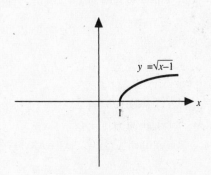

Fig. 45-2

45.5   Let $P(x_0, y_0)$ and $Q(x_0 + \Delta x, y_0 + \Delta y)$ be two distinct points on the parabola $y = x^2 - 3$. Compute $\Delta y / \Delta x$ and interpret.

Here

$$y_0 = x_0^2 - 3$$

$$y_0 + \Delta y = (x_0 + \Delta x)^2 - 3 = x_0^2 + 2x_0 \cdot \Delta x + (\Delta x)^2 - 3$$

$$\Delta y = [x_0^2 + 2x_0 \cdot \Delta x + (\Delta x)^2 - 3] - [x_0^2 - 3] = 2x_0 \cdot \Delta x + (\Delta x)^2$$

and

$$\frac{\Delta y}{\Delta x} = 2x_0 + \Delta x$$

In Fig. 45-3, PR is parallel to the x axis and QR is parallel to the y axis. If α denotes the inclination of the secant line PQ, $\tan \alpha = \Delta y / \Delta x$; thus, $\Delta y / \Delta x$ is the slope of the secant line PQ.

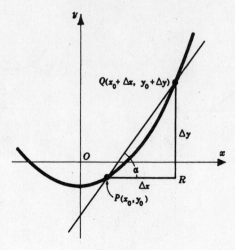

Fig. 45-3

45.6   If $s = 3t^2 + 10$ is the distance a body moving in a straight line is from a fixed point O of the line at time t, (a) find the change Δs is s when t changes from $t = t_0$ to $t = t_0 + \Delta t$, (b) find $\Delta s / \Delta t$ and interpret.

(a)   Here $s_0 = 3t_0^2 + 10$.   Then $s_0 + \Delta s = 3(t_0 + \Delta t)^2 + 10 = 3t_0^2 + 6t_0 \cdot \Delta t + 3(\Delta t)^2 + 10$ and $\Delta s = 6t_0 \cdot \Delta t + 3(\Delta t)^2$.

(b) $\dfrac{\Delta s}{\Delta t} = \dfrac{6t_0 \cdot \Delta t + 3(\Delta t)^2}{\Delta t} = 6t_0 + 3\Delta t$. Since Δs is the distance the body moves in time Δt, $\dfrac{\Delta s}{\Delta t}$ is the average rate of change of distance with respect to time or the average velocity of the body in the interval t_0 to $t_0 + \Delta t$.

45.7   Find

(a) $g'(x)$, given $g(x) = 5$ (c) $k'(x)$, given $k(x) = 4x^2$

(b) $h'(x)$, given $h(x) = 3x$ (d) $f'(x)$, given $f(x) = 4x^2 + 3x + 5$

Thus verify: If $f(x) = k(x) + h(x) + g(x)$, then $f'(x) = k'(x) + h'(x) + g'(x)$.

(a) $y = g(x) = 5$ (b) $y = h(x) = 3x$

$y + \Delta y = g(x + \Delta x) = 5$ $y + \Delta y = 3(x + \Delta x) = 3x + 3\Delta x$

$\Delta y = 0$ $\Delta y = 3\Delta x$

$\dfrac{\Delta y}{\Delta x} = 0$ $\dfrac{\Delta y}{\Delta x} = 3$

$g'(x) = \lim_{\Delta x \to 0} 0 = 0$ $h'(x) = \lim_{\Delta x \to 0} 3 = 3$

(c) $y = k(x) = 4x^2$

$y + \Delta y = 4(x + \Delta x)^2 = 4x^2 + 8x \cdot \Delta x + 4(\Delta x)^2$

$\Delta y = 8x \cdot \Delta x + 4(\Delta x)^2$

$\dfrac{\Delta y}{\Delta x} = 8x + 4\Delta x$

$k'(x) = \lim_{\Delta x \to 0} (8x + 4\Delta x) = 8x$

(d) $y = f(x) = 4x^2 + 3x + 5$

$y + \Delta y = 4(x + \Delta x)^2 + 3(x + \Delta x) + 5 = 4x^2 + 8x \cdot \Delta x + 4(\Delta x)^2 + 3x + 3\Delta x + 5$

$\Delta y = 8x \cdot \Delta x + 3\Delta x + 4(\Delta x)^2$

$\dfrac{\Delta y}{\Delta x} = 8x + 3 + 4\Delta x$

$f'(x) = \lim\limits_{\Delta x \to 0}(8x + 3 + 4\Delta x) = 8x + 3$

Thus $f'(x) = k'(x) + h'(x) + g'(x) = 8x + 3 + 0$.

45.8 Place a straight edge along PQ in Fig. 45-3. Keeping P fixed, let Q move along the curve toward P and thus verify that the straight edge approaches the tangent line PT as limiting position.

Now as Q moves toward $P, \Delta x \to 0$ and $\lim\limits_{\Delta x \to 0} \dfrac{\Delta y}{\Delta x} = \lim\limits_{\Delta x \to 0}(2x_0 + \Delta x) = 2x_0$. Thus the slope of the tangent line to $y = f(x) = x^2 - 3$ at the point $P(x_0, y_0)$ is $m = f'(x_0) = 2x_0$.

45.9 Find the slope and equation of the tangent line to the given curve $y = f(x)$ at the given point:

(a) $y = 2x^3$ at $(1, 2)$,   (b) $y = -3x^2 + 4x + 5$ at $(3, -10)$,   (c) $y = x^2 - 4x + 3$ at $(2, -1)$.

(a) By the five-step rule, $f'(x) = 6x^2$; then the slope $m = f'(1) = 6$. The equation of the tangent line at $(1,2)$ is $y - 2 = 6(x - 1)$ or $6x - y - 4 = 0$.

(b) Here $f'(x) = -6x + 4$ and $m = f'(3) = -14$. The equation of the tangent line at $(3, -10)$ is $14x + y - 32 = 0$.

(c) Here $f'(x) = 2x - 4$ and $m = f'(2) = 0$. The equation of the tangent line at $(2, -1)$ is $y + 1 = 0$. Identify the given point on the parabola.

45.10 Find the equation of the tangent line to the parabola $y^2 = 8x$ at (a) the point $(2, 4)$, (b) the point $(2, -4)$.

Let $P(x, y)$ and $Q(x + \Delta x, y + \Delta y)$ be two nearby points on the parabola. Then

$$y^2 = 8x \tag{1}$$

$$(y + \Delta y)^2 = 8(x + \Delta x) \tag{2}$$

or $y^2 + 2y \cdot \Delta y + (\Delta y)^2 = 8x + 8 \cdot \Delta x$

Subtracting (1) from (2),

$$2y \cdot \Delta y + (\Delta y)^2 = 8 \cdot \Delta x, \quad \Delta y(2y + \Delta y) = 8 \cdot \Delta x, \quad \text{and} \quad \frac{\Delta y}{\Delta x} = \frac{8}{2y + \Delta y}.$$

Now as Q moves along the curve toward $P, \Delta x \to 0$ and $\Delta y \to 0$. Thus

$$m = \lim_{\Delta x \to 0} \frac{\Delta y}{\Delta x} = \lim_{\Delta y \to 0} \frac{8}{2y + \Delta y} = \frac{8}{2y} = \frac{4}{y}$$

(a) At point $(2, 4)$ the slope of the tangent line (also called the slope of the curve) is $m = \frac{4}{4} = 1$ and the equation of the tangent line is $x - y + 2 = 0$.

(b) At point $(2, -4)$ the slope of the tangent line is $m = -1$ and the equation is $x + y + 2 = 0$.

45.11 Find the equation of the tangent line to the ellipse $4x^2 + 9y^2 = 25$ at (a) the point $(2, 1)$, (b) the point $(0, \frac{5}{3})$.

Let $P(x, y)$ and $Q(x + \Delta x, y + \Delta y)$ be two nearby points on the ellipse. (Why should P not be taken at an extremity of the major axis?) Then

$$4x^2 + 9y^2 = 25 \tag{1}$$

$$4x^2 + 8x \cdot \Delta x + 4(\Delta x)^2 + 9y^2 + 18y \cdot \Delta y + 9(\Delta y)^2 = 25 \tag{2}$$

Subtracting (*1*) from (*2*), $8x \cdot \Delta x + 4(\Delta x)^2 + 18y \cdot \Delta y + 9(\Delta y)^2 = 0$. Then

$$\Delta y(18y + 9\Delta y) = -\Delta x(8x + 4 \cdot \Delta x) \quad \text{and} \quad \frac{\Delta y}{\Delta x} = -\frac{8x + 4\Delta y}{18y + 9\Delta y}$$

When Q moves along the curve toward $P, \Delta x \to 0$ and $\Delta y \to 0$. Then $m = \lim\limits_{\Delta x \to 0} \frac{\Delta y}{\Delta x} = -\frac{4x}{9y}$.

(*a*) At point (2, 1), $m = -\frac{8}{9}$ and the equation of the tangent line is $8x + 9y - 25 = 0$.

(*b*) At point $(0, \frac{5}{3})$, $m = 0$ and the equation of the tangent line is $y - \frac{5}{3} = 0$.

45.12 The normal line to a curve at a point P on it is perpendicular to the tangent line at P. Find the equation of the normal line to the given curve at the given point of (*a*) Problem 45.9(*b*), (*b*) Problem 45.11.

(*a*) The slope of the tangent line is -14; the slope of the normal line is $\frac{1}{14}$. The equation of the normal line is $y + 10 = \frac{1}{14}(x - 3)$ or $x - 14y - 143 = 0$.

(*b*) The slope of the tangent line at (2, 1) is $-\frac{8}{9}$; the slope of the normal line is $\frac{9}{8}$. The equation of the normal line is $9x - 8y - 10 = 0$.
At the point $(0, \frac{5}{3})$ the normal line is vertical. Its equation is $x = 0$.

45.13 If $s = f(t)$ is the distance of a body, moving in a straight line, from a fixed point O of the line at time t, then (see Problem 45.6) $\frac{\Delta s}{\Delta t} = \frac{f(t + \Delta t) - f(t)}{\Delta t}$ is the average velocity of the body in the interval of time t to $t + \Delta t$ and

$$v = s' = \lim\limits_{\Delta t \to 0} \frac{\Delta s}{\Delta t} = \lim\limits_{\Delta t \to 0} \frac{f(t + \Delta t) - f(t)}{\Delta t}$$

is the *instantaneous velocity of the body* at time t. For $s = 3t^2 + 10$ of Problem 45.6, find the (instantaneous) velocity of the body at time (*a*) $t = 0$, (*b*) $t = 4$.

Here $v = s' = 6t$. (*a*) When $t = 0, v = 0$. (*b*) When $t = 4, v = 24$.

45.14 The height above the ground of a bullet shot vertically upward with initial velocity of 1152 ft/s is given by $s = 1152t - 16t^2$. Find (*a*) the velocity of the bullet 20 s after was fired and (*b*) the time required for the bullet to reach its maximum height and the maximum height attained.

Here $v = 1152 - 32t$.

(*a*) When $t = 20, v = 1152 - 32(20) = 512$ ft/s.

(*b*) At its maximum height, the velocity of the bullet is 0 ft/s. When $v = 1152 - 32t = 0, t = 36$ s. When $t = 36, s = 1152(36) - 16(36)^2 = 20\,736$ ft, the maximum height.

45.15 Find the derivative of each of the following polynomials:

(*a*) $f(x) = 3x^2 - 6x + 5$,   (*b*) $f(x) = 2x^3 - 8x + 4$,   (*c*) $f(x) = (x - 2)^2(x - 3)^2$.

(*a*) $f'(x) = 3 \cdot 2x^{2-1} - 6x^{1-1} + 0 = 6x - 6$.
(*b*) $f'(x) = 2 \cdot 3x^{3-1} - 8x^{1-1} + 0 = 6x^2 - 8$.
(*c*) Here $f(x) = x^4 - 10x^3 + 37x^2 - 60x + 36$. Then $f'(x) = 4x^3 - 30x^2 + 74x - 60 = (x - 2)(x - 3)(4x - 10)$.

45.16 For each of the following functions, find $f'(x), f''(x)$, and $f'''(x)$:

(a) $f(x) = 2x^2 + 7x - 5$,   (b) $f(x) = x^3 - 6x^2$,   (c) $f(x) = x^5 - x^3 + 3x$.

(a) $f'(x) = 4x + 7$, $f''(x) = 4$, $f'''(x) = 0$
(b) $f'(x) = 3x^2 - 12x$, $f''(x) = 6x - 12$, $f'''(x) = 6$
(c) $f'(x) = 5x^4 - 3x^2 + 3$, $f''(x) = 20x^3 - 6x$, $f'''(x) = 60x^2 - 6$

Supplementary Problems

45.17 Find all (real) values of x for which each of the following is defined:

(a) $x^2 - 3x + 4$ (d) $\dfrac{1}{(x-2)(x+3)}$ (g) $\dfrac{1}{x^2 + 4}$

(b) $\dfrac{1}{x^2}$ (e) $\dfrac{1}{x^2 - 4x + 3}$ (h) $\dfrac{x^2 - 9}{x - 3}$

(c) $\dfrac{1}{x - 2}$ (f) $\dfrac{1}{x^2 - 4}$ (i) $\dfrac{x - 3}{x^2 - 9}$

Ans. (a) all x (c) $x \neq 2$ (e) $x \neq 1, 3$ (g) all x (i) $x \neq \pm 3$

(b) $x \neq 0$ (d) $x \neq 2, -3$ (f) $x \neq \pm 2$ (h) $x \neq 3$

45.18 For each function $f(x)$ of Problem 45.17 evaluate $\lim\limits_{x \to 1} f(x)$, when it exists.

Ans. (a) 2   (b) 1   (c) -1 (d) $-\frac{1}{4}$ (f) $-\frac{1}{3}$ (g) $\frac{1}{5}$ (h) 4   (i) $\frac{1}{4}$

45.19 For each function $f(x)$ of Problem 45.17 evaluate $\lim\limits_{x \to 3} f(x)$, when it exists.

Ans. (a) 4   (b) $\frac{1}{9}$ (c) 1   (d) $\frac{1}{6}$ (f) $\frac{1}{5}$ (g) $\frac{1}{13}$ (h) 6   (i) $\frac{1}{6}$

45.20 Use the five-step rule to obtain $f'(x)$ or $f'(t)$, given   (a) $f(x) = 3x + 5$ (b) $f(x) = x^2 - 3x$
(c) $f(t) = 2t^2 + 8t + 9$ (d) $f(t) = 2t^3 - 12t^2 + 20t + 3$

Ans. (a) 3   (b) $2x - 3$ (c) $4t + 8$ (d) $6t^2 - 24t + 20$

45.21 Find the equation of the tangent and normal to each curve at the given point on it.

(a) $y = x^2 + 2, P(1, 3)$ Ans. $2x - y + 1 = 0, x + 2y - 7 = 0$

(b) $y = 2x^2 - 3x, P(1, -1)$ Ans. $x - y - 2 = 0, x + y = 0$

(c) $y = x^2 - 4x + 5, P(1, 2)$ Ans. $2x + y - 4 = 0, x - 2y + 3 = 0$

(d) $y = x^2 + 3x - 10, P(2, 0)$ Ans. $7x - y - 14 = 0, x + 7y - 2 = 0$

(e) $x^2 + y^2 = 25, P(4, 3)$ Ans. $4x + 3y - 25 = 0, 3x - 4y = 0$

(f) $y^2 = 4x - 8, P(3, -2)$ Ans. $x + y - 1 = 0, x - y - 5 = 0$

(g) $x^2 + 4y^2 = 8, P(-2, -1)$ Ans. $x + 2y + 4 = 0, 2x - y + 3 = 0$

(h) $2x^2 - y^2 = 9, P(-3, 3)$ Ans. $2x + y + 3 = 0, x - 2y + 9 = 0$

45.22 A particle moves along the x axis according to the law $s = 2t^2 + 8t + 9$ [see Problem 45.20(c)], where s (ft) is the directed distance of the particle from the origin O at time t (seconds). Locate the particle and find its velocity when (a) $t = 0$, (b) $t = 1$.

   Ans.   (a)   9 ft to the right of O, $v = 8$ ft/s   (b)   19 ft to the right of O, $v = 12$ ft/s

45.23 A particle moves along the x axis according to the law $s = 2t^3 - 12t^2 + 20t + 3$ [see Problem 45.20(d)], where s is defined as in Problem 45.22.

   (a)   Locate the particle and find its velocity when $t = 2$.   (b)   Locate the particle when $v = 2$ ft/s.

   Ans.   (a)   11 ft to the right of O, $v = -4$ ft/s

            (b) $t = 1$, 13 ft to the right of O; $t = 3$, 9 ft to the right of O

45.24 The height (s m) of a bullet shot vertically upwards is given by $s = 1280t - 16t^2$, with t measured in seconds. (a) What is the initial velocity? (b) For how long will it rise? (c) How high will it rise?

   Ans.   (a)   1280 m/s   (b)   40 s   (c)   25 600 m

45.25 Find the coordinates of the points for which the slope of the tangent to $y = x^3 - 12x + 1$ is 0.

   Ans. $(2, -15)$, $(-2, 17)$

45.26 At what point on $y = \frac{1}{2}x^2 - 2x + 3$ is the tangent perpendicular to that at the point $(1, 0)$?

   Ans. $(3, \frac{3}{2})$

45.27 Show that the equation of the tangent to the conic $Ax^2 + 2Bxy + Cy^2 + 2Dx + 2Ey + F = 0$ at the point $P_1(x_1, y_1)$ on it is given by $Ax_1x + B(x_1y + y_1x) + Cy_1y + D(x_1 + x) + E(y_1 + y) + F = 0$. Use this as a formula to solve Problem 45.21.

45.28 Show that the tangents at the extremities of the latus rectum of the parabola $y^2 = 4px$ (a) are mutually perpendicular and (b) intersect on the directrix.

45.29 Show that the tangent of slope $m \neq 0$ to the parabola $y^2 = 4px$ has equation $y = mx + p/m$.

45.30 Show that the slope of the tangent at either end of either latus rectum of the ellipse $b^2x^2 + a^2y^2 = a^2b^2$ is equal numerically to its eccentricity. Investigate the case of the hyperbola.

Chapter 46

Differentiation of Algebraic Expressions

DIFFERENTIATION FORMULAS

I. If $y = f(x) = kx^n$, where k and n are constants, then $y' = f'(x) = knx^{n-1}$. (See for example, Problem 46.1.)

II. If $y = f(x) \pm g(x)$, then $y' = f'(x) \pm g'(x)$ provided $f'(x)$ and $g'(x)$ exist.

III. If $y = k \cdot u^n$, where k and n are constants and u is a function of x, then $y' = knu^{n-1} \cdot u'$, provided u' exists. This is a form of the chain rule. (For a verification, see Problem 46.2.)

EXAMPLE 1. Find y', given (a) $y = 8x^{5/4}$, (b) $y = (x^2 + 4x - 1)^{3/2}$.

(a) Here $k = 8$, $n = \frac{5}{4}$. Then $y' = knx^{n-1} = 8 \cdot \frac{5}{4}x^{5/4-1} = 10x^{1/4}$.

(b) Let $u = x^2 + 4x - 1$ so that $y = u^{3/2}$. Then differentiating with respect to x, $u' = 2x + 4$ and

$$y' = \tfrac{3}{2}u^{1/2} \cdot u' = \tfrac{3}{2}\sqrt{x^2 + 4x - 1}(2x + 4) = 3(x + 2)\sqrt{x^2 + 4x - 1}$$

(See Problem 46.3.)

IV. If $y = f(x) \cdot g(x)$, then $y' = f(x) \cdot g'(x) + g(x) \cdot f'(x)$, provided $f'(x)$ and $g'(x)$ exist. (For the derivation, see Problem 46.4.)

EXAMPLE 2. Find y' when $y = (x^3 + 3x^2 + 1)(x^2 + 2)$.

Take $f(x) = x^3 + 3x^2 + 1$ and $g(x) = x^2 + 2$. Then $f'(x) = 3x^2 + 6x$, $g'(x) = 2x$, and

$$y' = f(x) \cdot g'(x) + g(x) \cdot f'(x)$$
$$= (x^3 + 3x^2 + 1)(2x) + (x^2 + 2)(3x^2 + 6x)$$
$$= 5x^4 + 12x^3 + 6x^2 + 14x$$

V. If $y = \dfrac{f(x)}{g(x)}$, then $y' = \dfrac{g(x) \cdot f'(x) - f(x) \cdot g'(x)}{[g(x)]^2}$, when $f'(x)$ and $g'(x)$ exist and $g(x) \neq 0$. (For a derivation, see Problem 46.6.)

EXAMPLE 3. Find y', given $y = \dfrac{x+1}{x^2+1}$.

Take $f(x) = x + 1$ and $g(x) = x^2 + 1$. Then

$$y' = \frac{g(x) \cdot f'(x) - f(x) \cdot g'(x)}{[g(x)]^2} = \frac{(x^2+1)(1) - (x+1)(2x)}{(x^2+1)^2} = \frac{1 - 2x - x^2}{(x^2+1)^2}.$$

Solved Problems

46.1 Use the five-step rule to obtain y' when $y = 6x^{3/2}$.

We have
$$y = 6x^{3/2}$$
$$y + \Delta y = 6(x + \Delta x)^{3/2}$$
$$\Delta y = 6(x + \Delta x)^{3/2} - 6x^{3/2} = 6[(x + \Delta x)^{3/2} - x^{3/2}]$$

and
$$\frac{\Delta y}{\Delta x} = 6 \cdot \frac{(x + \Delta x)^{3/2} - x^{3/2}}{\Delta x} = 6 \cdot \frac{(x + \Delta x)^{3/2} - x^{3/2}}{\Delta x} \cdot \frac{(x + \Delta x)^{3/2} + x^{3/2}}{(x + \Delta x)^{3/2} + x^{3/2}}$$

$$= 6 \cdot \frac{(x + \Delta x)^3 - x^3}{\Delta x[(x + \Delta x)^{3/2} + x^{3/2}]} = 6 \cdot \frac{3x^2 + 3x \cdot \Delta x + (\Delta x)^2}{(x + \Delta x)^{3/2} + x^{3/2}}$$

Then
$$y' = \lim_{\Delta x \to 0} 6 \cdot \frac{3x^2 + 3x \cdot \Delta x + (\Delta x)^2}{(x + \Delta x)^{3/2} + x^{3/2}} = 6 \cdot \frac{3x^2}{2x^{3/2}} = 9x^{1/2}$$

(NOTE: By Formula I, with $k = 6$ and $n = \frac{3}{2}$, we find $y' = knx^{n-1} = 6 \cdot \frac{3}{2}x^{1/2} = 9x^{1/2}$.)

46.2 Use the five-step rule to find y' when $y = (x^2 + 4)^{1/2}$. Solve also by using Formula II.

We have
$$y = (x^2 + 4)^{1/2}$$
$$y + \Delta y = [(x + \Delta x)^2 + 4]^{1/2}$$
$$\Delta y = [(x + \Delta x)^2 + 4]^{1/2} - (x^2 + 4)^{1/2}$$

and
$$\frac{\Delta y}{\Delta x} = \frac{[(x + \Delta x)^2 + 4]^{1/2} - (x^2 + 4)^{1/2}}{\Delta x} \cdot \frac{[(x + \Delta x)^2 + 4]^{1/2} + (x^2 + 4)^{1/2}}{[(x + \Delta x)^2 + 4]^{1/2} + (x^2 + 4)^{1/2}}$$

$$= \frac{(x + \Delta x)^2 + 4 - (x^2 + 4)}{\Delta x\{[(x + \Delta x)^2 + 4]^{1/2} + (x^2 + 4)^{1/2}\}} = \frac{2x + \Delta x}{[(x + \Delta x)^2 + 4]^{1/2} + (x^2 + 4)^{1/2}}$$

Then
$$y' = \lim_{\Delta x \to 0} \frac{2x + \Delta x}{[(x + \Delta x)^2 + 4]^{1/2} + (x^2 + 4)^{1/2}} = \frac{2x}{2(x^2 + 4)^{1/2}} = \frac{x}{(x^2 + 4)^{1/2}}$$

Let $u = x^2 + 4$ so that $y = u^{1/2}$. Then $u' = 2x$ and $y' = \dfrac{1}{2}u^{-1/2} \cdot u' = \dfrac{1}{2}(x^2 + 4)^{-1/2} \cdot 2x = \dfrac{x}{\sqrt{x^2 + 4}}$.

46.3 Find y', given (a) $y = (2x - 5)^3$, (b) $y = \frac{2}{3}(x^6 + 4x^3 + 5)^2$, (c) $y = \dfrac{4}{x^2}$, (d) $y = \dfrac{1}{\sqrt{x}}$, (e) $y = 2(3x^2 + 2)^{1/2}$.

(a) Let $u = 2x - 5$ so that $y = u^3$. Then, differentiating with respect to x, $u' = 2$ and $y' = 3u^2 \cdot u' = 3(2x - 5)^2 \cdot 2 = 6(2x - 5)^2$.

(b) Let $u = x^6 + 4x^3 + 5$ so that $y = \frac{2}{3}u^2$. Differentiating with respect to x, $u' = 6x^5 + 12x^2$ and $y' = \frac{4}{3}u \cdot u' = \frac{4}{3}(x^6 + 4x^3 + 5)(6x^5 + 12x^2) = 8(x^6 + 4x^3 + 5)(x^5 + 2x^2)$.

(c) Here $y = 4x^{-2}$ and $y' = 4(-2)x^{-3} = -8/x^3$.

(d) Here $y = x^{-1/2}$ and $y' = (-\frac{1}{2})x^{-3/2} = -1/(2x\sqrt{x})$.

(e) Since $y = 2(3x^2 + 2)^{1/2}$, $y' = 2(\frac{1}{2})(3x^2 + 2)^{-1/2}(6x) = 6x/(3x^2 + 2)^{1/2}$.

46.4 Derive: If $y = f(x) \cdot g(x)$, then $y' = f(x) \cdot g'(x) + g(x) \cdot f'(x)$, provided $f'(x)$ and $g'(x)$ exist.

Let $u = f(x)$ and $v = g(x)$ so that $y = u \cdot v$.

As x changes to $x + \Delta x$, let u change to $u + \Delta u$, v change to $v + \Delta v$, and y change to $y + \Delta y$. Then

$$y + \Delta y = (u + \Delta u)(v + \Delta v) = uv + u \cdot \Delta v + v \cdot \Delta u + \Delta u \cdot \Delta v$$

$$\Delta y = u \cdot \Delta v + v \cdot \Delta u + \Delta u \cdot \Delta v$$

and

$$\frac{\Delta y}{\Delta x} = u \cdot \frac{\Delta v}{\Delta x} + v \cdot \frac{\Delta u}{\Delta x} + \Delta u \cdot \frac{\Delta v}{\Delta x}$$

Then

$$y' = \lim_{\Delta x \to 0} \left(u \cdot \frac{\Delta v}{\Delta x} + v \cdot \frac{\Delta u}{\Delta x} + \Delta u \cdot \frac{\Delta v}{\Delta x} \right)$$

$$= u \cdot v' + v \cdot u' + 0 \cdot v' = u \cdot v' + v \cdot u' = f(x) \cdot g'(x) + g(x) \cdot f'(x)$$

46.5 Find y', given (a) $y = x^5(1 - x^2)^4$, (b) $y = x^2\sqrt{x^2 + 4}$, (c) $y = (3x + 1)^2(2x^3 - 3)^{1/3}$.

(a) Set $f(x) = x^5$ and $g(x) = (1 - x^2)^4$. Then $f'(x) = 5x^4$, $g'(x) = 4(1 - x^2)^3(-2x)$, and

$$y' = f(x) \cdot g'(x) + g(x) \cdot f'(x) = x^5 \cdot 4(1 - x^2)^3(-2x) + (1 - x^2)^4 \cdot 5x^4$$

$$= x^4(1 - x^2)^3[-8x^2 + 5(1 - x^2)] = x^4(1 - x^2)^3(5 - 13x^2)$$

(b)

$$y = x^2 \cdot \tfrac{1}{2}(x^2 + 4)^{-1/2} \cdot 2x + (x^2 + 4)^{1/2} \cdot 2x$$

$$= x^3(x^2 + 4)^{-1/2} + 2x(x^2 + 4)^{1/2} = \frac{x^3 + 2x(x^2 + 4)}{(x^2 + 4)^{1/2}} = \frac{3x^3 + 8x}{\sqrt{x^2 + 4}}$$

(c) Here $y = (3x + 1)^2(2x^3 - 3)^{1/3}$ and

$$y' = (3x + 1)^2 \cdot \tfrac{1}{3}(2x^3 - 3)^{-2/3} \cdot 6x^2 + (2x^3 - 3)^{1/3} \cdot 2(3x + 1) \cdot 3$$

$$= 2x^2(3x + 1)^2(2x^3 - 3)^{-2/3} + 6(3x + 1)(2x^3 - 3)^{1/3} = \frac{2x^2(3x + 1)^2 + 6(3x + 1)(2x^3 - 3)}{(2x^2 - 3)^{2/3}}$$

$$= \frac{2(3x + 1)[x^2(3x + 1) + 3(2x^3 - 3)]}{(2x^3 - 3)^{2/3}} = \frac{2(3x + 1)(9x^3 + x^2 - 9)}{(2x^3 - 3)^{2/3}}$$

46.6 Prove: If $y = \dfrac{f(x)}{g(x)}$, if $f'(x)$ and $g'(x)$ exist, and if $g(x) \neq 0$, then $y' = \dfrac{g(x) \cdot f'(x) - f(x) \cdot g'(x)}{[g(x)]^2}$.

Let $u = f(x)$ and $v = g(x)$ so that $y = u/v$. Then

$$y + \Delta y = \frac{u + \Delta u}{v + \Delta v} \qquad \Delta y = \frac{u + \Delta u}{v + \Delta v} - \frac{u}{v} = \frac{v \cdot \Delta u - u \cdot \Delta v}{v(v + \Delta v)}$$

and

$$\frac{\Delta y}{\Delta x} = \frac{v \cdot \Delta u - u \cdot \Delta v}{\Delta x \cdot v(v + \Delta v)} = \frac{v \cdot \dfrac{\Delta u}{\Delta x} - u \cdot \dfrac{\Delta v}{\Delta x}}{v(v + \Delta v)}$$

Then

$$y' = \lim_{\Delta x \to 0} \frac{v \cdot \dfrac{\Delta u}{\Delta x} - u \cdot \dfrac{\Delta v}{\Delta x}}{v(v + \Delta v)} = \frac{v \cdot u' - u \cdot v'}{v^2} = \frac{g(x) \cdot f'(x) - f(x) \cdot g'(x)}{[g(x)]^2}$$

46.7 Find y', given (a) $y = 1/x^3$, (b) $y = \dfrac{2x}{x - 3}$, (c) $y = \dfrac{x + 5}{x^2 - 1}$, (d) $y = \dfrac{x^3}{\sqrt{4 - x^2}}$, (e) $y = \dfrac{\sqrt{2x - 3x^2}}{x + 1}$.

(a) Take $f(x) = 1$ and $g(x) = x^3$; then

$$y' = \frac{g(x) \cdot f'(x) - f(x) \cdot g'(x)}{[g(x)]^2} = \frac{x^3 \cdot 0 - 1 \cdot 3x^2}{(x^3)^2} = -\frac{3}{x^4}$$

Note that it is simpler here to write $y = x^{-3}$ and $y' = -3x^{-4} = -3/x^4$.

(b) Take $f(x) = 2x$ and $g(x) = x - 3$; then

$$y' = \frac{g(x) \cdot f'(x) - f(x) \cdot g'(x)}{[g(x)]^2} = \frac{(x-3) \cdot 2 - 2x \cdot 1}{(x-3)^2} = \frac{-6}{(x-3)^2}$$

Note that $y = \dfrac{2x}{x-3} = 2 + \dfrac{6}{x-3} = 2 + 6(x-3)^{-1}$ and $y' = 6(-1)(x-3)^{-2} = -6/(x-3)^2$.

(c)
$$y' = \frac{(x^2-1)(1) - (x+5)(2x)}{(x^2-1)^2} = -\frac{1 + 10x + x^2}{(x^2-1)^2}$$

(d)
$$y' = \frac{(4-x^2)^{1/2} \cdot 3x^2 - x^3 \cdot \frac{1}{2}(4-x^2)^{-1/2}(-2x)}{4-x^2} = \frac{(4-x^2)(3x^2) + x^3 \cdot x}{(4-x^2)^{3/2}} = \frac{12x^2 - 2x^4}{(4-x^2)^{3/2}}.$$

The derivative exists for $-2 < x < 2$.

(e)
$$y' = \frac{(x+1) \cdot \frac{1}{2}(2x - 3x^2)^{-1/2}(2 - 6x) - (2x - 3x^2)^{1/2}(1)}{(x+1)^2}$$

$$= \frac{(x+1)(1-3x) - (2x - 3x^2)}{(x+1)^2(2x - 3x^2)^{1/2}} = \frac{1 - 4x}{(x+1)^2(2x - 3x^2)^{1/2}}$$

The derivative exists for $0 < x < \frac{2}{3}$.

46.8   Find y', y'', y''' given (a) $y = 1/x$; (b) $y = \dfrac{1}{x-1} + \dfrac{1}{x+1}$; (c) $y = \dfrac{2}{x^2-1}$.

(a) Here $y = x^{-1}$; then $y' = -1 \cdot x^{-2} = -x^{-2}$, $y'' = 2x^{-3}$, $y''' = -6x^{-4}$ or $y' = -1/x^2$, $y'' = 2/x^3$, $y''' = -6/x^4$.

(b) Here $y = (x-1)^{-1} + (x+1)^{-1}$; then

$$y' = -1(x-1)^{-2} + (-1)(x+1)^{-2} = -\frac{1}{(x-1)^2} - \frac{1}{(x+1)^2}$$

$$y'' = 2(x-1)^{-3} + 2(x+1)^{-3} = \frac{2}{(x-1)^3} + \frac{2}{(x+1)^3}$$

$$y''' = -6(x-1)^{-4} - 6(x+1)^{-4} = -\frac{6}{(x-1)^4} - \frac{6}{(x+1)^4}$$

(c) Here $y = 2(x^2 - 1)^{-1}$; then

$$y' = 2(-1)(x^2-1)^{-2} \cdot (2x) = -4x(x^2-1)^{-2} = \frac{-4x}{(x^2-1)^2}$$

$$y'' = -4(x^2-1)^{-2} - 4x(-2)(x^2-1)^{-3}(2x) = -4(x^2-1)^{-2} + 16x^2(x^2-1)^{-3} = \frac{12x^2 + 4}{(x^2-1)^3}$$

$$y''' = \frac{(x^2-1)^3(24x) - (12x^2+4) \cdot 3(x^2-1)^2(2x)}{(x^2-1)^6} = \frac{(x^2-1)(24x) - 6x(12x^2+4)}{(x^2-1)^4} = \frac{-48x(x^2+1)}{(x^2-1)^4}$$

Supplementary Problems

46.9   Use the differentiation formulas to find y', given

(a) $y = 2x^3 + 4x^2 - 5x + 8$ *Ans.* $y' = 6x^2 + 8x - 5$

(b) $y = -5 + 3x - \frac{3}{2}x^2 - 7x^3$ *Ans.* $y' = 3 - 3x - 21x^2$

(c) $y = (x-2)^4$ *Ans.* $y' = 4(x-2)^3$

(d) $y = (x^2 + 2)^3$ *Ans.* $y' = 6x(x^2 + 2)^2$

(e) $y = (4 - x^2)^{10}$ *Ans.* $y' = -20x(4 - x^2)^9$

(f) $y = (2x^2 + 4x - 5)^6$ *Ans.* $y' = 24(x+1)(2x^2 + 4x - 5)^5$

(g) $y = \frac{1}{5}x^{5/2} + \frac{1}{3}x^{3/2}$ *Ans.* $y' = \frac{1}{2}x^{1/2}(x+1)$

(h) $y = (x^2 - 4)^{3/2}$ *Ans.* $y' = 3x(x^2 - 4)^{1/2}$

(i) $y = (1 - x^2)^{1/2}$ *Ans.* $y' = -\dfrac{x}{(1 - x^2)^{1/2}}$

(j) $y = \dfrac{6}{x} + \dfrac{4}{x^2} - \dfrac{3}{x^3}$ *Ans.* $y' = -\dfrac{6}{x^2} - \dfrac{8}{x^3} + \dfrac{9}{x^4}$

(k) $y = x^3(x+1)^2$ *Ans.* $y' = x^2(x+1)(5x+3)$

(l) $y = (x+1)^3(x-3)^2$ *Ans.* $y' = (x+1)^2(x-3)(5x-7)$

(m) $y = (x+2)^2(2-x)^3$ *Ans.* $y' = -(x+2)(2-x)^2(5x-2)$

(n) $y = \dfrac{x+1}{x-1}$ *Ans.* $y' = -\dfrac{2}{(x-1)^2}$

(o) $y = \dfrac{x^2 + 2x - 3}{x^2}$ *Ans.* $y' = \dfrac{6 - 2x}{x^3}$

(p) $y = \dfrac{x^2 + 1}{x^2 + 2}$ *Ans.* $y' = \dfrac{2x}{(x^2 + 2)^2}$

(q) $y = \dfrac{1}{(2x+1)^3}$ *Ans.* $y' = -\dfrac{6}{(2x+1)^4}$

(r) $y = \dfrac{1}{(x^2 - 9)^{1/2}}$ *Ans.* $y' = -\dfrac{x}{(x^2 - 9)^{3/2}}$

(s) $y = \dfrac{1}{(16 - x^2)^{1/2}}$ *Ans.* $y' = \dfrac{x}{(16 - x^2)^{3/2}}$

(t) $y = \dfrac{x}{(x+1)^{1/2}}$ *Ans.* $y' = \dfrac{x+2}{2(x+1)^{3/2}}$

(u) $y = \dfrac{(x^2 + 2)^{1/2}}{x}$ *Ans.* $y' = \dfrac{-2}{x^2(x^2 + 2)^{1/2}}$

46.10 For each of the following, find $f'(x)$, $f''(x)$, and $f'''(x)$:

(a) $f(x) = 3x^4 - 8x^3 + 12x^2 + 5$ (c) $f(x) = \dfrac{1}{4 - x}$

(b) $f(x) = x^3 - 6x^2 + 9x + 18$ (d) $f(x) = (1 - x^2)^{3/2}$

Ans. (a) $f'(x) = 12x(x^2 - 2x + 2)$, $f''(x) = 12(3x^2 - 4x + 2)$, $f'''(x) = 24(3x - 2)$

(b) $f'(x) = 3(x^2 - 4x + 3)$, $f''(x) = 6(x - 2)$, $f'''(x) = 6$

(c) $f'(x) = \dfrac{1}{(4 - x)^2}$, $f''(x) = \dfrac{2}{(4 - x)^3}$, $f'''(x) = \dfrac{6}{(4 - x)^4}$

(d) $f'(x) = -3x(1 - x^2)^{1/2}$, $f''(x) = \dfrac{3(2x^2 - 1)}{(1 - x^2)^{1/2}}$, $f'''(x) = \dfrac{3x(3 - 2x^2)}{(1 - x^2)^{3/2}}$

46.11 In each of the following state the values of x for which $f(x)$ is continuous; also find $f'(x)$ and state the values of x for which it is defined.

(a) $f(x) = \dfrac{1}{x^2}$ (b) $f(x) = \dfrac{1}{x - 2}$ (c) $f(x) = (x - 2)^{4/3}$ (d) $f(x) = (x - 2)^{1/3}$

Ans. (a) $x \neq 0$; $f'(x) = \dfrac{-2}{x^3}$, $x \neq 0$ (c) all x; $f'(x) = \frac{4}{3}(x - 2)^{1/3}$; all x

(b) $x \neq 2$; $f'(x) = \dfrac{-1}{(x - 2)^2}$, $x \neq 2$ (d) all x; $f'(x) = \dfrac{1}{3(x - 2)^{2/3}}$, $x \neq 2$

[NOTE: Parts (a) and (b) verify: If $f(x)$ is not continuous at $x = x_0$, then $f'(x)$ does not exist at $x = x_0$. Parts (c) and (d) verify: If $f(x)$ is continuous at $x = x_0$, its derivative $f'(x)$ may or may not exist at $x = x_0$.]

Chapter 47

Applications of Derivatives

INCREASING AND DECREASING FUNCTIONS.   A function $y = f(x)$ is said to be an *increasing function* if y *increases* as x increases, and a *decreasing function* if y *decreases* as x increases.

Let the graph of $y = f(x)$ be as shown in Fig. 47-1. Clearly $y = f(x)$ is an increasing function from A to B and from C to D, and is a decreasing function from B to C and from D to E. At any point of the curve between A and B (also, between C and D), the inclination θ of the tangent line to the curve is acute; hence, $f'(x) = \tan \theta > 0$. At any point of the curve between B and C (also, between D and E), the inclination θ of the tangent line is obtuse; hence, $f'(x) = \tan \theta < 0$.

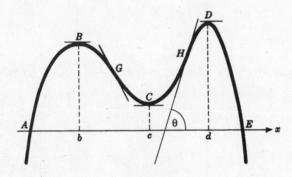

Fig. 47-1

Thus, for values of x for which $f'(x) > 0$, the function $f(x)$ is an increasing function; for values of x for which $f'(x) < 0$, the function is a decreasing function.

When $x = b, x = c$, and $x = d$, the function is neither increasing nor decreasing since $f'(x) = 0$. Such values of x are called *critical values* for the function $f(x)$.

EXAMPLE 1.   For the function $f(x) = x^2 - 6x + 8, f'(x) = 2x - 6$.

Setting $f'(x) = 0$, we find the critical value $x = 3$. Now $f'(x) < 0$ when $x < 3$, and $f'(x) > 0$ when $x > 3$. Thus, $f(x) = x^2 - 6x + 8$ is a decreasing function when $x < 3$ and an increasing function when $x > 3$. (See Problem 47.1.)

RELATIVE MAXIMUM AND MINIMUM VALUES. Let the curve of Fig. 47-1 be traced from left to right. Leaving A, the tracing point rises to B and then begins to fall. At B the ordinate $f(b)$ is greater than at any point of the curve near to B. We say that the point $B(b, f(b))$ is a *relative maximum point* of the curve or that the function $f(x)$ has a *relative maximum* $[= f(b)]$ when $x = b$. By the same argument $D(d, f(d))$ is also a relative maximum point of the curve or $f(x)$ has a relative maximum $[= f(d)]$ when $x = d$.

Leaving B, the tracing point falls to C and then begins to rise. At C the ordinate $f(c)$ is smaller than at any point of the curve near to C. We say that the point $C(c, f(c))$ is a *relative minimum point* of the curve or that $f(x)$ has a *relative minimum* $[= f(c)]$ when $x = c$.

Note that the relative maximum and minimum of this function occur at the critical values. While not true for all functions, the above statement is true for all of the functions considered in this chapter.

Test for relative maximum. If $x = a$ is a critical value for $y = f(x)$ and if $f'(x) > 0$ for all values of x less than but near to $x = a$ while $f'(x) < 0$ for all values of x greater than but near to $x = a$, then $f(a)$ is a relative maximum value of the function.

Test for relative minimum. If $x = a$ is a critical value for $y = f(x)$ and if $f'(x) < 0$ for all values of x less than but near to $x = a$ while $f'(x) > 0$ for all values of x greater than but near to $x = a$, then $f(a)$ is a relative minimum value of the function.

If as x increase, in value through a critical value, $x = a, f'(x)$ does not change sign, then $f(a)$ is neither a relative maximum nor a relative minimum value of the function.

EXAMPLE 2.   For the function of Example 1, the critical value is $x = 3$.

Since $f'(x) = 2(x - 3) < 0$ for $x < 3$ and $f'(x) > 0$ for $x > 3$, the given function has a relative minimum value $f(3) = -1$.

In geometric terms, the point $(3, -1)$ is a relative minimum point of the curve $y = x^2 - 6x + 8$. (See Problem 47.2.)

ANOTHER TEST FOR MAXIMUM AND MINIMUM VALUES.   At A on the curve of Fig. 47-1, the inclination θ of the tangent line is acute. As the tracing point moves from A to B, θ decreases; thus $f'(x) = \tan \theta$ is a decreasing function. At B, $f'(x) = 0$. As the tracing point moves from B to G, θ is obtuse and decreasing; thus $f'(x) = \tan \theta$ is a decreasing function. Hence, from A to G, $f'(x)$ is a decreasing function and its derivative $f''(x) < 0$. In particular, $f''(b) < 0$. Similarly, $f''(d) < 0$.

As the tracing point moves from G to C, θ is obtuse and increasing; thus $f'(x)$ is an increasing function. At $C, f'(x) = 0$. As the tracing point moves from C to H, θ is acute and increasing; thus $f'(x)$ is an increasing function. Hence, from G to $H, f'(x)$ is an increasing function and $f''(x) > 0$. In particular, $f''(c) > 0$.

Test for relative maximum.   If $x = a$ is a critical value for $y = f(x)$ and if $f''(a) < 0$, then $f(a)$ is a relative maximum value of the function $f(x)$.

Test for relative minimum.   If $x = a$ is a critical value for $y = f(x)$ and if $f''(a) > 0$, then $f(a)$ is a relative minimum value of the function $f(x)$.

The test fails when $f''(a) = 0$. When this occurs, the tests of the preceding section must be used. (See Problem 47.3.)

CONCAVITY.   Suppose that $f(x)$ is a differentiable function on (a, b). Then, if $f'(x)$ is increasing on (a, b), we call f concave upward on (a, b). See Fig. 47-2(a). If $f'(x)$ is decreasing on (a, b), we say f is concave downward on (a, b). See Fig. 47-2(b).

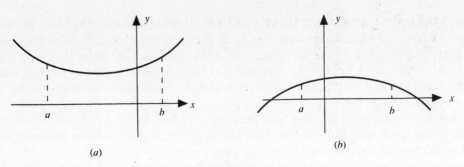

Fig. 47-2

INFLECTION POINT OF A CURVE. If at $x = a$, not necessarily a critical value for $f(x)$, the concavity changes from downward to upward or upward to downward, $(a, f(a))$ is an inflection point of $f(x)$. See Fig. 47-3.

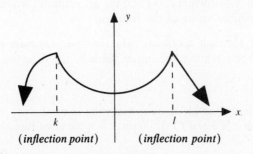

Fig. 47-3

In Fig. 47-1, G and H are inflection points of the curve. Note that at points between A and G the tangent lines to the curve lie above the curve, at points between G and H the tangent lines lie below the curve, and at points between H and E the tangent lines line above the curve. At G and H, the points of inflection, the tangent line *crosses* the curve. Thus, $f''(x)$ must be zero at an inflection point and change sign there.

EXAMPLE 3.   For the function $f(x) = x^2 - 6x + 8$ of Example 1, $f'(x) = 2x - 6$ and $f''(x) = 2$. At the critical value $x = 3$, $f''(x) > 0$; hence $f(3) = -1$ is a relative minimum value. Since $f''(x) = 2 \neq 0$, the parabola $y = x^2 - 6x + 8$ has no inflection point. (See Problem 47.4.)

EXAMPLE 4.   For the function $f(x) = x^3 + x^2 + x, f''(x) = 6x + 2 = 0$ when $x = -\frac{1}{3}$. Since $f'''(-\frac{1}{3}) \neq 0$, concavity must be changing when $x = -\frac{1}{3}$.

VELOCITY AND ACCELERATION.   Let a particle move along a horizontal line and let its distance (in feet) at time $t \geq 0$ (in seconds) from a fixed point O of the line be given by $s = f(t)$. Let the positive direction on the line be to the right (that is, the direction of increasing s). A complete description of the motion may be obtained by examining $f(t), f'(t)$, and $f''(t)$. It was noted in Chapter 45 that $f'(t)$ gives the velocity v of the particle. The *acceleration* of the particle is given by a $a = f''(t)$.

EXAMPLE 5.   Discuss the motion of a particle which moves along a horizontal line according to the equation $s = t^3 - 6t^2 + 9t - 2$.

When $t = 0$, $s = f(0) = 2$. The particle begins its motion from $A(s = 2)$ See Fig. 47-4.

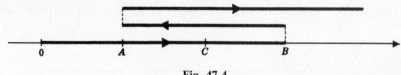

Fig. 47-4

Direction of Motion. Here $v = f'(t) = 3t^2 - 12t + 9 = 3(t-1)(t-3)$. When $t = 0, v = f'(0) = 9$); the particle leaves A with initial velocity 9 ft/s.

Now $v = 0$ when $t = 1$ and $t = 3$. Thus, the particle moves (from A) for 1 s, stops momentarily ($v = 0$, when $t = 1$), moves off for two more seconds, stops momentarily, and then moves off indefinitely.

On the interval $0 < t < 1, v > 0$. Now $v > 0$ indicates that s is increasing; thus the body leaves A with initial velocity 9 ft/s and moves to the right for 1 s to B [$s = f(1) = 6$] where it stops momentarily.

On the interval $1 < t < 3, v < 0$. Now $v < 0$ indicates that s is decreasing; thus the particle leaves B and moves to the left for 2 s to $A[s = f(3) = 2]$ where it stops momentarily.

On the interval $t > 3, v > 0$. The particle leaves A for the second time and moves to the right indefinitely.

Velocity and Speed. We have $a = f''(t) = 6t - 12 = 6(t-2)$. The acceleration is 0 when $t = 2$.

On the interval $0 < t < 2, a < 0$. Now $a < 0$ indicates that v is decreasing; thus the particle moves for the first 2 s with decreasing velocity. For the first second (from A to B) the velocity decreases from $v = 9$ to $v = 0$. The speed (numerical value of the velocity) decreases from 9 to 0; that is, the particle "slows up." When $t = 2$, $f(t) = 4$ (the particle is at C) and $f'(t) = -3$. Thus from B to $C(t = 1$ to $t = 2)$, the velocity decreases from $v = 0$ to $v = -3$. On the other hand, the speed increases from 0 to 3; that is, the particle "speeds up."

On the interval $t > 2, a > 0$; thus the velocity is increasing. From C to A ($t = 2$ to $t = 3$) the velocity increases from $v = -3$ to $v = 0$ while the speed decreases from 3 to 0. Thereafter ($t > 3$) both the velocity and speed increase indefinitely. (See Problem 47.9.)

DIFFERENTIALS. Let $y = f(x)$. Define dx (read, differential x) by the relation $dx = \Delta x$ and define dy (read, differential y) by the relation $dy = f'(x) \cdot dx$. Note $dy \neq \Delta y$.

EXAMPLE 6. If $y = f(x) = x^3$, then

$$\Delta y = (x + \Delta x)^3 - x^3 = 3x^2 \cdot \Delta x + 3x(\Delta x)^2 + (\Delta x)^3 = 3x^2 \cdot dx + 3x(dx)^2 + (dx)^3$$

while $dy = f'(x) \cdot dx = 3x^2 \cdot dx$. Thus, if dx is small numerically, dy is a fairly good approximation of Δy and simple to compute.

Suppose now that $x = 10$ and $dx = \Delta x = .01$. Then for the function above, $\Delta y = 3(10)^2(.01) + 3(10)(.01)^2 + (.01)^3 = 3.0031$ while $dy = 3(10)^2(.01) = 3$.

Solved Problems

47.1 Determine the intervals on which each of the following is an increasing function and the intervals on which it is a decreasing function:

(a) $f(x) = x^2 - 8x$

(b) $f(x) = 2x^3 - 24x + 5$

(c) $f(x) = x^3 + 3x^2 + 9x + 5$

(d) $f(x) = x^3 + 3x$

(e) $f(x) = (x-2)^3$

(f) $f(x) = (x-1)^3(x-2)$

(a) Here $f'(x) = 2(x - 4)$. Setting this equal to 0 and solving, we find the critical value to be $x = 4$. We locate the point $x = 4$ on the x axis and find that $f'(x) < 0$ for $x < 4$, and $f'(x) > 0$ when $x > 4$. See Fig. 47-5. Thus, $f(x) = x^2 - 8x$ is an increasing function when $x > 4$, and is a decreasing function when $x < 4$.

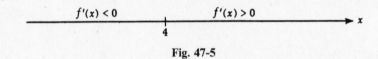

Fig. 47-5

(b) $f'(x) = 6x^2 - 24 = 6(x + 2)(x - 2)$; the critical values are $x = -2$ and $x = 2$. Locating these points and determining the sign of $f'(x)$ on each of the intervals $x < -2, -1 < x < 2$, and $x > 2$ (see Fig. 47-6), we find that $f(x) = 2x^3 - 24x + 5$ is an increasing function on the intervals $x < -2$ and $x > 2$, and is a decreasing function on the interval $-2 < x < 2$.

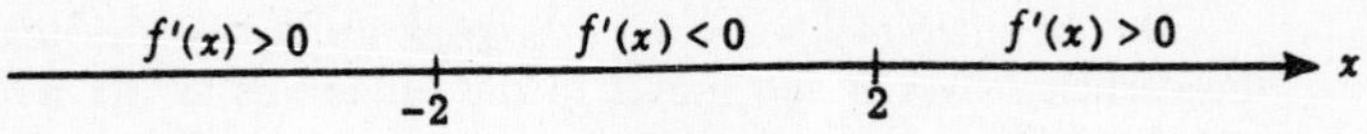

$$\underset{-2}{\underbrace{\qquad f'(x) > 0 \qquad}} \quad \underset{2}{\underbrace{\qquad f'(x) < 0 \qquad}} \quad \underbrace{\qquad f'(x) > 0 \qquad} \longrightarrow x$$

Fig. 47-6

(c) $f'(x) = -3x^2 + 6x + 9 = -3(x + 1)(x - 3)$; the critical values are $x = -1$ and $x = 3$. See Fig. 47-7. Then $f(x)$ is an increasing function on the interval $-1 < x < 3$, and a decreasing function on the intervals $x < -1$ and $x > 3$.

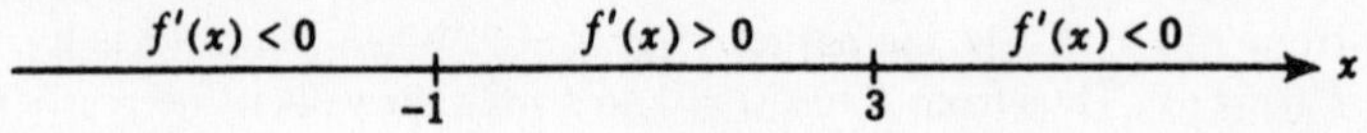

$$\underset{-1}{\underbrace{\qquad f'(x) < 0 \qquad}} \quad \underset{3}{\underbrace{\qquad f'(x) > 0 \qquad}} \quad \underbrace{\qquad f'(x) < 0 \qquad} \longrightarrow x$$

Fig. 47-7

(d) $f'(x) = 3x^2 + 3 = 3(x^2 + 1)$; there are no critical values. Since $f'(x) > 0$ for all values of x, $f(x)$ is everywhere an increasing function.

(e) $f'(x) = 3(x - 2)^2$; the critical value is $x = 2$. See Fig. 47-8. Then $f(x)$ is an increasing function on the intervals $x < 2$ and $x > 2$.

$$\underset{2}{\underbrace{\qquad f'(x) > 0 \qquad}} \quad \underbrace{\qquad f'(x) > 0 \qquad} \longrightarrow x$$

Fig. 47-8

(f) $f'(x) = (x - 1)^2(4x - 7)$; the critical values are $x = 1$ and $x = \frac{7}{4}$. See Fig. 47-9. Then $f(x)$ is an increasing function on the interval $x > \frac{7}{4}$ and is a decreasing function on the intervals $x < 1$ and $1 < x < \frac{7}{4}$.

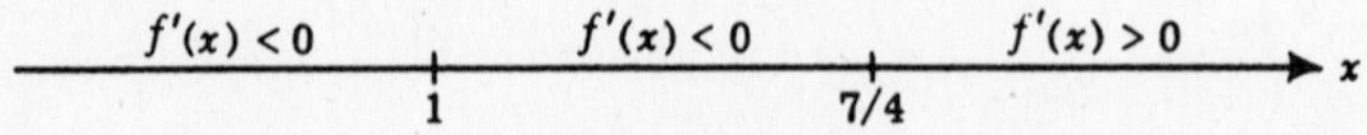

$$\underset{1}{\underbrace{\qquad f'(x) < 0 \qquad}} \quad \underset{7/4}{\underbrace{\qquad f'(x) < 0 \qquad}} \quad \underbrace{\qquad f'(x) > 0 \qquad} \longrightarrow x$$

Fig. 47-9

47.2   Find the relative maximum and minimum values of the functions of Problem 47.1

(a) The critical value is $x = 4$. Since $f'(x) < 0$ for $x < 4$ and $f'(x) > 0$ for $x > 4$, the function has a relative minimum vale $f(4) = -16$.

(b) The critical values are $x = -2$ and $x = 2$. Since $f'(x) > 0$ for $x < -2$ and $f'(x) < 0$ for $-2 < x < 2$, the function has a relative maximum value $f(-2) = 37$. Since $f'(x) < 0$ for $-2 < x < 2$ and $f'(x) > 0$ for $x > 2$, the function has a relative minimum value $f(2) = -27$.

(c) The critical values are $x = -1$ and $x = 3$. Since $f'(x) < 0$ for $x < -1$ and $f'(x) > 0$ for $-1 < x < 3$, $f(x)$ has a relative minimum value $f(-1) = 0$. Since $f'(x) > 0$ for $-1 < x < 3$ and $f'(x) < 0$ for $x > 3$, the function has a relative maximum value $f(3) = 32$.

(d) The function has neither a relative maximum nor a relative minimum value.

(e) The critical value is $x = 2$. Since $f'(x) > 0$ for $x < 2$ and $f'(x) > 0$ for $x > 2$, the function has neither a relative maximum nor minimum value.

(f) The critical values are $x = 1$ and $x = \frac{7}{4}$. The function has a relative minimum value $f(\frac{7}{4}) = -\frac{27}{256}$. The critical value $x = 1$ yields neither a relative maximum nor minimum value.

47.3 Find the relative maximum and minimum values of the functions of Problem 47.1, using the second derivative test.

(a) $f(x) = x^2 - 8x, f'(x) = 2x - 8, f''(x) = 2$.
   The critical value is $x = 4$. Since $f''(4) = 2 \neq 0, f(4) = -16$ is a relative minimum value of the function.

(b) $f(x) = 2x^3 - 24x + 5, f'(x) = 6x^2 - 24, f''(x) = 12x$.
   The critical values are $x = -2$ and $x = 2$. Since $f''(-2) = -24 < 0, f(-2) = 37$ is a relative maximum value of the function; since $f''(2) = 24 > 0, f(2) = -27$ is a relative minimum value.

(c) $f(x) = -x^3 + 3x^2 + 9x + 5, f'(x) = -3x^2 + 6x + 9, f''(x) = -6x + 6$.
   The critical values are $x = -1$ and $x = 3$. Since $f''(-1) > 0, f(-1) = 0$ is a relative minimum value of the function; since $f''(3) < 0, f(3) = 32$ is a relative maximum value.

(d) $f(x) = x^3 + 3x, f'(x) = 3x^2 + 3, f''(x) = 6x$.
   There are no critical values; hence the function has neither a relative minimum nor a relative maximum value.

(e) $f(x) = (x - 2)^3, f'(x) = 3(x - 2)^2, f''(x) = 6(x - 2)$.
   The critical value is $x = 2$. Since $f''(2) = 0$, the test fails. The test of Problem 47.2 shows that the function has neither a relative maximum nor a relative minimum value.

(f) $f(x) = (x - 1)^3(x - 2), f'(x) = (x - 1)^2(4x - 7), f''(x) = 6(2x - 3)(x - 1)$.
   The critical value are $x = 1$ and $x = \frac{7}{4}$. Since $f''(1) = 0$, the test fails; the test of Problem 47.2 shows that $f(1)$ is neither a relative maximum nor a relative minimum value of the function. Since $f''(\frac{7}{4}) > 0$, $f(\frac{7}{4}) = -\frac{27}{256}$ is a relative minimum value.

47.4 Find the inflection points and plot the graph of each of the given curves. In sketching the graph, locate the x and y intercepts when they can be found, the relative maximum and minimum points (see Problem 47.2), and the inflection points, if any. Additional points may be found if necessary.

(a) $y = f(x) = x^2 - 8x$ (c) $y = f(x) = -x^3 + 3x^2 + 9x + 5$ (e) $y = f(x) = (x - 2)^3$
(b) $y = f(x) = 2x^3 - 24x + 5$ (d) $y = f(x) = x^3 + 3x$ (f) $y = f(x) = (x - 1)^3(x - 2)$

(a) Since $f''(x) = 2$, the parabola does not have an inflection point. It is always concave upward.
   The x and y intercepts are $x = 0, x = 8$, and $y = 0; (4, -16)$ is a relative minimum point. See Fig. 47-10(a).

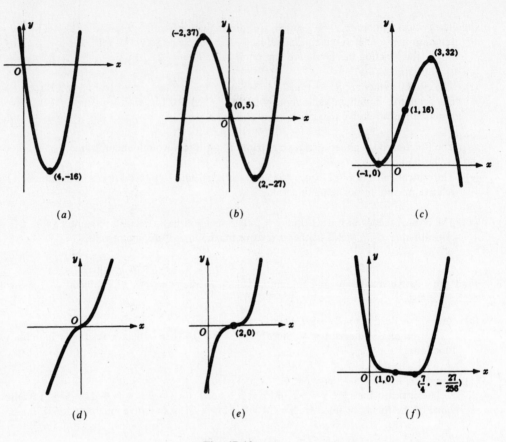

Fig. 47-10

(b) $f''(x) = 12x$ and $f'''(x) = 12$. Since $f''(x) = 0$ when $x = 0$ and $f'''(0) = 12 \neq 0, (0, 5)$ is an inflection point. Notice the change in concavity.

They y intercept is $y = 5$, the x intercepts cannot be determined; $(-2, 37)$ is a relative maximum point, $(2, -27)$ is a relative minimum point; $(0, 5)$ is an inflection point. See Fig. 47-10 (b).

(c) $f''(x) = -6x + 6$ and $f'''(x) = -6$. Since $f''(x) = 0$ when $x = 1$ and $f'''(1) = -6 \neq 0, (1, 16)$ is an inflection point.

The x and y intercepts are $x = -1, x = 5$, and $y = 5; (-1, 0)$ is a relative minimum point and $(3, 32)$ is a relative maximum point; $(1, 16)$ is an inflection point. See Fig. 47-10 (c).

(d) $f''(x) = 6x$ and $f'''(x) = 6$. The point $(0, 0)$ is an inflection point.

The x and y intercepts are $x = 0, y = 0; (0, 0)$ is an inflection point. The curve can be sketched after locating the points $(1, 4), (2, 14), (-1, -4)$, and $(-2, -14)$. See Fig. 47-10 (d).

(e) $f''(x) = 6(x - 2)$ and $f'''(x) = 6$. The point $(2, 0)$ is an inflection point.

The x and y intercepts are $x = 2, y = -8; (2, 0)$ is an inflection point. The curve can be sketched after locating the points $(3, 1), (4, 8)$, and $(1, -1)$. See Fig. 47-10 (e).

(f) $f''(x) = 6(2x - 3)(x - 1)$ and $f'''(x) = 6(4x - 5)$. The inflection points are $(1, 0)$ and $(\frac{3}{2}, -\frac{1}{16})$.

The x and y intercepts are $x = 1, \ x = 2$, and $y = 2; (\frac{7}{4}, -\frac{277}{256})$ is a relative minimum point; $(1, 0)$ and $(\frac{3}{2}, -\frac{1}{16})$ are inflection points. For the graph, see Fig. 47-10 (f).

47.5 Find two integers whose sum is 12 and whose product is a maximum.

Let x and $12 - x$ be the integers; their product is $P = f(x) = x(12 - x) = 12x - x^2$.

Since $f'(x) = 12 - 2x = 2(6 - x), x = 6$ is the critical value. Now $f''(x) = -2$; hence $f''(6) = -2 < 0$ and $x = 6$ yields a relative maximum. The integers are 6 and 6.

Note that we have, in effect proved that the rectangle of given perimeter has maximum area when it is a square.

47.6 A farmer wishes to enclose a rectangular plot for a pasture, using a wire fence on three sides and a hedge row as the fourth side. If he has 2400 ft of wiring what is the greatest area he can fence off?

Let x denote the length of the equal sides to be wired; then the length of the third side is $2400 - 2x$. See Fig. 47-11.

The area is $A = f(x) = x(2400 - 2x) = 2400x - 2x^2$. Now $f'(x) = 2400 - 4x = 4(600 - x)$ and the critical value is $x = 600$. Since $f''(x) = -4, x = 600$ yields a relative maximum $f(600) = 720\,000$ ft^2.

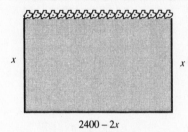

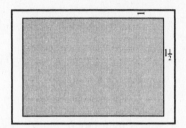

Fig. 47-11 **Fig. 47-12**

47.7 A page is to contain 54 square inches of printed material. If the margins are 1 in. at top and bottom and $1\frac{1}{2}$ in. at the sides, find the most economical dimensions of the page. See Fig. 47-12.

Let the dimensions of the printed material be denoted by x and y; then $xy = 54$.

The dimensions of the page are $x + 3$ and $y + 2$; the area of the page is $A = (x + 3)(y + 2)$.

Since $y = 54/x, A = f(x) = (x + 3)(54/x + 2) = 60 + 162/x + 2x$. Then $f'(x) = -162/x^2 + 3$ and the critical values are $x = \pm 9$. Since $f''(x) = 324/x^3$, the relative minimum is given by $x = 9$. The required dimensions of the page are 12 in. wide and 8 in. high.

47.8 A cylindrical container with circular base is to hold 64 cubic centimeters. Find the dimensions so that the amount (surface area) of metal required is a minimum when (a) the container is an open cup and (b) a closed can.

Let r and h respectively be the radius of the base and height in centimeters, V be the volume of the container, and A be the surface area of the metal required.

(a) $V = \pi r^2 h = 64$ and $A = 2\pi rh + \pi r^2$. Solving for $h = 64/\pi r^2$ in the first relation and substituting in the second, we have $A = 2\pi r\left(\dfrac{64}{\pi r^2}\right) + \pi r^2 = \dfrac{128}{r} + \pi r^2$.

Then $\dfrac{dA}{dr} = -\dfrac{128}{r^2} + 2\pi r = \dfrac{2(\pi r^3 - 64)}{r^2}$ and the critical value is $r = \dfrac{4}{\sqrt[3]{\pi}}$.

Now $h = \dfrac{64}{\pi r^2} = \dfrac{4}{\sqrt[3]{\pi}}$; thus, $r = h\dfrac{4}{\sqrt[3]{\pi}}$ cm.

(b) $V = \pi r^2 h = 64$ and $A = 2\pi rh + 2\pi r^2 = 2\pi r\left(\dfrac{64}{\pi r^2}\right) + 2\pi r^2 = \dfrac{128}{r} + 2\pi r^2$.

Then $\dfrac{dA}{dr} = -\dfrac{128}{r^2} + 4\pi r = \dfrac{4(\pi r^3 - 32)}{r^2}$ and the critical value is $r = 2\sqrt[3]{\dfrac{4}{\pi}}$.

Now $h = \dfrac{64}{\pi r^2} = 4\sqrt[3]{\dfrac{4}{\pi}}$; thus, $h = 2r = 4\sqrt[3]{\dfrac{4}{\pi}}$ cm.

47.9   Study the motion of a particle which moves along a horizontal line in accordance with

(a) $x = t^3 - 6t^2 + 3$,   (b) $x = t^3 - 5t^2 + 7t - 3$,   (c) $v = (t-1)^2(t-4)$,   (d) $v = (t-1)^4$.

(a)   Here $v = 3t^2 - 12t = 3t(t-4) = 0$ when $t = 0$ and $t = 4$; $a = 6t - 12 = 6(t-2) = 0$ when $t = 2$. The particle leaves $A(s = 3)$ with velocity 0 and moves to $B(-29)$ where it stops momentarily; thereafter it moves to the right. See Fig. 47-13.

The intervals of increasing and decreasing speed are shown in Fig. 47-14.

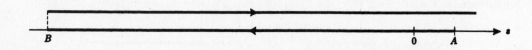

Fig. 47-13

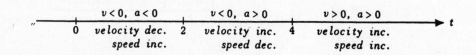

Fig. 47-14

(b)   Here $v = 3t^2 = 10t + 7 = (t-1)(3t-7) = 0$ when $t = 1$ and $t = \frac{7}{5}$; $a = 6t = 10 = 2(3t - 5) = 0$ when $t = \frac{5}{3}$. The particle leaves $A(s = -3)$ with velocity 7 ft/s and moves to 0 where it stops momentarily, then it moves to $B(-\frac{32}{27})$ where is stops momentarily. Thereafter it moves to the right. See Fig. 47-15.

The intervals of increasing and decreasing speed are shown in Fig. 47-16.

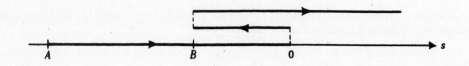

Fig. 47-15

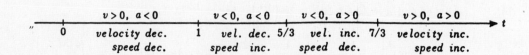

Fig. 47-16

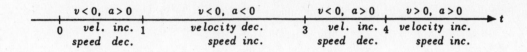

Fig. 47-17

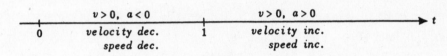

Fig. 47-18

(c) Here $v = 0$ when $t = 1$ and $t = 4$. Also $a = 3t^2 - 12t + 9 = 3(t-1)(t-3) = 0$ when $t = 1$ and $t = 3$. The intervals of increasing and decreasing speed are shown in Fig. 47-17.

Note that the particle stops momentarily at the end of 1 s but does not then reverse its direction of motion as, for example, in (b).

(d) Here $a = 4(t-1)^3$ and $v = a = 0$ when $t = 1$. The intervals of increasing and decreasing speed are shown in Fig. 47-18.

47.10 Find dy in terms of x and dx, given

(a) $y = f(x) = x^2 + 5x + 6$,   (b) $y = f(x) = x^4 - 4x^3 + 8$,   (c) $y = f(x) = x^2 + 1/x^2$.

(a) Since $f'(x) = 2x + 5$, $dy = f'(x)\,dx = (2x + 5)\,dx$.

(b) Since $f'(x) = 4x^3 - 12x^2$, $dy = (4x^3 - 12x^2)\,dx$.

(c) Since $f'(x) = 2x - 2/x^3$, $dy = (2x - 2/x^3)\,dx$.

47.11 Find the approximate displacement of a particle moving along the x axis in accordance with the law $s = t^4 - t^2$, from the time $t = 1.99$ to $t = 2$.

Here $ds = (4t^3 - 2t)dt$. We take $t = 2$ and $dt = -0.01$. Then $ds = (4 \cdot 8 - 2 \cdot 2)(-.01) = -0.28$ and the displacement is 0.28 unit.

47.12 Find using differentials the approximate area of a square whose side is 3.01 cm.

Here $A = x^2$ and $dA = 2x\,dx$. Taking $x = 3$ and $dx = 0.01$, we find $dA = 2 \cdot 3(.01) = 0.06$ cm^2. Now the area (9 cm^2) of a square 3 cm. on a side is increased approximately 0.06 cm^2 when the side is increased to 3.01 cm. Hence the approximate area is 9.06 cm^2. The true area is 9.0601 cm^2.

Supplementary Problems

47.13 Determine the intervals on which each of the following is an increasing function and the intervals on which it is a decreasing function.

(a) $f(x) = x^2$ *Ans.* Dec. for $x < 0$; inc. for $x > 0$
(b) $f(x) = 4 - x^2$ *Ans.* Inc. for $x < 0$; dec. for $x > 0$
(c) $f(x) = x^2 + 6x - 5$ *Ans.* Dec. for $x < -3$; inc. for $x > -3$
(d) $f(x) = 3x^2 + 6x + 18$ *Ans.* Dec. for $x < -1$; inc. for $x > -1$
(e) $f(x) = (x - 2)^4$ *Ans.* Dec. for $x < 2$; inc. for $x > 2$
(f) $f(x) = (x - 1)^3(x + 2)^2$ *Ans.* Inc. for $x < -2$; dec. for $-2 < x < -\frac{4}{5}$;
                                                 inc. for $-\frac{4}{5} < x < x1$ and for $x > 1$

47.14 Find the relative maximum and minimum values of the functions of Problem 47.13.

> *Ans.* (*a*) Min. $= 0$ (*c*) Min. $= -14$ (*e*) Min. $= 0$
>
> (*b*) Max. $= 4$ (*d*) Min. $= 15$ (*f*) Max. $= 0$, Min. $= -26\,244/3125$

47.15 Investigate for relative maximum (minimum) points and points of inflection. Sketch each locus.

> (*a*) $y = x^2 - 4x + 8$ (*b*) $y = (x-1)^3 + 5$ (*c*) $y = x^4 + 32x + 40$ (*d*) $y = x^3 - 3x^2 - 9x + 6$
>
> *Ans.* (*a*) Min. (2, 4)   (*c*) Min. $(-2, -8)$
>
> (*b*) I.P. (1, 5)   (*d*) Max, $(-1, 11)$, Min. $(3, -21)$, I.P. $(1, -5)$

47.16 The sum of two positive numbers is 12. Find the numbers

(*a*) If the sum of their squares is a minimum

(*b*) If the product of one and the square of the other is a maximum

(*c*) If the product of one and the cube of the other is a maximum

> *Ans.* (*a*) 6 and 6   (*b*) 4 and 8   (*c*) 3 and 9

47.17 Find the dimensions of the largest open box which can be made from a sheet of tin 24 in. square by cutting equal squares from the corners and turning up the sides.

> *Ans.* $16 \times 16 \times 4$ in.

47.18 Find the dimensions of the largest open box which can be made from a sheet of tin 60 in. by 28 in. by cutting equal squares from the corners and turning up the sides.

> *Ans.* $48 \times 16 \times 6$ in.

47.19 A rectangular field is to be enclosed by a fence and divided into two smaller plots by a fence parallel to one of the sides. Find the dimensions of the largest such field which can be enclosed by 1200 ft of fencing.

> *Ans.* 200×300 ft

47.20 If a farmer harvests his crop today, he will have 1200 kg worth $2.00 per kg. Every week he waits, the crop increases by 100 kg but the price drops 10¢ per kg. When should he harvest the crop?

> *Ans.* 4 weeks from today

47.21 The base of an isosceles triangle is 20 ft and its altitude is 40 ft. Find the dimensions of the largest inscribed rectangle if two of the vertices are on the base of the triangle.

> *Ans.* 10×20 ft

47.22 For each of the following compute $\Delta y, dy$, and $\Delta y - dy$.

> (*a*) $y = \frac{1}{2}x^2 + x; x = 2, \Delta x = \frac{1}{4}$ *Ans.* $\Delta y = \frac{25}{32}, dy = \frac{3}{4}, \Delta y - dy = \frac{1}{32}$
>
> (*b*) $y = x^2 - x; x = 3, \Delta x = .01$ *Ans.* $\Delta y = .0501, dy = .05, \Delta y - dy = .0001$

47.23 Approximate using differentials the volume of a cube whose side is 3.005 in.

> *Ans.* 27.135 in^3

47.24 Approximate using differentials the area of a circular ring whose inner radius is 5 in. and whose width is $\frac{1}{8}$ in.

> *Ans.* 1.25π in^2

Chapter 48

Integration

IF $F(x)$ IS A FUNCTION whose derivative $F'(x) = f(x)$, then $F(x)$ is called an *antiderivative* of $f(x)$.

For example, $F(x) = x^3$ is an antiderivative of $f(x) = 3x^2$ since $F'(x) = 3x^2 = f(x)$. Also, $G(x) = x^3 + 5$ and $H(x) = x^3 - 6$ are antiderivatives of $f(x) = 3x^2$. Why?

If $F(x)$ and $G(x)$ are two distinct antiderivatives of $f(x)$, then $F(x) = G(x) + C$, where C is a constant. (See Problem 48.1)

THE INDEFINITE INTEGRAL OF $f(x)$ is denoted by $\int f(x)\,dx$, and is the most general antiderivative of $f(x)$—that is

$$\int f(x)\,dx = F(x) + C$$

where $F(x)$ is any function such that $F'(x) = f(x)$ and C is an arbitrary constant. Thus the indefinite integral of $f(x) = 3x^2$ is $\int 3x^2 dx = x^3 + C$.

We shall use the following antidifferentiation formulas:

I. $\displaystyle\int x^n\,dx = \frac{x^{n+1}}{n+1} + C$, where $n \neq -1$

II. $\displaystyle\int cf(x)\,dx = c\int f(x)\,dx$, where c is a constant

III. $\displaystyle\int [f(x) + g(x)]\,dx = \int f(x)\,dx + \int g(x)\,dx$

EXAMPLE 1

(a) $\displaystyle\int x^5 dx = \frac{x^{5+1}}{5+1} + C = \frac{x^6}{6} + C$

(b) $\displaystyle\int 4x^3\,dx = 4\int x^3\,dx = 4\left[\frac{x^4}{4} + C\right] = x^4 + 4C$, but if we call $4C$ by the name C, then C_1 still represents an arbitrary constant; so we can simply write $\displaystyle\int 4x^3\,dx = x^4 + C$

(c) $\displaystyle\int 3x\,dx = 3\int x\,dx = 3 \cdot \frac{x^2}{2} + C = \frac{3}{2}x^2 + C$

(d) $\displaystyle\int \frac{dx}{x^3} = \int x^{-3}\,dx = \frac{x^{-2}}{-2} + C = -\frac{1}{2x^2} + C$

(e) $\displaystyle\int (x^5 + 4x^3 + 3x)\,dx = \int x^5\,dx + \int 4x^3\,dx + \int 3x\,dx = \frac{x^6}{6} + x^4 - \frac{3x^2}{2} + C$

(See Problems 48.2–48.8.)

AREA BY SUMMATION.   Consider the area A bounded by curve $y = f(x) \geq 0$, the x axis, and the ordinates $x = a$ and $x = b$, where $b > a$.

Let the interval $a \leq x \leq b$ be divided into n equal parts each of length Δx. At each point of subdivision, construct the ordinate, thus dividing the area into n strips, as in Fig. 48-1. Since the areas of the strips are unknown, we propose to approximate each strip by a rectangle whose area can be found. In Fig. 48-2, a representative strip and its approximating rectangle are shown.

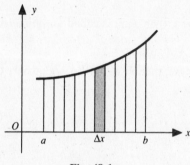

Fig. 48-1

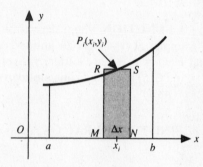

Fig. 48-2

Suppose the representative strip is the ith strip counting from the left, and let $x = x_i$ be the coordinate of the midpoint of its base. Denote by $y_i = f(x_i)$ the ordinate of the point P_i (on the curve) whose abscissa is x_i. Through P_i pass a line parallel to the x axis and complete the rectangle $MRSN$. This rectangle of area $y_i \Delta x$ is the approximating rectangle of the ith strip. When each strip is treated similarly, it seems reasonable to take

$$y_1 \Delta x + y_2 \Delta x + y_3 \Delta x + \cdots + y_n \Delta x = \sum_{i=1}^{n} y_i \Delta x$$

as an approximation of the area sought.

Now suppose that the number of strips (with approximating rectangles) is indefinitely increased so that $\Delta x \to 0$. It is evident from the figure that by so increasing the number of approximating rectangles the sum of their areas more nearly approximates the area sought; that is,

$$A = \lim_{n \to \infty} \sum_{i=1}^{n} y_i \Delta x$$

IF WE DEFINE $\displaystyle\int_a^b f(x)\,dx$ (read, the *definite integral* of $f(x)$ between $x = a$ and $x = b$) as

$$\int_a^b f(x)\,dx = F(x)\Big|_a^b = F(b) - F(a), \quad (F'(x) = f(x))$$

then the area bounded by $y = f(x) \geq 0$, the x axis, and the ordinates $x = a$ and $x = b$ $(b > a)$ is given by

$$A = \int_a^b f(x)\,dx$$

(See Problems 48.9–48.12.)

Solved Problems

48.1 Prove: If $F(x)$ and $G(x)$ are distinct integrals of $f(x)$, then $F(x) = G(x) + C$, where C is a constant.

Since $F(x)$ and $G(x)$ are integrals of $f(x)$, $F'(x) = G'(x) = f(x)$.

Suppose $F(x) - G(x) = H(x)$; differentiating with respect to x, $F'(x) - G'(x) = H'(x)$ and $H'(x) = 0$. Thus, $H(x)$ is a constant, say, C, and $F(x) = G(x) + C$.

48.2(a) $\displaystyle\int \sqrt{x}\,dx = \int x^{1/2}\,dx = \frac{x^{3/2}}{\frac{3}{2}} + C = \frac{2}{3}x^{3/2} + C$

(b) $\displaystyle\int (3x^2 + 5)\,dx = x^3 + 5x + C$

(c) $\displaystyle\int (5x^6 + 2x^3 - 4x + 3)\,dx = \frac{5}{7}x^7 + \frac{1}{2}x^4 - 2x^2 + 3x + C$

(d) $\displaystyle\int (80x^{19} - 32x^{15} - 12x^{-3})\,dx = 4x^{20} - 2x^{16} + \frac{6}{x^2} + C$

48.3 At every point (x, y) of a certain curve, the slope is equal to 8 times the abscissa. Find the equation of the curve if it passes through (1,3).

Since $m = \dfrac{dy}{dx} = 8x$, we have $dy = 8x\,dx$. Then $y = \int 8x\,dx = 4x^2 + C$, a family of parabolas. We seek the equation of the parabola of this family which passes through the point (1,3). Then $3 = 4(1)^2 + C$ and $C = -1$. The curve has equation $y = 4x^2 - 1$.

48.4 For a certain curve $y'' = 6x - 10$. Find its equation if it passes through point (1,1) with slope -1.

Since $y'' = 6x - 10$, $y' = 3x^2 - 10x + C_1$; since $y' = -1$ when $x = 1$, we have $-1 = 3 - 10 + C_1$ and $C_1 = 6$. Then $y' = 3x^2 - 10x + 6$.

Now $y = x^3 - 5x^2 + 6x + C_2$ and since $y = 1$ when $x = 1$, $1 = 1 - 5 + 6 + C_2$ and $C_2 = -1$. Thus the equation of the curve is $y = x^3 - 5x^2 + 6x - 1$.

48.5 The velocity at time t of a particle moving along the x axis is given by $v = x' = 2t + 5$. Find the position of the particle at time t, if $x = 2$ when $t = 0$.

Select a point on the x axis as origin and assume positive direction to the right. Then at the beginning of the motion ($t = 0$) the particle is 2 units to the right of the origin.

Since $v = \dfrac{dx}{dt} = 2t + 5$, $dx = (2t + 5)\,dt$. Then $x = \int (2t + 5)\,dt = t^2 + 5t + C$.

Substituting $x = 2$ and $t = 0$, we have $2 = 0 + 0 + C$ so that $C = 2$. Thus the position of the particle at time t is given by $x = t^2 + 5t + 2$.

48.6A body moving in a straight line has an acceleration equal to $6t^2$, where time (t) is measured in seconds and distance s is measured in feet. If the body starts from rest, how far will it move during the first 2 s?

Let the body start from the origin; then it is given that when $t = 0$, $v = 0$ and $s = 0$.

Since $a = \dfrac{dv}{dt} = 6t^2$, $dv = 6t^2\,dt$. Then $v \int 6t^2\,dt = 2t^3 + C_1$. When $t = 0$, $v = 0$; then $0 = 2 \cdot 0 + C_1$ and $C_1 = 0$. Thus $v = 2t^3$.

Now $v = \dfrac{ds}{dt} = 2t^3$; then $ds = 2t^3$ and $s = \int 2t^3\,dt = \frac{1}{2}t^4 + C_2$. When $t = 0$, $s = 0$; then $C_2 = 0$ and $s = \frac{1}{2}t^4$. When $t = 2$, $s = \frac{1}{2}(2)^4 = 8$. The body moves 8 ft during the first 2 s.

48.7A ball is thrown upward from the top of a building 320 ft high with initial velocity 128 ft/s. Determine the velocity with which the ball will strike the street below. (Assume acceleration is 32 ft/s, directed downward.)

First we choose an origin from which all distances are to be measured and a direction (upward or downward) which will be called positive.

First Solution.   Take the origin at the top of the building and positive direction as upward.

Then $a = \dfrac{dv}{dt} = -32$ and $v = -32t + C_1$.

When the ball is released, $t = 0$ and $v = 128$; then $128 = -32(0) + C_1$ and $C_1 = 128$.
    Now $v = ds/dt = -32t + 128$ and $s = -16t^2 + 128t + C_2$. When the ball is released, $t = 0$ and $s = 0$; then $C_2 = 0$ and $s = -16t^2 + 128t$.
    When the ball strikes the street, it is 320 ft below the origin, that is, $s = -320$; hence, $-320 = -16t^2 + 128t$, $t^2 - 8t - 20 = (t + 2)(t - 10) = 0$, and $t = 10$. Finally, when $t = 10$, $v = -32(10) + 128 = -129$ ft/s.

Second Solution.   Take the origin on the street and positive direction as before. Then $a = dv/dt = -32$ and $v = -32t + 128$ as in the first solution.
    Now $s = -16t^2 + 128t + C_2$ but when $t = 0$, $s = 320$. Thus $C_2 = 320$ and $s = -16t^2 + 128 + 320$. When the ball strikes the street $s = 0$; then $t = 10$ and $v = -192$ ft/s as before.

48.8   A ball was dropped from a balloon 640 ft above the ground. If the balloon was rising at the rate of 48 ft/s, find (*a*) the greatest distance above the ground attained by the ball, (*b*) the time the ball was in the air, and (*c*) the speed of the ball when it struck the ground.
    Assume the origin at the point where the ball strikes the ground and positive distance to be directed upward. Then

$$a = \frac{dv}{dt} = -32 \qquad \text{and} \qquad v = -32t + C_1$$

When $t = 0$, $v = 48$; hence, $C_1 = 48$. Then $v = ds/dt = -32 + 48$ and $s = -16t^2 + 48t + C_2$. When $t = 0$, $s = 640$; hence, $C_2 = 640$ and $s = -16t^2 + 48t + 640$.

(*a*)   When $v = 0, t = \frac{3}{2}$ and $s = -16(\frac{3}{2})^2 + 48(\frac{3}{2}) + 640 = 676$. The greatest height attained by the ball was 676 ft.

(*b*)   When $s = 0, -16t^2 + 48t + 640 = 0$ and $t = -5, 8$. The ball was in the air for 8 s.

(*c*)   When $t = 8$, $v = -32(8) + 48 = -208$. The ball struck the ground with speed 208 ft/s.

48.9   Find the area bounded by the line $y = 4x$, the x axis, and the ordinates $x = 0$ and $x = 5$.
    Here $y \geq 0$ on the interval $0 \leq x \leq 5$. Then

$$A = \int_0^5 4x\, dx = 2x^2 \Big|_0^5 = 50 \text{ sq units}$$

Note that we have found the area of a right triangle whose legs are 5 and 20 units. See Fig. 48-3. The area is $\frac{1}{2}(5)(20) = 50$ sq units.

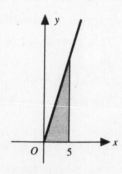

Fig. 48-3

48.10   Find the area bounded by the parabola $y = 8 + 2x - x^2$ and the x axis.

The x intercepts are $x = -2$ and $x = 4$; $y \geq 0$ on the interval $-2 \leq x \leq 4$. See Fig. 48-4. Hence

$$A = \int_{-2}^{4} (8 + 2x - x^2)\, dx = \left(8x + x^2 - \frac{x^3}{3}\right)\Big|_{-2}^{4}$$

$$= \left(8 \cdot 4 + 4^2 - \frac{4^3}{3}\right) - \left[8(-2) + (-2)^2 - \frac{(-2)^3}{3}\right] = 36 \text{ sq units}$$

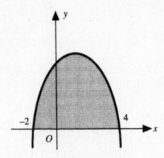

Fig. 48-4

48.11   Find the area bounded by the parabola $y = x^2 + 2x - 3$, the x axis, and the ordinates $c = -2$ and $x = 0$.
On the interval $-2 \leq x \leq 0$, $y \leq 0$. Here

$$A \int_{-2}^{0} (x^2 + 2x - 3)\, dx = \left(\frac{x^3}{3} + x^2 - 3x\right)\Big|_{-2}^{0} = 0 - \left[\frac{(-2)^3}{3} + (-2)^2 - 3(-2)\right] = -\frac{22}{3}$$

The negative sign indicates that the area lies entirely below the x axis. The area is $\frac{22}{3}$ sq units. See Fig. 48-5.

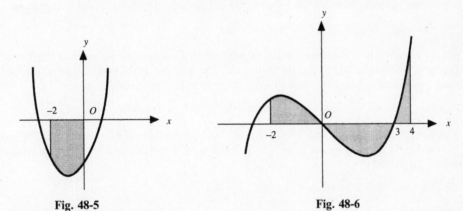

Fig. 48-5   **Fig. 48-6**

48.12   Find the area bounded by the curve $y = x^3 - 9x$, the x axis, and the ordinates $x = -2$ and $x = 4$.
The purpose of this problem is to show that the required area is *not* given by $\int_{-2}^{4}(x^3 - 9x)\, dx$.
From Fig. 48-6, we note that y changes sign at $x = 0$ and at $x = 3$. The required area consists of three pieces, the individual areas being given, apart from sign, by

$$A_1 = \int_{-2}^{0} (x^3 - 9x)\, dx = (\tfrac{1}{4}x^4 - \tfrac{9}{2}x^2)\Big|_{2}^{0} = 0 - (4 - 18) = 14$$

$$A_2 = \int_{0}^{3} (x^3 - 9x)\, dx = (\tfrac{1}{4}x^4 - \tfrac{9}{2}x^2)\Big|_{0}^{3} = (\tfrac{81}{4} - \tfrac{81}{2}) - 0 = -\tfrac{81}{4}$$

$$A_3 = \int_{3}^{4} (x^3 - 9x)\, dx = (\tfrac{1}{4}x^4 - \tfrac{9}{2}x^2)\Big|_{3}^{4} = (64 - 72) - (\tfrac{81}{4} - \tfrac{81}{2}) = \tfrac{49}{4}$$

Thus, $A = A_1 - A_2 + A_3 = 14 + \tfrac{81}{4} + \tfrac{49}{4} = \tfrac{93}{2}$ sq units.
Note that $\int_{-2}^{4}(x^3 - 9x)\, dx = 6 < A_1$, an absurd result.

Supplementary Problems

48.13 Find the following indefinite integrals.

(a) $4 \int dx = 4x + C$

(b) $\int \frac{1}{2} x \, dx = \frac{1}{4} x^2 + C$

(c) $\int (3x^2 + 4x - 5) \, dx = x^3 + 2x^2 - 5x + C$

(d) $\int x(1-x) \, dx = \frac{1}{2} x^2 - \frac{1}{3} x^3 + C$

(e) $\int 3(x+1)^2 \, dx = (x+1)^3 + C$

(f) $\int (x-1)(x+2) \, dx = \frac{1}{3} x^3 + \frac{1}{2} x^2 - 2x + C$

(g) $\int \frac{dx}{x^2} = -\frac{1}{x} + C$

(h) $\int \frac{x^2 - 2}{x^2} \, dx = x + \frac{2}{x} + C$

48.14 Find the equation of the family of curves whose slope is the given function of x. Find also the equation of the curve of the family passing through the given point.

(a) $m = 1, (1, -2)$ *Ans.* $y = x + C, \ y = x - 3$

(b) $m = -6x, (0, 0)$ *Ans.* $y = -3x^2 + C, \ y = -3x^2$

(c) $m = 3x^2 + 2x, (1, -3)$ *Ans.* $y = x^3 + x^2 + C, \ y = x^3 + x^2 - 5$

(d) $m = 6x^2, (0, 1)$ *Ans.* $y = 2x^3 + C, \ y = 2x^3 + 1$

48.15 For a certain curve $y'' = 6x + 8$. Find its equation if it passes through (1,2) with slope $m = 6$.

Ans. $y = x^3 + 4x^2 - 5x + 2$

48.16 A stone is dropped from the top of a building 400 ft high. Taking the origin at the top of the building and positive direction downward, find (a) the velocity of the stone at time t, (b) the position at time t, (c) the time it takes for the stone to reach the ground, and (d) the velocity when it strikes the ground.

Ans. (a) $v = 32t$ (b) $s = 16t^2$ (c) 5 s   (d) 160 ft/s

48.17 A stone is thrown downward with initial velocity 20 ft/s from the top of a building 336 ft high. Following the directions of Problem 48.16, find (a) the velocity and position of the stone 2 s later, (b) the time it takes to reach the ground, and (c) the velocity with which it strikes the ground.

Ans. (a) 84 ft/s, 232 ft above the ground   (b) 4 s   (c) 148 ft/s

48.18 A stone is thrown upward with initial velocity 16 ft/s from the top of a building 192 ft high. Find (a) the greatest height attained by the stone, (b) the total time in motion, and (c) the speed with which the stone strikes the ground.

Ans. (a) 196 ft   (b) 4 s   (c) 112 ft/s

48.19 A boy on top of a building 192 ft high throws a rock straight down. What initial velocity did he give it if it strikes the ground after 3 s?

Ans. 16 ft/s

48.20 Find the area bounded by the x axis, the given curve, and the indicated ordinates.

(a) $y = x^2$ between $x = 2$ and $x = 4$ *Ans.* $\frac{56}{3}$ square units

(b) $y = 4 - 3x^2$ between $x = -1$ and $x = 1$ *Ans.* 6 square units

(c) $y = x^{1/2}$ between $x = 0$ and $x = 9$ *Ans.* 18 square units

(d) $y = x^2 - x - 6$ between $x = 0$ and $x = 2$ *Ans.* $\frac{34}{3}$ square units

(e) $y = x^3$ between $x = -2$ and $x = 4$ *Ans.* 68 square units

(f) $y = x^3 - x$ between $x = -1$ and $x = 1$ *Ans.* $\frac{1}{2}$ square unit

Chapter 49

Infinite Sequences

GENERAL TERM OF A SEQUENCE. Frequently the law of formation of a given sequence may be stated by giving a representative or *general term* of the sequence. This general term is a function of n, where n is the number of the term in the sequence. For this reason, it is also called the nth term of the sequence.

When the general term is given, it is a simple matter to write as many terms of the sequence as desired.

EXAMPLE 1

(a) Write the first four terms and the tenth term of the sequence whose general term is $1/n$.

The first term ($n = 1$) is $\frac{1}{1} = 1$, the second term ($n = 2$) is $\frac{1}{2}$, and so on. The first four terms are $1, \frac{1}{2}, \frac{1}{3}, \frac{1}{4}$ and the tenth term is $\frac{1}{10}$.

(b) Write the first terms and the ninth term of the sequence whose general term is $(-1)^{n-1} \dfrac{2n}{n^2 + 1}$.

The first term ($n = 1$) is $(-1)^{1-1} \dfrac{2 \cdot 1}{1^2 + 1} = 1$, the second term ($n = 2$) is $(-1)^1 \dfrac{2 \cdot 2}{2^2 + 1} = -\dfrac{4}{5}$, and so on.

The first four terms are $1, -\frac{4}{5}, \frac{3}{5}, -\frac{8}{17}$ and the ninth term is $(-1)^8 \dfrac{2 \cdot 9}{9^2 + 1} = \dfrac{9}{41}$.

Note that the effect of the factor $(-1)^{n-1}$ is to produce a sequence whose terms have alternate signs, the sign of the first term being positive. The same pattern of signs is also produced by the factor $(-1)^{n+1}$. In order to produce a sequence whose terms alternate in sign, the first term being negative, the factor $(-1)^n$ is used.

When the first few terms of a sequence are given and they match an obvious pattern, the general term is obtained by inspection.

EXAMPLE 2. Obtain the general term for each of the sequences:

(a) $1, 4, 9, 16, 25, \ldots$.

The terms of the sequence are the squares of the positive integers; the general term is n^2.

(b) $3, 7, 11, 15, 19, 23, \ldots$.

This is an arithmetic progression having $a = 3$ and $d = 4$. The general term is $a + (n-1)d = 4n - 1$. Note, however, that the general term can be obtained about as easily by inspection.

(See Problems 49.1–49.3.)

LIMIT OF AN INFINITE SEQUENCE. From Example 4 of Chapter 8, the line $y = 2$ is a horizontal asymptote of $xy - 2x - 1 = 0$. To show this, let $P(x, y)$ move along the curve so that its abscissa takes on the values $10, 10^2, 10^3, \ldots, 10^n, \ldots$. Then the corresponding values of y are

$$2.1, 2.01, 2.001, \ldots, 2 + \frac{1}{10^n}, \ldots \qquad (49.1)$$

and we infer that, by proceeding far enough along in this sequence, the difference between the terms of the sequence and 2 may be made as small as we please. This is equivalent to the following: Let ε denote a positive number, as small as we please; then there is a term of the sequence such that the difference between it and 2 is less than ε, and the same is true for all subsequent terms of the sequence. For example, let $\varepsilon = 1/10^{25}$ then the difference between the term $2 + 1/10^{26}$ and 2, $2 + 1/10^{26} - 2 = 1/10^{26}$, is less then $\varepsilon = 1/10^{25}$ and the same is true for the terms $2 + 1/10^{27}$, $2 + 1/10^{28}$, and so on.

The behavior of the terms of the sequence (49.1) discussed above is indicated by the statement: *The limit of the sequence (49.1) is 2.* In general, if, for an infinite sequence

$$s_1, s_2, s_3, \ldots, s_n, \ldots \qquad (49.2)$$

and a positive number ε, however small, there exists a number s and a positive integer m such that for all $n > m$

$$|s - s_n| < \varepsilon,$$

then the limit of the sequence is s.

EXAMPLE 3.   Show, using the above definition, that the limit of sequence (49.1) is 2.

Take $\varepsilon = 1/10^p$, where p is a positive integer as large as we please; thus, ε is a positive number as small as we please. We must produce a positive integer m (in other words, a term s_m) such that for $n > m$ (that is, for all subsequent terms) $|s - s_n| < \varepsilon$. Now

$$\left| 2 - \left(2 + \frac{1}{10^n} \right) \right| < \frac{1}{10^p} \quad \text{or} \quad \frac{1}{10^n} < \frac{1}{10^p}$$

requires $n > p$. Thus, $m = p$ is the required value of m.

The statement that the limit of the sequence (49.2) is s describes the behavior of s_n as n increases without bound over the positive integers. Since we shall repeatedly be using the phrase "as n increases without bound" or the phrase "as n becomes infinite," which we shall take to be equivalent to the former phrase, we shall introduce the notation $n \to \infty$ for it. Thus the behavior of s_n may be described briefly by

$$\lim_{n \to \infty} s_n = s$$

(read: the limit of s_n, as n becomes infinite, is s).

We state, without proof, the following theorem:

If $\lim\limits_{n \to \infty} s_n = s$ and $\lim\limits_{n \to \infty} t_n = t$, then

(A) $\lim\limits_{n \to \infty} (s_n \pm t_n) = \lim\limits_{n \to \infty} s_n \pm \lim\limits_{n \to \infty} t_n = s \pm t$

(B) $\lim\limits_{n \to \infty} (s_n \cdot t_n) = \lim\limits_{n \to \infty} s_n \cdot \lim\limits_{n \to \infty} t_n = s \cdot t$

(C) $\lim\limits_{n \to \infty} \dfrac{s_n}{t_n} = \dfrac{\lim\limits_{n \to \infty} s_n}{\lim\limits_{n \to \infty} t_n} = \dfrac{s}{t}$,   provided $t \neq 0$

or, in words, if each of two sequences approaches a limit, then the limits of the sum, difference, product, and quotient of the two sequences are equal, respectively, to the sum, difference, product, and quotient of their limits provided only that, in the case of the quotient, the limit of the denominator is not zero.

This theorem makes it possible to find the limit of a sequence directly from its general term. In this connection, we shall need

$$\lim_{n\to\infty} a = a, \qquad \text{where } a \text{ is any constant} \tag{49.3}$$

$$\lim_{n\to\infty} \frac{1}{n^k} = 0, \qquad k > 0 \text{ (See Problem 49.4.)} \tag{49.4}$$

$$\lim_{n\to\infty} \frac{1}{b^n} = 0, \qquad \text{where } b \text{ is a constant} > 1 \text{ (See Problem 49.5.)} \tag{49.5}$$

(See Problem 49.6.)

THE FOLLOWING THEOREMS are useful in establishing whether or not certain sequences have a limit.

I. Suppose M is a fixed number, such that for all values of n,

$$s_n \le s_{n+1} \qquad \text{and} \qquad s_n \le M;$$

then $\lim\limits_{n\to\infty} s_n$ exists and is $\le M$.

   If, however, s_n eventually exceeds M, no matter how large M may be, $\lim\limits_{n\to\infty} s_n$ does not exist.

II. Suppose M is a fixed number such that, for all values of n,

$$s_n \ge s_{n+1} \qquad \text{and} \qquad s_n \ge M;$$

then $\lim\limits_{n\to\infty} s_n$ exists and is $\ge M$.

   If, however, s_n is eventually smaller than M, no matter how small M may be, $\lim\limits_{n\to\infty} s_n$ does not exist.

EXAMPLE 4

(a) For the sequence $\dfrac{5}{2}, 3, \dfrac{19}{6}, \dfrac{13}{4}, \ldots, \left(\dfrac{7}{2} - \dfrac{1}{n}\right), \ldots, s_n < s_{n+1}$ and $s_n < 4$, for all values of n; the sequence has a limit ≤ 4. In fact, $\lim\limits_{n\to\infty} s_n = \dfrac{7}{2}$.

(b) For the sequence $3, 5, 7, 9, \ldots, 2n+1, \ldots, s_n < s_{n+1}$ but s_n will eventually exceed any chosen M, however large (if $M = 2^{1000} + 1$, then $2n+1 > M$ for $n > 2^{999}$), and the sequences does not have a limit.

(See Problems 49.7–49.9.)

RECURSIVELY DEFINED SEQUENCES. Sequences can be defined recursively. For example, suppose that $a_1 = 1$ and $a_{n+1} = 2a_n$ for every natural number n. Then,

$$a_1 = 1, \qquad a_2 = a_{1+1} = 2a_1 = 2, \qquad a_3 = a_{2+1} = 2a_2 = 4, \qquad \text{etc.}$$

Thus, the sequence is $1, 2, 4, 8, \ldots$.

   One famous such sequence is the Fibonacci sequence:

$$a_1 = 1, \qquad a_2 = 1, \qquad a_{n+2} = a_{n+1} + a_n.$$

The sequence is $1, 1, 2, 3, 5, 8, 13, \ldots$.

Solved Problems

49.1   Write the first five terms and the tenth term of the sequence whose general term is

(a) $4n - 1$.

   The first term is $4 \cdot 1 - 1 = 3$, the second term is $4 \cdot 2 - 1 = 7$, the third term is $4 \cdot 3 - 1 = 11$, the fourth term is $4 \cdot 4 - 1 = 15$, the fifth term is $4 \cdot 5 - 1 = 19$; the tenth term is $4 \cdot 10 - 1 = 39$.

(b) 2^{n-1}.

The first term is $2^{1-1} = 2^0 = 1$, the second term is $2^{2-1} = 2$, the third is $2^{3-1} = 2^2 = 4$, the fourth is $2^3 = 8$, the fifth is $2^4 = 16$; the tenth is $2^9 = 512$.

(c) $\dfrac{(-1)^{n-1}}{n+1}$.

The first term is $\dfrac{(-1)^{1-1}}{1+1} = \dfrac{(-1)^0}{2} = \dfrac{1}{2}$, the second is $\dfrac{(-1)^{2-1}}{2+1} = -\dfrac{1}{3}$, the third is $\dfrac{(-1)^2}{3+1} = \dfrac{1}{4}$, the fourth is $-\frac{1}{5}$, the fifth is $\frac{1}{6}$; the tenth is $-\frac{1}{11}$.

49.2 Write the first four terms of the sequences whose general term is

(a) $\dfrac{n+1}{n!}$.

The terms are $\dfrac{1+1}{1!}, \dfrac{2+1}{2!}, \dfrac{3+1}{3!}, \dfrac{4+1}{4!}$ or $2, \dfrac{3}{2}, \dfrac{2}{3}, \dfrac{5}{24}$.

(b) $\dfrac{x^{2n-1}}{(2n+1)!}$.

The required terms are $\dfrac{x}{3!}, \dfrac{x^3}{5!}, \dfrac{x^5}{7!}, \dfrac{x^7}{9!}$.

(c) $(-1)^{n-1}\dfrac{x^{2n-2}}{(n-1)!}$.

The terms are $\dfrac{x^0}{0!}, -\dfrac{x^2}{1!}, \dfrac{x^4}{2!}, -\dfrac{x^6}{3!}$ or $1, -x^2, \dfrac{x^4}{2}, \dfrac{x^6}{6}$.

49.3 Write the general term for each of the following sequences:

(a) $2, 4, 6, 8, 10, 12, \ldots$.

The first term is $2 \cdot 1$, the second is $2 \cdot 2$, the third is $2 \cdot 3$, etc.; the general term is $2n$.

(b) $1, 3, 5, 7, 9, 11, \ldots$.

Each term of the given sequence is 1 less than the corresponding term of the sequences in (a); the general term is $2n - 1$.

(c) $2, 5, 8, 11, 14, \ldots$.

The first term is $3 \cdot 1 - 1$, the second term is $3 \cdot 2 - 1$, the third term is $3 \cdot 3 - 1$, and so on; the general term is $3n - 1$.

(d) $2, -5, 8, -11, 14, \ldots$.

This sequence may be obtained from that in (c) by changing the signs of alternate terms beginning with the second; the general term is $(-1)^{n-1}(3n - 1)$.

(e) $2.1, 2.01, 2.001, 2.0001, \ldots$.

The first term is $2 + 1/10$, the second term is $2 + 1/10^2$, the third is $2 + 1/10^3$, and so on; the general term is $2 + 1/10^n$.

(f) $\frac{1}{8}, -\frac{1}{27}, \frac{1}{64}, -\frac{1}{125}, \ldots$.

The successive denominators are the cubes of $2, 3, 4, 5 \ldots$ or of $1+1, 2+1, 3+1, 4+1, \ldots$; the general term is $\dfrac{(-1)^{n-1}}{(n+1)^3}$.

(g) $\dfrac{3}{1}, \dfrac{4}{1 \cdot 2}, \dfrac{5}{1 \cdot 2 \cdot 3}, \ldots$.

Rewriting the sequence as $\dfrac{1+2}{1!}, \dfrac{2+2}{2!}, \dfrac{3+2}{3!}, \ldots$, the general term is $\dfrac{n+2}{n!}$.

(h) $\dfrac{x}{2}, \dfrac{x^2}{6}, \dfrac{x^3}{24}, \dfrac{x^4}{120}, \ldots$.

The denominators are $2!, 3!, \ldots, (n+1)!, \ldots$; the general term is $\dfrac{x^n}{(n+1)!}$.

(i) $x, \dfrac{-x^3}{3!}, \dfrac{x^5}{5!}, \dfrac{-x^7}{7!}, \ldots$.

The exponents of x are $2\cdot 1 - 1, 2\cdot 2 - 1, 2\cdot 3 - 1, \ldots$; the general term is $(-1)^{n-1}\dfrac{x^{2n-1}}{(2n-1)!}$.

(j) $3, 4, \frac{5}{2}, 1, \frac{7}{24}, \ldots$.

Rewrite the sequence as $\dfrac{3}{0!}, \dfrac{4}{1!}, \dfrac{5}{2!}, \dfrac{6}{3!}, \ldots$; the general term is $\dfrac{n+2}{(n-1)!}$.

49.4 Show that $\lim\limits_{n\to\infty}\dfrac{1}{n^k} = 0$ when $k > 0$.

Take $\varepsilon = 1/p^k$, where p is a positive integer as large as we please. We seek a positive number m such that for $n > m$, $|0 - 1/n^k| = 1/n^k < 1/p^k$. Since this inequality is satisfied when $n > p$, it is sufficient to take for m any number equal to or greater than p.

49.5 Show that $\lim\limits_{n\to\infty}\dfrac{1}{b^n} = 0$ when $b > 1$.

Take $\varepsilon = 1/b^p$, where p is a positive integer. Since $b > 1$, $b^p > 1$ and $\varepsilon = 1/b^p < 1$. Thus, ε may be made as small as we please by taking p sufficiently large. We seek a positive number m such that for $n > m$, $|0 - 1/b^n| = 1/b^n < 1/b^p$. Since $n > p$ satisfies the inequality, it is sufficient to take for m any number equal to or greater than p.

49.6 Evaluate each of the following:

(a) $\lim\limits_{n\to\infty}\left(\dfrac{3}{n} + \dfrac{5}{n^2}\right) = \lim\limits_{n\to\infty}\dfrac{3}{n} + \lim\limits_{n\to\infty}\dfrac{5}{n^2} = 0 + 0 = 0$

(b) $\lim\limits_{n\to\infty}\dfrac{n}{n+1} = \lim\limits_{n\to\infty}\dfrac{n/n}{\dfrac{n+1}{n}} = \lim\limits_{n\to\infty}\dfrac{1}{1+1/n} = \dfrac{\lim\limits_{n\to\infty}1}{\lim\limits_{n\to\infty}(1+1/n)} = \dfrac{1}{1+0} = 1$

(c) $\lim\limits_{n\to\infty}\dfrac{n^2+2}{2n^2-3n} = \lim\limits_{n\to\infty}\dfrac{1+2/n^2}{2-3/n} = \dfrac{1+0}{2-0} = \dfrac{1}{2}$

(d) $\lim\limits_{n\to\infty}\left(4 - \dfrac{2^n-1}{2^{n+1}}\right) = 4 - \lim\limits_{n\to\infty}\dfrac{2^n-1}{2^{n+1}} = 4 - \lim\limits_{n\to\infty}\left(\dfrac{1}{2} - \dfrac{1}{2^{n+1}}\right)$

$\qquad\qquad = 4\dfrac{1}{2} - \lim\limits_{n\to\infty}\left(-\dfrac{1}{2^{n+1}}\right) = 4 - \dfrac{1}{2} - 0 = 3.5$

49.7 Show that every infinite arithmetic sequence fails to have a limit except when $d = 0$.

(a) If $d > 0$, then $s_n = a + (n-1)d < s_{n+1} = a + nd$; but s_n eventually exceeds any previously selected M, however large. Thus, the sequence has no limit.

(b) If $d < 0$, then $s_n > s_{n+1}$; but s_n eventually becomes smaller than any previously selected M, however small. Thus, the sequence has no limit.

(c) If $d = 0$, the sequence is $a, a, a, \ldots, a, \ldots$ with limit a.

49.8 Show that the infinite geometric sequence $3, 6, 12, \ldots, 3\cdot 2^{n-1}, \ldots$, does not have a limit.

Here $s_n < s_{n+1}$; but $3\cdot 2^{n-1}$ may be made to exceed any previously selected M, however, large. The sequence has no limit.

49.9 Show that the following sequence does not have a limit:

$$1, 1+\dfrac{1}{2}, 1+\dfrac{1}{2}+\dfrac{1}{3}, 1+\dfrac{1}{2}+\dfrac{1}{3}+\dfrac{1}{4}+\cdots, \qquad 1+\dfrac{1}{2}+\dfrac{1}{3}+\dfrac{1}{4}+\cdots+\dfrac{1}{n}, \qquad \cdots .$$

Here $s_n < s_{n+1}$. Let M, as large as we please, be chosen, Now

$$1+\tfrac{1}{2}+\tfrac{1}{3}+\tfrac{1}{4}+\cdots = 1+\tfrac{1}{2}+\left(\tfrac{1}{3}+\tfrac{1}{4}\right)+\left(\tfrac{1}{5}+\tfrac{1}{6}+\tfrac{1}{7}+\tfrac{1}{8}\right)+\left(\tfrac{1}{9}+\cdots+\tfrac{1}{16}\right)+\left(\tfrac{1}{17}+\cdots+\tfrac{1}{32}\right)+\cdots$$

and

$$\tfrac{1}{3}+\tfrac{1}{4}>\tfrac{1}{2}, \quad \tfrac{1}{5}+\tfrac{1}{6}+\tfrac{1}{7}+\tfrac{1}{8}>\tfrac{1}{2}, \quad \tfrac{1}{9}+\cdots+\tfrac{1}{16}>\tfrac{1}{2}, \quad \text{and so on.}$$

Since the sum of each group exceeds $\frac{1}{2}$ and we may add as many groups as we please, we can eventually obtain a sum of groups which exceeds M. Thus, the sequence has no limit.

Supplementary Problems

49.10 Write the first terms of the sequence whose general term is

(a) $\dfrac{1}{1+n}$ (c) $\dfrac{1}{3^n}$ (e) $\dfrac{n^2}{3n-2}$ (g) $(-1)^{n+1}\dfrac{1}{n!}$

(b) $\dfrac{1}{n+n\sqrt{n}}$ (d) $\dfrac{2n-1}{2n+3}$ (f) $(-1)^{n+1}\dfrac{2n+1}{2^{n+1}}$ (h) $\dfrac{n^2}{(2n)!}$

Ans.   (a) $\frac{1}{2},\frac{1}{3},\frac{1}{4},\frac{1}{5}$ (c) $\frac{1}{3},\frac{1}{9},\frac{1}{27},\frac{1}{81}$ (f) $\frac{3}{4},-\frac{5}{8},\frac{7}{16},-\frac{9}{32}$

　　　(b) $\dfrac{1}{2},\dfrac{1}{1+2\sqrt{2}},\dfrac{1}{1+3\sqrt{3}},\dfrac{1}{9}$ (d) $\frac{1}{5},\frac{3}{7},\frac{5}{9},\frac{7}{11}$ (g) $1,-\frac{1}{2},\frac{1}{6},-\frac{1}{24}$

　　　　　　　　　　　　　　　　　　　　　　　　　(e) $1,1,\frac{9}{7},\frac{8}{5}$ (h) $\frac{1}{2},\frac{1}{6},\frac{1}{80},\frac{1}{2520}$

49.11 Write the general term of each sequence.

(a) $1,\frac{1}{3},\frac{1}{5},\frac{1}{7},\ldots$ (c) $1,\frac{3}{5},\frac{2}{5},\frac{5}{17},\frac{3}{13},\ldots$ (f) $\frac{1}{3},-\frac{1}{15},\frac{1}{35},-\frac{1}{63},\ldots$

(b) $\dfrac{4}{1\cdot3},\dfrac{5}{2\cdot4},\dfrac{6}{3\cdot5},\dfrac{7}{4\cdot6},\cdots$ (d) $\frac{1}{2},\frac{3}{8},\frac{7}{24},\frac{15}{64},\cdots$ (g) $\frac{1}{2},\frac{3}{8},\frac{5}{16},\frac{9}{32},\ldots$

　　　　　　　　　　　　　　　　　　　　(e) $2,1,\frac{8}{9},1,\frac{32}{25},\frac{16}{9},\ldots$ (h) $\frac{1}{2},-x^2/4,x^4/6,-x^6/8,\ldots$

Ans.   (a) $\dfrac{1}{2n-1}$ (c) $\dfrac{n+1}{n^2+1}$ (e) $\dfrac{2^n}{n^2}$ (g) $\dfrac{1+2^{n-1}}{2^{n+1}}$

　　　(b) $\dfrac{n+3}{n(n+2)}$ (d) $\dfrac{2^n-1}{n\cdot2^n}$ (f) $\dfrac{(-1)^{n+1}}{(2n-1)(2n+1)}$ (h) $(-1)^{n+1}\dfrac{x^{2n-2}}{2n}$

49.12 Evaluate.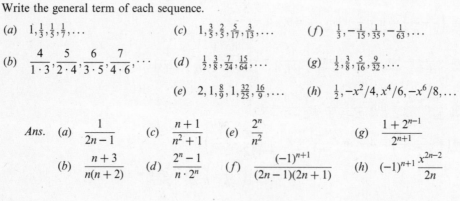

(a) $\displaystyle\lim_{n\to\infty}\left(2-\frac{1}{n}\right)$ (d) $\displaystyle\lim_{n\to\infty}\frac{2n^2+5n-6}{n^2+n-1}$ (g) $\displaystyle\lim_{n\to\infty}\frac{1}{2^n+1}$

(b) $\displaystyle\lim_{n\to\infty}\frac{3n+1}{3n-2}$ (e) $\displaystyle\lim_{n\to\infty}\frac{n}{n+1}$ (h) $\displaystyle\lim_{n\to\infty}\frac{2^n+1}{2^{n+1}+1}$

(c) $\displaystyle\lim_{n\to\infty}\frac{2n}{(n+1)(n+2)}$ (f) $\displaystyle\lim_{n\to\infty}\frac{(3n)!\,n}{(3n+1)!}$

Ans.   (a) 2   (b) 1   (c) 0   (d) 2   (e) 1   (f) $\frac{1}{3}$ (g) 0   (h) $\frac{1}{2}$

49.13 Explain why each of the following has no limit:

(a) $1,3,5,7,9,\ldots$ (c) $1,-2,4,-8,16,-32,\ldots$

(b) $1,0,1,0,1,0,\ldots$ (d) $\frac{1}{25},\frac{4}{25},\frac{9}{25},\frac{16}{25},\ldots$

49.14 Write out the next four terms given the recursive formula: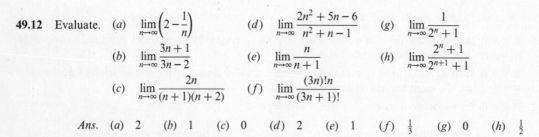

(a) $a_1=-1, a_{n+1}=\frac{1}{3}a_n$

(b) $a_1=2, a_2=3, a_{n+2}=a_{n+1}+\dfrac{a_n}{2}$

Ans.   (a) $-\frac{1}{3},-\frac{1}{9},-\frac{1}{27},-\frac{1}{81}$

　　　(b) $2+\frac{3}{2}, 3+\frac{7}{4}, \frac{7}{2}+\frac{19}{8}, \frac{19}{4}+\frac{47}{16}$

Chapter 50

Infinite Series

THE INDICATED SUM of the terms of an infinite sequence is called an *infinite series*. Let

$$s_1 + s_2 + s_3 + \cdots + s_n + \cdots \qquad (50.1)$$

be such a series and define the sequence of *partial sums*

$$S_1 = s_1, \quad S_2 = s_1 + s_2, \quad \ldots, \quad S_n = s_1, + s_2 + \cdots + s_n, \quad \ldots$$

If $\lim_{n\to\infty} S_n$ exists, the series *(50.1)* is called *convergent*; if $\lim_{n\to\infty} S_n = S$, the series is said to converge to S. If $\lim_{n\to\infty} S_n$ does not exist, the series is called *divergent*.

EXAMPLE 1

(*a*) Every infinite geometric series

$$a + ar + ar^2 + \cdots + ar^{n-1} + \cdots$$

is convergent if $|r| < 1$ and is divergent if $|r| \geq 1$. (See Problem 50.1.)

(*b*) The *harmonic series* $1 + \frac{1}{2} + \frac{1}{3} + \cdots + 1/n + \cdots$ is divergent. (See Problem 49.9 and Problem 50.2.)

A NECESSARY CONDITION THAT *(50.1)* BE CONVERGENT is $\lim_{n\to\infty} s_n = 0$; that is, if *(50.1)* is convergent then $\lim_{n\to\infty} s_n = 0$. However, the condition is *not sufficient* since the harmonic series is divergent although $\lim_{n\to\infty} s_n = \lim_{n\to\infty}(1/n) = 0$.

A SUFFICIENT CONDITION THAT *(50.1)* BE DIVERGENT is $\lim_{n\to\infty} s_n \neq 0$; that is, if $\lim_{n\to\infty} s_n$ exists and is different from 0, or if $\lim_{n\to\infty} s_n$ does not exist, the series is divergent. This, in turn, is not a necessary condition since the harmonic series is divergent although $\lim_{n\to\infty} s_n = 0$. (See Problem 50.3).

SERIES OF POSITIVE TERMS

COMPARISON TEST FOR CONVERGENCE of a series of positive terms.

   I.   If every term of a given series of positive terms is less than or equal to the corresponding term of a known convergent series from some point on in the series, the given series is convergent.

   II. If every term of a given series of positive terms is equal to or greater than the corresponding term of a known divergent series from some point on in the series, the given series is divergent.

The following series will be found useful in making comparison tests:

(a)   The geometric series $a + ar + ar^2 + \cdots + ar^n + \cdots$ which converges when $|r| < 1$ and diverges when $|r| \geq 1$

(b)   The p series $1 + \frac{1}{2^p} + \frac{1}{3^p} + \cdots + \frac{1}{n^p} + \cdots$ which converges for $p > 1$ and diverges for $p \leq 1$

(c)   Each new series tested

In comparing two series, it is not sufficient to examine the first few terms of each series. The *general terms must be compared*, since the comparison must be shown from some point on. (See Problems 50.4–50.6.)

THE RATIO TEST FOR CONVERGENCE.   If, in a series of positive terms, the *test ratio*

$$r_n = \frac{s_{n+1}}{s_n}$$

approaches a limit R as $n \to \infty$, the series is convergent if $R < 1$ and is divergent if $R > 1$. If $R = 1$, the test fails to indicate convergency or divergency. (See Problem 50.7.)

SERIES WITH NEGATIVE TERMS

A SERIES WITH ALL ITS TERMS NEGATIVE may be treated as the negative of a series with all of its terms positive.

ALTERNATING SERIES.   A series whose terms are alternately positive and negative, as

$$s_1 - s_2 + s_3 - \cdots + (-1)^{n-1} s_n + \cdots \qquad (50.2)$$

where each s is positive, is called an *alternating series*.

An alternating series (50.2) is convergent provided $s_n \geq s_{n+1}$, for every value of n, and $\lim\limits_{n \to \infty} s_n = 0$.

(See Problem 50.8.)

ABSOLUTELY CONVERGENT SERIES.   A series (50.1), $s_1 + s_2 + s_3 + \cdots + s_n + \cdots$ in which some of the terms are positive and some are negative is called *absolutely convergent* if the series of absolute values of the terms

$$|s_1| + |s_2| + |s_3| + \cdots + |s_n| + \cdots \qquad (50.3)$$

is convergent.

CONDITIONALLY CONVERGENT SERIES.   A series (50.1), where some of the terms are positive and some are negative, is called *conditionally convergent* if it is convergent but the series of absolute values of its terms is divergent.

EXAMPLE 2.   The series $1 - \frac{1}{2} + \frac{1}{3} - \frac{1}{4} + \cdots$ is convergent, but the series of absolute values of its terms $1 + \frac{1}{2} + \frac{1}{3} + \frac{1}{4} + \cdots$ is divergent. Thus, the given series is conditionally convergent.

THE GENERALIZED RATIO TEST.   Let (50.1) $s_1 + s_2 + s_3 + \cdots + s_n + \cdots$ be a series some of whose terms are positive and some are negative. Let

$$\lim_{n \to \infty} \frac{|s_{n+1}|}{|s_n|} = R$$

The series (50.1) is absolutely convergent if $R < 1$ and is divergent if $R > 1$. If $R = 1$, the test fails. (See Problem 50.9.)

Solved Problems

50.1 Examine the infinite geometric series $a + ar + ar^2 + \cdots + ar^n + \cdots$ for convergence and divergence.

The sum of the first n terms is $S_n = \dfrac{a(1 - r^n)}{1 - r} = \dfrac{a}{1 - r}(1 - r^n)$.

If $|r| < 1$, $\lim\limits_{n \to \infty} r^n = 0$; then $\lim\limits_{n \to \infty} S_n = \dfrac{a}{1 - r}$ and the series is convergent.

If $|r| > 1$, $\lim\limits_{n \to \infty} r^n$ does not exist and $\lim\limits_{n \to \infty} S_n$ does not exist; the series is divergent.

If $r = 1$, the series is $a + a + a + \cdots + a + \cdots$; then $\lim\limits_{n \to \infty} S_n = \lim\limits_{n \to \infty} na$ does not exist.

If $r = -1$, the series is $a - a + a - a + \cdots$; then $S_n = a$ or 0 according as n is odd or even, and $\lim\limits_{n \to \infty} S_n$ does not exist.

Thus, the infinite geometric series convergence to $\dfrac{a}{1 - r}$ when $|r| < 1$, and diverges when $|r| \geq 1$.

50.2 Show that the following series are convergent:

(a) $1 + \dfrac{2}{3} + \dfrac{4}{9} + \dfrac{8}{27} + \cdots + \dfrac{2^{n-1}}{3^{n-1}} + \cdots$.

This is a geometric series with ratio $r = \frac{2}{3}$; then $|r| < 1$ and the series is convergent.

(b) $2 - \frac{3}{2} + \frac{9}{8} - \frac{27}{32} + \cdots + (-1)^{n-1}2\left(\frac{3}{4}\right)^{n-1} + \cdots$.

This is a geometric series with ratio $r = -\frac{3}{4}$; then $|r| < 1$ and the series is convergent.

(c) $1 + \dfrac{1}{2^p} + \dfrac{1}{2^p} + \dfrac{1}{4^p} + \dfrac{1}{4^p} + \dfrac{1}{4^p} + \dfrac{1}{4^p} + \dfrac{1}{8^p} + \cdots, p > 1$.

This series may be rewritten as $1 + \dfrac{2}{2^p} + \dfrac{4}{4^p} + \dfrac{8}{8^p} + \cdots$, a geometric series with ratio $r = 2/2^p$. Since $|r| < 1$, when $p > 1$, the series is convergent.

50.3 Show that the following series are divergent:

(a) $2 + \dfrac{3}{2} + \dfrac{4}{3} + \cdots + \dfrac{n+1}{n} + \cdots$.

Since $\lim\limits_{n \to \infty} s_n = \lim\limits_{n \to \infty} \dfrac{n+1}{n} = \lim\limits_{n \to \infty}\left(1 + \dfrac{1}{n}\right) = 1 \neq 0$, the series is divergent.

(b) $\dfrac{1}{2} + \dfrac{3}{8} + \dfrac{5}{16} + \dfrac{9}{32} + \cdots + \dfrac{2^{n-1} + 1}{4 \cdot 2^{n-1}} + \cdots$.

Since $\lim\limits_{n \to \infty} s_n = \lim\limits_{n \to \infty} + \dfrac{2^{n-1} + 1}{4 \cdot 2^{n-1}} = \lim\limits_{n \to \infty} \dfrac{1 + 1/2^{n-1}}{4} = \dfrac{1}{4} \neq 0$, the series is divergent.

50.4 Show that $1 + \dfrac{1}{2^p} + \dfrac{1}{3^p} + \dfrac{1}{4^p} + \cdots + \dfrac{1}{n^p} + \cdots$ is divergent for $p \leq 1$ and convergent for $p > 1$.

For $p = 1$, the series is the harmonic series and is divergent.

For $p < 1$, including negative values, $\dfrac{1}{n^p} \geq \dfrac{1}{n}$, for every n. Since every term of the given series is equal to or greater than the corresponding term of the harmonic series, the given series is divergent.

For $p > 1$, compare the series with the convergent series

$$1 + \frac{1}{2^p} + \frac{1}{2^p} + \frac{1}{4^p} + \frac{1}{4^p} + \frac{1}{4^p} + \frac{1}{4^p} + \frac{1}{8^p} + \cdots \qquad (1)$$

of Problem 50.2 (c). Since each term of the given series is less than or equal to the corresponding term of series (1), the given series is convergent.

50.5 Use Problem 50.4 to determine whether the following series are convergent or divergent:

(a) $1 + \frac{1}{2\sqrt{2}} + \frac{1}{3\sqrt{3}} + \frac{1}{4\sqrt{4}} + \cdots$. The general term is $\frac{1}{n\sqrt{n}} = \frac{1}{n^{3/2}}$.

This is a p series with $p = \frac{3}{2} > 1$; the series is convergent.

(b) $1 + 4 + 9 + 16 + \cdots$. The general term is $n^2 = \frac{1}{n^{-2}}$.

This is a p series with $p = -2 < 1$; the series is divergent.

(c) $1 + \frac{\sqrt[4]{2}}{4} + \frac{\sqrt[4]{3}}{9} + \frac{\sqrt[4]{4}}{16} + \cdots$. The general term is $\frac{\sqrt[4]{n}}{n^2} = \frac{1}{n^{7/4}}$.

The series is convergent since $p = \frac{7}{4} > 1$.

50.6 Use the comparison test to determine whether each of the following is convergent or divergent:

(a) $1 + \frac{1}{2!} + \frac{1}{3!} + \frac{1}{4!} + \cdots$.

The general term $\frac{1}{n!} \le \frac{1}{n^2}$. Thus, the terms of the given series are less than or equal to the corresponding terms of the p series with $p = 2$. The series is convergent.

(b) $\frac{1}{2} + \frac{1}{3} + \frac{1}{5} + \frac{1}{9} + \cdots$.

The general term $\frac{1}{1 + 2^{n-1}} \le \frac{1}{2^{n-1}}$. Thus, the terms of the given series are less than or equal to the corresponding terms of the geometric series with $a = 1$ and $r = \frac{1}{2}$. The series is convergent.

(c) $\frac{2}{1} + \frac{3}{4} + \frac{4}{9} + \frac{5}{16} + \cdots$.

The general term $\frac{n+1}{n^2} = \frac{1}{n} + \frac{1}{n^2} > \frac{1}{n}$. Thus, the terms of the given series are equal to or greater than the corresponding terms of the harmonic series. The series is divergent.

(d) $\frac{1}{3} + \frac{1}{12} + \frac{1}{27} + \frac{1}{48} + \cdots$.

The general term $\frac{1}{3 \cdot n^2} \le \frac{1}{n^2}$. Thus, the terms of the given series are less than or equal to the corresponding terms of the p series with $p = 2$. The series is convergent.

(e) $1 + \frac{1}{2} + \frac{1}{3^2} + \frac{1}{4^3} + \frac{1}{5^4} + \cdots$.

The general term $\frac{1}{n^{n-1}} \le \frac{1}{n^2}$ for $n \ge 3$. Thus, neglecting the first two terms, the given series is term by term less than or equal to the corresponding terms of the p series with $p = 2$. The given series is convergent.

50.7   Apply the ratio test to each of the following. If it fails, use some other method to determine convergency or divergency.

   (a) $\frac{1}{2} + \frac{1}{2} + \frac{3}{8} + \frac{1}{4} + \cdots$ or $\frac{1}{2} + \frac{2}{2^2} + \frac{3}{2^3} + \frac{4}{2^4} + \cdots$.

   For this series $s_n = \dfrac{n}{2^n}$, $s_{n+1} = \dfrac{n+1}{2^{n+1}}$, and $r_n = \dfrac{s_{n+1}}{s_n} = \dfrac{n+1}{2^{n+1}} \cdot \dfrac{2^n}{n} = \dfrac{n+1}{2n}$. Then $R = \lim\limits_{n \to \infty} r_n =$

   $\lim\limits_{n \to \infty} \dfrac{n+1}{2n} = \lim\limits_{n \to \infty} \dfrac{1 + 1/n}{2} = \dfrac{1}{2} < 1$ and the series is convergent.

   (b) $3 + \frac{9}{2} + \frac{9}{2} + \frac{27}{8} + \frac{81}{40} + \cdots$ or $\frac{3}{1!} + \frac{3^2}{2!} + \frac{3^3}{3!} + \cdots$.

   Here $s_n = \dfrac{3^n}{n!}$, $s_{n+1} = \dfrac{3^{n+1}}{(n+1)!}$, and $r_n = \dfrac{3^{n+1}}{(n+1)!} \cdot \dfrac{n!}{3^n} = \dfrac{3}{n+1}$. Then $R = \lim\limits_{n \to \infty} \dfrac{3}{n+1} = 0$ and the series is convergent.

   (c) $\frac{1}{1 \cdot 1} + \frac{1}{2 \cdot 3} + \frac{1}{3 \cdot 5} + \frac{1}{4 \cdot 7} + \cdots$.

   Here $s_n = \dfrac{1}{n(2n-1)}$, $s_{n+1} = \dfrac{1}{(n+1)(2n+1)}$, and $r_n = \dfrac{n(2n-1)}{(n+1)(2n+1)}$.

   Then $R = \lim\limits_{n \to \infty} \dfrac{n(2n-1)}{(n+1)(2n+1)} = \lim\limits_{n \to \infty} \dfrac{2 - 1/n}{(1 + 1/n)(2 + 1/n)} = 1$ and the test fails.

   Since $\dfrac{1}{n(2n-1)} \leq \dfrac{1}{n^2}$, the given series is term by term less than or equal to the convergent p series, with $p = 2$. The given series is convergent.

   (d) $\frac{2}{1^2 + 1} + \frac{2^3}{2^2 + 2} + \frac{2^5}{3^2 + 3} + \cdots$.

   Here $s_n = \dfrac{2^{2n-1}}{n^2 + n}$, $s_{n+1} = \dfrac{2^{2n+1}}{(n+1)^2 + (n+1)}$, and $r_n = \dfrac{2^{2n+1}}{(n+1)(n+2)} \cdot \dfrac{n(n+1)}{2^{2n-1}} = \dfrac{4n}{n+2}$. Then $R =$

   $\lim\limits_{n \to \infty} \dfrac{4}{1 + 2/n} = 4$ and the series is divergent.

   (e) $\frac{1}{5} + \frac{2}{25} + \frac{6}{125} + \frac{24}{625} + \cdots$.

   In this series $s_n = \dfrac{n!}{5^n}$, $s_{n+1} = \dfrac{(n+1)!}{5^{n+1}}$, and $r_n = \dfrac{(n+1)!}{5^{n+1}} \cdot \dfrac{5^n}{n!} = \dfrac{n+1}{5}$. Now $\lim\limits_{n \to \infty} r_n$ does not exist.

   However, since $s_n \to \infty$ as $n \to \infty$, the series is divergent.

50.8   Test the following alternating series for convergence:

   (a) $1 - \frac{1}{3} + \frac{1}{5} - \frac{1}{7} + \cdots$.

   $s_n > s_{n+1}$, for all values of n, and $\lim\limits_{n \to \infty} s_n = \lim\limits_{n \to \infty} \dfrac{1}{2n-1} = 0$. The series is convergent.

   (b) $\frac{1}{2^3} - \frac{2}{3^3} + \frac{3}{4^3} - \frac{4}{5^3} + \cdots$.

   $s_n > s_{n+1}$, for all values of n, and $\lim\limits_{n \to \infty} s_n = \lim\limits_{n \to \infty} \dfrac{n}{(n+1)^3} = 0$. The series is convergent.

50.9   Investigate the following for absolute convergence, conditional convergence, or divergence:

   (a) $1 - \frac{1}{2} + \frac{1}{4} - \frac{1}{8} + \cdots$.

   Here $|s_n| = \dfrac{1}{2^{n-1}}$, $|s_{n+1}| = \dfrac{1}{2^n}$, and $R = \lim\limits_{n \to \infty} \dfrac{|s_{n+1}|}{|s_n|} = \lim\limits_{n \to \infty} \dfrac{2^{n-1}}{2^n} = \dfrac{1}{2} < 1$. The series is absolutely convergent.

(b) $1 - \dfrac{4}{1!} + \dfrac{4^2}{2!} - \dfrac{4^3}{3!} + \cdots$.

Here $|s_n| = \dfrac{4^{n-1}}{(n-1)!}$, $|s_{n+1}| = \dfrac{4^n}{n!}$, and $R = \lim\limits_{n \to \infty} \dfrac{4}{n} = 0$. The series is absolutely convergent.

(c) $\dfrac{1}{2 - \sqrt{2}} - \dfrac{1}{3 - \sqrt{3}} + \dfrac{1}{4 - \sqrt{4}} - \dfrac{1}{5 - \sqrt{5}} + \cdots$.

The ratio test fails here.

Since $\dfrac{1}{n + 1 - \sqrt{n+1}} > \dfrac{1}{n + 2 - \sqrt{n+2}}$ and $\lim\limits_{n \to \infty} \dfrac{1}{n + 1 - \sqrt{n+1}} = 0$, the series is convergent.

Since $\dfrac{1}{n + 1 - \sqrt{n+1}} > \dfrac{1}{n + 1}$ for all values of n, the series of absolute values is term by term greater

than the harmonic series, and thus is divergent. The given series is conditionally convergent.

Supplementary Problems

50.10 Investigate each of the following series for convergence or divergence:

(a) $\dfrac{1}{3} + \dfrac{1}{6} + \dfrac{1}{11} + \cdots + \dfrac{1}{2^n + n} + \cdots$

(f) $1 + \dfrac{1}{\sqrt[3]{2}} + \dfrac{1}{\sqrt[3]{3}} + \cdots + \dfrac{1}{\sqrt[3]{n}} + \cdots$

(b) $\dfrac{1}{2} + \dfrac{1}{4} + \dfrac{1}{6} + \cdots + \dfrac{1}{2n} + \cdots$

(g) $\dfrac{1}{2} + \dfrac{1}{2 \cdot 2^2} + \dfrac{1}{3 \cdot 2^3} + \cdots + \dfrac{1}{n \cdot 2^n} + \cdots$

(c) $1 + \dfrac{1}{3} + \dfrac{1}{5} + \cdots + \dfrac{1}{2n - 1} + \cdots$

(h) $\dfrac{1}{2} + \dfrac{2}{3} + \dfrac{3}{4} + \cdots + \dfrac{n}{n + 1} + \cdots$

(d) $\dfrac{2}{1!} + \dfrac{2^2}{2!} + \dfrac{2^3}{3!} + \cdots + \dfrac{2^n}{n!} + \cdots$

(i) $\dfrac{1}{1 \cdot 3} + \dfrac{1}{3 \cdot 5} + \dfrac{1}{5 \cdot 7} + \cdots + \dfrac{1}{(2n - 1)(2n + 1)} + \cdots$

(e) $2 + \dfrac{1}{2} + \dfrac{8}{27} + \dfrac{1}{4} + \cdots + \dfrac{2^n}{n^3} + \cdots$

(j) $1 + \dfrac{2^2 + 1}{2^3 + 1} + \dfrac{3^2 + 1}{3^3 + 1} + \cdots + \dfrac{n^2 + 1}{n^3 + 1} + \cdots$

Ans. (a) Convergent (c) Divergent (e) Divergent (g) Convergent (i) Convergent
(b) Divergent (d) Convergent (f) Divergent (h) Divergent (j) Divergent

50.11 Investigate the following alternating series for convergence or divergence:

(a) $\frac{1}{4} - \frac{1}{10} + \frac{1}{28} - \frac{1}{82} + \cdots$

(f) $\frac{2}{3} - \frac{3}{4} \cdot \frac{1}{2} + \frac{4}{5} \cdot \frac{1}{3} - \frac{5}{6} \cdot \frac{1}{4} + \cdots$

(b) $2 - \frac{3}{2} + \frac{4}{3} - \frac{5}{4} + \cdots$

(g) $\dfrac{2}{2 \cdot 3} - \dfrac{2^2}{3 \cdot 4} + \dfrac{2^3}{4 \cdot 5} - \dfrac{2^4}{5 \cdot 6} + \cdots$

(c) $1 - \frac{1}{2} + \frac{1}{3} - \frac{1}{4} + \cdots$

(h) $2 - \dfrac{2^3}{3!} + \dfrac{2^5}{5!} - \dfrac{2^7}{7!} + \cdots$

(d) $\frac{1}{2} - \frac{2}{3} + \frac{3}{4} - \frac{4}{5} + \cdots$

(e) $\frac{1}{4} - \frac{3}{6} + \frac{5}{8} - \frac{7}{10} + \cdots$

Ans. (a) Abs. Conv. (c) Cond. Conv. (e) Divergent (g) Divergent
(b) Divergent (d) Divergent (f) Cond. Conv. (h) Abs. Conv.

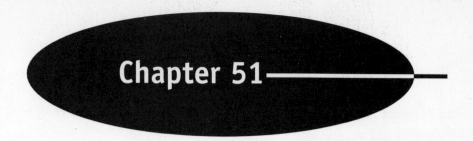

Chapter 51

Power Series

INFINITE SERIES OF THE FORM

$$c_0 + c_1 x + c_2 x^2 + \cdots + c_{n-1}(x)^{n-1} + \cdots \qquad (51.1)$$

and

$$c_0 + c_1(x-a) + c_2(x-a)^2 + \cdots + c_{n-1}(x-a)^{n-1} + \cdots, \qquad (51.2)$$

where $a, c_0, c_1, c_2, \ldots$ are constants, are called *power series*. The first is called a power series in x and the second a power series in $(x-a)$.

The power series (51.1) converges for $x = 0$ and (51.2) converges for $x = a$. Both series may converge for other values of x but not necessarily for every finite value of x. Our problem is to find for a given power series all values of x for which the series converges. In finding this set of values, called the *interval of convergence* of the series, the generalized ratio test of Chapter 50 will be used.

EXAMPLE 1. Find the interval of convergence of the series

$$x + x^2/2 + x^3/3 + \cdots$$

Since

$$|s_n| = \left|\frac{x^n}{n}\right| \quad \text{and} \quad |s_{n+1}| = \left|\frac{x^{n+1}}{n+1}\right|,$$

$$R = \lim_{n \to \infty} \frac{|s_{n+1}|}{|s_n|} = \lim_{n \to \infty} \left|\frac{x^{n+1}}{n+1} \cdot \frac{n}{x^n}\right| = \lim_{n \to \infty} \left|\frac{n}{n+1} x\right| = |x|.$$

Then, by the ratio test, the given series is convergent for all values of x such that $|x| < 1$, that is, for $-1 < x < 1$; the series is divergent for all values of x such that $|x| > 1$, that is, for $x < -1$ and $x > 1$; and the test fails for $x = \pm 1$.

But, when $x = 1$ the series is $1 + \frac{1}{2} + \frac{1}{3} + \frac{1}{4} + \cdots$ and is divergent, and when $x = -1$ the series is $-1 + \frac{1}{2} - \frac{1}{3} + \frac{1}{4} - \cdots$ and is convergent.

Thus, the series converges on the interval $-1 \le x < 1$. This interval may be represented graphically as in Fig. 51-1. The solid line represents the interval on which the series converges, the thin lines the intervals on which the series diverges. The solid circle represents the end point for which the series converges, the open circle the end point at which the series diverges. (See Problems 51.1–51.8.)

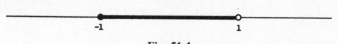

Fig. 51-1

389

Solved Problems

In Problems 51.1–51.6, find the interval of convergence *including* the end points.

51.1 $1 + \dfrac{x}{1!} + \dfrac{x^2}{2!} + \dfrac{x^3}{3!} + \cdots$.

For this series

$$|s_n| = \left|\frac{x^{n-1}}{(n-1)!}\right|, \qquad |s_{n+1}| = \left|\frac{x^n}{n!}\right|, \qquad \text{and} \qquad R = \lim_{n \to \infty} \frac{|s_{n+1}|}{|s_n|} = \lim_{n \to \infty}\left|\frac{x^n}{n!} \cdot \frac{(n-1)!}{x^{n-1}}\right| = \lim_{n \to \infty}\left|\frac{x}{n}\right| = 0.$$

The series is *everywhere convergent*; that is, it is convergent for all finite values of x.

51.2 $1 + x + 2x^2 + 3x^3 + \cdots$.

Here $|s_n| = |(n-1)x^{n-1}|, |s_{n+1}| = |nx^n|$, and $R = \lim_{n \to \infty}\left|\frac{nx^n}{(n-1)x^{n-1}}\right| = \lim_{n \to \infty}\left|\frac{n}{n-1}x\right| = |x|$.

The series converges on the interval $-1 < x < 1$ and diverges on the intervals $x < -1$ and $x > 1$.

When $x = 1$, the series is $1 + 1 + 2 + 3 + \cdots$ and is divergent.

When $x = -1$, the series is $1 - 1 + 2 - 3 + \cdots$ and is divergent.

The interval of convergence $-1 < x < 1$ is indicated in Fig. 51-2.

Fig. 51-2

51.3 $1 + \dfrac{x}{2} + \dfrac{x^2}{4} + \dfrac{x^3}{8} + \cdots$.

Here $|s_n| = \left|\dfrac{x^{n-1}}{2^{n-1}}\right|, |s_{n+1}| = \left|\dfrac{x^n}{2^n}\right|$, and $R = \lim_{n \to \infty}\left|\dfrac{x^n}{2^n} \cdot \dfrac{2^{n-1}}{x^{n-1}}\right| = \lim_{n \to \infty}\left|\dfrac{x}{2}\right| = \dfrac{1}{2}|x|$.

The series converges for all values of x such that $\frac{1}{2}|x| < 1$, that is, for $-2 < x < 2$, and diverges for $x < -2$ and $x > 2$.

For $x = 2$, the series is $1 + 1 + 1 + 1 + \cdots$ and is divergent.

For $x = -2$, the series is $1 - 1 + 1 - 1 + \cdots$ and is divergent.

The interval of convergence $-2 < x < 2$ is indicated in Fig. 51-3.

Fig. 51-3

51.4 $\dfrac{1!}{x+1} - \dfrac{2!}{(x+1)^2} + \dfrac{3!}{(x+1)^3} + \cdots$.

For this series $|s_n| = \left|\dfrac{n!}{(x+1)^n}\right|, |s_{n+1}| = \left|\dfrac{(n+1)!}{(x+1)^{n+1}}\right|$, and $R = \lim_{n \to \infty}\left|\dfrac{(n+1)!}{(x+1)^{n+1}} \cdot \dfrac{(x+1)^n}{n!}\right| =$ $\lim_{n \to \infty}\left|\dfrac{n+1}{x+1}\right|$ does not exist.

Thus, the series diverges for every value of x.

51.5 $\dfrac{x+3}{1\cdot 4}+\dfrac{(x+3)^2}{2\cdot 4^2}+\dfrac{(x+3)^3}{3\cdot 4^3}+\cdots.$

For this series $|s_n|=\left|\dfrac{(x+3)^n}{n\cdot 4^n}\right|$, $|s_{n+1}|=\left|\dfrac{(x+3)^{n+1}}{(n+1)\cdot 4^{n+1}}\right|$, and

$$R=\lim_{n\to\infty}\left|\dfrac{(x+3)^{n+1}}{(n+1)4^{n+1}}\cdot\dfrac{n4^n}{(x+3)^n}\right|=\lim_{n\to\infty}\left|\dfrac{n}{n+1}\cdot\dfrac{x+3}{4}\right|=\dfrac{1}{4}|x+3|.$$

The series converges for all values of x such that $\frac{1}{4}|x+3|<1$, that is, for $-4<x+3<4$ or $-7<x<1$, and diverges for $x<-7$ and $x>1$.

For $x=-7$, the series is $-1+\frac{1}{2}-\frac{1}{3}+\cdots$ and is convergent.

For $x=1$, the series is $1+\frac{1}{2}+\frac{1}{3}+\cdots$ and is divergent.

The interval of convergence $-7\le x<1$ is indicated in Fig. 51-4.

Fig. 51-4

51.6 $\dfrac{1}{1\cdot 2\cdot 3}-\dfrac{(x+2)^2}{2\cdot 3\cdot 4}+\dfrac{(x+2)^4}{3\cdot 4\cdot 5}-\cdots.$

Here $|s_n|=\left|\dfrac{(x+2)^{2n-2}}{n(n+1)(n+2)}\right|$, $|s_{n+1}|=\left|\dfrac{(x+2)^{2n}}{(n+1)(n+2)(n+3)}\right|$, and

$$R=\lim_{n\to\infty}\left|\dfrac{(x+2)^{2n}}{(n+1)(n+2)(n+3)}\cdot\dfrac{n(n+1)(n+2)}{(x+2)^{2n-2}}\right|=\lim_{n\to\infty}\left|\dfrac{n}{n+3}(x+2)^2\right|=(x+2)^2.$$

The series converges for all values of x such that $(x+2)^2<1$, that is, for $-3<x<-1$, and diverges for $x<-3$ and $x>-1$.

For $x=-3$ and $x=-1$ the series is $\dfrac{1}{1\cdot 2\cdot 3}-\dfrac{1}{2\cdot 3\cdot 4}+\dfrac{1}{3\cdot 4\cdot 5}-\cdots$ and is convergent.

The interval of convergence $-3\le x\le-1$ is indicated in Fig. 51-5.

Fig. 51-5

51.7   Expand $(1+x)^{-1}$ as a power series in x and examine for convergence.

By division, $\dfrac{1}{x+1}=1-x+x^2-x^3+x^4-\cdots$. Then $R=\lim_{n\to\infty}\left|\dfrac{x^{n+1}}{x^n}\right|=|x|.$

The series converges for $-1<x<1$, and diverges for $x<-1$ and $x>1$.

For $x=1$, the series is $1-1+1-1+\cdots$ and is divergent.

For $x=-1$, the series is $1+1+1+1+\cdots$ and is divergent.

The interval of convergence $-1<x<1$ is indicated in Fig. 51-6.

Fig. 51-6

Thus, the series $1 - x + x^2 - x^3 + x^4 - \cdots$ *represents* the function $f(x) = (1 + x)^{-1}$ for all x such that $|x| < 1$. It does not represent the function for, say, $x = -4$. Note that $f(-4) = -\frac{1}{3}$, while for $x = -4$ the series is $1 + 4 + 16 + 64 + \cdots$.

51.8   Expand $(1 + x)^{1/2}$ in a power series in x and examine for convergence.

By the binomial theorem

$$(1 + x)^{1/2} = 1 + \frac{1}{2}x + \frac{(\frac{1}{2})(-\frac{1}{2})}{1 \cdot 2}x^2 + \frac{(\frac{1}{2})(-\frac{1}{2})(-\frac{3}{2})}{1 \cdot 2 \cdot 3}x^3 + \frac{(\frac{1}{2})(-\frac{1}{2})(-\frac{3}{2})(-\frac{5}{2})}{1 \cdot 2 \cdot 3 \cdot 4}x^4 + \cdots$$

Except for $n = 1$,

$$|s_n| = \left| \frac{(\frac{1}{2})(-\frac{1}{2})(-\frac{3}{2}) \cdots (-2n + 5)/2}{(n - 1)!}x^{n-1} \right|,$$

$$|s_{n+1}| = \left| \frac{(\frac{1}{2})(-\frac{1}{2})(-\frac{3}{2}) \cdots (-2n + 3)/2}{n!}x^n \right|, \quad \text{and} \quad R = \lim_{n \to \infty} \left| \frac{-2n + 3}{2n}x \right| = |x|.$$

The series is convergent for $-1 < x < 1$, and divergent for $x < -1$ and $x > 1$. An investigation at the end points is beyond the scope of this book.

Supplementary Problems

In Problems 51.9–51.21 find the interval of convergence including the end points.

51.9 $1 + x^2 + x^4 + \cdots + x^{2n-2} + \cdots$

   Ans. $-1 < x < 1$

51.10 $\dfrac{x}{1 \cdot 2} + \dfrac{x^2}{2 \cdot 3} + \dfrac{x^3}{3 \cdot 4} + \cdots + \dfrac{x^n}{n(n + 1)} + \cdots$

   Ans. $-1 \leq x \leq 1$

51.11 $\dfrac{x}{1 \cdot 3} + \dfrac{x^2}{2 \cdot 3^2} + \dfrac{x^3}{3 \cdot 3^3} + \cdots + \dfrac{x^n}{n \cdot 3^n} + \cdots$

   Ans. $-3 \leq x < 3$

51.12 $\dfrac{x}{1^2 + 1} + \dfrac{x^2}{2^2 + 1} + \dfrac{x^3}{3^2 + 1} + \cdots + \dfrac{x^n}{n^2 + 1} + \cdots$

   Ans. $-1 \leq x \leq 1$

51.13 $\dfrac{x^2}{4} - \dfrac{x^4}{8} + \dfrac{x^6}{16} - \cdots + (-1)^{n-1}\dfrac{x^{2n}}{2^{n+1}} + \cdots$

   Ans. $-\sqrt{2} < x < \sqrt{2}$

51.14 $\dfrac{1 \cdot x}{2 \cdot 1} - \dfrac{1 \cdot 3 \cdot x^3}{2 \cdot 4 \cdot 3} + \dfrac{1 \cdot 3 \cdot 5 \cdot x^5}{2 \cdot 4 \cdot 6 \cdot 5} - \dfrac{1 \cdot 3 \cdot 5 \cdot 7 \cdot x^7}{2 \cdot 4 \cdot 6 \cdot 8 \cdot 7} + \cdots$

   Ans. $-1 \leq x \leq 1$

51.15 $\dfrac{1\cdot 1}{3\cdot 2}x+\dfrac{3\cdot 2}{5\cdot 3}x^2+\dfrac{5\cdot 3}{7\cdot 4}x^3+\dfrac{7\cdot 4}{9\cdot 5}x^4+\cdots$

      Ans. $-1<x<1$

51.16 $(x-2)+\tfrac{1}{2}(x-2)^2+\tfrac{1}{3}(x-2)^3+\tfrac{1}{4}(x-2)^4+\cdots$

      Ans. $1\le x<3$

51.17 $\dfrac{x+1}{1\cdot 2}-\dfrac{(x+1)^2}{3\cdot 2^2}+\dfrac{(x+1)^3}{5\cdot 2^3}-\dfrac{(x+1)^4}{7.2^4}+\cdots$

      Ans. $-3<x\le 1$

51.18 $\dfrac{x-a}{b}+\dfrac{(x-a)^2}{b^2}+\dfrac{(x-a)^3}{b^3}+\dfrac{(x-a)^4}{b^4}+\cdots$

      Ans. $a-b<x<a+b$

51.19 $\dfrac{x-2}{x}+\dfrac{1}{2}\left(\dfrac{x-2}{x}\right)^2+\dfrac{1}{3}\left(\dfrac{x-2}{x}\right)^3+\dfrac{1}{4}\left(\dfrac{x-2}{x}\right)^4+\cdots$

      Ans. $x\ge 1$

51.20 $x+x^2/2^2+x^3/3^3+x^4/4^4+\cdots+x^n/n^n+\cdots$

      Ans.   All values of x.

51.21 $x-2^2x^2+3^3x^3-4^4x^4+\cdots$

      Ans. $x=0$

Chapter 52

Polar Coordinates

IN THE POLAR COORDINATE SYSTEM a point in the plane is located by giving its position relative to a fixed point and a fixed line (direction) through the fixed point. The fixed point O (see Fig. 52-1) is called the *pole* and the fixed half line OA is called the *polar axis*.

Let θ denote the smallest positive angle measured counterclockwise in degrees or radians from OA to OB, and let r denote the (positively) directed distance OP. Then P is uniquely determined when r and θ are known. These two measures constitute the *polar coordinates* of P and we write $P(r, \theta)$. The quantity r is called the *radius vector* and θ is called the *vectorial angle* of P. Note that a positive direction, indicated by the arrow, has been assigned on the half line $\overrightarrow{OB}$.

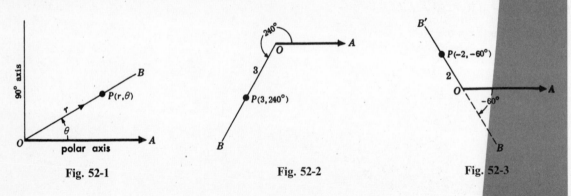

| Fig. 52-1 | Fig. 52-2 | Fig. 52-3 |

EXAMPLE 1. Locate the point $P(3, 240°)$ or $P(3, 4\pi/3)$. Refer to Fig. 52-2.

Lay off the vectorial angle $\theta = m\angle AOB = 240°$, measured counterclockwise from $\overrightarrow{OA}$, and on $\overrightarrow{OB}$ locate P such that $r = OP = 3$.

In the paragraph above we have restricted r and θ so that $r \geq 0$ and $0° \leq \theta < 360°$. In general, these restrictions will be observed; however, at times it will be more convenient to permit r and θ to have positive or negative values. If θ is negative and r is positive, we lay off the angle $\theta = \angle AOB$, measured clockwise from $\overrightarrow{OA}$, and locate P on $\overrightarrow{OB}$ so that $OP = r$. If r is negative, we lay off $\theta = \angle AOB$, extend $\overrightarrow{OB}$ through the pole to B$'$, and locate P on $\overrightarrow{OB'}$ a distance $|r|$ from O.

EXAMPLE 2. Locate the point $P(-2, 60°)$ or $P(-2, -\pi/3)$. Refer to Fig. 52-3.

Lay off the vectorial angle $\theta = \angle AOB = 60°$, measured clockwise from OA, extend OB through the pole to B$'$, and on $\overrightarrow{OB'}$ locate P a distance 2 units from O. (See Problems 52.1–52.2.)

394

Although not a part of the polar system, it will be helpful at times to make use of a half line, called the 90° axis, which issues from the pole perpendicular to the polar axis.

TRANSFORMATIONS BETWEEN POLAR AND RECTANGULAR COORDINATES.
If the pole and polar axis of the polar system coincide respectively with the origin and positive x axis of the rectangular system, and if P has rectangular coordinates (x, y) and polar coordinates (r, θ), then the following relations hold:

(1) $x = r\cos\theta$ (4) $\theta = \arctan y/x$
(2) $y = r\sin\theta$ (5) $\sin\theta = y/r$ and $\cos\theta = x/r$
(3) $r = \sqrt{x^2 + y^2}$

If relations (3)–(5) are to yield the restricted set of coordinates of the section above, θ is to be taken as the smallest positive angle satisfying (5) or, what is equivalent, θ is the smallest positive angle satisfying (4) and terminating in the quadrant in which $P(x, y)$ lies.

EXAMPLE 3. Find the rectangular coordinates of $P(3, 300°)$.
Here $r = 3$ and $\theta = 300°$; then $x = r\cos\theta = 3\cos 300° = 3(\frac{1}{2}) = \frac{3}{2}$, $y = r\sin\theta = 3\sin 300° = 3(-\frac{1}{2}\sqrt{3}) = -3\sqrt{3}/2$, and the rectangular coordinates are $(\frac{3}{2}, -3\sqrt{3}/2)$.

EXAMPLE 4. Find the polar equation of the circle whose rectangular equation is $x^2 + y^2 - 8x + 6y - 2 = 0$.
Since $x = r\cos\theta$, $y = r\sin\theta$, and $x^2 + y^2 = r^2$, the polar equation is $r^2 - 8r\cos\theta + 6r\sin\theta - 2 = 0$. (See Problems 52.3–52.5.)

CURVE SKETCHING IN POLAR COORDINATES.
Preliminary to sketching the locus of a polar equation, we discuss symmetry, extent, etc., as in the case of rectangular equations. However, there are certain complications at times due to the fact that in polar coordinates a given curve may have more than one equation.

EXAMPLE 5. Let $P(r, \theta)$ be an arbitrary point on the curve $r = 4\cos\theta - 2$. Now P has other representations: $(-r, \theta + \pi), (-r, \theta - \pi), (r, \theta - 2\pi), \dots$.
Since (r, θ) satisfies the equation $r = 4\cos\theta - 2$, $(-r, \theta + \pi)$ satisfies the equation $-r = 4\cos(\theta + \pi) - 2 = -4\cos\theta - 2$ or $r = 4\cos\theta + 2$. Thus, $r = 4\cos\theta - 2$ and $r = 4\cos\theta + 2$ are equations of the same curve. Such equations are called *equivalent*. The reader will show that $(-r, \theta - \pi)$ satisfies $r = 4\cos\theta + 2$ and $(r, \theta - 2\pi)$ satisfies $r = 4\cos\theta - 2$.

EXAMPLE 6. Show that point $A(-1, \pi/6)$ is on the ellipse $r = \dfrac{3}{4 + 2\sin\theta}$.
Note that the given coordinates do not satisfy the given equation.

First Solution. Another set of coordinates for A is $(1, 7\pi/6)$. Since these coordinates satisfy the equation, A is on the ellipse.

Second Solution. An equivalent equation for the ellipse is

$$-r = \frac{3}{4 + 2\sin(\theta - \pi)} \quad \text{or} \quad r = \frac{-3}{4 - 2\sin\theta}$$

Since the given coordinates satisfy this equation, A is on the ellipse.

SYMMETRY.   A locus is symmetric with respect to the polar axis if an equivalent equation is obtained
when

(*a*) θ is replaced by $-\theta$, or
(*b*) θ is replaced by $\pi - \theta$ and r by $-r$ in the given equation.

A locus is symmetric with respect to the 90° axis if an equivalent equation is obtained when

(*a*) θ is replaced by $\pi - \theta$, or
(*b*) θ is replaced by $-\theta$ and r by $-r$ in the given equation.

A locus is symmetric with respect to the pole if an equivalent equation is obtained when

(*a*) θ is replaced by $\pi + \theta$, or
(*b*) r is replaced by $-r$ in the given equation.

EXTENT.   The locus whose polar equation is $r = f(\theta)$ is a closed curve if r is real and finite for all
values of θ, but is not a closed curve if there are values of one variable which make the
other become infinite.

The equation should also be examined for values of one variable which make the other
imaginary.

At times, as in the equation $r = a(1 + \sin\theta)$, the values of θ which give r its maximum
values can be readily determined. Since the maximum value of $\sin\theta$ is 1, the maximum value
of r is $2a$ which it assumes when $\theta = \frac{1}{2}\pi$.

DIRECTIONS AT THE POLE.   Unlike all other points, the pole has infinitely many pairs of coordinates
$(0, \theta)$ when θ is restricted to $0° \leq \theta < 360°$. While two such pairs $(0, \theta_1)$
and $(0, \theta_2)$ define the pole, they indicate different directions (measured
from the polar axis) there. Thus, the values of θ for which $r = f(\theta) = 0$
give the directions of the tangents to the locus $r = f(\theta)$ at the pole.

POINTS ON THE LOCUS.   We may find as many points on a locus as desired by assigning values to θ
in the given equation and solving for the corresponding values of r.

EXAMPLE 7.   Discuss and sketch the locus of the cardioid $r = a(1 - \sin\theta)$.

Symmetry.   An equivalent equation is obtained when θ is replaced by $\pi - \theta$; the locus is symmetric with respect to
the 90° axis.

Extent.   Since r is real and $\leq 2a$ for all values of θ, the locus is a closed curve, lying within a circle of radius $2a$ with
center at the pole. Since $\sin\theta$ is of period 2π, the complete locus is described as θ varies from 0 to 2π.

Direction at the pole.   When $r = 0$, $\sin\theta = 1$ and $\theta = \frac{1}{2}\pi$. Thus, the locus is tangent to the 90° axis at the pole.

After locating the points in Table 52.1 and making use of symmetry of the locus with respect to the 90° axis, we
obtain the required curve as shown in Fig. 52-4. (See Problems 52.11–52.17.)

Table 52.1

θ	$\frac{1}{2}\pi$	$2\pi/3$	$3\pi/4$	$5\pi/6$	π	$7\pi/6$	$5\pi/4$	$4\pi/3$	$3\pi/2$
r	0	$0.13a$	$0.29a$	$0.5a$	a	$1.5a$	$1.71a$	$1.87a$	$2a$

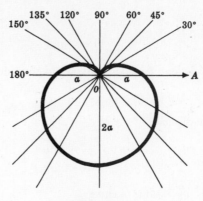

Fig. 52-4

INTERSECTIONS OF POLAR CURVES. It is to be expected that in finding the points of intersection of two curves with polar equations $r = f_1(\theta)$ and $r = f_2(\theta)$, we set $f_1(\theta) = f_2(\theta)$ and solve for θ. However, because of the multiplicity of representations both of the coordinates of a point and the equation of a curve, this procedure will fail at times to account for all of the intersections. Thus, it is a better policy to determine from a figure the exact number of intersections before attempting to find them.

EXAMPLE 8.   Since each of the circles $r = 2\sin\theta$ and $r = 2\cos\theta$ passes through the pole, the circles intersect in the pole and in one other point. See Fig. 52-5. Since each locus is completely described on the interval 0 to π, we set $2\sin\theta = 2\cos\theta$ and solve for θ on this interval. The solution $\theta = \frac{1}{4}\pi$ yields the point $(\sqrt{2}, \frac{1}{4}\pi)$.

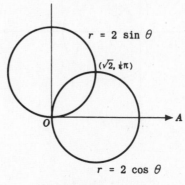

Fig. 52-5

Analytically we may determine whether or not the pole is a point of intersection by setting $r = 0$ in each of the equations and solving for θ. Setting $\sin\theta = 0$ we find $\theta = 0$, and setting $\cos\theta = 0$ we find $\theta = \frac{1}{2}\pi$. Since both equations have solutions, the pole is a point of intersection. The procedure above did not yield this solution since the coordinates of the pole $(0, 0)$ satisfy $r = 2\sin\theta$ while the coordinates $(0, \frac{1}{2}\pi)$ satisfy $r = 2\cos\theta$. (See Problems 52.18–52.19.)

SLOPE OF A POLAR CURVE.   We state the following results without proof:

Given a polar function, $r = f(\theta)$,

$$\frac{dy}{dx} = \frac{r'\sin\theta + r\cos\theta}{r'\cos\theta - r\sin\theta} \qquad (1)$$

If $f(x) = \sin x$, then

$$f'(x) = \cos x \qquad (2)$$

and if $g(x) = \cos x$, then

$$g'(x) = -\sin x \tag{3}$$

Using (1), (2), and (3) above, we can find derivatives and thus slopes for polar curves.

EXAMPLE 9.   Find the slope of the cardioid

$$r = 2(1 + \cos \theta) \ \text{ at } \ \theta = \frac{\pi}{3}.$$

$$\frac{dy}{dx} = \text{slope of curve}$$

$$= \frac{r'\sin \theta + r\cos \theta}{r'\cos \theta - r\sin \theta}$$

$$= \frac{(-2\sin \theta)\,(\sin \theta) + 2(1 + \cos \theta)\,(\cos \theta)}{(-2\sin \theta)\,(\cos \theta) - 2(1 + \cos \theta)\,(\sin \theta)}$$

At $\theta = \dfrac{\pi}{3}, \dfrac{dy}{dx} = 0$. This indicates that $r = 2(1 + \cos \theta)$ has a horizontal tangent line at $\theta = \dfrac{\pi}{3}$. See Fig. 52-6 and see Problem 52.20.

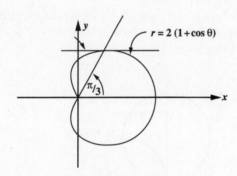

Fig. 52-6

Solved Problems

52.1   Locate the following points and determine which coincide with $P(2, 150°)$ and which with $Q(2, 30°)$:

(a) $A(2, 750°)$ (b) $B(-2, -30°)$ (c) $C(-2, 330°)$ (d) $D(-2, -150°)$ (e) $E(2, -210°)$

The points B, C, and E coincide with P; the points A and D coincide with Q. See Figs. 52-7(a)–(e).

52.2   Find the distance between the points

(a) $P_1(5, 20°)$ and $P_2(3, 140°)$ (b) $P_1(4, 50°)$ and $P_2(3, 140°)$ (c) $P_1(r_1, \theta_1)$ and $P_2(r_2, \theta_2)$

In any triangle $OP_1P_2, (P_1P_2)^2 = (OP_1)^2 + (OP_2)^2 - 2(OP_1)(OP_2)\cos \angle P_1OP_2$.

(a)   From Fig. 52-8 (a), $(P_1P_2)^2 = (5)^2 + (3)^2 - 2 \cdot 5 \cdot 3 \cos 120° = 49$; hence, $P_1P_2 = 7$.

(b)   From Fig. 52-8 (b), $P_1P_2 = \sqrt{(4)^2 + (3)^2 - 2 \cdot 4 \cdot 3 \cos 90°} = 5$.

(c)   From Fig. 52-8 (c), $P_1P_2 = \sqrt{r_1^2 + r_2^2 - 2r_1r_2 \cos(\theta_1 - \theta_2)}$.

52.3 Find the set of polar coordinates, satisfying $r \geq 0, 0° \leq \theta < 360°$, of P whose rectangular coordinates are

(a) $(2, -2\sqrt{3})$, (b) (a, a), (c) $(-3, 0)$, (d) $(0, 2)$. Find two other sets of polar coordinates for each point.

(a) We have $r = \sqrt{x^2 + y^2} = \sqrt{(2)^2 + (-2\sqrt{3})^2} = 4$ and $\theta = \arctan y/x = \arctan(-\sqrt{3})$. Since the point is in the first quadrant, we take $\theta = 300°$. The polar coordinates are $(4, 300°)$ or $(4, 5\pi/3)$. Equivalent sets of polar coordinates are $(4, -60°)$ and $(-4, 2\pi/3)$.

(b) Here $r = \sqrt{a^2 + a^2} = a\sqrt{2}$ and $\theta = \arctan 1$, when $a > 0$. Since the point is in the first quadrant, we take $\theta = \frac{1}{4}\pi$. The polar coordinates are $(a\sqrt{2}, \frac{1}{4}\pi)$. Equivalent sets are $(a\sqrt{2}, -7\pi/4)$ and $(-a\sqrt{2}, -3\pi/4)$.

(c) Here $r = \sqrt{(-3)^2 + (0)^2} = 3$. Since the point is on the negative x axis, we take $\theta = \pi$ and the polar coordinates are $(3, \pi)$. Equivalent sets are $(-3, 0)$ and $(3, -\pi)$.

(d) Here $r = \sqrt{(0)^2 + (2)^2} = 2$. Since the point is on the positive y axis, we take $\theta = \pi/2$ and the polar coordinates are $(2, \pi/2)$. Equivalent sets are $(2, -3\pi/2)$ and $(-2, 3\pi/2)$.

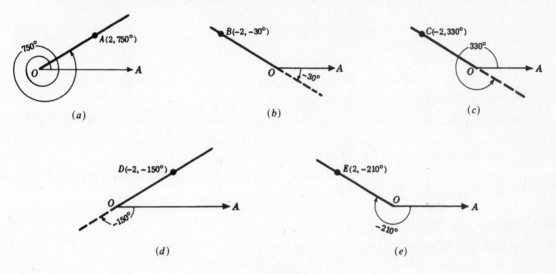

(a)   (b)   (c)

(d)   (e)

Fig. 52-7

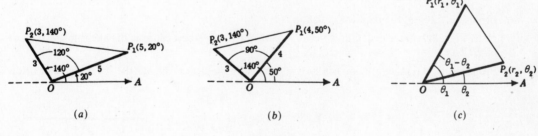

(a)   (b)   (c)

Fig. 52-8

52.4 Transform each of the following rectangular equations into their polar form:

(a) $x^2 + y^2 = 25$ (c) $3x - y = 0$ (e) $(x^2 + y^2 - ax)^2 = a^2(x^2 + y^2)$

(b) $x^2 - y^2 = 4$ (d) $x^2 + y^2 = 4x$ (f) $x^3 + xy^2 + 6x^2 - 2y^2 = 0$

We make use of the transformation: $x = r\cos\theta, y = r\sin\theta, x^2 + y^2 = r^2$.

(a) By direct substitution we obtain $r^2 = 25$ or $r = \pm 5$. Now $r = 5$ and $r = -5$ are equivalent equations since they represent the same locus, a circle with center at the origin and radius 5.

(b) We have $(r\cos\theta)^2 - (r\sin\theta)^2 = r^2(\cos^2\theta - \sin^2\theta) = r^2\cos 2\theta = 4$.

(c) Here $3r\cos\theta - r\sin\theta = 0$ or $\tan\theta = 3$. The polar equation is $\theta = \arctan 3$.

(d) We have $r^2 = 4r\cos\theta$ or $r = 4\cos\theta$ as the equation of the circle of radius 2 which passes through the origin and has its center on the polar axis.

(e) Here $(r^2 - ar\cos\theta)^2 = a^2 r^2$; then $(r - a\cos\theta)^2 = a^2$ and $r - a\cos\theta = \pm a$. Thus we may take $r = a(1 + \cos\theta)$ or $r = -a(1 - \cos\theta)$ as the polar equation of the locus.

(f) Writing it as $x(x^2 + y^2) + 6x^2 - 2y^2 = 0$, we have $r^3\cos\theta + 6r^2\cos^2\theta - 2r^2\sin^2\theta = 0$. Then $r\cos\theta = 2\sin^2\theta - 6\cos^2\theta - 2(\sin^2\theta + \cos^2\theta) - 8\cos^2\theta = 2 - 8\cos^2\theta$ and $r = 2(\sec\theta - 4\cos\theta)$ is the polar equation.

52.5 Transform each of the following equations into its rectangular form:

(a) $r = -2$ (c) $r\cos\theta = -6$ (e) $r = 4(1 + \sin\theta)$ (f) $r = \dfrac{4}{2 - \cos\theta}$

(b) $\theta = 3\pi/4$ (d) $r = 2\sin\theta$

In general, we attempt to put the polar equation in a form so that the substitutions $x^2 + y^2$ for r^2, x for $r\cos\theta$, and y for $r\sin\theta$ can be made.

(a) Squaring, we have $r^2 = 4$; the rectangular equation is $x^2 + y^2 = 4$.

(b) Here $\theta = \arctan y/x = 3\pi/4$; then $y/x = \tan 3\pi/4 = -1$ and the rectangular equation is $x + y = 0$.

(c) The rectangular form is $x = -6$.

(d) We first multiply the given equation by r to obtain $r^2 = 2r\sin\theta$. The rectangular form is $x^2 + y^2 = 2y$.

(e) After multiplying by r, we have $r^2 = 4r + 4r\sin\theta$ or $r^2 - 4r\sin\theta = 4r$; then $(r^2 - 4r\sin\theta)^2 = 16r^2$ and the rectangular equation is $(x^2 + y^2 - 4y)^2 = 16(x^2 + y^2)$.

(f) Here $2r - r\cos\theta = 4$ or $2r = r\cos\theta + 4$; then $4r^2 = (r\cos\theta + 4)^2$ and the rectangular form of the ellipse is $4(x^2 + y^2) = (x + 4)^2$ or $3x^2 + 4y^2 - 8x - 16 = 0$.

52.6 Derive the polar equation of the straight line:

(a) Passing through the pole with vectorial angle k

(b) Perpendicular to the polar axis and $p > 0$ units from the pole

(c) Parallel to the polar axis and $p > 0$ units from the pole

Let $P(r, \theta)$ be an arbitrary point on the line.

(a) From Fig. 52-9 (a) the required equation is $\theta = k$.

(b) From Fig. 52-9 (b) the equation is $r\cos\theta = p$ or $r\cos\theta = -p$ according as the line is to the right or left of the pole.

(c) From Fig. 52-9 (c) the equation is $r\sin\theta = p$ or $r\sin\theta = -p$ according as the line is above or below the pole.

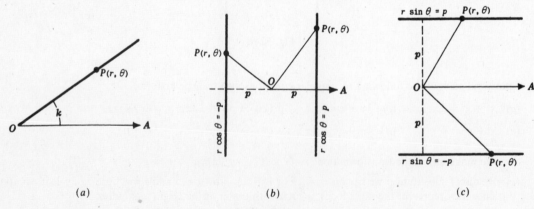

Fig. 52-9

52.7 Derive the polar equivalent of the normal form of the rectangular equation of the straight line not passing through the pole.

Let $P(r, \theta)$ be an arbitrary point on the line. Then the foot of the normal from the pole has coordinates $N(p, w)$. Using triangle ONP, the required equation is $r \cos(\theta - \omega) = p$. See Fig. 52-10.

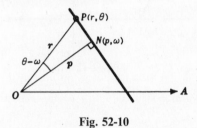

Fig. 52-10

52.8 Derive the polar equation of the circle of radius a whose center is at (c, γ).

Let $P(r, \theta)$ be an arbitrary point on the circle. See Fig. 52-11. Then [see Problem 52-2(c)]

$$r^2 + c^2 - 2rc \cos(\gamma - \theta) = a^2$$

or
$$r^2 - 2rc \cos(\gamma - \theta) + c^2 - a^2 = 0 \tag{1}$$

is the required equation.

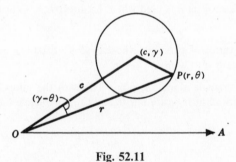

Fig. 52.11

The following special cases are of interest:

(a) If the center is at the pole, (1) becomes $r^2 = a^2$. Then $r = a$ or $r = -a$ is the equation of the circle of radius a with center at the pole.

(b) If $(c, \gamma) = (\pm a, 0°)$, (1) becomes $r = \pm 2a \cos \theta$. Thus, $r = 2a \cos \theta$ is the equation of the circle of radius a passing through the pole and having its center on the polar axis; $r = -2a \cos \theta$ is the equation of the circle of radius a passing through the pole and having its center on the polar axis extended.

(c) Similarly if $(c, \gamma) = (\pm a, 90°)$, we obtain $r = \pm 2a \sin \theta$ as the equation of the circle of radius a passing through the pole and having its center on the 90° axis or the 90° axis extended.

52.9 Derive the polar equation of a conic of eccentricity e, having a focus at the pole and p units from the corresponding directrix, when the axis on which the focus lies coincides with the polar axis.

In Fig. 52-12 a focus is at O and the corresponding directrix DD' is to the right of O. Let $P(r, \theta)$ be an arbitrary point on the conic. Now

$$\frac{OP}{PM} = e$$

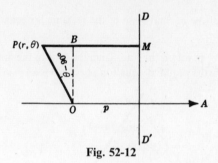

Fig. 52-12

where $OP = r$ and $PM = PB + BM = r\sin(\theta - 90°) + p = p - r\cos\theta$. Thus

$$\frac{r}{p - r\cos\theta} = e, \qquad r(1 + e\cos\theta) = ep, \qquad \text{and} \qquad r = \frac{ep}{1 + e\cos\theta}.$$

It is left for the reader to derive the equation $r = \dfrac{ep}{1 - e\cos\theta}$ when the directrix DD' lies to the left of O.

Similarly it may be shown that the polar equation of a conic of eccentricity e, having a focus at the pole and p units from the corresponding directrix, is

$$r = \frac{ep}{1 \pm e\sin\theta}$$

where the positive sign (negative sign) is used when the directrix lies above (below) the pole.

52.10 Find the locus of the third vertex of a triangle whose base is a fixed line of length $2a$ and the product of the other two sides is the constant b^2.

Take the base of the triangle along the polar axis with the midpoint of the base at the pole. The coordinates of the end points of the base are $B(a, 0)$ and $C(a, \pi)$. Denote the third (variable) vertex by $P(r, \theta)$. See Fig. 52-13.

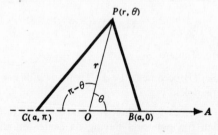

Fig. 52-13

From the triangle BOP, $(BP)^2 = r^2 + a^2 - 2ar\cos\theta$ and from the triangle COP, $(CP)^2 = r^2 + a^2 - 2ar\cos(\pi - \theta) = r^2 + a^2 + 2ar\cos\theta$. Now $(BP)(CP) = b^2$; hence

$$(r^2 + a^2 - 2ar\cos\theta)(r^2 + a^2 + 2ar\cos\theta) = (b^2)^2 = b^4$$

Then
$$(r^2 + a^2)^2 - 4a^2 r^2 \cos^2\theta = b^4$$

$$r^4 + 2a^2 r^2 (1 - 2\cos^2\theta) = r^4 - 2a^2 r^2 \cos 2\theta = b^4 - a^4$$

$$r^4 - 2a^2 r^2 \cos 2\theta + a^4 \cos^2 2\theta = b^4 - a^4 + a^4 \cos^4 2\theta = b^4 - a^4 \sin^2 2\theta$$

and the required equation is $r^2 = a^2 \cos 2\theta \pm \sqrt{b^4 - a^4 \sin^2 2\theta}$.

52.11   Sketch the conic $r = \dfrac{3}{2 - 2\sin\theta}$.

To put the equation in standard form, in which the first term in the denominator is 1, divide numerator and denominator by 2 and obtain $r = \dfrac{\frac{3}{2}}{1 - \sin\theta}$. The locus is a parabola ($e = 1$) with focus at the pole. It opens upward ($\theta = \frac{1}{2}\pi$ makes r infinite).

When $\theta = 0, r = \frac{3}{2}$. When $\theta = 3\pi/2, r = \frac{3}{4}$; the vertex is on the 90° axis extended $\frac{3}{4}$ unit below the pole. With these facts the parabola may be sketched readily as in Fig. 52-14.

The equation in rectangular coordinates is $4x^2 = 12y + 9$.

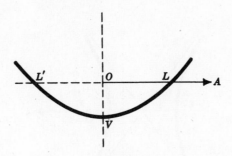

Fig. 52-14

52.12   Sketch the conic $r = \dfrac{18}{5 + 4\sin\theta}$.

After dividing numerator and denominator by 5, we have

$$r = \frac{\frac{18}{5}}{1 + \frac{4}{5}\sin\theta}$$

The locus is an ellipse ($e = \frac{4}{5}$) with a focus at the pole.

Since an equivalent equation is obtained when θ is replaced by $\pi - \theta$, the ellipse is symmetric with respect to the 90° axis; thus, the major axis is along the 90° axis. Since $ep = \frac{18}{5}$ and $ep = \frac{4}{5}, p = \frac{9}{2}$; the directrix is $\frac{9}{2}$ units above the pole. When $\theta = \frac{1}{2}\pi, r = 2$; when $\theta = 3\pi/2, r = 18$. Thus the vertices are 2 units above and 18 units below the pole. Since $a = \frac{1}{2}(2 + 18) = 10, b = \sqrt{a^2(1 - e^2)} = 6$. With these facts the ellipse may be readily sketched as in Fig. 52-15.

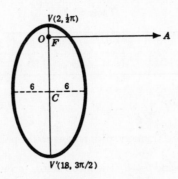

Fig. 52-15

In rectangular coordinates, the equation is $25x^2 + 9y^2 + 144y - 324 = 0$.

52.13   Sketch the conic $r = \dfrac{8}{3 - 5\cos\theta}$.

After dividing the numerator and denominator by 3, we have $r = \dfrac{\frac{8}{3}}{1 - \frac{5}{3}\cos\theta}$. The locus is a hyperbola $(e = \frac{5}{3})$ with a focus at the pole.

An equivalent equation is obtained when θ is replaced by $-\theta$; hence, the hyperbola is symmetric with respect to the polar axis and its transverse axis is on the polar axis. When $\theta = 0, r = -4$ and when $\theta = \pi, r = 1$; the vertices are respectively 4 units and 1 unit to the left of the pole. Then $a = \frac{1}{2}(4 - 1) = \frac{3}{2}$ and $b = \sqrt{a^2(e^2 - 1)} = 2$. The asymptotes, having slopes $\pm b/a = \pm \frac{4}{3}$, intersect at the center $\frac{1}{2}(1 + 4) = \frac{5}{2}$ units to the left of the pole. Since $ep = \frac{8}{3}$ and $e = \frac{5}{3}$, $p = \frac{8}{5}$; the directrix is $\frac{8}{5}$ units to the left of the pole.

In rectangular coordinates, the equation is $16x^2 - 9y^2 + 80x + 64 = 0$. See Fig. 52-16.

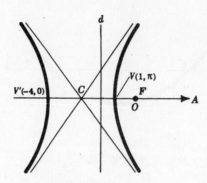

Fig. 52-16

52.14   Sketch the limacon $r = 2a\cos\theta + b$ when (a) $a = 2, b = 5$,   (b) $a = 2, b = 4$,   (c) $a = 2, b = 3$.

(a)   The equation is $r = 4\cos\theta + 5$.

   Symmetry.   An equivalent equation is obtained when θ is replaced by $-\theta$; the locus is symmetric with respect to the polar axis.

   Extent.   Since r is real and finite for all values of θ, the locus is a closed curve. Since $\cos\theta$ is of period 2π, the complete locus is described as θ varies from 0 to 2π.

   Directions at the Pole.   When $r = 0, \cos\theta = -\frac{5}{4}$; the locus does not pass through the pole. After locating the points in Table 52.2 and making use of symmetry with respect to the polaris axis, we obtain the required curve shown in Fig. 52-17(a). The equation in rectangular coordinates is $(x^2 + y^2 - 4x)^2 = 25(x^2 + y^2)$.

Table 52.2

θ	0	$\pi/6$	$\pi/4$	$\pi/5$	$\pi/2$	$2\pi/3$	$3\pi/4$	$5\pi/6$	π
r	9.00	8.48	7.84	7.00	5.00	3.00	2.16	1.52	1.00

(b)   The equation is $r = 4(1 + \cos\theta)$.

   The locus a closed curve, symmetric with respect to the polar axis, and is completely described as θ varies from 0 to 2π.

   When $r = 0, \cos\theta = -1$ and $\theta = \pi$. The locus passes through the pole and is tangent to the polar axis there.

   After locating the points in Table 52.3 and making use of symmetry, we obtain the required curve shown in Fig. 52.17(b).

   In rectangular coordinates the equation of the cardioid is $(x^2 + y^2 - 4x)^2 = 16(x^2 + y^2)$.

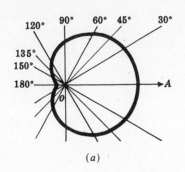

(a)

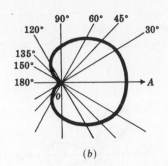

(b)

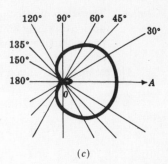

(c)

Fig. 52-17

Table 52.3

θ	0	$\pi/6$	$\pi/4$	$\pi/3$	$\pi/2$	$2\pi/3$	$3\pi/4$	$5\pi/6$	π
r	8.00	7.48	6.84	6.00	4.00	2.00	1.16	0.52	0

(c) The equation is $r = 4 \cos\theta + 3$.

The locus is a closed curve, symmetric with respect to the polar axis, and is completely described as θ varies from 0 to 2π.

When $r = 0, \cos\theta = -\frac{3}{4} = -0.750$ and $\theta = 138°40', 221°20'$. The locus passes through the pole with tangents $\theta = 138°40'$ and $\theta = 221°20'$.

After putting in the these tangents as guide lines, locating the points in Table 52.4, and making use of symmetry, we obtain the required curve shown in Fig. 52.17(c).

The equation in rectangular coordinates is $(x^2 + y^2 - 4x)^2 = 9(x^2 + y^2)$.

Table 52.4

θ	0	$\pi/6$	$\pi/4$	$\pi/3$	$\pi/2$	$2\pi/3$	$3\pi/4$	$5\pi/6$	π
r	7.00	6.48	5.84	5.00	3.00	1.00	0.16	−0.48	−1.00

52.15 Sketch the rose $r = a \cos 3\theta$.

The locus is a closed curve, symmetric with respect to the polar axis. When $r = 0, \cos 3\theta = 0$ and $\theta = \pi/6, \pi/2, 5\pi/6, 7\pi/5, \ldots$; the locus passes through the pole with tangent lines $\theta = \pi/6, \theta = \pi/2$, and $\theta = 5\pi/6$ there.

The variation of r as θ changes is shown in Table 52.5.

Table 52.5

θ	3θ	r
0 to $\pi/6$	0 to $\pi/2$	a to 0
$\pi/6$ to $\pi/3$	$\pi/2$ to π	0 to $-a$
$\pi/3$ to $\pi/2$	π to $3\pi/2$	$-a$ to 0
$\pi/2$ to $2\pi/3$	$3\pi/2$ to 2π	0 to a
$2\pi/3$ to $5\pi/6$	2π to $5\pi/2$	a to 0
$5\pi/6$ to π	$5\pi/2$ to 3π	0 to $-a$

Caution.   The values plotted are (r, θ) not $(r, 3\theta)$. The curve starts at a distance a to the right of the pole on the polar axis, passes through the pole tangent to the line $\theta = \pi/6$, reaches the tip of a loop when $\theta = \pi/3$, passes through the pole tangent to the line $\theta = \pi/2$, and so on. The locus is known as a three-leaved rose.

The rectangular equation is $(x^2 + y^2)^2 = ax(x^2 + 3y^2)$. See Fig. 52-18.

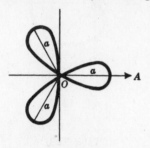

Fig. 52-18

In general, the roses $r = a \sin n\theta$ and $r = a \cos n\theta$ consist of n leaves when n is an *odd* integer.

52.16   Sketch the rose $r = a \sin 4\theta$.

The locus is a closed curve, symmetric with respect to the polar axis (an equivalent equation is obtained when θ is replaced by $\pi - \theta$ and r by $-r$), with respect to the 90° axis (an equivalent equation is obtained when θ is replaced by $-\theta$ and r by $-r$), and with respect to the pole (an equivalent equation is obtained when θ is replaced by $\pi + \theta$).

When $r = 0$, $\sin 4\theta = 0$ and $\theta = 0, \pi/4, \pi/2, 3\pi/4, \ldots$; the locus passes through the pole with tangent lines $\theta = 0, \theta = \pi/4, \theta = \pi/2$, and $\theta = 3\pi/4$ there.

The variation of r as θ changes from 0 to $\pi/2$ is shown in Table 52.6.

Table 52.6

θ	4θ	r
0 to $\pi/8$	0 to $\pi/2$	0 to a
$\pi/8$ to $\pi/4$	$\pi/2$ to π	a to 0
$\pi/4$ to $3\pi/8$	π to $3\pi/2$	0 to $-a$
$3\pi/8$ to $\pi/2$	$3\pi/2$ to 2π	$-a$ to 0

The complete curve, consisting of 8 leaves, can be traced by making use of the symmetry. See Fig. 52-19.

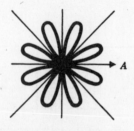

Fig. 52-19

In rectangular coordinates, the equation of the locus is $(x^2 + y^2)^5 = 16a^2(x^3y - xy^3)^2$.

In general, the roses $r = a \sin n\theta$ and $r = a \cos n\theta$ consist of $2n$ leaves when n is an *even* integer.

52.17 Sketch the locus of $r = \cos\frac{1}{2}\theta$.

Other equations of the locus are $-r = \cos\frac{1}{2}(\theta + \pi) = -\sin\frac{1}{2}\theta$ or $r = \sin\frac{1}{2}\theta, -r = \cos\frac{1}{2}(\theta - \pi) = \sin\frac{1}{2}\theta$ or $r = -\sin\frac{1}{2}\theta$, and $r = \cos\frac{1}{2}(\theta - 2\pi) = -\cos\frac{1}{2}\theta$.

The locus is a closed curve, symmetric with respect to the polar axis, the 90° axis, and the pole. It is completely described as θ varies from 0 to 4π.

When $r = 0, \theta = \pi, 3\pi, \ldots$; the line $\theta = \pi$ is tangent to the locus at the pole.

The curve is traced by locating the points (Table 52.7) and making use of symmetry. See Fig. 52-20.

Table 52.7

θ	$\frac{1}{2}\theta$	r
0	0	1.00
$\pi/6$	$\pi/12$	0.97
$\pi/3$	$\pi/6$	0.87
$\pi/2$	$\pi/4$	0.71
$2\pi/3$	$\pi/3$	0.50
$5\pi/6$	$5\pi/12$	0.26
π	$\pi/2$	0.00

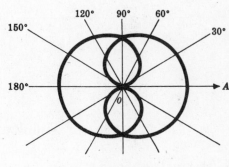

Fig. 52-20

52.18 Find the points of intersection of the limacon $r = 2\cos\theta + 4$ and the circle $r = 8\cos\theta$.

From Fig. 52-21 there are two points of intersection.

Setting $2\cos\theta + 4 = 8\cos\theta$, we obtain $\cos\theta = \frac{2}{3}$; then $\theta = 48°10'$ and $311°50'$. (We solve for θ on the range $0 \le \theta < 2\pi$ since the limacon is completely described on this range.) The points of intersection are $(\frac{16}{3}, 48°10')$ and $(\frac{16}{3}, 311°50')$.

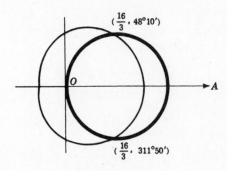

Fig. 52-21

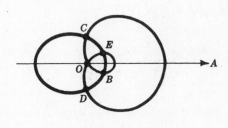

Fig. 52-22

52.19 Find the points of intersection of the ellipse $r = \dfrac{4}{2 + \cos\theta}$ and the limacon $r = 4\cos\theta - 2$.

From Fig. 52-22 there are four points of intersection.

Setting $\dfrac{4}{2 + \cos\theta} = 4\cos\theta - 2$, we have $2\cos^2\theta + 3\cos\theta - 4 = 0$. Then $\cos\theta = \dfrac{-3 \pm \sqrt{41}}{4} = 0.851$ or -2.351, and $\theta = 31°40'$ and $328°20'$. The corresponding points are $E(-5 + \sqrt{41}, 31°40')$ and $B(-5 + \sqrt{41}, 328°20')$.

To obtain the other two points, we solve the equation of the ellipse with another equation $r = 4\cos\theta + 2$ (see Example 5) of the limacon. From $4\cos\theta + 2 = \dfrac{4}{2 + \cos\theta}$, we obtain $2\cos^2\theta + 5\cos\theta = (\cos\theta)$

$(2 \cos \theta + 5) = 0$. Then $\cos \theta = 0$ and $\theta = \frac{1}{2}\pi$ and $3\pi/2$. The corresponding points are $C(2, \pi/2)$ and $D(2, 3\pi/2)$.

[NOTE: When sketching $r = 4 \cos \theta - 2$, the coordinates of C were found as $(-2, 3\pi/2)$ and those of D as $(-2, \pi/2)$.]

52.20   Find $\dfrac{dy}{dx}$ at $\theta = \dfrac{\pi}{6}$ for $r = 2(1 + \cos \theta)$.

$$\frac{dy}{dx} = \frac{(-2 \sin \theta)(\sin \theta) + 2(1 + \cos \theta)(\cos \theta)}{(-2 \sin \theta)(\cos \theta) - 2(1 + \cos \theta)(\sin \theta)} \text{ when } \theta = \tfrac{\pi}{6}.$$

Ans. -1

Supplementary Problems

52.21   Find the rectangular coordinates of P whose polar coordinates are

(a) $(-2, 45°)$ (b) $(3, \pi)$ (c) $(2, \pi/2)$ (d) $(4, 2\pi/3)$

Ans.   (a) $(-\sqrt{2}, -\sqrt{2})$ (b) $(-3, 0)$ (c) $(0, 2)$ (d) $(-2, 2\sqrt{3})$

52.22   Find a set of polar coordinates of P whose rectangular coordinates are

(a) $(1, \sqrt{3})$ (b) $(0, -5)$ (c) $(1, -1)$ (d) $(-12, 5)$

Ans.   (a) $(2, \pi/3)$ (b) $(5, 3\pi/2)$ (c) $(\sqrt{2}, 7\pi/4)$ (d) $(13, \pi - \text{Arctan} \tfrac{5}{12})$

52.23   Transform each of the following rectangular equations into polar form:

(a) $x^2 + y^2 = 16$ (b) $y^2 - x^2 = 9$ (c) $x = 4$ (d) $y = \sqrt{3}x$ (e) $xy = 12$
(f) $(x^2 + y^2)x = 4y^2$

Ans.   (a) $r = 4$ (c) $r \cos \theta = 4$ (e) $r^2 \sin 2\theta = 24$
(b) $r^2 \cos 2\theta + 9 = 0$ (d) $\theta = \pi/3$ (f) $r = 4 \tan \theta \sin \theta$

52.24   Transform each of the following polar equations into rectangular form:

(a) $r \sin \theta = -4$ (c) $r = 2 \cos \theta$ (e) $r = 1 - 2 \cos \theta$ (f) $r = \dfrac{4}{1 - 2 \sin \theta}$
(b) $r = -4$ (d) $r = \sin 2\theta$

Ans.   (a) $y = -4$ (c) $x^2 + y^2 - 2x = 0$ (e) $(x^2 + y^2 + 2x)^2 = x^2 + y^2$
(b) $x^2 + y^2 = 16$ (d) $(x^2 + y^2)^3 = 4x^2y^2$ (f) $x^2 - 3y^2 - 16y - 16 = 0$

52.25   Derive the polar equation $r = \dfrac{ep}{1 \pm e \sin \theta}$ of the conic of eccentricity e with a focus at the pole and with corresponding directrix p units from the focus.

52.26   Write the polar equation of each of the following:

(a) Straight line bisecting the second and fourth quadrants

(b) Straight line through $(4, 2\pi/3)$ and perpendicular to the polar axis

(c) Straight line through $N(3, \pi/6)$ and perpendicular to the radius vector of N

(d) Circle with center at $C(4, 3\pi/2)$ and radius $=4$

(e) Circle with center at $C(-4, 0)$ and radius $=4$

(f) Circle with center at $C(4, \pi/3)$ and radius $=4$

(g) Parabola with focus at the pole and directrix $r = -4 \sec \theta$

(h) Parabola with focus at the pole and vertex at $V(3, \pi/2)$

(i) Ellipse with eccentricity $\frac{3}{4}$, one focus at the pole, and the corresponding directrix 5 units above the polar axis

(j) Ellipse with one focus at the pole, the other focus at $(8, \pi)$, and eccentricity $= \frac{2}{3}$

(k) Hyperbola with eccentricity $= \frac{3}{2}$, one focus at the pole, and the corresponding directrix 5 units to the left of the 90° axis

(l) Hyperbola, conjugate axis $= 24$ parallel to and below the polar axis, transverse axis $= 10$, and one focus at the pole

Ans. (a) $\theta = 3\pi/4$	(c) $r\cos(\theta - \pi/6) = 3$	(e) $r = -8\cos\theta$	
(b) $r\cos\theta = -2$	(d) $r = -8\sin\theta$	(f) $r = 8\cos(\theta - \pi/3)$	
(g) $r = \dfrac{4}{1 - \cos\theta}$	(i) $r = \dfrac{15}{4 + 3\sin\theta}$	(k) $r = \dfrac{15}{2 - 3\cos\theta}$	
(h) $r = \dfrac{6}{1 + \sin\theta}$	(j) $r = \dfrac{10}{3 + 2\cos\theta}$	(l) $r = \dfrac{144}{5 - 13\sin\theta}$	

52.27 Discuss and sketch:

(a) $r = \sin(\theta - 45°) = -2$

(b) $r = 10 \sin\theta$

(c) $r = -6\cos\theta$

(d) $r = \dfrac{8}{2 - \sin\theta}$

(e) $r = \dfrac{6}{1 + 2\cos\theta}$

(f) $r = \dfrac{2}{1 - \cos\theta}$

(g) $r = 2 - 4\cos\theta$

(h) $r = 4 - 2\cos\theta$

(i) $r^2 = 9\cos 2\theta$

(j) $r^2 = 16\sin 2\theta$

(k) $r = 2\cos 2\theta$

(l) $r = 4\sin 2\theta$

(m) $r = 2a\tan\theta \sin\theta$

(n) $r = 4\tan^2\theta \sec\theta$

(o) $r = \cos \frac{3}{2}\theta$

(p) $r = 2\theta$

(q) $r = a/\theta$

52.28 Find the complete intersection of

(a) $r = 2\cos\theta, r = 1$ (b) $r^2 = 4\cos 2\theta, r = 2\sqrt{2}\sin\theta$ (c) $r = 1 + \sin\theta, r = \sqrt{3}\cos\theta$

Ans. (a) $(1, \pi/3), (1, 5\pi/3)$ (b) $(0, 0), (\sqrt{2}, \pi/6), (\sqrt{2}, 5\pi/6)$ (c) $(0, 0), (\frac{3}{2}, \pi/6)$

52.29 Find $\dfrac{dy}{dx}$ for $r = \sin\theta$ at $\theta = \pi/2$.

Ans. 0

Introduction to the Graphing Calculator

IN THIS TEXT, a number of opportunities to utilize calculators have been provided to the reader. In many of these cases, a traditional handheld calculator would be sufficient. However, handheld algebraic-entry graphing calculators have become quite prevalent, and although some of the applications of such calculators are well above the level of

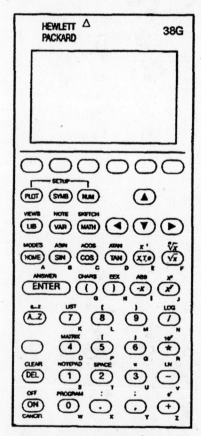

Fig. A1-1
(Reprinted with the permission of the Hewlett-Packard Company.)

the mathematics in this book, it would be worthwhile to introduce these devices at this point. The goal here is to make you familiar with such calculators and to make you at ease with them. In that way, when you engage in more advanced mathematics such as the calculus, it will be easier for you to use graphing calculators.

First, what does such a calculator look like? An example is indicated in Fig. A1-1. This diagram illustrates the keyboard of the Hewlett-Packard HP 38G calculator.

While there are other such calculators available to you on the market, I have found that this particular one offers the easiest retrieval of answers and meshes particularly well with the needs of the mathematics student at both the precalculus level and more advanced levels as well. In this appendix, you will be presented with a brief overview of the calculator and of its use in solving equations and graphing functions.

You turn the calculator on by pressing the $\boxed{ON}$ key. Note that the $\boxed{ON}$ key is also the $\boxed{CANCEL}$ key. Also, if you wish to lighten or darken your calculator's screen, simply hold down the $\boxed{ON}$ key while you press the $\boxed{-}$ or $\boxed{+}$ keys. (Note: Please refer to the *Reference Manual* provided with the HP 38G for a more detailed description of the many uses of the calculator.)

The HOME screen is the main area in which you will work. Press $\boxed{HOME}$ to find this area. Note that the $\boxed{HOME}$ screen is divided into two main parts: The large rectangular area is the space in which entries and answers are indicated; the smaller rectangular area is the Editline. For example, if you press the keystrokes

$\boxed{7}$ $\boxed{5}$ $\boxed{ENTER}$

then you will find the following on the HOME screen:

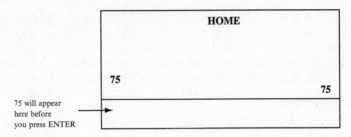

If you press the keystrokes

$\boxed{7}$ $\boxed{5}$ $\boxed{/}$ $\boxed{5}$ $\boxed{ENTER}$

then you will find the following on the $\boxed{HOME}$ screen:

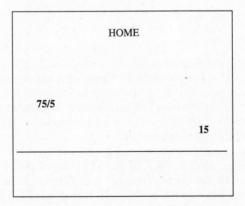

Notice that the HP 36G uses algebraic-entry notation. Thus, to perform a calculation, you first press the operations required, and then press $\boxed{ENTER}$. Again, if you wish to find $689-231$, press the keystrokes

$\boxed{6}$ $\boxed{8}$ $\boxed{9}$ $\boxed{-}$ $\boxed{2}$ $\boxed{3}$ $\boxed{1}$ $\boxed{ENTER}$

To find 5 squared (5^2), press the keystrokes

$\boxed{5}$ $\boxed{x^y}$ $\boxed{2}$ $\boxed{\text{ENTER}}$

Now perform the following calculations using the HP 36G or similar graphing calculator:

1. $483 + 286$

2. $47 - 81$

3. $843 * 35$

4. $75 / 21$

5. $45x^y2$, or $(45)^2$

6. $\sqrt{15}$

7.   The reciprocal of 21

8.   The absolute value of -45

The HP 38G can be used to solve simple equations and to graph elementary functions. Let's look at some examples. Please refer to the HP 38G *Reference Manual* for more detailed instructions.

Let us solve the equation $X - 2 = 9$. At any one of the equation lines (marked E1, E2, etc.), enter the equation $X - 2 = 9$. Note that "X" is entered by pressing the $\boxed{\text{A...Z}}$ key, and then pressing the $\boxed{\bullet}$ key. Also note that the "=" sign is entered by pressing the key under the "=" sign on the lower, darkened area of the screen. After the equation appears correctly, press $\boxed{\text{ENTER}}$. A check mark should appear next to the equation you have entered. This check mark indicates that when you attempt to solve, the equation checked is the one you will be solving. Next, press the $\boxed{\text{NUM}}$ key, and then the key under the word "SOLVE" in the lower, darkened area of the screen. The number 11 should appear next to the symbol X.

Now try to solve the equation $2X - 11 = 13$ using the HP 38G and the following series of steps. Press $\boxed{\text{LIB}}$, use the arrow keys to scroll to Solve, press $\boxed{\text{ENTER}}$, go to any of the "E" lines, and enter the equation $2X - 11 = 13$. Press $\boxed{\text{ENTER}}$. Press the key under the $\boxed{\sqrt{\text{CHK}}}$ box on the lower, darkened area of the screen. Press the $\boxed{\text{NUM}}$ key and then press the key under the word SOLVE in the lower, darkened area of the screen. You should see the answer 12 appear next to the symbol X.

Next, solve each of the following using the HP 38G or similar graphing calculator:

1. $2X - 15 = 26$

2. $3X = 4X + 4$

3. $5Y = 6Y + 12.5$

Let's now investigate the graph of the equation $Y = X - 4$. Press the $\boxed{\text{LIB}}$ key, and use the arrow keys to locate "FUNCTION." Press $\boxed{\text{ENTER}}$, and at any one of the "F" lines, enter the expression $X - 4$. Make certain that you press $\boxed{\text{ENTER}}$. Note that you may enter the "X" by pressing the key under the symbol X in the lower, darkened area of the screen. Make certain that the equation you have entered is "checked," and then press the $\boxed{\text{PLOT}}$ key. The graph of the line will appear on a set of coordinate axes. Consult the *Reference Manual* for more details concerning the use of the HP 38G for graphing purposes.

Now try to graph the equation $Y = 2X - 11$ using the HP 38G and the following series of steps. Press $\boxed{\text{LIB}}$, use the arrow keys to scroll to Function, and press $\boxed{\text{ENTER}}$. Go to any of the "F" lines, and enter $2X - 11$. Press $\boxed{\text{ENTER}}$. Make certain that the expression is "checked" and that no other expressions are "checked." All expressions checked will be plotted when you press $\boxed{\text{PLOT}}$. Now press $\boxed{\text{PLOT}}$. The graph will appear on the coordinate axes on the calculator's screen.

Now graph each of the following using the HP 38G or similar calculator:

1. $Y = 2X - 5$

2. $Y = -5X = 13$

3. $2Y = 3X - 7$ (*Hint*: Divide both sides of the equation by 2.)

Supplementary Problems

1.   Solve the equations

$$X + 2Y = 7$$

$$X + 3Y = 10$$

using the graphing calculator.

2.   Graph each on the calculator:

(a) $y = x^2 + 7$ (b) $y = 2x^3 - 8$ (c) $y = \sqrt{x}$

(d) $y = 2 \sin x - 4$ (e) $y = \exp(x^2)$

3.   Find the absolute minimum value for $y = 3x^2 - 8$ using the calculator.

4.   Graph $x^2 - y^2 = 4$ using the calculator.

5.   Graph $x^2 - 2y^2 = 16$.

6.   Find dy/dx, where $y = x^4 - 4x^3 + 6x + 11$.

7.   Find $\int (x^5 - \sin x)\, dx$, where $y = x^4 - 4x^3 + 6x + 11$.

8.   Graph $y = \sin(1/x)$, where $y = x^4 - 4x^3 + 6x + 11$.

9.   Graph $y = \exp(\sin x)$, where $y = x^4 - 4x^3 + 6x + 11$.

10.   Find where $y = x^2 - \sin x$.

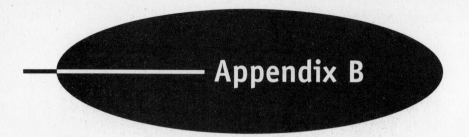

The Number System of Algebra

ELEMENTARY MATHEMATICS is concerned mainly with certain elements called *numbers* and with certain operations defined on them.

The unending set of symbols $1, 2, 3, 4, 5, 6, 7, 8, 9, 10, 11, 12, \ldots$ used in counting are called *natural numbers*.

In adding two of these numbers, say, 5 and 7, we begin with 5 (or with 7) and count to the *right* seven (or five) numbers to get 12. The sum of two natural numbers is a natural number; that is, the sum of two members of the above set is a member of the set.

In subtracting 5 from 7, we begin with 7 and count to the *left* five numbers to 2. It is clear, however, that 7 cannot be subtracted from 5, since there are only four numbers to the left of 5.

INTEGERS. In order that subtraction be always possible, it is necessary to increase our set of numbers. We prefix each natural number with a + sign (in practice, it is more convenient not to write the sign) to form the *positive integers*, we prefix each natural number with a − sign (the sign must always be written) to form the *negative integers*, and we create a new symbol 0, read *zero*. On the set of *integers*

$$\ldots, -8, \; -7, \; -6, \; -5, \; -4, \; -3, \; -2, \; -1, \; 0, \; +1, \; +2, \; +3, \; +4, \; +5, \; +6, \; +7, \; +8, \ldots$$

the operations of addition and subtraction are possible without exception.

To add two integers such as $+7$ and -5, we begin with $+7$ and count to the left (indicated by the sign of -5) five numbers to $+2$, or we begin with -5 and count to the right (indicated by the sign of $+7$) seven numbers to $+2$. How would you add -7 and -5?

To subtract $+7$ from -5, we begin with -5 and count to the left (opposite to the direction indicated by $+7$) seven numbers to -12. To subtract -5 from $+7$ we begin with $+7$ and count to the right (opposite to the direction indicated by -5) five numbers to $+12$. How would you subtract $+7$ from $+5$? -7 from -5? -5 from -7?

If one is to operate easily with integers, it is necessary to avoid the process of counting. To do this, we memorize an addition table and establish certain rules of procedure. We note that each of the numbers $+7$ and -7 is seven steps from 0 and indicate this fact by saying that the *numerical value* of each of the numbers $+7$ and -7 is 7. We may state:

Rule 1. To add two numbers having like signs, add their numerical values and prefix their common sign.

Rule 2. To add two numbers having unlike signs, subtract the smaller numerical value from the larger, and prefix the sign of the number having the larger numerical value.

Rule 3. To subtract a number, change its sign and add.

Since $3 \cdot 2 = 2 + 2 + 2 = 3 + 3 = 6$, we assume

$$(+3)\,(+2) = +6 \qquad (-3)\,(+2) = (+3)\,(-2) = -6 \text{ and } (-3)\,(-2) = +6$$

414

Rule 4.   To multiply or divide two numbers (never divide by 0!), multiply or divide the numerical values, prefixing a + sign if the two numbers have like signs and a − sign if the two numbers have unlike signs.

If m and n are integers, then $m+n$, $m-n$, and $m \cdot n$ are integers but $m \div n$ may do not be an integer. (Common fractions will be treated in the next section.) Moreover, there exists a unique integer x such that $m+x=n$. If $x=0$, then $m=n$; if x is positive $(x>0)$, then m is less than $n(m<n)$; if x is negative $(x<0)$, then m is greater than $n(m>n)$.

The integers may be made to correspond one-to-one with equally spaced points on a straight line as in Fig. A2-1. Then $m>n$ indicates that the point on the scale corresponding to m lies to the right of the point corresponding to n. There will be no possibility of confusion if we write the point m rather than the point which corresponds to m, and we shall do so hereafter. Then $m<n$ indicates that the point m lies to the left of n.

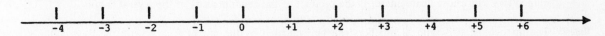

Fig. A2-1

Every positive integer m is divisible by ± 1 and $\pm m$. A positive integer $m>1$ is called a *prime* if its only factors or divisors are ± 1 and $\pm m$; otherwise, m is called *composite*. For example, 2, 7, 19 are primes, while $6=2 \cdot 3, 18=2 \cdot 3 \cdot 3$, and $30=2 \cdot 3 \cdot 5$ are composites. In these examples, the composite numbers have been expressed as products of prime factors, that is, factors which are prime numbers. Clearly, if $m=r \cdot s \cdot t$ is such a factorization of m, then $-m=(-1) \, r \cdot s \cdot t$ is such a factorization of m.

THE RATIONAL NUMBERS.   The set of rational numbers consists of all numbers of the form m/n, where m and $n \neq 0$ are integers. Thus, the rational numbers include the integers and common fractions.

Every rational number has an infinitude of representations; for example, the integer 1 may be represented by $1/1, 2/2, 3/3, 4/4 \ldots$ and the fraction 2/3 may be represented by $4/6, 6/9, 8/12, \ldots$. A fraction is said to be expressed in lowest terms by the representation m/n, where m and n have no common prime factor. The most useful rule concerning rational numbers is, therefore:

Rule 5.   The value of a rational number is unchanged if both the numerator and denominator are multiplied or divided by the same nonzero number.

Caution: We use Rule 5 with division to reduce a fraction to lowest terms. For example, we write $15/21 = \cancel{3}/\cancel{3} \cdot 5/7 = 5/7$ and speak of canceling the 3s. Now canceling is not an operation on numbers. We cancel or strike out the 3s as a safety measure, that is, to be sure that they will not be used in computing the final result. The operation is division and Rule 5 states that we may divide the numerator by 3 provided we also divide the denominator by 3. This point is belabored here because of the all too common error:

$$\frac{12\cancel{a}-5}{7\cancel{a}}$$

The fact is the $\dfrac{12a-5}{7a}$ cannot be further simplified for if we divide $7a$ by a we must also divide $12a$ and 5 by a. This would lead to the more cumbersome

$$\frac{12-5/a}{7}$$

The rational numbers may be associated in a one-to-one manner with points on a straight line as in Fig. A2-2. Here the point associated with the rational number m is m units from the point (called the origin) associated with 0, the distance between the points 0 and 1 being the unit of measure.

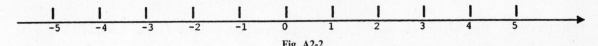

Fig. A2-2

If two rational numbers have representations r/n and s/n, where n is a positive integer, then $r/n > s/n$ if $r > s$, and $r/n < s/n$ if $r < s$. Thus, in comparing two rational numbers it is necessary to express them with the same denominator. Of the many denominators (positive integers) there is always a least one, called the *least common denominator*. For the fractions 3/5 and 2/3, the least common denominator is 15. We conclude that $3/5 < 2/3$ since $3/5 = 9/15 < 10/15 = 2/3$.

Rule 6. The sum (difference) of two rational numbers expressed with the same denominator is a rational number whose denominator is the common denominator and whose numerator is the sum (difference) of the numerators.

Rule 7. The product of two or more rational numbers is a rational number whose numerator is the product of the numerators and whose denominator is the product of the denominators of the several factors.

Rule 8. The quotient of two rational numbers can be evaluated by the use of Rule 5 with the least common denominator of the two numbers as the multiplier.

If a and b are rational numbers, $a + b$, $a - b$, and $a \cdot b$ are rational numbers. Moreover, if a and b are $\neq 0$, there exists a rational number x, unique except for its representation, such that

$$ax = b \qquad\qquad (A2.1)$$

When a or b or both are zero, we have the following situations:

$b = 0$ and $a \neq 0$: $\quad$ (*A2.1*) becomes $a \cdot x = 0$ and $x = 0$, that is, $0/a = 0$ when $a \neq 0$.

$a = 0$ and $b \neq 0$: $\quad$ (*A2.1*) becomes $0 \cdot x = b$; then $b/0$, when $b \neq 0$, is without meaning since $0 \cdot x = 0$.

$a = 0$ and $b = 0$: $\quad$ (*A2.1*) becomes $0 \cdot x = 0$; then $0/0$ is indeterminate since every number x satisfies the equation.

In brief: $0/a = 0$ when $a \neq 0$, but division by 0 is never permitted.

DECIMALS.   In writing numbers we use a positional system, that is, the value given any particular digit depends upon its position in the sequence. For example, in 423 the positional value of the digit 4 is 4 (100), while in 234 the positional value of the digit 4 is 4 (1). Since the positional value of a digit involves the number 10, this system of notation is called the *decimal system*. In this system, the number 4238.75 means

$$4(1000) + 2(100) + 3(10) + 8(1) + 7(1/10) + 5(1/100)$$

It is interesting to note that from this example certain definitions to be made in a later study of exponents may be anticipated. Since $1000 = 10^3, 100 = 10^2, 10 = 10^1$, it would seem natural to define $1 = 10^0$, $1/10 = 10^{-1}, 1/100 = 10^{-2}$.

By the process of division, any rational number can be expressed as a decimal; for example, $70/33 = 2.121212\ldots$. This is termed a *repeating decimal*, since the digits 12, called the *cycle*, are repeated without end. It will be seen later that every repeating decimal represents a rational number.

In operating with decimals, it may be necessary to "round off" a decimal representation to a prescribed number of decimal places. For example, $1/3 = 0.3333\ldots$ is written as 0.33 to two decimal places and $2/3 = 0.6666\ldots$ is written as 0.667 to three decimal places. In rounding off, use will be made of the Computer's Rule:

(*a*) Increase the last digit retained by 1 if the digits rejected exceed the sequence 50000.... For example, 2.384629... becomes 2.385 to three decimal places.

(*b*) Leave the last digit retained unchanged if the digits rejected are less than 5000.... For example, 2.384629... becomes 2.38 to two decimal places.

(*c*) Make the last digit retained even if the digit rejected is exactly 5; for example, to three decimal places 11.3865 becomes 11.386 and 9.3815 becomes 9.382.

PERCENTAGE.   The symbol %, read percent, means per hundred; thus; 5% is equivalent to 5/ 100 or 0.05.

Any number, when expressed in decimal notation, can be written as a percent by multiplying by 100 and adding the symbol %. For example, $0.0125 = 100(0.0125)\% = 1.25\% = 1\frac{1}{4}\%$, and $7/20 = 0.35 = 35\%$.

Conversely, any percentage may be expressed in decimal form by dropping the symbol % and dividing by 100. For example, $42.5\% = 42.5/100 = 0.425, 3.25\% = 0.0325$, and $2000\% = 20$.

When using percentages, express the percent as a decimal and, when possible, as a simple fraction. For example, $4\frac{1}{4}\%$ of $48 = 0.0425 \cdot 48 = 2.04$ and $12\frac{1}{2}\%$ of $5.28 = 1/8$ of $5.28 = 0.66$. (See Problems.)

THE IRRATIONAL NUMBERS. The existence of numbers other than the rational numbers may be inferred from either of the following considerations:

(a) We may conceive of a nonrepeating decimal constructed in endless time by setting down a succession of digits chosen at random.

(b) The length of the diagonal of a square of side 1 is not a rational number; that is, there exists no rational number a such that $a^2 = 2$. Numbers such as $\sqrt{2}, \sqrt[3]{2}, \sqrt[3]{-3}$, and π (but not $\sqrt{-3}$ or $\sqrt[4]{5}$) are called *irrational numbers*. The first three of these are called *radicals*. The radical $\sqrt[n]{a}$ is said to be of order n; n is called the *index*, and a is called the *radicand*.

THE REAL NUMBERS. The set of *real numbers* consists of the rational and irrational numbers. The real numbers may be ordered by comparing their decimal representations. For example, $\sqrt{2} = 1.4142\ldots$; then $7/5 = 1.4 < \sqrt{2}, 3/2 = 1.5 > \sqrt{2}$, etc.

We assume that the totality of real numbers may be placed in one-to-one correspondence with the totality of points on a straight line. See Fig. A2-3.

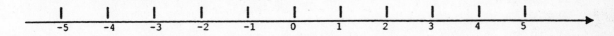

Fig. A2-3

The number associated with a point on the line, called the *coordinate* of the point, gives its distance and direction from the point (called the *origin*) associated with the number 0. If a point A has coordinate a, we shall speak of it as the point $A(a)$.

The directed distance from point $A(a)$ to point $B(b)$ on the real number scale is given by $AB = b - a$. The midpoint of the segment AB has coordinate $\frac{1}{2}(a + b)$.

THE COMPLEX NUMBERS. In the set of real numbers there is no number whose square is -1. If there is to be such a number, say, $\sqrt{-1}$, then by definition $(\sqrt{-1})^2 = -1$. Note carefully that $(\sqrt{-1})^2 = \sqrt{-1}\sqrt{-1} = \sqrt{(-1)(-1)} = \sqrt{1} = 1$ is incorrect. In order to avoid this error, the symbol i with the following properties is used:

$$\text{If } a > 0, \sqrt{-a} = i\sqrt{a}; \quad i^2 = -1$$

Then

$$(\sqrt{-2})^2 = \sqrt{-2}\sqrt{-2} = (i\sqrt{2})(i\sqrt{2}) = i^2 \cdot 2 = -2$$

and

$$\sqrt{-2}\sqrt{-3} = (i\sqrt{2})(i\sqrt{3}) = i^2\sqrt{6} = -\sqrt{6}$$

Numbers of the form $a + bi$, where a and b are real numbers, are called *complex numbers*. In the complex number $a + bi$, a is called the *real part* and bi is called the *imaginary part*. Numbers of the form ci, where c is real, are called *imaginary numbers* or sometimes *pure imaginary numbers*.

The complex number $a + bi$ is a real number when $b = 0$ and a pure imaginary number when $a = 0$.
Only the following operations will be considered here:

To add (subtract) two complex numbers, add (subtract) the real parts and add (subtract) the pure imaginary parts; that is, $(a + ib) + (c + id) = (a + b) + (b + d)i$.

To multiply two complex numbers, form the product treating i as an ordinary number and then replace i^2 by -1; that is, $(a + ib)(c + id) = (ac - bd) + (bc + ad)i$.

Supplementary Problems

1. Arrange each of the following so that they may be separated by $<$.

    (a) $2/3, -3/4, 5/6, -1, 4/5, -4/3, -1/4$

    (b) $3/2, 2, 7/5, 4/3, 3$

    (c) $3/2, \sqrt{3}, -1/2, \sqrt{5}, 0$

2. Determine the greater of each pair.

    (a) $|4 + (-2)|$ and $|-4| + |(-2)|$

    (b) $|4 + (-2)|$ and $|4| + |(-2)|$

    (c) $|4 - (-2)|$ and $|4| - |(-2)|$

3. Convert each of the following fractions into equivalent fractions having the indicated denominator:

    (a) $3/5, 15$

    (b) $-3/5, 20$

    (c) $7/3, 42$

    (d) $5/7, 35$

    (e) $12/13, 156$

4. Perform the indicated operations.

    (a) $(-2)(3)(-5)$

    (b) $3(-2)(4) + (-5)(2)(0)$

    (c) $-8 - (-6) + 2$

    (d) $3/4 + 2/3$

    (e) $3/4 - 2/3$

    (f) $5/6 - 1/2 - 2/3$

    (g) $3/4 - 7/12 - 1/3$

    (h) $(1/2)(8/9)(6/5)$

    (i) $3/8 \times 5\frac{1}{3}$

    (j) $2\frac{1}{4} \times 2\frac{2}{3} \times 1\frac{2}{5} \times 2\frac{1}{7}$

    (k) $25/32 \div 35/64$

    (l) $3\frac{1}{3} \div 7/10$

    (m) $(1\frac{1}{2} \times 2\frac{1}{4}) \div 1\frac{1}{8}$

    (n) $\dfrac{3 - 2/3}{5 + 5/6}$

    (o) $\dfrac{2/3 + 3/4}{5/6 - 7/8}$

    (p) $\dfrac{1\frac{1}{2} - 2\frac{2}{3}}{3\frac{1}{5} - 1\frac{1}{4}}$

5. Perform the indicated operations.

   (a) $5\sqrt{3} + 2\sqrt{3} - 8\sqrt{3}$

   (b) $5\sqrt{2} + \sqrt{32} - 3\sqrt{8}$

   (c) $\sqrt[3]{12} \cdot \sqrt[3]{36}$

   (d) $(1 + \sqrt{2})(3 - \sqrt{2})$

   (e) $(2\sqrt{3} + 3\sqrt{2})(2\sqrt{3} - 3\sqrt{2})$

   (f) $(4\sqrt{3} - 3\sqrt{5})(2\sqrt{3} + \sqrt{5})$

   (g) $\dfrac{4 - 2\sqrt{3}}{5\sqrt{3}}$

   (h) $\dfrac{2\sqrt{5} - 3\sqrt{2}}{3\sqrt{5} + 4\sqrt{2}}$

   (i) $\dfrac{3\sqrt{2} - 4\sqrt{3}}{4\sqrt{2} + 3\sqrt{3}}$

6. Perform the indicated operations.

   (a) $i\sqrt{12} + i\sqrt{75} - i\sqrt{108}$

   (b) $i\sqrt{50} - i\sqrt{32} - i\sqrt{8}$

   (c) $\sqrt[4]{-27} + 2\sqrt{-12} - 5\sqrt{-48}$

   (d) $3\sqrt{-20} + 5\sqrt{-80} - \sqrt{-45}$

   (e) $\sqrt{-9} \cdot \sqrt{-16}$

   (f) $\sqrt{-12} \cdot \sqrt{-27}$

   (g) $3i(2 + i)$

   (h) $(-3 + 5i) + (4 - 2i)$

   (i) $(3 + 5i) - (-4 - 2i)$

   (j) $(3 + 5i)(2 - 7i)$

   (k) $(2\sqrt{3} + i\sqrt{2})(3\sqrt{3} - 5i\sqrt{2})$

   (l) $(3 - 2i)(1 + 5i)(-2 - i)$

   (m) i^{73}

Answers to Supplementary Problems

1.   (a) $-4, 3, -1, -3/4, -1/4, 2/3, 4/5, 5/6$

     (b) $4/3, 7/5, 3/2, 2, 3$

     (c) $-\sqrt{5}, -1\sqrt{2}, 0, 3/2, \sqrt{3}$

2.   (a) Second

     (b) Second

     (c) First

3.   (a) $9/15$

     (b) $-12/20$

     (c) $98/42$

     (d) $25/35$

     (e) $144/156$

4. (a) 30
    (b) −24
    (c) 0
    (d) 17/12
    (e) 1/12
    (f) −1/3
    (g) −1/6
    (h) 8/15
    (i) 2
    (j) 18
    (k) 10/7
    (l) 100/21
    (m) 3
    (n) 2/5
    (o) −34
    (p) −70/117

5. (a) $-\sqrt{3}$
    (b) $3\sqrt{2}$
    (c) $6\sqrt[3]{2}$
    (d) $1 + 2\sqrt{2}$
    (e) −6
    (f) $9 - 2\sqrt{15}$
    (g) $\dfrac{4\sqrt{3} - 6}{15}$
    (h) $\dfrac{54 - 17\sqrt{10}}{13}$
    (i) $\dfrac{12 + 7\sqrt{6}}{5}$

6. (a) $i\sqrt{3}$
    (b) $3i\sqrt{2}$
    (c) $-4i\sqrt{3}$
    (d) $-17i\sqrt{5}$
    (e) −12
    (f) −18
    (g) $-3 + 6i$
    (h) $1 + 3i$
    (i) $7 + 7i$
    (j) $41 - 11i$
    (k) $28 - 7i\sqrt{6}$
    (l) $-13 - 39i$
    (m) i

Appendix C

Mathematical Modeling

ONE OF THE MOST IMPORTANT CHANG... the precalculus curriculum over the last 10 years is the introduction into that curriculum of mathematica... Certainly it is the case that problems and problem solving have been a significant part of that curricul... is modeling different?

According to the National Council of Teachers... matics' publication, *Mathematical Modeling in the* *Secondary School Curriculum*, a mathematical model is... hematical structure that approximates the features of phenomenon. The process of devising a mathemat... is called mathematical modeling." (Swetz and Hartzler, NCTM, Reston, Virginia 1991.) Thus, one can s... hematical modeling does not in any way replace problem solving in the curriculum. Instead, it is a kind o... lving.

EXAMPLE. One of the most significant applications of modeling i... is in the area of population growth. Table A3.1 gives the population for a culture of bacteria from time $t = 0$ until t...

Table A3.1

Time (t)	Population (p)
0	0
1	1
2	2
3	4
4	5
5	7

See Fig. A3-1 for the graph of these data with x axis representing t and y axis...

Notice that for the times from 0 to 2, the graph in Fig. A3-1 is linear. In fact, ... $t = 4$. Let us find the equation of the line that approximates these data: Since the d... $(2, 2)$, a reasonable model for these data is the equation $y = x$. This equation, $y = $ ... linear through

$$p(x) = x$$

... nts $(1, 1)$ and ... rm,

is a linear model. We can use this linear model to predict the population of this commu...

421

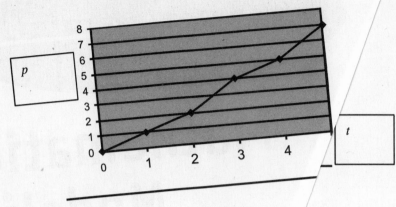

Fig. A3-1

For example, $p(x) = x$ predicts that

$$p(3) = 3.$$

Since $p(3) = 4$, our model is off by 25%. $P(x) = x$ predicts that

$$p(4) = 4$$

but $p(4) = 5$, so the model is off by 20%.
The reader should ask herself/himself whether the quadr

is a better model for these data. For example, what is the
predict in terms of growth for various larger values of
Using the example above, we see that the critic

or for $t = 2, 3, 4$, etc.? What do these two models
ulation at time 10 is 35, which is the better model?
modeling are as follows:
observed).

(a) Conjecture what model best fits the data

(b) Analyze the model mathematically.   is in (b).

(c) Draw reasonable conclusions based

easonable fit for given data.

In the example above, we

(a) Conjectured that a linear mode matically.

(b) Conjectured a model and an an additional model.

(c) Drew conclusions which in

ementary Problems

Population of U.S.
150,697,000
131,669,000
122,775,000
105,711,000
91,972,000

1. For the following data fro

0

1900	75,995,000
1890	62,480,000
1880	50,156,000
1870	38,558,000
1860	31,443,000
1850	23,192,000
1840	17,069,000
1830	12,866,000
1820	9,638,000
1810	7,240,000
1800	5,308,000
1790	3,929,000

(a) Graph these data, using the vertical axis for population (in millions).

(b) For which years is the graph almost linear? *Ans.* From 1880 to 1900

(c) Find the equation of an approximating linear model for these data.

(d) Use your model in (c) to predict the population in 1980.
     (NOTE: The actual population in 1980 was 227 million.)

(e) What is the percent error in your model for 1980?

(f) Construct a quadratic model for these data.
     (**Hint**: Review Chapter 35 before attempting this.)

(g) Which is the better model for the above data: a linear function or a quadratic function? Why?

Index